Soil Biology

Volume 29

Series Editor
Ajit Varma, Amity Institute of Microbial Technology,
Amity University Uttar Pradesh, Noida, UP, India

For further volumes:
http://www.springer.com/series/5138

Bhupinder Pal Singh • Annette L. Cowie
K. Yin Chan
Editors

Soil Health and Climate Change

Springer

Editors
Dr. Bhupinder Pal Singh
Forest Science Centre
NSW Department of Primary Industries
PO Box 100
Beecroft, NSW 2119
Australia
BP.Singh@sf.nsw.gov.au;
Bhupinderpal.Singh@industry.nsw.gov.au

Dr. Annette L. Cowie
National Centre for Rural Greenhouse Gas Research
A partnership between University of New England and NSW Department of Primary Industries
University of New England
Armidale, NSW 2351
Australia
Annette.Cowie@une.edu.au

Dr. K. Yin Chan (retired)
NSW Department of Primary Industries
Locked Bag 4
Richmond, NSW 2753
Australia
yin.chan@mq.edu.au

ISSN 1613-3382
ISBN 978-3-642-20255-1 e-ISBN 978-3-642-20256-8
DOI 10.1007/978-3-642-20256-8
Springer Heidelberg Dordrecht London New York

Library of Congress Control Number: 2011933561

Cover design: SPi Publisher Services

Printed on acid-free paper

Springer is part of Springer Science+Business Media (www.springer.com)

Preface

There is growing consensus among the scientific community that global warming caused by increased concentration of greenhouse gases in the atmosphere is one of the most serious environmental problems facing the world today. It is estimated that global mean surface temperature could rise over 6°C by 2100. In addition to global warming, increased greenhouse gas concentrations may increase the occurrence of precipitation extremes: greater precipitation is expected in already-wet areas and increased drought in already-dry areas. Furthermore, widespread expansion of industry and agricultural activities may increase atmospheric nitrogen deposition to unprecedented levels, which will modify climate change impacts. Climate change is also expected to increase the severity and frequency of wildfire, floods, and pest and pathogen attacks. These global environmental changes will pose serious consequences for the overall functioning of terrestrial ecosystems, particularly for agriculture and forestry.

Soil is an important component of terrestrial ecosystems that support life on the earth; it serves as a buffer medium for perturbations to biogeochemical processes of importance to global climate change, acts as a source or a sink for greenhouse gases, and thus underpins social, economic and environmental well-being of humans. Many of the ecosystem services provided by soils are reliant on organic matter. It is therefore critical that land is managed to increase soil organic matter, which will assist in halting rising atmospheric [CO_2], improving soil structure, and decreasing soil erosion and land degradation. In view of changing climate, it is vital that soil health is maintained because a healthy soil is able to sustain physical, chemical and biological functions, and recover following perturbations, due to inherent resilience. A healthy soil enhances plant productivity, promotes plant, animal and human health, maintains water and air quality, supports a diverse community of soil organisms, and resists stresses of human impact and climatic perturbations, so resists environmental degradation. A healthy soil is a complex dynamic living resource that is resilient as a result of its capacity for self-organisation. "Soil health" is thus a broad term, encompassing physical, chemical and biological characteristics, which may be assessed through quantitative measures and also qualitatively expressed indicators. The terms "soil health" and "soil quality" are both used in the literature to describe the capacity of a soil to contribute to ecosystem functions, meet human needs and bear stresses. In this book, we use these two terms synonymously.

Several books on the topic of soil health or soil quality have been published over the last two decades, mainly with a focus on assessing soil health/quality indicators or soil functions in relation to managing soil health in terrestrial ecosystems under existing climatic regimes. The principal objectives of this book are to: (i) present a comprehensive overview of responses of key soil properties or processes to potential impacts of climate change; (ii) highlight the importance, for major conventional and emerging land use systems, of maintaining soil health to mitigate and adapt to climate change impacts; and (iii) describe soil-related feedback processes with implications for plant productivity and climate change. A better understanding of the influences of global environmental changes and land management on soil health is important for ensuring sustainable agro-ecosystems, developing adaptive strategies and sustaining the capacity of soil to meet demands for food, fibre, fodder, timber, and fuel for present and future generations.

Part I provides an overview of the concept of soil health, highlighting the role of soil carbon sequestration for improving soil health and mitigating and adapting to potential impacts of climate change. This section also provides a review of current knowledge about physical, chemical and biological indicators of soil health within the context of climate change and their significance for monitoring impacts of land management and climate change on soil health. Part II focuses on important soil attributes and processes including soil structure, soil pH, soil organic matter, nitrogen cycling, soil respiration and soil biota, and their responses and/or their role in sustaining the environmental functions of soil ecosystems under future climate change scenarios. Part III considers a range of conventional land use systems such as cropping, pastoral, forestry and rangeland, as well as rehabilitated mine-sites, with focus on managing soil health and the processes in these systems that can help to mitigate and adapt to climate change impacts. In Part IV, special attention is given to describing emerging management systems such as organic farming, biochar and bioenergy, and the impact of these systems on soil health and climate sustainability.

With contributions from internationally renowned experts, this book will be a great knowledge resource on the topical area of "soil health and climate change". We believe the book will interest students and researchers in soil, plant and environmental sciences, as well as policy makers and industry stakeholders involved in natural resource management, agricultural development and climate change mitigation through land use management. The compiled information is expected to generate stimulating discussions among scientists and will assist in formulating research aiming to tackle knowledge gaps identified by the contributors.

We thank the authors for their contribution to this volume and appreciate their diligence in responding to reviewers' comments, thereby ensuring high standards. All chapters have been peer-reviewed as per the standards of international scientific journals, and we are thankful to reviewers for providing critical assessment and suggestions that helped in improving the chapters. Last but not least, we also express our thanks to the series editor, Prof. Ajit Varma, and the publisher for providing us the opportunity to edit this book.

Sydney, NSW, Australia

Bhupinder Pal Singh
Annette L. Cowie
K. Yin Chan

Contents

Part I
Soil Health Indicators for the Climate Change

Chapter 1
Soil Health and Climate Change: An Overview

Rattan Lal

1.1 Introduction

Soil health refers to the capacity of soil to perform agronomic and environmental functions. Important among these functions are: agronomic/biomass productivity, response to management and inputs, and resistance to biotic and abiotic stresses. With reference to agricultural land use, soil health refers to its capacity to sustain and support growth of crops and animals while also maintaining and improving the environment. Such definitions of soil health imply an integrated holistic or system-level approach, and are based on the concept that the whole is bigger than sum of its components. Key components include soil properties, processes, and synergistic interactions among them. An integrated approach considers soil as a living system which responds to managerial interventions as does an organism (Kibblewhite et al. 2008). It is in this context that the concept of soil health is similar to that of human health (Magdoff 2001), and is determined by maintenance at an optimum level of key soil properties and processes. Key soil properties important to maintaining good soil health include favorable soil texture and structure or tilth, good internal drainage, optimal water, and nutrient retention capacities and soil reaction. Relevant soil processes include good aeration, low susceptibility to erosion, and strong nutrient cycling. An optimal level of soil organic matter (SOM) content is essential to all key soil properties and processes, which are strong determinants of soil health. To be in good health, a soil must also be relatively free from pests and pathogens including nematodes and weeds, and have adequate nutrient reserves and suitable elemental concentrations and balance. A healthy soil must also have strong resistance to degradation processes and able to recover following a perturbation because of inherent resilience (Magdoff 2001). The term "soil health" is primarily used by farmers, land managers, extension agents, and other practicing professionals.

R. Lal (✉)
Carbon Management and Sequestration Center, The Ohio State University, Columbus, OH 43210, USA
e-mail: lal.1@osu.edu

B.P. Singh et al. (eds.), *Soil Health and Climate Change*, Soil Biology 29,
DOI 10.1007/978-3-642-20256-8_1, © Springer-Verlag Berlin Heidelberg 2011

Table 1.1 Parameters to measure soil quality/soil health and express soil health

Quantifiable soil quality parameters	Qualitative soil health characteristics
Particle size distribution	Texture. feel
Water stake aggregation, mean weight diameter	Tilth, cloddiness
Pore size distribution and total porosity	Internal drainage
Water retention capacity	Droughtiness, inundation
Erodibility	Prone to erosion
Infiltration capacity/rate	Time to ponding
pH	Taste, smell
Cation/anion exchange capacity	Buffering
Electrical conductivity	Salinity
Nutrient concentration and availability	Fertility
Soil organic carbon concentration	Color, smell
Microbial biomass carbon	Biodiversity
Time to recover/restore following disturbance	Resilience

In comparison, the term "soil quality" is used by soil scientists and ecologists (Herrick et al. 1999; Karlen et al. 2003). In this book, are used these two terms synonymously. Soil health denotes numerous functions and ecosystem services provided by a soil. Important among ecosystem services are net primary production (NPP), denaturing, and filtering of pollutants to purify water, improving air quality by scrubbing contaminants, enhancing the environment, and moderating climate at local, regional, and global scales. Therefore, soil quality is assessed by identifying and measuring some key parameters (Table 1.1; see Chap. 2). Parameters used as indicators of soil health are similar to those used for assessing soil quality. However, parameters are assessed and characterized qualitatively for soil health and quantitatively for soil quality (Table 1.1). Because of a strong similarity and often interchangeable use of these terms, several indices of assessing soil quality are also used to assess soil health (Table 1.2). Soil health is assessed using a composite soil health index and several biological indicators (Table 1.2). These key parameters are specific to three distinct but interrelated components: physical, chemical, and biological (Fig. 1.1). The strong interaction among these components (i.e., biophysical, biochemical, and physicochemical) determines soil quality/soil health. However, the same parameters are also relevant to denote soil health, although used in somewhat descriptive and qualitative terms. The term "soil quality" is used under both natural and managed ecosystems, while "soil health" is used for soils managed to grow crops and pastures. Management, characterization, and knowledge about SOM pool are equally important in describing soil health or assessing soil quality. Cycling of carbon (C) through atmosphere–plant–soil continuum, and its transformation and retention among these components (Fig. 1.1), is important to soil health. The C cycle is also important to gaseous composition of the atmosphere, the global climate change, and quality of water as moderated by fate and transport of pollutants and sediments.

Most of the indicators of soil health listed in Table 1.2 are specific to land use or soil management functions. There are only few indices of generic application.

Table 1.2 Assessment of soil health at different scales

Soil	quality/health index	Scale	Application	Reference
1.	Composite soil health index (CSHI)	Field plots	Assess impact of tillage, rotation	Idowu et al. (2009)
2.	Soil biological quality	Field plots	Response to crop management	Gil et al. (2009a, b)
3.	Holistic assessment of soil quality	Landscape	Urban ecosystems	Schindelbeck et al. (2008)
4.	Soil quality index (SQI)	Watershed	Impact of conservation practices	Karlen et al. (2008)
5.	Short-term indicators	Landscape	Mulch management for vegetables	Mochizuki et al. (2008)
6.	Soil biological quality	Field plot	Impacts of pesticides	Korthals et al. (2005)
7.	Soil health dynamic	Field plot	Rotation and cover crop impacts	Carter et al. (2003)
8.	Biological indicators	Field	Organic farming impact	Stockdale and Watson (2009)
9.	Biological indicators	Landscape	Rainfed farming, tree crops	Moreno et al. (2009)
10.	Chemical and biological parameters	Field	Impact of organic farming	Van Diepeningen et al. (2009)
11.	Soil organic carbon	Field	Agricultural land use	Farquharson et al. (2003)

The soil quality index (SQI) by Karlen et al. (2008) is of a wider application. It is based on integration of soil physical, chemical, and biological properties and processes. Several biological indicators (Farquharson et al. 2003; Stockdale and Watson 2009) are specific to organic farming. The index by Schindelbeck et al. (2008) is specifically designed for urban ecosystems. The Cornell Soil Health Index (Gugino et al. 2009) is applicable to arable and urban ecosystems. Other soil quality indices have been proposed by Andrews et al. (2004) and Lal (1994).

The objective of this chapter is to: (1) describe the processes, factors, and causes influencing soil health, (2) discuss the importance of soil organic C (SOC) pool and its management on soil health, and (3) explain the significance of soil C sequestration in off-setting anthropogenic emissions, mitigating climate change by atmospheric enrichment of CO_2 and other greenhouse gases (GHGs), and adapting to possible climatic disruptions through enhancement of soil's buffering capacity.

1.2 Soil Health and Human Health

The strong link between soil health and human health, although known to ancient civilizations, is neither widely recognized nor adequately understood. Indian and Chinese literature has documented the impact of iodine (I) on human health for four to five millennia. Yet, the topic is undervalued by policy makers and planners, often

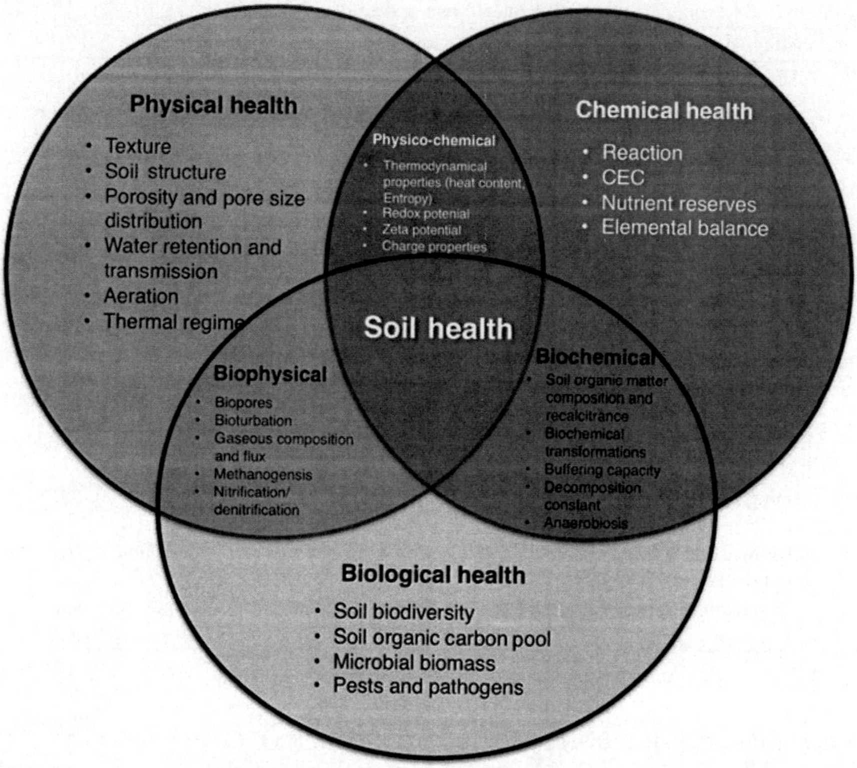

Fig. 1.1 Components of soil health

with severe consequences. For example, food insecurity perpetuates in regions with the highest rate of soil degradation such as in Sub-Saharan Africa and South Asia (Abrahams 2002). In addition to the amount of food production (e.g., cereals, food legumes, roots and tubers, milk and animal products), the quality of food is also strongly impacted by soil health. Important aspects linking soil health with human health through quality of the food produced are the concentrations of micronutrients (i.e., I, Fe, Se, CO, Cr, Cu, F, Mn, Mo, Zn, Ni, Si, V), protein, and essential amino acids in agronomic produce. Soil health can affect human health through deficiency, excess, or imbalance of some of these elements in soils. Soil health also affects human health through its effects on the dietary intake and the environment (Marlow et al. 2009).

The close interaction between soil health, human heath, and environmental quality is depicted in Fig. 1.2. Environmental impacts of soil health are related to its interaction with the hydrosphere that alters water quality, biosphere that moderates NPP and nutrient/elemental uptake, lithosphere that affects stability of the landscape (seismic activities, landslides), and the atmosphere that affects global climate change through alterations in the concentrations of GHGs and other airborne contaminants (e.g., soot, particles, pathogens). The strong link between soil

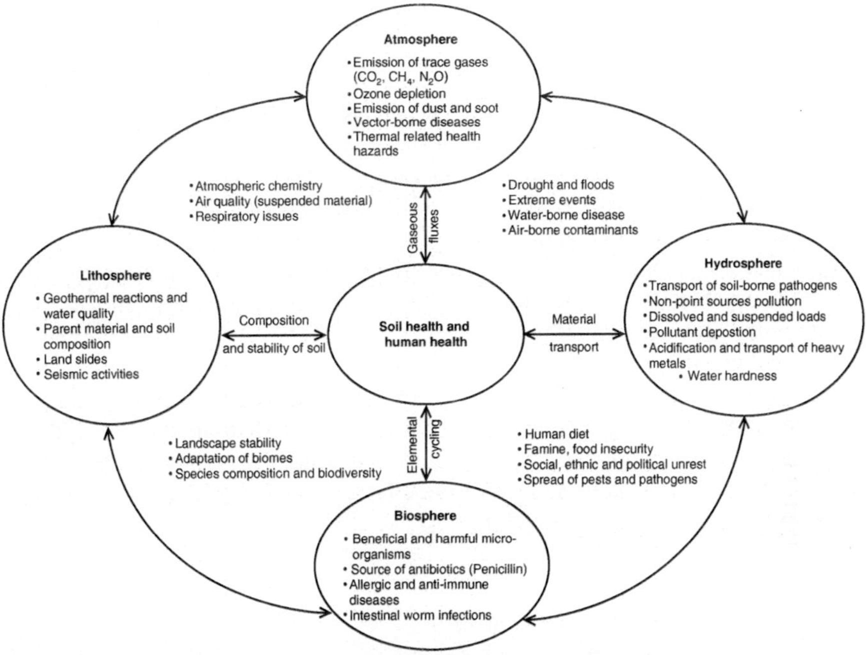

Fig. 1.2 Indirect effects of soil health on human health through interactions with the environment

health and global climate is moderated through storage and emission of carbonaceous (CO_2, CH_4, soot), nitrogenous (N_2O, NO_x), and other organic and inorganic compounds. How humans manage soils influences the production and emission of these gases. The magnitude of gaseous emission is also influenced by the prevailing climate. Thus, there is a positive feedback relating gaseous emissions from soils to changing climate. Climate change and global warming may affect human health through shifts in the geography of vector-borne diseases such as malaria, dengue, and chistosomiasis (Abrahams 2002). Changes in weather patterns may also exacerbate food and water shortages, increase the thermal-related mortality, and aggravate respiratory problems.

1.3 Soil Erosion and Gaseous Emissions

Soil erosion is a four-step process: detachment, transport, redistribution, and deposition. Soil detachment is caused by the kinetic energy of raindrop, overland flow, and wind velocity. The breakdown of aggregates and the detachment of particles are relatively more by wind-driven rain than by drops impacting soil in a windless rain. The kinetic energy from these and other agents of erosion disrupts aggregates and exposes the SOC/SOM, hitherto encapsulated and protected, to microbial/

enzymatic reactions. Thus, breakdown of aggregates increases the rate of SOM decomposition. Aggregate breakdown is also exacerbated by transport and the attendant rolling action along with the shearing force of overland flow and blowing wind. The rate of decomposition is also accentuated by erosion-induced changes in soil temperature and moisture regimes, especially on summit and shoulder slope landscape positions. Under these conditions, erosion accentuates emissions of CO_2 (and also of CH_4 and N_2O) from soil to the atmosphere (Lal 2003). In contrast to the increase in decomposition of SOM during the first three phases of the erosional process (i.e., detachment, transport, and redistribution), SOC transported into aquatic ecosystems and buried in depressional sites is protected against decompositions and is sequestered (Stallald 1998; Van Oost et al. 2007). The net effect of accelerated erosion, as influenced by the four-step process, is increase in emission of GHGs from soil to the atmosphere. Lal (2003) estimated the erosion-induced CO_2 emission of 1.1 Pg C/year globally and 15 Tg C/year in the USA. Thus, adoption of conservation-effective measures and restoration of eroded/degraded/desertified soils (see Chaps. 5 and 9) have a large technical potential to sequester C and mitigate the greenhouse effect.

1.4 Restoring Soil Health in Managed and Disturbed Ecosystems

Choice of strategies for managing soil health depends on the land use, antecedent soil properties, the desired function of interest to humans, and the required ecosystem services. Key soil properties and processes that impact soil health in agricultural, urban, and mine land ecosystems are outlined in Fig. 1.3. Three components of soil health (i.e., physical, chemical, and biological) are also important to sustainable management of croplands and grazing/pasture lands (see Chap. 2). Soil physical health depends on texture, structure, rooting depth, drainage, available water capacity, erodibility, and heat capacity which moderates soil temperature. Determinants of soil chemical health include soil pH, nutrient reserves, surface charge properties, salt concentration and electrical conductivity, and elemental balance. Soil biological health is fundamental to microbial transformations of biomass C into humus and fluxes of GHGs into the atmosphere. Key determinants of soil biological health are magnitude of the SOC pool and its composition, microbial biomass C, soil biodiversity, and prevalence of soil-borne pathogens. In agricultural ecosystems, the goal of soil health management is to maintain and enhance agronomic productivity and economic profitability, and the strategy may differ among land uses and cropping systems (Bell et al. 2007).

Urban ecosystems are gaining importance because of the strong increase in urban populations at global scale. Urban agriculture is also gaining importance because of the need for growing more food locally and creating productive landscapes (Pearson et al. 2010). The multifunctional use of urban green space is

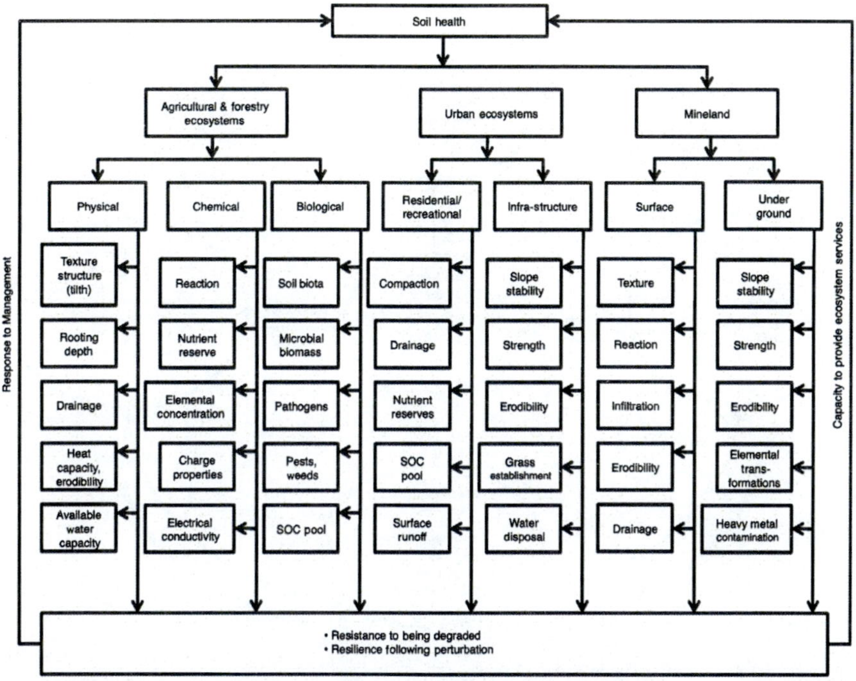

Fig. 1.3 Concepts of soil health for specific land uses and functions

widely recognized (Van Leeuwen et al. 2010). Home lawns and recreational lands, although intensively managed with relatively high inputs, have a large technical potential of soil C sequestration of ~1 Mg C/ha/year (Qian and Follett 2002). Management of landscape for infrastructure (roads, parking areas, airports, etc.) is also important to environments and especially to water quality (Fig. 1.3). Strategies of soil health management comprise sustainable management options for resources including soil, crop, water, and micro-climate (Fig. 1.4).

In addition to advancing food security by growing more grains (cereals and legumes) and roots/tubers (cassava, yam, sweet potato, taro), there is also an increasing interest in producing organic food (Lynch 2009). The goal of organic farming is to develop farm enterprises that are sustainable and harmonious with the environment (Heckman 2006; Srivastava et al. 2007; Watson et al. 2002; see Chap. 14). Organic farming involves adoption of production systems based on reducing external inputs and prohibiting use of synthetic fertilizers, chemical pesticides, and genetically modified organisms (Mason and Spaner 2006; see Chap. 14). Thus, there is a strong reliance on the use of organic nutrient sources for crop production (Rosen and Allan 2007; Srivastava et al. 2009), such as animal manure (Alves et al. 2006; Evanylo et al. 2008), green manure (Chaphale and Badole 1999; Fageria 2007; Khan et al. 2000), residue management and mulch farming (Govaerts et al. 2007; Iqbal et al. 2008; Kachroo et al. 2006; Kornecki et al. 2006;

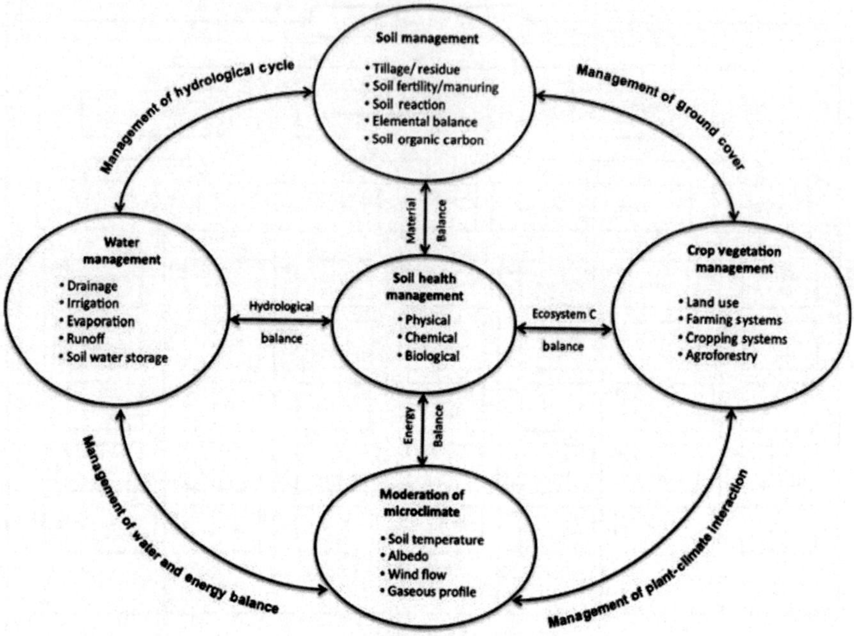

Fig. 1.4 Strategies of soil health management

Lal et al. 2004; Wuest et al. 2000), and use of crop rotations and forages in the rotation cycle (Entz et al. 2002; Katsvairo et al. 2007; Oddino et al. 2008). There is also a growing interest in conservation agriculture (CA) as a system to improve soil health (Burman et al. 2004; Govaerts et al. 2007, 2008; Kassam et al. 2009; Paday et al. 2008; Peiretti 2004; Rainford 2008; Thomas et al. 2007). The goal is to improve soil health through environmentally based practices (Fageria 2002; Narwal 2002). Assessing the environmental impact of organic farming entails measurement of: (1) SOC pools and its dynamics, (2) soil and plant biodiversity, (3) use efficiency of energy, water, and nutrients, (4) nutrient cycling, and (5) emission of GHGs and the attendant impact on climate change (Lynch 2009). In view of the growing population and increasing global food demand, principal constraints to adoption of organic farming are low crop yields due to severe soil nutrient deficiency and high competition from weeds. Yet, organic farming may have specific niches even in densely populated countries of South Asia (Ramesh et al. 2005) because of an increasing public awareness of environmental and health issues. Furthermore, incorporating organic amendments in soil (e.g., manure, compost) and use of residue mulch can suppress soil-borne pathogens, nematodes, and root diseases (Abawi and Widmer 2000). Bailey and Lazarovits (2003) observed that these practices influence pathogen viability and disruption and release of those biologically active substances (from both crop residues and soil microorganisms) that suppress diseases (e.g., root rot in cereals). In this context, long-term use of organic amendments can lead to development of disease-suppressive soils because of improvements in soil's biological health. Thus, some relevant indicators of soil's biological health within organic farming systems include: (1) high biological activity and soil biodiversity, (2) amount and quality of SOC pool, and (3) soil resilience against tillage and mechanical weed control measures. Stockdale and Watson (2009) outlined four principles of organic farming: (1) sustain and enhance the health of the soil, plant animal, human, and planet, (2) strengthen living ecological systems and cycles, and work with them, emulate them, and sustain them, (3) ensure fairness with regard to the common environment and life opportunities, and (4) protect the health and well being of current and future generations and the environment (see Chap. 14).

1.5 Agronomic Strategies of Soil Health Management

In addition to the use of organic amendments (mulching with crop residues), other agronomic strategies of soil health management include conversion to no-till (NT) farming and CA, adoption of integrated nutrient management (INM) involving liberal use of manures and compost, and frequent use of cover crops and green manures in complex crop rotations (Table 1.3). These strategies have proven useful for a range of soils, crops, and eco-regions throughout the world, especially when these have been specifically adapted under site-specific situations. A notable example of improved agronomic practices are those developed for the rice–wheat system (Balasubramanian et al. 2007; Channabasavnna et al. 2002) which were the basis of the "green revolution" in Asia that is continuing to deliver productivity benefits (Table 1.4). Rice, a semi-aquatic plant usually cultivated under submerged conditions, is now increasingly being grown under aerobic environments (Bouman et al. 2007). Water scarcity, especially under dryland farming in arid and semi-arid environments, has necessitated the use of waste and brackish/saline water for irrigation (Tabatabaei and Najafi 2004, 2009), and management of saline soils (Rengasamy 2005). High salinity waste water can also be used for irrigation (Kahlown and Azam 2003) and with drip irrigation technology (Ali 1997).

1.6 Soil Health and Climate Change

Atmospheric concentration of CO_2 has increased by 39% from 280 ppm in the preindustrial era to 390 ppm in 2010, and the present concentration exceeds the range observed over the last 65,000 years (IPCC 2007). Increase in the concentration of CO_2 and other GHGs (CH_4, N_2O) has increased the global mean temperature by $0.76 \pm 0.19°C$ over the twentieth century (IPCC 2007), and with the business as usual, increase in global temperature by the end of the twenty-first century may be $4 \pm 2°C$ with severe adverse impacts on biomes and the ecosystems services.

Table 1.3 Agronomic strategies for soil health management

Technique	Country	Soil	Crop	Impact	References
1. Mulching/residue management	Canada	–	Potato	Suppressing soil-borne diseases	Bailey and Lazarovits (2003)
	Pakistan (Punjab)	Sandy clay loam	Corn	Increase SOC concentration, soil moisture storage and crop yield	Pervaiz et al. (2009)
	India (AP)	Alfisols	Sorghum	Improved soil biological quality (microbial biomass)	Sharma et al. (2008)
	India (Meghalaya)	Inceptisols	Vegetables	Improved soil biological quality (earthworms)	Das et al. (2008)
2. No-till/conservation agriculture	Australia	Limited resource areas	Rice	Improve SOC pool and soil quality	So et al. (2001)
	Australia (Qld)	Vertisols	Sugarcane	Improve sugarcane yield, and decreased detrimental soil biota	Pankhurst et al. (2003)
	Australia (Qld)	Vertisols	Grain crops	Erosion control, water quality, improvement	Silburn et al. (2007)
	Mexico (Central)	Cumulic Phaeozen	Maize, wheat	Increase soil C, infiltration rate, and soil microflora	Govaerts et al. (2007)
	Chile	Ultisols	Wheat–lupin	Improve microbial biomass C and N concentrations	Alvear et al. (2005)
	Canada	Orthic Podzol	Potato	Suppress soil-borne pathogens, improve soil structure	Carter et al. (2009)
	USA (Washington)	Aridisols	Grain crops	Reduce soil erosion, conserve water, improve soil health	Huggins and Reganold (2008)
	India	Alluvial	Wheat	Raised bed system, furrow irrigation	Singh et al. (2009)
3. Manuring/organic farming	Canada	–	Grain crops	Lower crop yields	Lynch (2009)
	India	Vertisols	Soybean, wheat	High microbial biomass C	Behera (2009)
	India	Vertisols/alfisols	Grain crops	Higher SOC pool and microbial biomass C	Vineela et al. (2008)
	USA (IA)	Loess	Vegetables	Better soil quality, higher profit	Delate (2002)
	Turkey	–	Lettece	Lower weeds, higher yields	Isik et al. (2009)
	Holland	–	Grain crops	Higher SOC pool and better soil quality	Van Diepeningen et al. (2006)

4. Forages, cover crops/green manure, rotations	West Africa	Alfisols	Grain legumes	Enhance soil fertility	Odion et al. (2007)
	India	Inceptisols	Wheat–cowpeas	High N-use efficiency	Yadav et al. (2003)
	Italy	–	Meadows, fodders	Weed control	Tomasoni et al. (2003)
	Japan	–	Green pea/wheat	Higher microbial activity	Zai et al. (2008a, b)
	India	Inceptisols	Sebsania in rice	Improved SOC and N pools	Ramesh and Chandrasekaran (2004)
	Philippines	Mahas clay	Mung bean in rice	Better soil health and more labile C	Shrestha et al. (2002)

Table 1.4 Agronomic practices to improve soil health and productivity

Technique	Cropping system	Country	Impact	References
1. Mulching	Rice–wheat	India	Increase water use efficiency, and yield of wheat, weed control; enhanced SOC pool, microbial biomass C, and microflora; improved soil microbial count and soil biological quality	Singh et al. (2008) Pal and Jat (2004) Jat et al. (2004, 2009)
2. Waste water	Sugarcane	Colombia	Better physicochemical quality parameters	Madera et al. (2009)
	Grain crops	India	Lower microbial biomass C and higher respiration rate	Masto et al. (2009)
	Lettuce	Ghana	Increased soil contaminants and health risks	Seidu et al. (2008)
	–	Japan	Increase in denitrification, high salt, and NO_3 concentrations	Shigemastu et al. (2008)
	Maize	Pakistan	Altered soil chemical, bacterial and VAM population	Faryal et al. (2007)
	Cotton, cereals (mixed)	Syria	Waste water treatment is recommended to reduce risks	Ryan et al. (2006)
3. Saline/ brackish water	Fodder	India	Yield reduction, increase in Na^+ uptake	Yadav et al. (2003, 2007)
	Tomato	Iran	Drip irrigation system is recommended	Aminipouri and Ghoddousi (1997) Gil et al. (2009a, b)

Soils have been a major source of atmospheric CO_2 and other GHGs (i.e., CH_4, N_2O) ever since the dawn of settled agriculture (Ruddiman 2003, 2005). The magnitude of CO_2-C emission from soil to the atmosphere since the industrial revolution (~1750 AD) is estimated at 78 ± 12 Pg (Lal 1999). Most soils under agricultural land use contain lower SOC pool than their counterpart under natural/undisturbed ecosystems because of: (1) lower amount of biomass and detritus material returned, (2) higher decomposition rate attributed to changes in soil temperature and moisture regimes, (3) more leaching losses of the dissolved organic C (DOC), and (4) severe losses by accelerated wind and water erosion. Indeed, there is a large flux of CO_2 from the oxidation of SOM from agricultural soils. Thus, most cropland soils have lost 25–75% of their original SOC pool. In Australia, SOC losses of up to 60% have been reported (Bell et al. 2007; see Chap. 9). The loss of SOC pool is more from soils of the tropics than temperate or boreal climates, with coarse (sandy) than fine (clayey) texture, characterized by high internal drainage than those of slow permeability, and low resilience and high susceptibility to degradation (e.g., erosion, nutrient depletion, salinization) than those of high resilience and low vulnerability. The magnitude of loss is also more from soils managed by extractive farming practices (i.e., residue removal, low or no external input of organic or inorganic fertilizers) than those managed by sustainably intensive methods of crop and animal production. Soil health in managed ecosystems is strongly impacted by the magnitude of SOC loss due to historic land use and soil/crop/pasture management practices. Soil health is adversely affected when the SOC level declines below the critical/threshold range. The range of critical level, varying among soils and climates, may be 1.0–2.0% in the topsoil layer (Aune and Lal 1998).

The magnitude of SOC pool and the rate of its decomposition depend on temperature and precipitation (see Chaps. 5 and 7). Therefore, the projected global warming and precipitation extremes may decrease the global SOC pool. Some soils of the boreal and arctic regions (i.e., Cryosols, Histosols) and peat lands which are now a net sink of atmospheric CO_2 may become a major source because of the positive feedback (Baird et al. 2009). In addition to CO_2, agricultural ecosystems are also source of CH_4 (i.e., rice paddies, livestock, manure management) and N_2O (i.e., fertilizers, manuring, biomass burning). These gases have more radiative forcing and higher global warming potential (GWP) than CO_2. Future climate warming may increase emissions of these gases from soils to the atmosphere.

Sustainable management of soil health is an important strategy for climate risk management through adaptation to climate variability (Baethgen 2010). Good soil health can moderate climatic disruptions through (1) reducing emission of CO_2 and other GHGs, and (2) sequestering CO_2 and oxidizing CH_4. The strategy is to convert agriculturally marginal soils to a restorative land use, and adopt recommended management practices (RMPs) on good soils to create a positive C budget such that $C_{input} > C_{output}$. Restorative land uses include establishing a perennial vegetative cover through afforestation, conversion of cropland to pastures with low stocking rate and controlled grazing, and reclamation and rehabilitation of degraded and desertified soils/ecosystems. Because of the severe depletion of the SOC pool, degraded soils have the highest sink capacity for sequestering atmospheric CO_2. Strategies of C_{input} include use of conservation agriculture with crop residue mulch and cover cropping, integrated nutrient management with liberal use of compost and manure in conjunction with chemical fertilizers and organic amendments, and complex cropping/farming systems involving forages and agroforestry. Processes leading to C_{output} from the soil involve losses by accelerated erosion, mineralization or decomposition, and leaching. The projected climate change may exacerbate the losses by erosion, decomposition, and leaching. Therefore, adoption of RMPs must be actively promoted to off-set the losses by enhancing C_{input} into the system (see Chap. 5). In addition to off-setting CO_2 emissions by sequestering C in the pedosphere, soils of a good health may also cause greater oxidation of CH_4 and reduce N_2O emission by moderating both processes of nitrification and denitrification.

The rate of soil C sequestration differs widely among soils, eco-regions, land uses, and management practices. Examples of some proven management practices to enhance SOC pool and improve soil health are listed in Table 1.5. The potential of SOC sequestration is high in desertified and degraded soils (Lal 2001).

Table 1.5 Management practices to enhance soil organic pool and improve soil health and mitigate climate change

Country	Soil	Management practices	Reference
India	Inceptisols	Crop residue management, organic amendments	Mandal et al. (2007)
USA	–	Corn stalk return, no-till	Hooker et al. (2005)
	Alfisols	Mulch	Duiker and Lal (1999)
Australia	Vertisols	Stubble management, fertilization, no-tillage	Farquharson et al. (2003)
China	Black soils (Mollisols)	Manure, rotations, straw management	Liu et al. (2003)
	Alluvial (Yangtze Plains)	Manure, crop residues, conservation tillage	Rui and Zhamg (2010)
UK	–	Conservation agriculture, straw management	Hazarika et al. (2009)

In general, the rate of SOC sequestration is higher in cooler than warmer climates and humid than arid regions. The rate of SOC sequestration is also more in soils of heavier than lighter texture, and in those with slow or poor than rapid or excessive internal drainage. On a global scale, the rate of SOC sequestration through adoption of RMPs is 300–500 kg/ha/year on croplands and 100–300 kg/ha/year on grazing lands or rangelands (see Chap. 5). The rate of SOC sequestration through afforestation of degraded soils (eroded, salinized, depleted of nutrients), especially in humid and subhumid regions, may be as much as 1,000 kg/ha/year. The net rate of sequestration must be adjusted for C emissions from farm operations and inputs (Lal 2004b). Technical potential of SOC sequestration in world soils is 3–4 Pg/year over ~50 years (Pacala and Socalow 2006), which has a drawdown capacity of reducing atmospheric CO_2 concentration by 50 ppm by 2150 AD (Hansen et al. 2008). Furthermore, C sequestration in soils and terrestrial ecosystems is highly cost-effective and an economic alternative (McKinsey and Co 2009). However, some have argued that estimates of SOC sequestration are highly optimistic (Schlesinger 1999, 2000).

The SOC concentration also improves adaptation to climate change by enhancing soil health and its buffering capacity. Enhancement of the SOC concentration, beyond the critical or threshold range, enhances soil health by: (1) improving soil structure and tilth, (2) enhancing plant available water capacity and reducing droughtiness, (3) increasing soils resistance to erosion and reducing erodibility, (4) increasing nutrient retention and availability, (5) increasing water infiltration rate and reducing surface runoff, (6) improving water quality and reducing nonpoint source pollution, (7) increasing soil biodiversity by providing food and habitat for soil biota, (8) reducing sedimentation in waterways and reservoirs, (9) improving use efficiency of inputs, and (10) increasing crop/plant growth and yield. It is also the improvement in soil health through SOC sequestration that is essential to increasing agronomic production (Lal 2004a) and advancing food security (Lal 2006).

1.7 Soil Resilience and Soil Health

Resilience is defined as the capacity to recover from a perturbation while retaining structure, function, identity, and feedbacks (Brand and Jax 2007; Holling et al. 2002; Lal 1997; Walker et al. 2004, 2010; Walker and Salt 2006). A resilient system is characterized by limits or thresholds and tipping points. Change beyond a threshold level in key properties and processes can lead to a tipping point with an attendant change in state. A soil with low resilience can degrade into another state because of the alterations in key soil properties and processes beyond the critical threshold. For example, decline in soil depth by accelerated erosion, decrease in SOC concentration to below the critical range of 1.1%, increase in electrical conductivity beyond 4 ds/m, and decrease in air porosity to <0.1 (Aune and Lal 1998) can drastically alter soil health with adverse impacts on agronomic production and other ecosystem services.

Resilience and transformability, essential attributes of a good soil, depend on soil health. In the context of changing climate and other perturbations, building soil resilience is essential to coping with external changes and meeting growing demands of increasing world population. Soils of good health, with high SOC concentration and favorable properties and processes (physical, chemical, biological) have high resilience. The strategy is to enable soils to recover and adapt rather than resist a change. Adaptability is essential for a soil to be resilient. Defining threshold levels of key soil properties, specific to land uses and ecosystems, is essential to managing and enhancing soil resilience (Lal 1997).

1.8 Eco-Efficiency and Soil Health

The term eco-efficiency, first used at the 1992 Rio Earth Summit for implementing Agenda 21, implies achieving more with less (Keating et al. 2010). With reference to agricultural ecosystems, the term eco-efficiency implies enhancing quantity and quality of agricultural produce with lower input of land, nutrients, water, energy, labor, etc. It also means slower depletion of soil's inherent fertility comprising the SOC pool, nutrient reserve, topsoil depth, etc. In agronomic terms, eco-efficiency is measured as crop productivity per unit area, time and input. It is also measured in terms of productivity per unit decline in SOC pool, topsoil depth, CEC, nutrient reserve, microbial biomass C, etc.

Considering the need of feeding 9.2 billion people by 2050 with limited availability of soil and water and competing uses for natural resources, it is important to identify, develop, validate, and use eco-efficient approaches to enhance agronomic production. The strong relationship between agronomic yield and SOC concentration in the root zone is widely recognized (Lal 2010). Such a relationship between agronomic yield and SOC concentration is especially strong in low-input (extractive farming) agriculture practiced by resource-poor and small-land holders of the tropics and subtropics. Increasing SOC pool in these systems enhances eco-efficiency through improvements in: (1) soil structure and tilth, (2) water retention, (3) nutrient retention, (4) biotic activity and species diversity, (5) erosion control, etc. (Lal 2010). The eco-efficiency can be improved by minimizing soil erosion, conserving water in the root zone, recycling plant nutrients, optimizing soil temperature regime, creating positive C budget, and enhancing soil resilience. Two among several examples of eco-efficient systems are presented by Kapkiyai et al. (1999) from Kenya and Wani et al. (2003) from Central India. In Kenya, Kapkiyai and colleagues demonstrated that agronomic productivity was enhanced from 1.4 Mg/ha/year in traditional systems to 6.0 Mg/ha/year with RMP, which also enhanced the SOC pool in the root zone from 23.6 Mg/ha to 28.7 Mg/ha over 18-year period. In Central India, Wani et al. (2003) reported from a 30-year study that adoption of RMPs increased productivity from 1.1 Mg/ha/year to 5.1 Mg/ha/year along with SOC sequestration of 330 kg/ha/year. The yield gap, difference in agronomic yield under on-station vis-à-vis under on-farm conditions, that exists throughout the developing countries can be abridged by enhancing eco-efficiency of production systems. Improving soil health, through enhancing SOC concentrations and quality, can also enhance eco-efficiency of production systems (Lal 2010).

1.9 Conclusions

Soil health, capacity of a soil to produce agronomic and economic goods and services while also maintaining the environment quality, is a term used by farmers and land managers. In comparison, the term soil quality is used by soil scientists, agronomists, and pedologist. Key indicators of soil health, similar to that of soil quality, are soil structure, soil organic carbon concentration and quality, water retention and intake rate, and soil biodiversity (see Chap. 2). Thus, maintaining and enhancing these soil properties above the threshold/critical levels are essential to sustaining/improving soil health. Enhancing the soil organic carbon pool also improves agro-ecosystem resilience, eco-efficiency, and adaptation to climate change. Technical potential of soil C sequestration through improvement in soil health is ~3 Pg/year for about 50 years with a drawdown capacity of reducing atmospheric CO_2 concentration by 50 ppm over the twenty-first century. Improving soil health, through restoration of degraded/desertified soil and adoption of RMPs, is also a necessity to feeding the world population of 9.2 billion by 2050.

References

Abawi GS, Widmer TL (2000) Impacts of soil health management practices on soilborne pathogens, nematodes and root diseases of vegetable crops. Appl Soil Ecol 15:37–47

Abrahams PW (2002) Soils: their implications to human health. Sci Total Environ 291:1–32

Ali I (1997) Agricultural reuse of high salinity wastewater through drip irrigation. In: Aminipouri B, Ghoddousi J (eds) Proceedings of the 8th international conference of rainwater catchment systems, vol 1 and 2 – rainwater catchment for survival. Ministry of Jihad-e-Sazandegi, Tehran, pp 1069–1075

Alvear M, Rosas A, Rouanet JL, Borie F (2005) Effects of three soil tillage systems on some biological activities in an Ultisol from southern Chile. Soil Till Res 82:195–202

Alves MC, Gonzalez AP, Colodro G, Perecin H, Vazquez EV (2006) Influence of biosolids rate on chemical properties of an oxisol in Sao Paulo, Brazil. Commun Soil Sci Plant Anal 37:2481–2493

Aminipouri B, Ghoddousi J (1997) Agricultural use of high salinity waste water through drip irrigation. In: Proceedings of the 8th international conference of rain water catchment systems, Tehran, pp 1069–1075

Andrews SS, Karlen DL, Camberdella CA (2004) The soil management assessment framework: a quantitative soil quality evaluation method. Soil Sci Soc Am J 68:1945–1962

Aune J, Lal R (1998) Agricultural productivity in the tropics and critical limits of properties of Oxisols, Ultisols and Alfisols. Trop Agric 74:96–103

Baethgen WE (2010) Climate risk management for adaptation to climate variability and change. Crop Sci 50:S70–S76

Bailey KL, Lazarovits G (2003) Suppressing soil-borne diseases with residue management and organic amendments. Soil Till Res 72:169–180

Baird AJ, Belyea LR, Comas X, Reeve AS, Slater LD (2009) Carbon cycling in northern peatlands, vol 184, Geophysical monograph. American Geophysical Union, Washington

Balasubramanian V, Sie M, Hijman RJ, Otsuka K (2007) Increasing rice production in Sub-Sarharan Africa: challenges and opportunities. Adv Agron 94:55–133

Behera UK (2009) Organic manuring for soil biological health and productivity of a wheat-soybean cropping system in the Vertisols of central India. Crop Pasture Sci 60:382–393

Bell MJ, Stirling GR, Pankhurst CE (2007) Management impacts on health of soils supporting Australian grain and sugarcane industries. Soil Till Res 97:256–271

Bouman BAM, Lampayan RM, Toung TP (2007) Water management in irrigated rice: coping with water scarcity. IRRI, Los Banos

Brand FS, Jax K (2007) Focusing the meanings(s) of resilience: resilience as a descriptive concept and a boundary object. Ecol Soc 12:23

Burman RA, Alesii BA, Hatfield JL, Karlen DL (2004) Profit, yield and soil quality effects of tillage systems in corn-soybean rotations. J Soil Wat Conserv 59:260–270

Carter MR, Kunelius HT, Sanderson JB, Kimpinski J, Platt HW, Bolinder MA (2003) Productivity parameters and soil health dynamics under long-term 2-year potato rotations in Atlantic Canada. Soil Till Res 72:153–168

Carter MR, Peters RD, Noronha C, Kimpinski J (2009) Influence of 10 years of conservation tillage on some biological properties of a fine sandy loam in the potato phase of two crop rotations in Atlantic Canada. Can J Soil Sci 89:391–402

Channabasavnna AS, Itnal CJ, Patil SG (2002) Productivity, economic analysis and changes in physico-chemical properties of soil as influenced by integrated rice (*Oryza sativa*)-based farming systems. Indian J Agron 47:1–5

Chaphale SD, Badole WP (1999) Effect of green manuring and NPK combinations on soil health and yield of rice (*Oryza sativa*). Indian J Agron 44:448–451

Das A, Patel DP, Munda GC, Hazarika UK, Bordoloi J (2008) Nutrient recycling potential in rice-vegetable cropping sequences under in situ residue management at mid-altitude subtropical Meghalaya. Nutr Cycl Agroecosys 82:251–258

Delate K (2002) Using an agroecological approach to farming systems research. HortTechnology 12:435–354

Duiker SJ, Lal R (1999) Crop residues and tillage effects on carbon sequestration in a Luvisol in central Ohio. Soil Till Res 52:73–81

Entz MH, Baron VS, Carr PM, Meyer DW, Smith SR, McCaughey WP (2002) Potential of forages to diversify cropping systems in the northern Great Plains. Agro J 94:240–250

Evanylo G, Shrony C, Spargo J, Starner D, Brosius M, Haering K (2008) Soil and water environmental effects on fertilizer-, manure-, and compost-based fertility practices in an organic vegetable cropping system. Agr Ecosyst Environ 127:50–58

Fageria NK (2002) Soil quality vs. environmentally-based agricultural management practices. Commun Soil Sci Plant Anal 33:2301–2329

Fageria NK (2007) Green manuring in crop production. J Plant Nutr 30:691–719

Farquharson RJ, Schwenke GD, Mullen JD (2003) Should we manage soil organic carbon in Vertosols in the northern grains region of Australia? Aust J Exp Agr 43:261–270

Faryal R, Rahir F, Hameed A (2007) Effect of wastewater treatment system: historical development. Pak J Bot 39:193–204

Gil SV, Becker A, Oddino C, Auaz M, Marinelli A, March G (2009a) Field assessment of biological, chemical and physical responses of soil to tillage intensity, fertilization and grazing. Environ Manag 44:378–386

Gil SV, Meriles J, Confornta C, Figoni G, Basanta M, Lovera E, March GJ (2009b) Field assessment of soil biological and chemical quality in response to crop management practices. World J Microbiol Biotechnol 25:439–448

Govaerts B, Mezzalama M, Unno Y, Sayre KD, Luna-Guideo M, Vanherck K, Dendooven L, Deckers J (2007) Influence of tillage, residue management and crop rotation on soil microbial biomass and catabolic diversity. Appl Soil Ecol 37:18–30

Govaerts B, Mezzalama M, Sayre KD, Crossa J, Lichter K, Troch V, Vanherck K, De Corte P, Deckers J (2008) Long-term consequences of tillage, residue management, and crop rotation on selected soil micro-flora groups in the subtropical highlands. Appl Soil Ecol 38:197–210

Gugino BK, Idowu OJ, Schindelbeck RR, Van Es HM, Wolfe DW, Moebius-Clume BN, Thies JE, Abawi GS (2009) Cornell soil health assessment training manual, 2nd edn. Cornell University, Ithaca

Hansen JE, Sato M, Kharecha P, Beerling D, Berner R, Masson-Delmotte V, Pagani M, Raymo M, Royer DL, Zachos JC (2008) Target atmosphere CO_2: where should humanity aim? Open Atmos Sci J 2:217–231

Hazarika S, Parkinson R, Bol R, Dixon L, Russel P, Donovan S, Allen D (2009) Effect of tillage system and straw management on organic matter dynamics. Agro Sustain Dev 29:525–533

Heckman J (2006) A history of organic farming: transitions form Sir Albert Howard's what in the soil to USDA national organic program. Renew Agric Food Syst 21:143–150

Herrick JE, Brown JR, Tugel AJ, Dhaver PL, Havstad KM (1999) Application of soil quality to monitoring and management: paradigms form rangeland ecology. Agro J 94:3–11

Holling CS, Gunderson LH, Peterson GD (2002) Sustainability and panarchies. In: Gunderson LH, Holling CS (eds) Panarchy: understanding transformations in human and natural systems. Island Press, Washington, pp 63–102

Hooker BA, Morris TF, Peters R, Cardon ZG (2005) Long-term effects of tillage and corn stalk return on soil carbon dynamics. Soil Sci Soc Am J 69:188–196

Huggins DR, Reganold JP (2008) No-till: the quiet revolution. Sci Am 299:70–77

Idowu OJ, van Es HM, Abawi GS, Wolfe DW, Schindelbeck RR, Moebius-Clune BN, Gugino BK (2009) Use of an integrative soil health test for evaluation of soil management impacts. Renew Agric Food Sys 24:214–224

IPCC (2007) The climate change 2007: the physical science basis. Cambridge University Press, Cambridge

Iqbal M, Hassan AU, Ibrahim M (2008) Effects of tillage systems and mulch on soil physical quality parameters and maize (*Zea mays* L.) yield in semi-arid Pakistan. Biol Agric Hortic 25:311–325

Isik D, Kaya E, Ngouajio M, Mennan H (2009) Summer cover crops for weed management and yield improvement in organic lettuce (*Lactuca sativa*) production. Phytoparasitica 37:193–203

Jat ML, Pal SS, Shukla L, Mathur JMS, Singh M (2004) Rice (*Oryza sativa*) residue management using cellulolytic fungi and its effect on wheat (*Triticum aestivum*) yield and soil health in rice-wheat cropping system. Indian J Agric Sci 74:117–120

Jat ML, Gathala MK, Ladha JK, Daharawat YS, Jat AS, Kumar V, Sharma SK, Kumar V, Gupta R (2009) Evaluation of precision land leveling and double zero-till systems in the rice-wheat rotation: water use, productivity, profitability and soil physical properties. Soil Till Res 105:112–121

Kachroo D, Dixit AK, Bali AS (2006) Influence of crop residue, fly ash and varying started dosages on growth, yield and soil characteristics in rice (*Oryza sativa*)-wheat (*Triticum aestivum*) cropping system under irrigated conditions of Jammu region. Indian J Agric Sci 76:3–6

Kahlown MA, Azam M (2003) Effect of saline drainage effluent on soil health and crop yield. Agric Water Manage 62:127–138

Kapkiyai JJ, Laranja NK, Qureshi JN, Smithson PC, Woomer PL (1999) Soil organic matter and nutrient dynamics in a Kenyon Nitosol under long-term fertilizers and organic input management. Soil Biol Biochem 31:1773–1782

Karlen DL, Ditzer CA, Andrews SS (2003) Soil quality: why and how? Geoderma 114:145–156

Karlen DL, Tomer MD, Deppel J, Cambardella CA (2008) A preliminary watershed scale soil quality assessment in north central Iowa, USA. Soil Till Res 99:291–299

Kassam A, Friedrich T, Shaxson F, Pretty J (2009) The spread of conservation agriculture: justification, sustainability and uptake. Int J Agric Sustain 7:292–320

Katsvairo TW, Wright DL, Marois JJ, Rich JR (2007) Transition from conventional farming to organic farming using bahiagrass. J Sci Food Agric 87:2751–2756

Keating B, Carberry PS, Bindragen PS, Asseng S, Meinke H, Dixon J (2010) Eco-efficient agriculture: concepts, challenges and opportunities. Crop Sci 50:S109–S119

Khan AR, Ghori AK, Singh SR (2000) Improvement of crop and soil sustainability through green manuring in a rainfed lowland rice ecosystem. Agrochimica 44:21–29

Kibblewhite MG, Ritz K, Swift MJ (2008) Soil health in agricultural systems. Philos Trans R Soc A 363:685–701

Kornecki TS, Grigg BC, Fouss JL, Southwick LM (2006) Effectiveness of post-harvest sugarcane residue and polyacrylamide on reducing deposition in quarter-drains. Appl Eng Agric 22:857–865

Korthals GW, Vlsser JHM, Molendijk LPG (2005) Improvement and monitoring soil health. In: Vanachter A (Ed) Proceedings of the 6th international symposium on chemical and non-chemical soil and substrate disinfestations, Corfu (698), pp 279–283

Lal R (1994) Methods and guidelines for assessing sustainable use of soil and water resources in the tropics. USDA/SMSS Bull. 21, Washington, 78 pp

Lal R (1997) Degradation and resilience of soils. Philos Trans R Soc B 353:997–1010

Lal R (1999) Soil management and restoration for carbon sequestration to mitigate the greenhouse effect. Prog Env Sci 1:307–326

Lal R (2001) Potential of desertification control to sequester carbon and mitigate the greenhouse effect. Clim Change 15:35–72

Lal R (2003) Soil erosion and the global carbon budget. Environ Int 29:437–450

Lal R (2004a) Soil carbon sequestration impact on global climate change and good security. Science 304:1623–1627

Lal R (2004b) Carbon emissions from farm operations. Environ Int 30:981–990

Lal R (2006) Enhancing crop yield in developing countries through restoration of soil organic carbon pool in agricultural lands. Land Degrad Dev 17:197–209

Lal R (2010) Beyond Copenhagen: mitigating climate change and achieving food security through soil carbon sequestration. Food Secur 2:169–177

Lal B, Rajput DS, Singh SD, Gupta HP, Suman M (2004) Tillage and residue management impacts on fodder sorghum production: yield, energy, and soil properties. Trop Agric 81:229–235

Liu XB, Han XZ, Song CY, Herbert SJ, Xing BS (2003) Soil organic carbon dynamics in black soils of China under different agricultural management systems. Commun Soil Sci Plant Anal 34:973–984

Lynch D (2009) Environmental impacts of organic agriculture: a Canadian perspective. Can J Plant Sci 89:621–628

Madera CA, Silva J, Mara DD, Torres P (2009) Wastewater use in agriculture: irrigation of sugarcane with effluents from the Canaveralejo wastewater treatment plant in Cali, Colombia. Environ Technol 30:1011–1015

Magdoff F (2001) Concept, components and strategies of soil health in agroecosystems. J Nematol 33:169–172

Mandal B, Majumder B, Bandyopadhyay PK, Hazra GC, Gangopadhyay A, Samantaray RN, Mishra AK, Chaudhury J, Saha MN, Kundu S (2007) The potential of cropping systems and soil amendments for carbon sequestration in soils under long-term experiments in subtropical India. Global Change Biol 13:357–369

Marlow HJ, Hayes WK, Soret S, Carter RL, Schwab ER, Sabate J (2009) Diet and the environment: does what you eat matter? Am J Clin Nutr 89:S1699–S1703

Mason HE, Spaner D (2006) Competitive ability of wheat in conventional and organic management systems: a review of the literature. Can J Plant Sci 86:333–343

Masto RE, Chhonkar PK, Singh D, Patra AK (2009) Changes in soil quality indicators under long-term sewage irrigation in a sub-tropical environment. Environ Geol 56:1237–1243

McKinsey & Co (2009) Pathways to a low-carbon economy, version 2 of the global greenhouse abatement cost curve. Mckinsey & Co, New York

Mochizuki MJ, Rangarajan A, Bellinder RR, van Es HM, Bjorkman T (2008) Rye mulch management affects short-term indicators of soil quality in the transition to conservation tillage for cabbage. Hortscience 43:862–867

Moreno B, Garcia-Rodriguez S, Canizaers R, Castro J, Benitez E (2009) Rainfed olive farming in south-eastern Spain: long-term effect of soil management on biological indicators of soil quality. Agric Ecosyst Environ 131:333–339

Narwal SS (2002) Crop allelopathy for weed management in sustainable agriculture. In: Reigosa MJ, Pedrol N (eds) Allelopathy: from molecules to ecosystems. Scientific Publishers Inc, Enfield, pp 209–229

Oddino CM, Marinelli AD, Auza M, March GJ (2008) Influence of crop rotation and tillage on incidence of brown root rot of peanut caused by Fusarium solani in Argentina. Can J Plant Pathol 30:575–580

Odion EC, Asiribo OE, Ogunlela VB, Singh BB, Tarawali SA (2007) Strategies to improve and sustain food production capacity in the savanna: the role of leguminous fodder crops in maintaining soil fertility and health. J Food Agric Environ 5:338–344

Pacala S, Socalow R (2006) Stabilization wedges: solving the climate problem for the next 50 years with current technologies. Science 305:968–972

Paday SC, Singh RD, Sahai S, Singh KP, Prakashi V, Kumar A, Kumar M, Srivastava AK (2008) Effect of tillage and irrigation on yield, profitability, water productivity and soil health in rice (*Oryza sativa*) wheat (*Triticum aestivum*) cropping system in north-west Himalayas. Indian J Agric Sci 78:1018–1022

Pal SS, Jat ML (2004) Enhanced decomposition of rice (*Oryza sativa*) straw and its effect on soil health and crop yield in rice-wheat (*Triticum aestivum*) cropping system in an Ustochrept of Indo-Gangetic plain. Indian J Agric Sci 74:409–414

Pankhurst CE, Magarey RC, Stirling GR, Blair BL, Bell MJ, Garside AL (2003) Management practices to improve soil health and reduce the effects of detrimental soil biota associated with yield decline of sugarcane in Queensland, Australia. Soil Till Res 72:125–137

Pearson LJ, Pearson L, Pearson CJ (2010) Sustainable urban agriculture: stock take and opportunities. Int J Agric Sustain 8:7–19

Peiretti RA (2004) No till cropping system and its evolution toward the achievement of the moshpa model principles (modern sustainable high productivity agricultural model). In: Moscardi F, HoffmannCampo CB, Saraiva OF, Calerani PR, Krzyanowski FC, CarraoPanizzi MC (eds) 7th world soybean research conference – 6th international soybean processing and utilization conference – 3rd Brazilian soybean congress, Foz do Iguacu

Pervaiz MA, Iqbal M, Shahzad K, Hassan AU (2009) Effect of mulch on soil physical properties and N, P, K concentration in maize (*Zea mays*) shoots under two tillage systems. Int J Agric Biol 11:119–124

Qian Y, Follett RF (2002) Assessing soil carbon sequestration in turfgress systems using long-term soil testing data. Agron J 94:930–935

Rainford C (2008) Soil health and productivity benefits of low-tillage organic systems: the common ground between organic and no-tilt farming practices. J Soil Water Conserv 63:19A–21A

Ramesh K, Chandrasekaran B (2004) Soil organic carbon build-up and dynamics in rice-rice cropping system. J Agron Crop Sci 190:21–27

Ramesh P, Singh M, Rao AS (2005) Organic farming: its relevance to the Indian context. Curr Sci 88:561–568

Rengasamy P (2005) World salinization with emphasis on Australia. J Exp Bot 57:1017–1023

Rosen CJ, Allan DL (2007) Exploring the benefits of organic nutrient sources for crop production and soil quality. HortTechnology 17:422–430

Ruddiman WF (2003) The anthropogenic greenhouse era began thousands of years ago. Clim Change 61:262–292

Ruddiman WF (2005) How did human first alter global climate. Sci Am 292:429–436

Rui WY, Zhamg WJ (2010) Effect size and duration of recommended management practices on carbon sequestration in paddy field in Yangtze delta plain of China: a meta-analysis. Agric Ecosyst Environ 135:199–205

Ryan J, Masrj S, Qadir M (2006) Nutrient monitoring of sewage water irrigation: impacts for soil quality and crop nutrition. Commun Soil Sci Plant Anal 37:2513–2521

Schindelbeck RR, van Es HM, Abawi GS, Wolf DW, Whitlow TL, Gugino BK, Idowu OJ, Moebius-Clune BN (2008) Comprehensive assessment of soil quality for landscape and urban management. Landsc Urban Plan 88:73–80

Schlesinger WH (1999) Carbon sequestration in soil. Science 284:2095

Schlesinger WH (2000) Carbon sequestration in soils: some cautions amidst optimism. Agric Ecosyst Environ 82:121–127

Seidu R, Heistad A, Amoah P, Drechsel P, Jenssen PD, Stenstrom TA (2008) Quantification of the health risk associated with wastewater reuse in Accra, Ghana: a contribution towards local guidelines. J Water Health 6:461–471

Sharma KL, Gajbhiye PN, Mandal UK, Korwar GR, Grace JK, Madhavi M, Yadav SK, Srinivas K, Ramesh V, Rani KU (2008) Effects of residue recycling on biological soil health indicators and surface CO_2 emissions in rainfed semi-arid tropical Alfisol under minimum tillage conditions. J Agrometeorol 10:46–50

Shigemastu T, Dayanthi WKCN, Yamashita N, Tana H, Tamashita T (2008) The first introduction of reclaimed wastewater to dry-farming field in Okinawa Island, Japan. Water Sci Technol 57:1161–1167

Shrestha RK, Ladha JK, Lefroy RDB (2002) Carbon management for sustainability of an intensively managed rice-based cropping system. Biol Fertil Soils 36:215–223

Silburn DM, Freebairn DM, Rattray DJ (2007) Tillage and the environment in sub-tropical Australia-Tradeoffs and challenges. Soil Till Res 97:306–317

Singh B, Shan YH, Johson-Beebout SE, Singy Y, Buresh RJ (2008) Crop residue management for lowland rice-based cropping systems in Asia. Adv Agron 98:117–119

Singh Y, Humphreys E, Kukal SS, Singh B, Kaur A, Thaman S, Prashar A, Yadav S, Timisina J, Shillon SS, Kaur N, Smith DJ, Gajari PR (2009) Crop performance in permanent raised bed rice-wheat cropping system in Punjab, India. Field Crops Res 110:1–20

So HB, Kirchhof G, Bakker R, Smith GD (2001) Low input tillage/cropping systems for limited resource areas. Soil Till Res 61:109–123

Srivastava R, Roseti D, Sharma AK (2007) The evaluation of microbial diversity in a vegetable based cropping system under organic farming practices. Appl Soil Ecol 36:116–123

Srivastava TK, Lal M, Singh KP, Suman A, Kumar P (2009) Enhancing soil health and sugarcane productivity in a plant-ratoon system through organic nutrient module in sub-tropics. Indian J Agric Sci 79:346–350

Stallald RF (1998) Terrestrial sedimentation and the carbon cycle: coupling weathering and erosion to carbon burial. Glob Biogeochem Cycles 12:231–257

Stockdale EA, Watson CA (2009) Biological indicators of soil quality in organic farming systems. Renew Agri Food Sys 24:308–318

Tabatabaei SH, Najafi P (2004) Effects of treated municipal wastewater injection with different irrigation methods on soil quality parameter. In: G. Huang and L.S. Pereira (eds) Land and water management: decision tools and practices, vol 1 and 2. pp 959–965, CGIAR Int. Conf., Beijing, China, 11–14 Oct. 2004

Tabatabaei SH, Najafi P (2009) Effects of irrigation with treated municipal wastewater on soil properties in arid and semi-arid regions. Irrig Drain 58:551–560

Thomas GA, Titmarsh GW, Freebairn DM, Radford BJ (2007) No-tillage and conservation farming practices in grain growing areas of Queensland – a review of 40 years of development. Aust J Exp Agr 47:887–898

Tomasoni C, Borrelli L, Pecetti L (2003) Influence of fodder crop rotations on the potential weed flora in the irrigated lowlands of Lombardy, Italy. Eur J Agron 19:439–451

Van Diepeningen AD, de Vos OJ, Korthals GW, van Bruggen AHC (2006) Effects of organic versus conventional management on chemical and biological parameters in agricultural soils. Appl Soil Eco 32:120–135

Van Leeuwen E, Nijkamp P, Vag TDN (2010) The multifunctional use of urban green space. Int J Agric Sustain 8:20–25

Van Oost K, Quine TA, Govers G, De Gryze S, Six J, Harden JW, Ritchie JC, McCarty GW, Heckrath G, Kosmas C, Giraldez JV, Marques da Silva JR, Merckx R (2007) The impact of agricultural soil erosion on the global carbon cycle. Science 318:626–629

Vineela C, Wani SP, Srinivasarao C, Padmaja B, Vittal KPR (2008) Microbial properties of soils as affected by cropping and nutrient management practices in several long-term manorial experiments in the semi-arid tropics of India. Appl Soil Ecol 40:165–173

Walker B, Salt D (2006) Resilience thinking: sustaining ecosystems and people in a changing world. Island Press, Washington

Walker BH, Holling CS, Carpenter SC, Kinzig AP (2004) Resilience adaptability and transformability. Ecol Soc 9:5

Walker B, Sayer J, Andrew NL, Campbell B (2010) Should enhanced resilience be an objective of natural resource management research for developing countries. Crop Sci 50:510–519

Wani SP, Pathak P, Jangawad LS, Eswaran H, Singh P (2003) Improved management of Vertisols in the semi-arid tropics for increased productivity and soil carbon sequestration. Soil Use Manag 19:217–222

Watson CA, Atkinson D, Gosling P, Jackson LR, Rayns FW (2002) Managing soil fertility in organic farming systems. Soil Use Manag 18:239–247

Wuest SB, Albrecht SL, Skirvin KW (2000) Crop residue position and interference with wheat seedling development. Soil Till Res 55:175–182

Yadav RL, Singh VK, Dwivedi BS, Shukla AK (2003) Wheat productivity and N use-efficiency as influenced by inclusion of cowpea as a grain legume in a rice-wheat system. J Agri Sci Part 2 (141):213–220

Yadav RK, Singh SP, Lal D, Kumar A (2007) Fodder production and soil health with conjunctive use of saline and good quality water in ustipsamments of a semi-arid region. Land Degrad Dev 18:153–161

Zai AKE, Horiuchi T, Matsui T (2008a) Effects of compost and green manure of pea and their combinations with chicken manure and rapeseed oil residue on soil fertility and nutrient uptake in wheat-rice cropping system. Afr J Agric Res 3:633–639

Zai AKE, Horiuchi T, Matsui T (2008b) Effects of green manure and compost of pea plant on wheat. Compost Sci Util 16:275–284

Chapter 2
Soil Health Indicators Under Climate Change: A Review of Current Knowledge

Diane E. Allen, Bhupinder Pal Singh, and Ram C. Dalal

2.1 Soil Health Indicators, Soil Health and Climate Change: Background

Soil health indicators are a composite set of measurable physical, chemical and biological attributes which relate to functional soil processes and can be used to evaluate soil health status, as affected by management and climate change drivers. Defining soil health in relation to climate change should consider the impacts of a range of predicted global change drivers such as rising atmospheric carbon dioxide (CO_2) levels, elevated temperature, altered precipitation (rainfall) and atmospheric nitrogen (N) deposition, on soil chemical, physical and biological functions (French et al. 2009). Many studies have progressed our understanding of relationships between particular soil properties and climate change drivers, e.g. responses to temperature, CO_2 or rainfall; however, Wixon and Balser (2009) note that "a comprehensive explanation of the factors at the heart of the issue is currently lacking". Determination of how predicted changes in climate relate to soil health will thus depend on our capacity to clearly define soil health properties and their relationship with specific soil functions, including complexity associated with interactive effects of climate change.

Note that both the terms "soil health" and "soil quality" are used synonymously in this chapter, although it is realised that the former term gives greater emphasis on soil biodiversity and ecological functions that make soil a dynamic living resource with capacity for self-organisation.

D.E. Allen (✉) • R.C. Dalal
Department of Environment and Resource Management, Environment and Resource Sciences, 41 Boggo Road, Dutton Park, QLD 4102, Australia
e-mail: Diane.Allen@derm.qld.gov.au

B.P. Singh
Forest Science Centre, NSW Department of Primary Industries, PO Box 100, Beecroft, NSW 2119, Australia

B.P. Singh et al. (eds.), *Soil Health and Climate Change*, Soil Biology 29,
DOI 10.1007/978-3-642-20256-8_2,

Indicators, calculated values or estimated statistics *relative* to a threshold level are being increasingly used across biological, environmental, economical, social, institutional and political disciplines to assess current condition or trend of soil health (Dalal et al. 2003a, b; Riley 2001c). Indicators may be used as an indirect measure of soil function, serving to assess soil quality or health and its direction of change with time, by linking functional relationships among measurable attributes and monitoring for sustainable land management, including environmental impacts (Dalal et al. 2003a, b; Doran 2002; Doran and Zeiss 2000).

Kinyangi (2007) proposes that soil health assessment involves an evaluation process consisting of a series of actions: (1) selection of soil health indicators, (2) determination of a minimum data set, (3) development of an interpretation scheme of indices, and (4) on-farm assessment and validation. While the use of integrative soil health tests is increasing (e.g. Idowu et al. 2009), limited information exists to evaluate the applicability of soil health indicators for monitoring soil functions within the context of climate change.

A range of frameworks identifying soil health indicators to measure change and implement strategies to adapt to climate change exist (Dalal et al. 1999; Dalal and Moloney 2000; Doran 2002; Kibblewhite et al. 2008; Lal 1999; Nuttall 2007; Schjønning et al. 2004; Stenberg 1999). Potential linkages between soil health indicators, land management and climate change from these reviews are summarised in Fig. 2.1. Elevated CO_2 concentration, increasing temperature, atmospheric N deposition and changes in total and seasonal distribution of rainfall and extreme events such as droughts and floods will impact on soil biological processes, C and N cycling, and consequently on soil structure and erosion events, nutrient availability and plant diseases, and hence on ecosystem functionality and agricultural productivity.

Major soil physical, chemical and biological properties which may indicate the status of soil health in relation to climate change impacts are listed in Table 2.1 and outlined below.

2.2 Soil Physical Properties, Soil Health and Climate Change

Soil physical properties provide information related to water and air movement through soil, as well as conditions affecting germination, root growth and erosion processes. Many soil physical properties thus form the foundation of other chemical and biological processes, which may be further governed by climate, landscape position and land use. A range of soil physical properties are highlighted as potential soil health indicators, and key soil physical indicators in relation to climate change include soil structure, water infiltration, bulk density, rooting depth, and soil surface cover, which are discussed below.

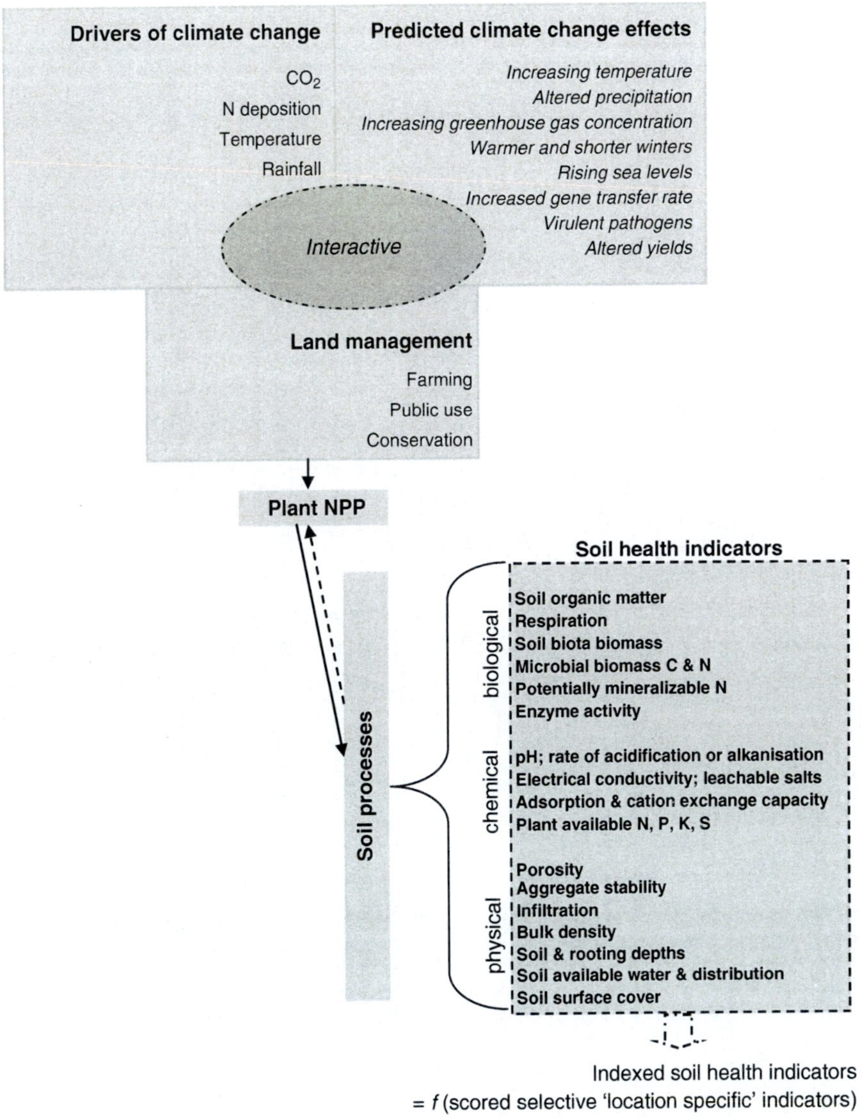

Fig. 2.1 Schematic representation of the potential links between climate change, land use and management change, and soil health indicators (modified from Dalal and Moloney 2000; French et al. 2009; Karlen et al. 2003; Nuttall 2007), including indexing of location-specific indicators to assess soil health

2.2.1 Soil Structure (Aggregate Stability, Porosity)

Aggregate stability, the resistance of soil aggregates to external energy such as high intensity rainfall and cultivation, is determined by soil structure, as well as a range

Table 2.1 Soil health indicators and relations to processes and functions under projected climate change scenarios[a]

Soil health indicators		Soil processes affected	Landscape scale (direct determination or estimated from pedotransfer functions)	Relevance to assess climate change impacts	Inclusion in a minimum data set
Physical	Soil structure	Aggregate stability, organic matter turnover	Aggregation, surface seal, indication of water and chemical retention and transportation	Medium	Frequent
	Porosity	Air capacity, plant available water capacity, relative field capacity	Soil crusting, reduced seed germination, aeration, water entry	High	Occasional/ frequent
	Infiltration	Soil water availability and movement	Potential for leaching, productivity, erosion	High	Occasional
	Bulk density	Soil structural condition; compaction	Volumetric basis for soil reporting	Low	Frequent
	Soil depth and rooting	Plant available water capacity, subsoil salinity	Productivity potential; uncertain whether trends can be discerned over long time periods	Medium	Occasional
	Soil/plant available water and distribution	Field capacity, permanent wilting point, macropore flow, texture	Water and chemical retention and transportation; yield	High	Frequent
	Soil protective cover	Soil water and nutrient movement, soil stabilisation, C and N fixation	Soil physical movement, organic matter input and movement	Medium	Frequent
Chemical	pH	Biological and chemical activity thresholds	Soil acidification, salinisation, electrical conductivity, soil structural stability	Medium	Frequent
	EC	Plant and microbial activity thresholds	Soil structural decline; leachable salts	Medium	Frequent
	Plant available N, P, K	Plant available nutrients and potential for loss	Capacity for crop growth and yield; environmental hazard (e.g. algal blooms)	Medium	Frequent

Biological	Soil organic matter				
	Light fraction or Macro-organic matter	Plant residue decomposition, organic matter storage and quality, macroaggregate formation	Loss of organic matter, soil aggregate formation	High	Frequent
	Mineralisable C and N	Metabolic activity of soil organisms, net inorganic N flux from mineralisation and immobilisation	Total organic C, soil respiration rate, nutrient supply	High	Occasional
			Microbial activity, nutrient supply		
	Soil total C and N	C and N mass and balance	Soil structure, nutrient supply	High	Occasional/ frequent
	Soil respiration	Microbial activity	Microbial activity	High	Occasional
	Microbial biomass C and N	Microbial activity	Soil structure, nutrient supply, pesticide degradation	High	Frequent/ Occasional
	Microbial quotients	Substrate use efficiency	Substrate quality	High	Occasional
	Microbial diversity	Nutrient cycling and availability		High	Occasional
	Other microbiological indicators, enzyme activity	Soil structure, labile carbon, K_m, V_{max}, K_i, Q_{10}	Biochemical activity, nutrient supply	High	Occasional

[a]Adapted from Dalal and Moloney 2000; Gregorich et al. 1994; Haynes 2008; Idowu et al. 2009; Kinyangi 2007; Reynolds et al. 2009; Stenberg 1999

of chemical and biological properties and management practices (Dalal and Moloney 2000; Moebius et al. 2007). It is considered a useful soil health indicator since it is involved in maintaining important ecosystem functions in soil including organic carbon (C) accumulation, infiltration capacity, movement and storage of water, and root and microbial community activity; it can also be used to measure soil resistance to erosion and management changes (Arias et al. 2005; Blanco-Canqui and Lal 2004; Lal 1999; Moebius et al. 2007; Rimal and Lal 2009; Weil and Magdoff 2004; see Chap. 3). Because of its association with the storage of soil organic carbon (SOC) and water, its measurement can be useful to guide climate adaptation strategies, especially in areas that are likely to experience high and intense rainfall and consequently increased erosion events. Since aggregate stability is measured in many different ways, standardised procedures are required within a soil health monitoring framework under climate change scenarios (Dalal and Moloney 2000; Salvador Sanchis et al. 2008).

Porosity, a measure of the void spaces in a material as a fraction (volume of voids to that of the total volume), and pore size distribution provide a direct, quantitative estimate of the ability of a soil to store root-zone water and air necessary for plant growth (Reynolds et al. 2002). Pore characteristics are strongly linked to soil physical quality; bulk density and macroporosity are functions of pore volume, while soil porosity and water release characteristics directly influence a range of soil physical indices including soil aeration capacity, plant available water capacity and relative field capacity (Reynolds et al. 2009). Recent studies to model soil water balance and ecosystem conditions under present-day and projected climatic scenarios use porosity as a model parameter (Porporato et al. 2005). Since root development and soil enzyme activities are closely related to soil porosity and pore size distribution (Piglai and De Nobili 1993) and because future climate change scenarios (e.g. elevated CO_2 and temperature, and variable and extreme rainfall events) may alter root development and soil biological activities (see Chaps. 7 and 8), soil porosity and pore size distribution and consequently soil functions are likely to be affected in unexpected directions; this aspect needs attention in future studies on the relationship of soil health and climate change. Moreover, pore size distribution and aeration status, besides other factors, are the key factors in governing methane (CH_4) fluxes, both CH_4 emission and uptake (Dalal et al. 2008), and nitrous oxide (N_2O) emissions from soil (Dalal et al. 2003a, b). However, data on relationships of greenhouse gas emissions and soil porosity and pore size distribution in response to climate change are limited and hence urgently required to guide development of climate adaptive strategies.

2.2.2 *Infiltration, Soil Available Water and Distribution*

Soil water infiltration, the rate at which water enters the soil surface and moves through soil depth, is gaining increasing interest within soil water modelling, as field-based measurements continue to improve (Dalal and Moloney 2000;

Joel and Messing 2001). Since infiltration rate may change significantly with soil use, management and time, it has been included as an indicator of soil health for assessments of land use change impacts (Arias et al. 2005; O'Farrell et al. 2010).

The availability of water for plant growth and important soil processes is governed by a range of soil properties including porosity, field capacity, lower limit of plant available water (thus excluding osmotic potential) and hence plant available water capacity, macropore flow and texture (Jarvis 2007; Reynolds et al. 2002). Plant available water capacity has been used as part of integrative soil health tests to assess management impacts, although variation in methodology may complicate the interpretation of results (Idowu et al. 2009). Furthermore, the soil available water and distribution may respond rapidly to climate change, especially to variable and high intensity rainfall or drought events, and thus management strategies, such as the planting of cover crops, conservation tillage and incorporation of organic matter, that maintain or even enhance water infiltration and available water in soil may help in mitigating the impacts of severe rainfall and drought events or severe erosion events (Lal 1995; Salvador Sanchis et al. 2008).

2.2.3 *Bulk Density*

Bulk density is routinely assessed in agricultural systems to characterise the state of soil compactness in response to land use and management (Håkansson and Lipiec 2000). It is considered as a useful indicator for the assessment of soil health with respect to soil functions such as aeration and infiltration (e.g. Dalal and Moloney 2000; Pattison et al. 2008; Reynolds et al. 2009). Since bulk density is in general negatively correlated with soil organic matter (SOM) or SOC content (Weil and Magdoff 2004), loss of organic C from increased decomposition due to elevated temperatures (Davidson and Janssens 2006) may lead to increase in bulk density and hence making soil more prone to compaction via land management activities and climate change stresses, for example, from variable and high intensity rainfall and drought events (Birkás et al. 2009).

2.2.4 *Rooting Depth*

Rooting depth is considered an important indicator of soil health, since changes in this property is likely to affect plant available water capacity, subsoil salinity, SOC content or other properties to indicate physicochemical constraints in the soil profile (Arias et al. 2005; Birkás et al. 2009; Dalal and Moloney 2000). Under prolonged drought, the impact of subsoil constraints such as salinity and high chloride concentrations (Dang et al. 2008; Rengasamy 2010) is likely to be greater on plant available water and hence plant productivity. Also, Birkás et al. (2008)

included rooting depth as a soil health parameter for monitoring of soil condition and plant growth under extreme drought and variable rainfall events to indicate the potential for adaptability and mitigation of climate stresses through alteration of rooting depth.

2.2.5 *Soil Surface Cover*

Soil surface cover (e.g. a layer of crop residues or biological soil crust) provides a range of important ecological functions including protection of soil surface by dissipating raindrop impact energy, soil stabilisation, reduction in erodible surface area, water and nutrient retention, C fixation and, in some instances, N fixation and support of native seed germination (Box and Bruce 1996; see Chap. 11). Dalal and Moloney (2000) note that soil surface cover provides an integrated indicator of soil physical management, organic matter input and effects associated with erosion and runoff, although correct timing of monitoring of soil surface cover change is required to evaluate impacts of management and climate change on soil health.

Soil structural conditions such as soil crust and soil seal formation, primarily related to sodicity, are also indicators that may be used to characterise soil health under climate change. The formation of soil crusts and seals can affect a range of soil processes, including water infiltration, oxygen diffusion, runoff, surface water evaporation and wind erosion. A range of methods exist to measure their thickness and strength, although research effort is needed to relate these properties with soil processes (Assouline 2004) affecting ecosystem functions and plant productivity, as well as to evaluate their role in mitigating adverse climate change impacts, thereby assisting in climate change adaptation.

2.3 Soil Chemical Properties, Soil Health and Climate Change

2.3.1 *pH*

Soil pH, a function of parent material, time of weathering, vegetation and climate, is considered as one of the dominant chemical indicators of soil health, identifying trends in change for a range of soil biological and chemical functions including acidification, salinisation, crop performance, nutrient availability and cycling and biological activity (Dalal and Moloney 2000; see Chap. 4). Soil pH has thus been included in integrative soil health tests to assess impacts of land use change and agricultural practices (Gil et al. 2009; Idowu et al. 2009; Pattison et al. 2008; Schindelbeck et al. 2008). While Brinkman and Sombroek (1999) suggested that most soils would not be subjected to rapid pH changes resulting from drivers of climate change such as elevated temperatures, CO_2 fertilisation, variable precipitation and atmospheric N deposition (DeVries and Breeuwsma 1987; McCarthy et al. 2001), it is likely, however, that these drivers of climate change will affect organic matter status, C and nutrient cycling, plant available water and hence plant productivity, which in turn will affect soil pH (Reth et al. 2005, see Chap. 4).

2.3.2 *Electrical Conductivity*

Soil electrical conductivity (EC), a measure of salt concentration, is considered an easily measured, reliable indicator of soil quality/health (Arnold et al. 2005). It can inform trends in salinity, crop performance, nutrient cycling (particularly nitrate) and biological activity and, along with pH, can act as a surrogate measure of soil structural decline especially in sodic soils (Arnold et al. 2005; Dalal and Moloney 2000). Electrical conductivity has been used as a chemical indicator to inform soil biological quality in response to crop management practices (Gil et al. 2009). Using elevation gradient as a surrogate for increasing temperatures and decreasing precipitation under climate change scenarios, Smith et al. (2002) found that EC decreased and pH increased in a semi-arid environment. Pariente (2001) examined the dynamics of soluble salts concentration in soils from four climatic regions (Mediterranean, semi-arid, mildly arid and arid) and found a non-linear relationship between the soluble salts content and rainfall, with sites that received $<$200 mm rainfall contained significantly high soluble contents and vice versa. Clearly, there is a need for comprehensive assessment of the influence of drivers of climate change on soil EC as an important soil health indicator in different ecosystems.

2.3.3 *Sorption Capacity and Cation Exchange Capacity*

Sorption capacity and cation exchange capacity (CEC) are considered important determinants of soil chemical quality, particularly the retention of major nutrient cations Ca, Mg and K, and immobilisation of potentially toxic cations Al and Mn; these properties can thus be useful indicators of soil health, informing of a soil's capacity to absorb nutrients, as well as pesticides and chemicals (Dalal and Moloney 2000; Ross et al. 2008). This may qualify CEC as a useful soil health indicator that is also required to be monitored in response to climate change. Since CEC of coarse-textured soils and low-activity clay soils is attributed to that of SOM (Weil and Magdoff 2004), the increasing decomposition and loss of SOM due to elevated temperatures (Davidson and Janssens 2006) may lead to the loss of CEC of these soils. Low CEC of soil may result in increased leaching of base cations in response to high and intense rainfall events, thus transporting alkalinity from soil to waterways (see Chap. 4).

2.3.4 Plant Available Nutrients

In their identification of basic soil properties to meet requirements of indicators for screening soil quality/health, Doran et al. (1999) list extractable nutrients N, phosphorus, potassium, since "they provide information on plant available nutrients and potential loss from soil. . .providing indication of productivity and environment quality". Measurement of extractable nutrients may provide indication of a soil's capacity to support plant growth; conversely, it may identify critical or threshold values for environmental hazard assessment (Dalal and Moloney 2000). Nutrient cycling, especially N, is intimately linked with soil organic C cycling (Weil and Magdoff 2004), and hence drivers of climate change such as elevated temperatures, variable precipitation and atmospheric N deposition are likely to impact on N cycling (see Chap. 6) and possibly the cycling of other plant available nutrients such as phosphorus and sulphur, although direction and exact magnitude of change in plant available nutrients need to be investigated in detail.

2.4 Soil Biological Properties, Soil Health and Climate Change

"While the chemistry (and physics) of the soil system provides the context. . . it is the soil biota which is adaptive to changes in environmental circumstances" (Kibblewhite et al. 2008). Under conditions of climate change, biological indicators form an integral component in soil health assessment, since, by virtue, they involve complex *adaptive* systems (i.e. the biota) by integrating key soil processes in ways that other indicators do not (Ritz et al. 2009; see Chap. 8 for a detailed overview).

Ritz et al. (2009) note that published information on potential biological indicators has shown an almost exponential increase since the 1970s; however, many of these publications cover a range of processes, for example, microbial, invertebrate or ecological processes. Consequently, a comprehensive explanation of the factors including formal criteria identifying a collective set of biological indicators is limited. Key biological indicators selected for the scope of this study include SOM and its constituents, soil C, respiration and soil microbial biomass. Other microbiological indicators will also be briefly discussed.

2.4.1 Soil Organic Matter

SOM comprises an extensive range of living and non-living components; it has been widely acknowledged that SOM is one of the most complex and heterogeneous components of soils, which vary in their properties, functions and turnover rates (Weil and Magdoff 2004). It is considered an important attribute of soil quality (or soil health) due to the many functions it provide and/or support, including the contribution to the charge characteristics of soils, a sink for and source of C and N, and to a variable extent regulates phosphorus and sulphur cycling, possesses an ability to complex with multivalent ions and organic compounds, provides microbial and faunal habitat and substrates, as well as affecting aggregate stability, trafficability, water retention and hydraulic properties (Haynes 2008; Weil and Magdoff 2004). Main indicators for evaluating SOM status include: SOC, since it comprises about 50% of SOM; organic N, since it is closely associated with organic C and is the most important nutrient for plant productivity; and readily mineralisable C and N (Gregorich et al. 1994; Haynes 2008; see Chaps. 5 and 6).

As SOM drives the majority of soil functions, decreases in SOM can lead to a decrease in fertility and biodiversity, as well as a loss of soil structure, resulting in reduced water holding capacity, increased risk of erosion and increased bulk density and hence soil compaction (Weil and Magdoff 2004). Land use and management practices that lead to build up of SOM will help in absorbing CO_2 from the atmosphere, thus mitigating global warming. By increasing water storage, SOM can play an important role in the mitigation of flooding impacts following extreme rainfall events, while storing water in the event of droughts thus increasing soil resilience.

SOM indicators have been used in long-term soil experiments for the assessment of climate change (Richter et al. 2007; Rinnan et al. 2007), although the response of SOM to elevated temperature remains scientifically controversial and without consensus (Ågren and Wetterstedt 2007; see Chap. 5). In general, increases in temperature have been reported to enhance decomposition of SOM, but rising temperature and precipitation, CO_2 fertilisation and atmospheric N deposition may support high plant productivity and organic matter input to soil and consequently increase SOM. According to Davidson and Janssens (2006) and Kuzyakov and Gavrichkova (2010), it is the accessibility and availability of SOM to microorganisms that govern SOM losses rather than the rate-modifying climate factor (i.e. temperature).

2.4.2 *Soil Carbon*

Soil C can be used as an indicator of change for a number of reasons: it is familiar, direct, linked to ecosystem performance and it has "memory", that is, changes across time; however, it may not be reliable as a solitary indicator of soil quality/health, as it may not encompass all ecosystem traits (Janzen 2005). While soil contains C in diverse forms and residence times, considerable research attention has focused on the SOC form, since it (1) has been largely modified through human activities, and (2) is predicted to decline with increase in mean global temperatures, which would have adverse effects on important soil functions and processes and soil quality/health (Lal et al. 2007). However, in his review on the role of soil C in the changing global C cycle, Janzen (2005) proposed that "if soil C is to be used as an indicator of ecosystem performance, we may need to rethink how it is measured and interpreted... since, although we have been measuring soil C for a long time, our methods were not developed to measure changing C stocks". Since then substantial progress has been made, especially in geospatial and geostatistical methodologies to measure soil C across the landscape and over time (Allen et al. 2010; Worsham et al. 2010).

2.4.3 Light Fraction and Macro-Organic Matter (Labile Organic Matter)

Light (or low-density) fraction and macro-organic components of SOM consist mainly of mineral-free particulate plant and animal residues, which serve as readily decomposable substrate for soil micro-organisms, as well as a labile nutrient reservoir (Gregorich et al. 1994; Post and Kwon 2000; Wagai et al. 2009). Since light fraction and macro-organic matter are responsive to management practices, they may act as early indicators to measure the effectiveness of changing management practice in adaptive response to climate change (Gregorich et al. 1994). For example, the labile soil organic C is rapidly depleted as the temperature rises (Davidson and Janssens 2006; Knorr et al. 2005). In addition, elevated CO_2 in the future may reduce sequestration of root-derived soil C, a major source of labile, light fraction C (Heath et al. 2005).

2.4.4 Potentially Mineralisable C and N

The amount of mineralisable organic matter in soil is an indicator of organic matter quality, acting as the interface between autotrophic and heterotrophic organisms during the nutrient cycling process (Gregorich et al. 1994). While potentially mineralisable C and N may be measured in both the field and laboratory, it is generally treated as a relative rather than an absolute value due to inconsistencies in methods (Haynes 2008). However, mineralisable organic matter may be a useful indicator to assess soil health under climate change, since it affects nutrient dynamics within single growing seasons, and may be used to compare management regimes and C sequestration over extended periods of time (Gregorich et al. 1994).

2.4.5 Soil Respiration

Soil respiration is often used as a biological indicator for soil health (see Chap. 7), since it is positively correlated with SOM content (and often with microbial biomass and activity) and can be determined as either CO_2 production or O_2 consumption, e.g. "soil" or "basal" respiration, using a range of in situ or laboratory methods (Dalal and Moloney 2000; Arias et al. 2005; Haynes 2008). Soil respiration, particularly its temperature response (often expressed as the Q_{10} function), is widely acknowledged to be a critical link between climate change and the global C cycle (Wixon and Balser 2009), although the nature of this relationship is under current scientific debate (Ågren and Wetterstedt 2007; Balser et al. 2006; Kuzyakov and Gavrichkova 2010; also see Chap. 7). Recent studies have also shown that soil respiration is relatively responsive to changes in the seasonal timing of rainfall (see Chap. 7), which is predicted to change according to global and regional climate models (Chou et al. 2008).

2.4.6 *Soil Microbial Biomass*

Microbial biomass, the living component of SOM, is considered the most labile C pool in soils and a sensitive indicator of changes in soil processes, with links to soil nutrient and energy dynamics, including mediating the transfer between SOC fractions (Haynes 2008; Post and Kwon 2000; Saha and Mandal 2009; Weil and Magdoff 2004). While it has been compared to related soil properties to determine change in SOM quality (Gregorich et al. 1994), its use as an indicator of soil quality or soil health is currently limited due to unavailability of benchmark values, difficulty in interpretation and cost-effectiveness of measurement procedures (Dalal 1998). However, soil microbial biomass, similar to labile C, has been shown to be responsive to short-term environmental changes (Haynes 2008), with recent studies revealing significant decline in the soil microbial biomass during long-term simulated climatic warming experiments (Rinnan et al. 2007). When combined with ^{13}C isotope labelling technique, the shift in microbial biomass ^{13}C may provide a more sensitive measure of changes in soil C processes in response to climate and land use changes than the total microbial biomass C (Paterson et al. 2009).

2.4.7 *The Use of Microbial and Metabolic "Quotients"*

In addition to individual biological indicators, ecophysiological indices or "quotients" have also been used to assess environmental change. Indices of physiological performance (e.g. respiration, growth/death, C uptake) against the total organic C or total microbial biomass C per unit time are expressed as microbial quotient (microbial C/organic C), and respiratory or metabolic quotient (CO_2-C respired/hr/microbial C), respectively, and have been used to assess management impacts on organic C dynamics (Moscatelli et al. 2005).

While the effectiveness of these quotients to assess ecosystem disturbance and development has been questioned (Moscatelli et al. 2005), there are cases where these quotients have provided useful information on substrate quality and ecosystem response to stresses across different soil types and environments (Haynes 2008; Sparling 1997; Weil and Magdoff 2004). Their potential for use as soil health indicators to assess climate change impacts is also promising, with research under controlled Free Air CO_2 Enrichment and N fertilisation trials, reporting that both microbial and metabolic quotients were sensitive to changes due to elevated CO_2 and N fertilisation (Moscatelli et al. 2005).

2.4.8 *Enzyme Activity*

Soil enzyme activities may serve to indicate change within the plant-soil system, since these (1) are closely linked to the cycling of nutrients and soil biology, (2) are easily measured, (3) integrate information on both the microbial status and the physicochemical soil conditions, and (4) show rapid response to changes in soil management (Aon et al. 2001; García-Ruiz et al. 2009). Studies of individual enzyme activities report strong temporal and spatial variability, thus often leading to conflicting results (Aon et al. 2001; García-Ruiz et al. 2009). Furthermore, Dorodnikov et al. (2009) showed that by altering the quantity and quality of below-ground C input by plants, elevated CO_2 may stimulate microbial enzyme activities, abundance of microbial enzymes and C turnover possibly affecting microbial community functioning in soil, and furthermore, the extent of stimulation of microbial enzyme activities may depend on soil aggregate size. In addition, atmospheric N deposition may affect extracellular enzymes, which are involved in SOC decomposition and nutrient cycling processes (Frey et al. 2004). It is still to be known how soil microbial enzyme activities involved in organic C turnover, nutrient cycling and greenhouse gas emissions will respond to the interactive effects of multiple global change drivers (such as climate change, land use change), thereby supporting the generalised view of their use as indicators of changes in soil health.

2.4.9 *Other Microbiological Indicators as Integrative Indicators of Soil Health*

Many other soil microbial properties may be considered with respect to soil health indicators; their inclusion within research literature is gaining popularity as more holistic definitions of soil health include soil and plant attributes and functionality, and as advances in microbiological techniques offer promising developments in the detection of soil health status (see Chap. 8). Other microbial indicators noted in the literature, although not included in the scope of this chapter, include microbial autotrophic nitrification, arbuscular mycorrhiza, community studies of soil macro-

and microfauna, and soil disease suppressiveness and plant disease incidence (for extended reviews, see Arias et al. 2005; Doran and Zeiss 2000; French et al. 2009; Kibblewhite et al. 2008; Pankhurst et al. 1997; Ritz et al. 2009).

2.5 Genetic and Functional Biodiversity of Soils, Soil Health and Climate Change

Soil community structure and function is a current theme for debate among soil scientists, particularly whether ecosystem functioning is influenced by biodiversity, and how the loss of microbial functional groups influences ecosystem functioning (Hunt and Wall 2002; Kibblewhite et al. 2008). Hunt and Wall (2002) note that much of the focus has been on aboveground systems to assess potential effects of climate change on plant biodiversity and biochemistry, although recent studies suggest that climatic stresses, such as warming, will have a profound effect on the rhizosphere, soil heterotrophic community structure and soil processes, including soil respiration, N mineralisation and ecosystem C functioning (Bardgett et al. 2008; Briones et al. 2009; see Chaps. 7 and 8). Changing climatic factors is also of concern due to possible evolutionary changes, which allow the spread of virulence factors and genes that aid in environmental survival (French et al. 2009). As molecular technique advancements continue to expand our understanding of microbial diversity and the conditions driving changes in microbial diversity and community structure (Arias et al. 2005; Prosser 2002), it is likely that the adoption of genetic and functional biodiversity indicators of soil health will increase, particularly within food web and nutrient cycling models, to improve predictions of climate change impacts on soil health (see Chap. 8).

2.6 Selection of Soil Health Key Indicators: Requirements and Conceptual Framework for In Situ Soil Health Assessment Under Climate Change

Many soil health indicators discussed above need prioritisation with regard to assessment of soil health under sustainable management practices and predicted climate change scenarios. Recent reviews have focused upon measurement of individual indicators or a suite of indicators to assess changes in soil health within a soil physical, chemical and/or biological context (Moebius et al. 2007; Zagal et al. 2009), although Stockdale and Watson (2009) note that work is needed to test existing indicator frameworks to guide management decisions, rather than simply to compare with the status quo.

The indicators of soil health are interlinked (Fig. 2.1) and so are the drivers of global change (land use change, elevated temperatures, elevated atmospheric

CO_2 concentration, increasing atmospheric N deposition, variability in the amount, intensity and distribution of rainfall, extreme climatic events, and their interactions). Conceptualisation of interactions between climate change, land management and soil health indicators, as presented in Fig. 2.1, further highlights the complexity and interdependence of many of these drivers to influence soil health, primarily through their influence on plant primary productivity, microbial and faunal biomass diversity and activity, including their products (intracellular and extracellular enzymes), as well as the overall C and N cycles (Bardgett et al. 2008; Kibblewhite et al. 2008; Wixon and Balser 2009; see Chaps. 6 and 8).

Table 2.1 presents a summary of major soil health indicators, their processes and functions under projected climate change scenarios, their relevance to climate change impacts and their frequency of inclusion within a minimum data set for soil health assessment. Most studies agree that a minimum data set for the assessment of soil health should include key indicators that: (1) are sensitive to changes due to management and climate variations, (2) integrate soil physical, chemical and biological properties, (3) are relatable to important soil functions, (4) are applicable to field conditions and (5) are accessible to many users (Idowu et al. 2009; Karlen et al. 2003; Lagomarsino et al. 2009; Pattison et al. 2008; Rao and Siddaramappa 2008; Schindelbeck et al. 2008). The selection of key indicators of soil health under climate change scenarios, as discussed in this chapter, should be able to reflect the mitigation and adaptive capability of soil and its resilience to climate change over short to long term.

2.7 Summary and Future Directions

Understanding soil health impacts in relation to climate change is possible through the use of indicators (measurable attributes or values) which relate soil physical, chemical and biological properties to ecological functions and which can be monitored in the context of sustainable land management and climate change. Key soil health indicators affected by climate change include aggregate stability, SOM, carbon and nitrogen cycling, microbial biomass and activity, and microbial fauna and flora diversity. Selection of indicators within a minimum data set depends on their sensitivity to management and climate changes, capacity to integrate and relate to other soil functions, ease of use, repeatability and cost of measurement. Soil health "tests" recommending a minimum data set of soil health indicators are being promoted within research and government organisations for agricultural management (e.g. Nelson et al. 2009; Pattison et al. 2008) and to assist monitoring efforts and policy development (Ritz et al. 2009; Schindelbeck et al. 2008). This minimum dataset could be used to assess the effect of climate change on soil health. Although tentative steps are underway, greater efforts are required to explore individual and interactive effects of drivers of global change (e.g. land use change, increasing temperatures, elevated CO_2 concentration, variability in the amount, intensity and distribution of rainfall, and increasing atmospheric N deposition) using controlled environment and long-term research experiments to assess soil health indicators that can be responsive to such treatment variations over wider spatiotemporal scales, and consequently, their monitoring and inclusion in a minimum data set can assist us in devising greenhouse gas mitigation and climate adaptive strategies.

Acknowledgements We thank the two anonymous reviewers for providing helpful comments on an earlier draft version of this chapter.

References

Ågren GI, Wetterstedt M (2007) What determines the temperature response of soil organic matter decomposition? Soil Biol Biochem 39:1794–1798

Allen DE, Pringle MJ, Page KL, Dalal RC (2010) A review of sampling designs for the measurement of soil organic carbon in Australian grazing lands. Rangel J 32:227–246

Aon MA, Cabello MN, Sarena DE, Colaneri AC, Franco MG, Burgos JL, Cortassa S (2001) I. Spatio-temporal patterns of soil microbial and enzymatic activities in agricultural soils. Appl Soil Ecol 18:239–254

Arias ME, González-Pérez FJ, González-Vila ASB (2005) Soil health – a new challenge for microbiologists and chemists. Int Microbiol 8:13–21

Arnold SL, Doran JW, Schepers J, Wienhold B (2005) Portable probes to measure electrical conductivity and soil quality in the field. Commun Soil Sci Plant Anal 36:2271–2287

Assouline S (2004) Rainfall-induced soil surface sealing: a critical review of observations, conceptual models, and solutions. Vadose Zone J 3:570–591

Balser TC, McMahon KD, Bart D, Bronson D, Coyle DR, Craig N, Flores-Mangual ML, Forshay K, Jones SE, Kent AE, Shade AL (2006) Bridging the gap between micro – and macro-scale perspectives on the role of microbial communities in global change ecology. Plant Soil 289: 59–70

Bardgett RD, Freeman C, Ostle NJ (2008) Microbial contributions to climate change through carbon cycle feedbacks. ISME J 2:805–814

Birkás M, Stingli A, Szemők A, Kalmár T, Bottlik L (2008) Soil condition and plant interrelations in dry years. Cereal Res Commun 36(Suppl):15–18

Birkás M, Dexter A, Szemők A (2009) Tillage-induced soil compaction, as a climate threat increasing stressor. Cereal Res Commun 37:379–382

Blanco-Canqui H, Lal R (2004) Mechanisms of carbon sequestration in soil aggregates. Crit Rev Plant Sci 23:481–504

Box JE, Bruce RR (1996) The effect of surface cover on infiltration and soil erosion. In: Agassi M (ed) Soil erosion, conservation and rehabilitation. Marcel Dekker Inc, New York

Brinkman R, Sombroek W (1999) Chapter 3. The effects of global change on soil conditions in relation to plant growth and food production. In: Bazzaz F, Sombroek W (eds) Global climate change and agricultural production. Food and Agriculture Organization of the United Nations and John Wiley and Sons Ltd., Rome, Italy. Accessed online from: http://www.fao.org. Accessed 15 Feb 2010

Briones MJI, Ostle NJ, McNamara NP, Poskitt J (2009) Functional shifts of grassland soil communities in response to soil warming. Soil Biol Biochem 41:315–322

Chou WW, Silver WL, Jackson RD, Thompson AW, Allen-Diaz B (2008) The sensitivity of annual grassland carbon cycling to the quantity and timing of rainfall. Glob Chang Biol 14: 1382–1394

Dalal RC (1998) Soil microbial biomass – what do the numbers really mean? Aust J Exp Agric 38:649–665

Dalal RC, Moloney D (2000) Sustainability indicators of soil health and biodiversity. In: Hale P, Petrie A, Moloney D, Sattler P (eds) Management for sustainable ecosystems. Centre for Conservation Biology, Brisbane, pp 101–108

Dalal RC, Lawrence P, Walker J, Shaw RJ, Lawrence G, Yule D, Doughton JA, Bourne A, Duivenvoorden L, Choy S, Moloney D, Turner L, King C, Dale A (1999) A framework to monitor sustainability in the grains industry. Aust J Exp Agric 39:605–620

Dalal RC, Eberhard R, Grantham T, Mayer DG (2003a) application of sustainability indicators, soil organic matter and electrical conductivity, to resource management in the northern grains region. Aust J Exp Agric 43:253–259

Dalal RC, Wang WJ, Robertson GP, Parton WJ (2003b) Nitrous oxide emission from Australian agricultural lands and mitigation options: a review. Aust J Soil Res 41:165–195

Dalal RC, Allen DE, Livesley SJ, Richards G (2008) Magnitude and biophysical regulators of methane emission and consumption in the Australian agricultural, forest, and submerged landscapes: a review. Plant Soil 309:43–76

Dang YP, Dalal RC, Mayer D, McDonald M, Routley R, Schwenke GD, Buck SR, Daniells IG, Singh DK, Manning W, Ferguson N (2008) High subsoil chloride concentrations reduce soil water extraction and crop yield on Vertisol in northeastern Australia. Aust J Agric Res 59:321–330

Davidson EA, Janssens IA (2006) Temperature sensitivity of soil carbon decomposition and feedbacks to climate change. Nature 440:165–173

DeVries W, Breeuwsma A (1987) The relationship between soil acidification and element cycling. Water Air Soil Pollut 35:293–310

Doran JW (2002) Soil health and global sustainability: translating science into practice. Agric Ecosyst Environ 88:119–127

Doran JW, Zeiss MR (2000) Soil health and sustainability: managing the biotic component of soil quality. Appl Soil Ecol 15:3–11

Doran JW, Jones AJ, Arshad MA, Gilley JE (1999) Determinants of soil quality and health. In: Pankhurst CE, Doube BM, Gupta VVSR (eds) Biological indicators of soil health. CAB International, Oxon

Dorodnikov M, Blagodatskaya E, Blagodatsky S, Marhan S, Fangmeier A, Kuzyakov Y (2009) Stimulation of microbial extracellular enzyme activities by elevated CO_2 depends on soil aggregate size. Glob Change Biol 15:1603–1614

French S, Levy-Booth D, Samarajeewa A, Shannon KE, Smith J, Trevors JT (2009) Elevated temperatures and carbon dioxide concentrations: effects on selected microbial activities in temperate agricultural soils. World J Microbiol Biotechnol 25:1887–1900

Frey SD, Knorr M, Parrent JL, Simpson RT (2004) Chronic nitrogen enrichment affects the structure and function of the microbial community in temperate hardwood and pine forests. For Ecol Manage 196:159–171

García-Ruiz R, Ochoa V, Vinegla B, Hinojosa MB, Pena-Santiago R, Liébanas G, Linares JC, Carreira JA (2009) Soil enzymes, nematode community and selected physico-chemical properties as soil quality indicators in organic and conventional olive-oil farming: influence of seasonality and site features. Appl Soil Ecol 41:305–314

Gil SV, Meriles J, Conforto C, Figoni G, Basanta M, Lovera E, March GJ (2009) Field assessment of soil biological and chemical quality in response to crop management practices. World J Microbiol Biotechnol 25:439–448

Gregorich EG, Carter MR, Angers DA, Monreal CM, Ellert BH (1994) Towards a minimum data set to assess soil organic matter quality in agricultural soils. Can J Soil Sci 74:367–385

Hadas A (1997) Soil tilth – the desired soil structural state obtained through proper soil fragmentation and reorientation processes. Soil Tillage Res 43:7–40

Håkansson I, Lipiec J (2000) A review of the usefulness of relative bulk density values in studies of soil structure and compaction. Soil Tillage Res 53:71–85

Haynes RJ (2008) Soil organic matter quality and the size and activity of the microbial biomass: their significance to the quality of agricultural soils. In: Huang Q, Huang PM, Violante A (eds) Soil mineral-microbe-organic interactions: theories and applications. Springer, Berlin, pp 201–230
Heath J, Ayres E, Possell M, Bardgett RD, Black HIJ, Grant H, Ineson P, Kerstiens G (2005) Rising atmospheric CO_2 reduces sequestration of root-derived soil carbon. Science 309:1711–1713
Hunt HW, Wall DH (2002) Modelling the effects of loss of soil biodiversity on ecosystem function. Glob Change Biol 8:33–50
Idowu OJ, van Es HM, Abawi GS, Wolfe DW, Schindelbeck RR, Moebius-Clune BN, Gugino BK (2009) Use of an integrative soil health test for evaluation of soil management impacts. Renew Agric Food Syst 24:214–224
Janzen HH (2005) Soil carbon: a measure of ecosystem response in a changing world? Can J Soil Sci 85:467–480
Jarvis NJ (2007) A review of non-equilibrium water flow and solute transport in soil macropores: principles, controlling factors and consequences for water quality. Eur J Soil Sci 58:523–546
Joel A, Messing I (2001) Infiltration rate and hydraulic conductivity measured with rain simulator and disc permeameter on sloping arid land. Arid Land Res Manage 15:371–384
Karlen DL, Doran JW, Andrews SS, Wienhold BJ (2003) Soil Quality – Humankinds' foundation for survival. J Soil Water Conserv 58:171–179
Kibblewhite MG, Ritz K, Swift MJ (2008) Soil health in agricultural systems. Philos Transact R Soc 363:685–701
Kinyangi J (2007) Soil health and soil quality: a review. Draft publication. Available on: http://www.cornell.edu.org; Accessed on: www.worldaginfo.org. Accessed 15 Feb 2010
Knorr W, Prentice IC, House JI, Holland EA (2005) Long-term sensitivity of soil organic carbon turnover to warming. Nature 433:298–301
Kuzyakov Y, Gavrichkova O (2010) Time lag between photosynthesis and carbon dioxide efflux from soil: a review of mechanisms and controls. Global Change Biology 16:3386–3406
Lagomarsino A, Moscatelli MC, Tizio AD, Mancinelli R, Grego S, Marinari S (2009) Soil biological indicators as a tool to assess the short-term impact of agricultural management on changes in organic C in a Mediterranean environment. Ecol Indic 9:518–527
Lal R (1995) Tillage systems in the tropics: management options and sustainability implications. Food and Agriculture Organization of the United Nations Soils Bulletin 71, FAO, Rome
Lal R (ed) (1999) Soil quality and soil erosion. Soil and Water Conservation Society, Ankeny
Lal R, Follett RF, Stewart BA, Kimble JM (2007) Soil carbon sequestration to mitigate climate change and advance food security. Soil Sci 172:943–956
McCarthy JJ, Canziani OF, Leary NA, Dokken DJ, White KS (2001) Climate Change 2001: impacts, adaptation, and vulnerability. Contribution of Working Group II to the Third Assessment Report of the Intergovernmental Panel on Climate Change. Cambridge University Press, Cambridge
Moebius BN, van Es HM, Schindelbeck RR, Idowu OJ, Clune DJ, Thies JE (2007) Evaluation of laboratory-measured soil properties as indicators of soil physical quality. Soil Sci 172:895–912
Moscatelli MC, Lagomarsino A, Marinari S, De Angelis P, Grego S (2005) Soil microbial indices as bioindicators of environmental changes in a poplar plantation. Ecol Indic 5:171–179
Nelson KL, Lynch DH, Boiteau G (2009) Assessment of changes in soil health throughout organic potato rotation sequences. Agric Ecosyst Environ 131:220–228
Nuttall JG (2007) Climate change – identifying the impacts on soil and soil health. Department of Primary Industries, Future Farming Systems Research Division, Victoria
O'Farrell PJO, Donaldson JS, Hoffman MT (2010) Vegetation transformaion, functional compensation, and soil health in a semi-arid environment. Arid Land Res Manage 24:12–30
Pankhurst CE, Doube BM, Gupta VVSR (eds) (1997) Biological indicators of soil health. CAB International, Oxon

Pariente S (2001) Soluble salts dynamics in the soil under different climatic conditions. Catena 43:307–321

Paterson E, Midwood AJ, Millard P (2009) Through the eye of the needle: a review of isotope approaches to quantify microbial processes mediating soil carbon balance. New Phytol 184:19–33

Pattison AB, Moody PW, Badcock KA, Smith LJ, Armour JA, Rasiah V, Cobon JA, Gulino L-M, Mayer R (2008) Development of key soil health indicators for the Australian banana industry. Appl Soil Ecol 40:155–164

Piglai M, De Nobili M (1993) Relationship between soil porosity and root development and soil enzyme activity in cultivated soils. Geoderma 56:243–256

Porporato A, Daly E, Rodriguez-Iturbe I (2005) Soil water balance and ecosystem response to climate change. Am Nat 164:625–632

Post WM, Kwon KC (2000) Soil carbon sequestration and land-use change: processes and potential. Glob Change Biol 6:317–328

Prosser JI (2002) Molecular and functional diversity in soil micro-organisms. Plant Soil 244:9–17

Rao BKR, Siddaramappa R (2008) Evaluation of soil quality parameters in a tropical paddy soil amended with rice residues and tree litters. Eur J Soil Biol 44:334–340

Rengasamy P (2010) Soil processes affecting crop production in salt-affected soils. Funct Plant Biol 37:613–620

Reth S, Reichstein M, Falge E (2005) The effect of soil water content, soil temperature, soil pH-value and root mass on soil CO_2 efflux. Plant Soil 268:21–33

Reynolds WD, Bowman BT, Drury CF, Tan CS, Lu X (2002) Indicators of good soil physical quality: density and storage parameters. Geoderma 110:131–146

Reynolds WD, Drury CF, Tan CS, Fox CA, Yang XM (2009) Use of indicators and pore volume-function characteristics to quantify soil physical quality. Geoderma 152:252–263

Richter DD, Hofmockel M, Callaham MA, Powlson DS, Smith P (2007) Long-term soil experiments: keys to managing Earth's rapidly changing ecosystems. Soil Sci Soc Am J 71:266–279

Riley J (2001c) Multidisciplinary indicators of impact and change: key issues for identification and summary. Agric Ecosyst Environ 87:245–259

Rimal BK, Lal R (2009) Soil and carbon losses from five different land management areas under simulated rainfall. Soil Tillage Res 106:62–70

Rinnan R, Michelsen A, Bååth E, Jonasson S (2007) Fifteen years of climate change manipulations alter soil microbial communities in a subarctic heath ecosystem. Glob Change Biol 13:28–39

Ritz K, Black HIJ, Campbell CD, Harris JA, Wood C (2009) Selecting biological indicators for monitoring soils: a framework for balancing scientific and technical opinion to assist policy development. Ecol Indic 9:1212–1221

Ross DS, Matschonat G, Skyllberg U (2008) Cation exchange in forest soils: the need for a new perspective. Eur J Soil Sci 59:1141–1159

Saha N, Mandal B (2009) Soil health – a precondition for crop production. In: Khan MS, Zaidi A, Musarrat J (eds) Microbial strategies for crop improvement. Springer, Heidelberg, pp 161–168

Salvador Sanchis MP, Torri D, Borselli L, Poesen J (2008) Climate effects on soil erodibility. Earth Surf Process Landf 33:1082–1097

Schindelbeck RR, van Es HM, Abawi GS, Wolfe DW, Whitlow TL, Gugino BK, Idowu OJ, Moebius-Clune BN (2008) Comprehensive assessment of soil quality for landscape and urban management. Landsc Urban Plan 88:73–80

Schjønning P, Elmholt S, Christensen BT (2004) Soil quality management – concepts and terms. In: Schjønning P, Elmholt S, Christensen BT (eds) Managing soil quality: challenges in modern agriculture. CAB International, Oxon

Smith JL, Halvoson JJ, Bolton H Jr (2002) Soil properties and microbial activity across a 500 m elevation gradient in a semi-arid environment. Soil Biol Biochem 34:1749–1757

Sparling GP (1997) Soil microbial biomass, activity and nutrient cycling as indicators of soil health. In: Pankhurst CE, Doube BM, Gupta VVSR (eds) Biological indicators of soil health. CAB International, Oxon, pp 79–96

Stenberg B (1999) Monitoring soil quality of arable land: microbiological indicators. Acta Agriculturae Scandinavica., Sect. B. Soil Plant Sci 49:1–24

Stockdale EA, Watson CA (2009) Biological indicators of soil quality in organic farming systems. Renew Agric Food Syst 24:308–318

Wagai R, Mayer LM, Kitayama K (2009) Nature of the "occluded" low-density fraction in soil organic matter studies: a critical review. Soil Sci Plant Nutr 55:13–25

Weil RR, Magdoff F (2004) Significance of soil organic matter to soil quality and health. In: Weil RR, Magdoff F (eds) Soil organic matter in sustainable agriculture. CRC press, Florida, pp 1–43

Wixon DL, Balser TC (2009) Complexity, climate change and soil carbon: a systems approach to microbial temperature response. Syst Res Behav Sci 26:601–620

Worsham L, Markewitz D, Nibblelink N (2010) Incorporating spatial dependence into estimates of soil carbon contents under different land covers. Soil Sci Soc Am J 74:635–646

Zagal E, Munoz C, Quiroz M, Córdova C (2009) Sensitivity of early indicators for evaluating quality changes in soil organic matter. Geoderma 151:191–198

Part II
Soil Attributes and Soil Processes in Response to the Climate Change

Chapter 3
Climate Change on Soil Structure and Soil Health: Impacts and Adaptation

Kwong Yin Chan

3.1 Introduction

In terrestrial ecosystems and agro-ecosystems, soil structure determines soil physical fertility which is the capacity of the soil to support and sustain plant production in relation to its physical properties. Hence, soil structure plays a fundamental role in controlling physical soil health, productivity and environmental quality (on-site as well as off-site) of agricultural land. However, soil structure in agro-ecosystems is sensitive to human disturbance and therefore to management practices. Historically, collapse of agriculture and large-scale land degradation problems can often be linked to mismanagement under episodes of drastic changes in climate, e.g. "dust bowl" of USA in 1930s (Follett 1993) and collapse of pasture industries in Australia in 1900s (Chartres et al. 1992). Today, many parts of the world are still faced with food shortages and land degradation problems (Lal 2000). Better knowledge of the impacts of hotter and drier regimes due to the predicted climate change on soil structure and its associated functions will help human kind to better prepare for the likely impacts on agriculture and food security and environmental quality.

This chapter begins with an outline of the three aspects of soil structure, namely form, stability and resilience and their relationship to soil functioning and crop production. This provides basic information on the underlying processes and factors affecting the sensitivity of soil structure to management and environmental changes. The impact of climate change on these fundamental aspects of soil structure and hence the consequent soil health and productivity changes will be elucidated both from first principles and from existing evidence. This is then followed by exploring the potential of soil structure (via maintenance and improvement) in adapting to (even benefiting from) climate changes in agro-ecosystems.

K.Y. Chan (✉)
NSW Department of Primary Industries, Locked Bag 4, Richmond, NSW 2753, Australia
e-mail: yin.chan@mq.edu.au

B.P. Singh et al. (eds.), *Soil Health and Climate Change*, Soil Biology 29,
DOI 10.1007/978-3-642-20256-8_3,

Finally, management practices that can help in improving soil structure and soil health and minimising the impact of climate changes are discussed.

3.2 Soil Structure and Soil Physical Fertility

3.2.1 The Three Aspects of Soil Structure

3.2.1.1 Structural Form

This refers to the architecture and hence the arrangement of the solid phase of the soil and of the pore space located between its constituent particles. This architecture of soil (the structural form) determines not only the density of packing of the solid phase (bulk density) and hence total porosity, but more importantly also pore size distribution and pore continuity of the soil profiles. Functionally, it is more relevant to consider the "pores" in the soil because pores of different sizes are important for different ecological processes operating within the soil. For example, macropores (>75 μm) predominantly controls flow of water and gases through soil, whereas smaller pores (<30–0.2 μm) are important in soil water storage (Kay 1990). Pore size distribution and pore continuity together control soil hydraulic properties by determining both soil moisture characteristics (soil water content–potential function) and soil permeability (hydraulic conductivity–water content function). These two functions determine important physical and biological processes such as infiltration, water holding capacity and drainage. Soil structure through the packing of the soil particles and their cementation (bonding) also controls mechanical properties such as soil strength, workability and friability (Dexter 1988). For maximum yield of crops other than rice, the structure of a soil should maximise the storage capacity of the soil for plant available water and optimise soil aeration in the root zone, with stable and porous aggregates at the surface (Emerson 1991). The importance of surface soil structure in controlling processes at the interphase of atmosphere, hydrosphere and geosphere, such as the processes of infiltration, and runoff, cannot be over-emphasised (Kay 1990). The basic relationships between soil structure and crop yield are illustrated in Fig. 3.1.

Importance of surface soil structure in controlling processes at the interphase of atmosphere, hydrosphere and geosphere, such process as infiltration, runoff cannot be over-emphasised (Kay and Angers 2000).

3.2.1.2 Stability and Resiliency: Soil Vulnerability to Degradation

In the present discussion of impacts of climate change on soil structure, it is pertinent to recognise two other aspects of soil structure, namely stability and resilience (Kay 1990). Stability refers to the ability of the soil to retain its structural

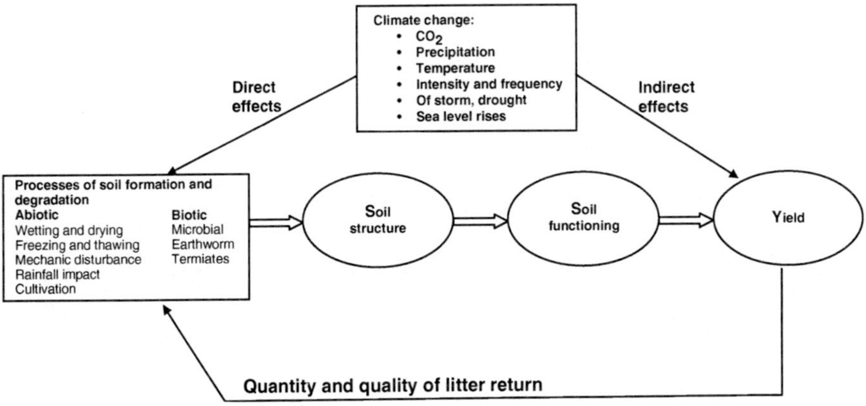

Fig. 3.1 The basic relationships between soil structure and crop yield under climate change scenarios

form, i.e. the arrangement of solid and void space, when exposed to different stresses. Soil structural stability is governed by the strength of inter-particle bonding holding the aggregates together. As such it is a function of the levels and locations of organic and inorganic substances (sesquioxides and carbonates) acting as effective cements in promoting aggregation. Importance of organic carbon on soil structural stability and its effect on the different levels of soil structural form are well documented. Labile forms control macro-aggregate stability, whereas more stable forms stabilise the micro-aggregates (Oades 1984; Tisdall and Oades 1982). Any impact of climate change on total organic carbon as well as its different forms (see Chap. 5) will affect different levels of soil structural stability and therefore soil structure and its associated functions.

Resilience refers to the ability of the soil to recover its structural form through natural processes when the applied stresses are reduced or removed, i.e. after degradation. Resiliency can arise from a number of processes such as wetting/drying, freezing/thawing and biological activities (e.g. soil faunal activities and root development). It is interesting to note that a stable soil is not necessarily a resilient soil (Kay 1990). For example, Oxisols (krasnozems) are often structurally very stable soils, but structural breakdown, once it occurs, especially in the subsoil, is difficult to recover because of low organic carbon and non-swelling properties. On the other hand, many self-mulching Vertisols, while structurally not very stable, tend to slake easily on wetting and are readily compacted, but are very resilient and can restructure after a couple of cycles of wetting and drying. Factors controlling resilience are not clearly understood, but clay mineralogy and SOC can both have an effect (Kay and Angers 2000).

Stability and resilience together determine the vulnerability of soil structure to external stresses (Schjonning et al. 2004). Soils that are least stable under a given stress and that do not recover when the stress is reduced or removed are most vulnerable to stress. Conversely, soils that have high stability and are very resilient

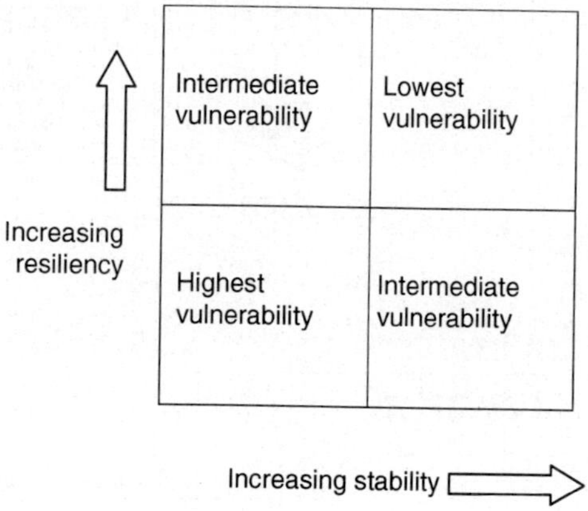

Fig. 3.2 Soil structural vulnerability in relation to stability and resiliency

are the least vulnerable. Any long-term impact of climate change on soil structure and all the consequent impacts on the associated soil functions will be dependent on its vulnerability. Therefore, it is worthwhile to rank soil in terms of vulnerability (such as in Fig. 3.2) as this will help to identify soils the more vulnerable soils and devise management practices for their management.

3.2.2 *Dynamic Nature of Soil Structure: Importance of Soil Organisms*

It is important to realise that soil structure is affected by a range of internal as well as external abiotic and biotic processes/forces. These include both formative and destructive forces, and at any time soil structure (with its three aspects) is the resultant of these processes (Harris et al. 1966; Kay 1990, also Fig. 3.1). The abiotic processes include drying, wetting, freezing, thawing, consolidation, as well as compression, shear and brittle fracture by equipment. As soil is a living system, soil structure is also continuously being modified by its inhabitants, including roots, microbes, earthworms and termites. Important roles played by different soil organisms on soil structure have been reviewed (e.g. Lavelle et al. 1997; Lee and Foster 1991). Soil structural degradation occurs when the effects of destructive forces exceed that of the formative forces, resulting in deleterious changes in form, stability and resilience of the soil structure. The aim of soil management is to promote soil forming processes and minimise the deleterious effects of soil structure destructive processes. In assessing the possible impact of climate change on soil structure and therefore the associated soil functions, it is necessary to evaluate the effect of climate change on these processes.

In summary, soil structure, in its three aspects, dynamically determines the physical fertility and hence a range of soil functions as mentioned above which in turn influences the agro-ecosystem health and productivity of the farming systems.

It is important to explore how soil structure can be affected by consequences of climate change, and how these attributes of soil structure can potentially be improved to help adapting future farming systems to climate changes.

3.3 Climate Change: Changes that Impact Soil Structure

Climate change scenarios considered by Intergovernmental Panel on Climate Change (IPCC) prediction include increase in atmospheric CO_2 concentration, increases in air temperature, changes in precipitation and prevalence of extreme climate events. For instance, global temperature change of 1.6–6.4°C by 2100, atmospheric CO_2 concentration increases by up to 550 ppm and precipitation change by at least 20% have been predicted (IPCC 2007). However, the predicted changes vary geographically and with future greenhouse gas (GHG) emission control. Therefore, the actual magnitude of changes of these parameters and consequences of these changes will therefore be location specific and be dependent on the extent of future success in reducing emission of GHG.

Changes in precipitation are likely to be different for different parts of the world. Over the twentieth century, precipitation has mostly increased in high northern latitudes, while decreases have dominated from 10°S to 30°N since the 1970s (IPCC 2007). There is high confidence that many areas in the subtropics and temperate regions, namely Australia, Mediterranean Basin, western USA, northwest and south-west Africa, and north-eastern Brazil will suffer a decrease in precipitation due to climate change (Garnaut 2008; IPCC 2007). For Australia, by 2050, the best estimate of precipitation change is: little change in the far north but grading southwards to a decrease of 5% (from -15 to 0%).

Of particular significant impacts on soil structure are the extreme climate events that include increasing intensity and frequency of drought, floods, storms and fires (IPCC 2007). Models show an increase in daily precipitation intensity and the number of dry days (CSIRO 2007). Even for locations where precipitation might increase or remain the same, a higher proportion of it will occur as intense storms (IPCC 2007).

Trends of many of these changes have already been recorded in different parts of the world in the last 30 years (Alpert et al. 2002; IPCC 2007; Kertesz and Mika 1999). Globally, the area of land classified as very dry has more than doubled since the 1970s and the trend is expected to continue. In the Mediterranean area, between 1951 and 1995, there was an average reduction of 10–20% in total precipitation, but torrential rainfall (those with intensity $>$128 mm/day) has increased from 1 to 4–5% with fewer rainy days (Alpert et al. 2002). In parts of Asia, Africa and Australia, the frequency and intensity of drought have been observed to increase in the recent decades (Dore 2005; Garnaut 2008).

All these changes can directly and indirectly affect soil structure and in turn the soil functional processes controlled by it.

3.4 Impact of Climate Change on Soil Structure

In principle, climate changes can affect all the three aspects of soil structure (form, stability and resilience) both directly and indirectly (Fig 3.1).

3.4.1 *Direct Effects*

In areas where reduction in precipitation have been predicted, i.e. in a hotter and drier environment, with more extreme event of drought, rainfall and fire, soils are subjected to a number of processes all conducive to structural degradation (Fig. 3.1).

3.4.1.1 Processes of Soil Structural Degradation

Slaking. Slaking is the macroscopic breakdown of soil aggregates into smaller fragments when immersed in water. The extent of incipient failure is dependent on the antecedent soil water potential, rate of wetting and aggregate stability of the soils (Chan and Mullins 1994; Emerson 1991; Quirk and Murray 1991). Degree of slaking increases with decreases in soil water potential (more breakdown as the soil becomes drier) and increases with rate of wetting (rainfall intensity). For the same soil with different soil organic carbon levels, slaking tends to be more severe in the lower organic carbon soils (Chan and Mullins 1994). Therefore, a dry soil with low organic carbon content when subjected to quick wetting is most vulnerable to structural breakdown via slaking. For the hardsetting soils (Mullins et al. 1990), rapid increase in slaking commences when the soil water potential is <6 MPa and more importantly, slaking results in irreversible breakdown of soil structure. This is because on slaking, unlike the other soils that tend to produce micro-aggregates, the hardsetting soils tend to result in fine particles of varying sizes, including primary particles of silt and clay which tend not to re-aggregate on drying (Chan and Mullins 1994). Repeated wetting and drying of soil results in reduction in aggregate stability (Kay et al. 1994). Slaking results in exposure of physically protected soil organic carbon to more rapid decomposition, and this in turn leads to further decreases in aggregate stability. The impact of repeated wetting and drying on the cracking clays may be less severe because of their self-mulching properties, a form of structural resilience (Kay and Angers 2000).

Dispersion. While slaking is the macroscopic breakdown of aggregates into smaller fragments, dispersion is the complete breakdown of aggregates into primary particles of clay, silt, sand and organic materials when immersed in water, and this can be due solely to osmotic stress (due to high sodicity and/or low electrolyte levels) or to osmotic stress-combined mechanical stresses exerted by raindrops or working the soil while wet (Emerson 1991). In low sodicity soils such as the hardsetting soils, collapse in soil structure can occur if the soils are mechanically disturbed while wet, e.g. cultivation or stock trampling.

Mechanical disturbance. This can result from both tillage and rainfall impacts. In agriculture, tillage is often the operation that causes the largest impact on soil structure, very often adversely (Russell 1977). The effect of tillage on soil structure is moisture dependent, and damage to soil structure can occur when the soil water content is either too dry (resulting in pulverisation) or too high (resulting in shearing and dispersion as discussed above) (Mullins et al. 1990). For a particular soil, there is a limited range of moisture content over which tillage can result in soil structural improvement. Hardsetting soils are notorious for the narrow range of soil water content for optimal tillage (Chan and Mullins 1994; Mullins et al. 1990).

Compaction. Compaction is the densification of soil due to external stresses such as trafficking, resulting in loss in porosity, particularly macropores and harder and stronger soil. These are undesirable changes in soil structure (form) with resulting adverse effects on infiltration and root growth. Compaction can occur in the surface as well as in subsoil. While compaction of surface soil can be removed by tillage, subsoil compaction tends to be more difficult to ameliorate and hence is more permanent. The extent of compaction is moisture dependent and is most vulnerable when the soil is worked or traffic when too wet.

3.4.1.2 Impacts of Climate Change

As shown in Fig. 3.1, with the predicted climate change, soils are expected to be exposed to drier environment for longer period of time and, on the other hand, more likely to be subjected to more intense rate of wetting and flooding. Under these conditions, soils are therefore more prone to structural breakdown due to slaking and dispersion, the latter is exacerbated by the mechanical rainfall impact which is expected to increase with increasing intensity in rain storms. Furthermore, as a result of climate change, there will be also higher incidence of tilling the soil at inappropriate soil water content, resulting in compaction and smearing (when too wet) and pulverisation (when too dry). The actual impact on structure under predicted climate change will be dependent on soil types, particularly as discussed above, on their vulnerability.

Under climate change, higher incidences of fire and flooding are also expected. Fire usually reduces soil aggregate stability and can induce, enhance or destroy water repellency depending on the temperature reached and its duration (Shakesby and Doerr 2006). Reduction in aggregate stability is caused by the removal of soil organic matter and fine root mat. These changes have implications for infiltration, overland flow and rain splash detachment (Greene et al. 1990). Reduced infiltration is often observed in burnt area relative to unburnt area due to induced water repellency and sealing of pores due to collapse of soil surface aggregates. On the other hand, fire (>270°C) can remove water repellence and increase infiltration of hydrophobic soils. DeBano (1971) reported that the infiltration capacity of a water-repellent soil was 25 times lower than for a similar soil rendered hydrophilic by heating.

Flooding of soils will occur more often due to more intense storms as well as the predicted sea water level rises. It has been predicted globally that sea water can rise by 26–59 cm (IPCC 2007). This together with the more prevalence of flooding incidences will result in submergence and loss of much of low lying prime agricultural land in many parts of the world. Periodic flooding can also drastically change soil structure and the physical properties associated with it directly via slaking and dispersion (Sedgley 1962) or indirectly via adverse impact on soil fauna, e.g. earthworms (Thonon and Klok 2007). Furthermore, flooding by sea water will result in salinity and sodicity problems leading to deterioration of soil structure and crop yield losses.

3.4.2 Indirect Impacts

Climate change can impact soil structure indirectly via possible effects on soil biology. With the different environmental changes associated with climate changes, direct and indirect changes on soil biology can occur (Fig. 3.1).

Soil organisms, particularly the "soil engineers", namely earthworm and termites, can have profound effects on soil structure and associated soil physical properties such as hydraulic conductivity (Lavelle et al. 1997; Lee and Foster 1991).

Earthworm and termite activities affect soil structure through: (1) the selective ingestion of mineral and organic particles; (2) their egestion as faecal pellets and organo-mineral aggregates; (3) digestion processes that may modify the colloidal properties of organic matter; and (4) building of long-lasting galleries, burrows and chambers. Absence of these macro-engineers results in strong profile differentiation because of the lack of mixing activities (Lavelle et al. 1997; Lee 1985).

The dominant role of the burrows (macropores) created by earthworms in controlling hydraulic properties of soil has been reported by many (e.g. Chan 2004; Chan and Heenan 1993; Edwards and Bohlen 1966). Chan (2004) reported that the infiltration rate of one single burrow created by an Australian anecic earthworm (*Anisochaeta chani*) was equivalent to 1.9 times that via the rest of soil matrix over 1 m^2 area, and there were 157/m^2 of these burrows present in the soil under pastures. Loss or reduction in the population of these earthworms with resulting reduction in these transmitting macropores can therefore have important consequences on the hydrology of the soils. Loss of continuous earthworm channels was the cause of the increased runoff and soil erosion observed in the adjacent cropping soils (Chan 2004). The potential impact of earthworm on soil structure and plant productivity can also be estimated from previous research on earthworm introduction. In New Zealand, Springett (1985) reported that the introduction of *Aporrectodea longa*, a deeper burrowing species, to low yield pasture with poor structure significantly increased porosity and doubled water infiltration.

The improved soil structure and soil–water relationship observed as a result of earthworm introduction were responsible for the observed increased root growth and pasture yield.

Possible direct responses to climate change of soil inhabitants can include changes in their activity, composition, abundance as well as the pattern of migration. It is expected that the predicted changes in water regime due to climate change, leading to longer period of summer drought and increases in the frequency of extreme events would differentially affect the activity and survival of the soil organisms. Increases in temperature can also have direct effect on length of life cycles of the soil organisms, and the effects are likely to be species specific, with resulting impact on their activity, abundance and diversity of the ecosystems. Large changes in activities and even survival of earthworms can result in response to quite small change in soil environmental factors. Daniel (1991) observed that for earthworm (*Lumbricus terrestris*), food consumption increased with temperature up to 20°C but declined above 22°C. The food consumption also declined when the soil water potential was below −7 kPa. Furthermore, burying rate of litter by earthworm, a measure of rate of decomposition, was also temperature dependent (Couteauz and Bolger 2000). Baker and Whiby (2003) observed that very few cocoon of *A. longa* survived 2 months in air-dried soil. They suggested that the more rapid drying out of soil and therefore disruption to their natural life cycle (by 1 month earlier than northern Tasmania) could be the reason for their inability to colonise mainland southern Australia. Migration of soil organisms in response to climate change can be vertical within the profile, e.g. downward migration to soil drying (Warburg et al. 1984).

The expected changes in temperature will expand the latitudinal distribution of termites and favour humivorous termites and endogeic earthworm species that feed in the soil (Lavelle et al. 1997). Invasion by peregrine exotic species can have dramatic effects on soil structure and the associated soil functions. This is well demonstrated in the case of forest clearing in Brazil (Barros et al. 2004). Soil structural degradation occurred in the pasture soils after forest clearing and the resulting reduction in soil macro-faunal diversity. Invasion of degraded pasture by earthworm *Pontoscolex corethrurus* in Brazilian Amazonia resulted in significant soil compaction (reduction in total porosity by 20–30%) resulting in reduction in infiltration (Barros et al. 2004).

Soil transportation experiment with varying temperature and moisture conditions showed that different species of enchytraeids responded differently to elevated soil temperature and moisture conditions (Briones et al. 1997).

Indirect effect of climate change on soil organisms occurs via changes in both the quantity and quality of food source (litter) (Couteauz and Bolger 2000). Yield reduction of agricultural crops has been predicted for large parts of the world, particularly those under rain-fed situation (IPCC 2007), and this will result in less cover and input of soil ecosystems. Both of the latter can affect soil structural stability and soil biology. In some African countries, yield from rain-fed agriculture could be reduced by up to 50% by 2020, and in southern and eastern Australia, production is projected to decline by 2030 (IPCC 2007).

Even with predicted increases in crop yield due to elevated CO_2 and higher precipitation, the higher quantity of litter will tend to have reduced nitrogen concentration but increased phenol concentration (Couteauz and Bolger 2000). These changes in the quality of the litter (food sources) will have effects on the abundance of some faunal species. Many aspects of the consequence of climate change on soil fauna are still poorly understood (Couteauz and Bolger 2000).

As outlined above, direct and indirect impacts of climate change on soil biology can result in profound changes in soil structure. However, the extent of soil structural changes will be dependent on the soil organisms, the magnitude of climate changes and location/soil specific. Improved understanding of the impact will require further monitoring and research.

3.4.3 Consequences: Implications on Soil Structure

There is increasing evidence in recent years that climate change is having adverse impacts on soil structure and soil functions, particularly in those regions of the world where water is already a limiting factor affecting agricultural production, viz. subtropical, temperate and semi-arid regions where precipitation is predicted to decline. Cerda (2000) studied the effect of climate and agriculture on soil aggregate stability along climatological gradients in Bolivia. Aggregate stability was measured by a number of methods with a range of energy application including the modified Emerson water dispersion test, drop impacts and ultrasonic disruption tests. Results indicated a declining aggregate stability with decreasing rainfall for both agricultural soils and scrubland soils, but for the same rainfall location, aggregate stability was consistently higher for the scrubland soil than the agricultural soils. The author also established that organic matter was the key factor determining soil aggregate stability on agricultural land. Similar trend of decreasing aggregate stability with decreasing rainfall was also established by Sarah (2005) under arid and Mediterranean climate conditions. Furthermore, a step-like threshold (reduction in aggregate stability) exists in the semi-arid zone, meaning that this area is threatened by land degradation even with a small decrease in mean annual rainfall (Sarah 2005). According to Kertesz and Mika (1999), aridification manifested as an increase in mean annual temperature (+0.0105°C yearly) and at the same time a decrease in the yearly precipitation (0.917 mm/year) has been documented in SE Europe during the last 110 years. The consequences are reduction in infiltration and recharge of the groundwater (fallen 2–4 m since 1960s), resulting in a shift of land use from cropping to plantations starting with orchards.

Under scenario of future climate change, with the combined reduction in aggregate stability (increase in soil erodibility) and more intense rainstorm events (increases in erosivity), the surface soil structure will be more prone to degradation with the formation of seal/crust. This has serious implication on soil water entry and water use efficiency (WUE). Soil surface seals and crusts resulting from aggregate breakdown reduce soil infiltration rate and may induce erosion by increasing runoff.

Table 3.1 Reduction in soil hydraulic conductivity due to seal/crust formation compared to undisturbed soils

Reference	Soil	Hydraulic conductivity (cm/h)		K/K′
		Crust/seal soil (K′)	Undisturbed soil (K)	
McIntyre (1958)	Red brown earth	0.0144	2.88	200
Bristow et al. (1994)		0.0616	1.396	22.7
Bohl and Roth (1993)	Loess soils (Haplic Luvisol, Eutric Gelysol)	16.2–60.0	70.0–97.8	1.6–5.0
Nishimura et al. (1993)	Andisol, clay loam	0.0226	0.36	15.9

As indicated in Table 3.1, presence of a thin seal/crust can dramatically reduce hydraulic conductivity by an order of magnitude (up to 200-folds). The actual reduction in hydraulic conductivity is dependent on rainfall intensity, soil types and soil structural stability (McIntyre 1958; Bohl and Roth 1993). Aggregate stability is a good indicator of erodibility and potential erosion risks (Sanchis et al. 2008). Erosion is predicted to increase under scenario of climate change, even in areas where precipitation is predicted to decrease (Aveyard 1993; Nearing et al. 2004) and the increase in erosion and therefore loss of invaluable topsoils is exacerbated by deterioration in structural forms and stability. The consequences are reduction in WUE, declining crop yield and increased risks of further erosion – thus establishing a vicious cycle of widespread land degradation.

Particularly vulnerable are the hardsetting soils because of their low stability and resilience, i.e. high vulnerability. Such soils are very widespread in Australia, Africa, Mediterranean areas of Europe and America [Luvisols, Planosols and Solonetz of World Soil Map (Mullins et al. 1990)]. Many are already in degraded state under current management practices producing low crop yield (Greacen 1983), and in many instances they are also located in regions where precipitation is predicted to decline due to future climate change. With the predicted climate change, management of these soils therefore presents challenges to agronomists and land managers.

3.5 Soil Structure on Climate Change: Adaptation and Mitigation

3.5.1 Climate Change Adaptation

The predicted changes in climate are likely to have adverse effect on agriculture in many parts of the world (IPCC 2007). In large parts of the world, e.g. Africa and southern Australia, significant declines in crop yield have been predicted due to lower rainfall and hence reduced water availability. However, with the projected food demand due to world population growth, agriculture will be under increasing pressure to maintain and even to increase production.

Agricultural systems in these regions will have to perform under hotter and drier environment where water will be more limiting. One logical approach to achieve this is to improve water capture from rain and storage of infiltrated water in soil within root zone with minimal evaporation losses. Thus, increasing WUE in agriculture under future climate change is crucial. As discussed above, these processes are controlled by hydraulic conductivity and water holding capacity of the soil, which in turn is determined by soil structural, form, stability and resilience. There are many examples of the benefits of improved soil structure on improving WUE and maintaining/improving yield of crops in semi-arid areas where water is limiting factor in crop production (Table 3.2).

Increased water capture and storage will help farming systems to adapt to more extreme rainfall events. As a consequence, during heavy precipitation on such improved systems, there will be less runoff and hence more infiltration which will improve the chance of producing a crop. In Australia, application of gypsum on sodic soils has increased crop yield under rain-fed conditions by greater than twofold, and this has been attributed to increases in infiltration, soil water storage and improved drainage (Howell 1987; Table 3.2).

Adoption of conservation tillage systems in grain production under rain-fed conditions in different parts of the world has also resulted in crop yield increases, which can be related to increases in WUE (Table 3.2). The latter was the result of increases in soil water storage due to higher infiltration rate (therefore less runoff) and the consequences of improved surface soil structure as a result of adoption of conservation tillage practices (Huang et al. 2008; So et al. 2008; Wani et al. 2007; Zhang et al. 2008).

Improvement in infiltration has also been reported in organic farming systems compared to conventional system (Colla et al. 2000). The latter authors reported a three-times higher amount of water infiltration in the organic tomato production systems compared to conventional system (Fig. 3.3). These authors attributed the reduced water use in organic cropping system compared to conventional system to the improvement in soil structure and resulting increases in infiltration.

The improvement in soil structure in these improved systems should also help to mitigate some of the undesirable consequences under climate change. In particular, it could reduce soil erosion and the associated deleterious environmental impacts.

3.5.2 Climate Change Mitigation: Soil Structure and GHGs

It is well documented that improvement in soil structure and increases in soil carbon tend to go hand-in-hand (e.g. Tisdall and Oades 1982; Chaney and Swift 1984; Chan et al. 1992). Therefore, management practices that lead to better soil structure

Table 3.2 Examples of improved water use efficiency and crop yield increases due to soil structural improvement

Method	Soil structural improvements	Soil water changes	Crop and yield improvements	Reference
Gypsum	Better infiltration	Improved water availability	20–50% increases in wheat yield	Howell (1987)
Gypsum	Improved hydraulic conductivity, improved drainage	137% increased soil water storage to 1.2 m depth	Better establishment, tiller production and less disease Up to 230% increase in wheat yield	McKenzie and So (1989)
Gypsum	Increased hydraulic conductivity	Increased water entry; higher subsoil water	Doubled WUE of pastures	Bridge and Tunny (1982)
Conservation tillage	Higher macroporosity, structural stability and infiltration, higher soil carbon	Increased plant available water	28% higher soybean yield than conventional tillage	So et al. 2008
Conservation tillage	Higher hydraulic conductivity, improved aggregate stability and higher organic carbon	Increased soil water storage, increased WUE	18% wheat and 20% pea yield than conventional	Zhang et al. (2008); Huang et al. (2008)
Conservation tillage	Higher infiltration rate; higher organic carbon; higher earthworm activity	Higher WUE	Mean grain yield over 21 seasons was 15% higher	Lal (1982)
Conservation agriculture	Higher infiltration; reduced draught	Higher WUE	Yield increases from 20 to 120% in East Africa	Wani et al. (2007)

often also result in higher soil carbon level, i.e. higher soil carbon sequestration and GHG mitigation benefits, e.g. conservation tillage (Hobb 2007). Soil structure can indirectly influence emission of GHGs such as nitrous oxide (N_2O) and methane (CH_4) via its ability to determine soil physical properties such as soil moisture characteristics, aeration and gaseous movement (Table 3.2). Previous soil aeration research has demonstrated the importance of the soil air/water balance upon aerobic and anaerobic microbial activities in soils (Linn and Doran 1984). Aerobic well-drained soils often found in well-structured soils are usually a sink for CH_4 due to the high rate of CH_4 diffusion into such soils and subsequent oxidation by methanotrophic micro-organisms (Dalal et al. 2008). While CH_4 tends to be favoured under anaerobic soil conditions, N_2O can be produced via nitrification (aerobic conditions) and denitrification (anaerobic conditions) (Dalal et al. 2003). Anaerobic conditions are often exacerbated under poor soil structure, e.g. compacted soils. Teepe et al. (2004) reported that soil compaction in the middle of wheel track in beech forest sites in Germany led to a 40-times increase in N_2O emission and reduction in CH_4 consumption by up to 90%. The changes in GHG

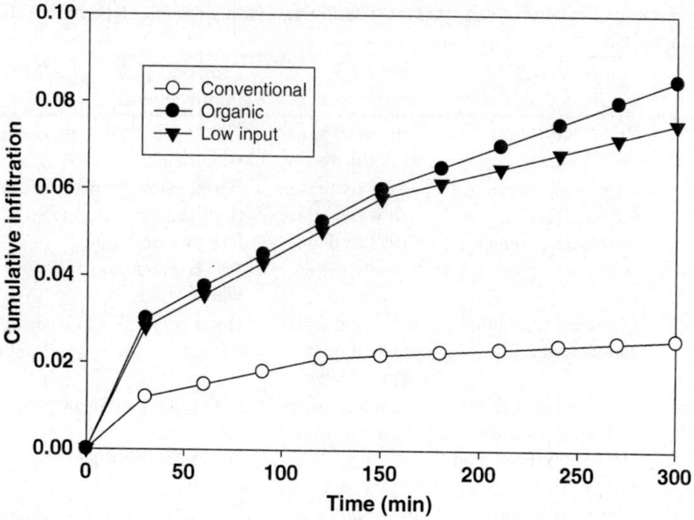

Fig. 3.3 Infiltration under conventional, organic and low input systems (adapted from Colla et al. 2000)

emissions could be related to the reduction in macroporosity resulting from soil compaction. Therefore, it is anticipated that better soil structure that promotes more aerobic soil conditions should predispose the agricultural systems to lower emission of CH_4 and N_2O, but the actual magnitude of reduction will be dependent on a number of interacting factors such as climate, soil types and other management practices which require further research.

3.6 Management Strategies for Minimising Impacts of Climate Change on Soil Structure

These can be grouped under three broad approaches.

3.6.1 Land Use Planning-Based Soil Structural Vulnerability

At present, in many land use capability classification systems, soil structural vulnerability is not a factor used. However, to minimise the likely adverse impact of soil structural degradation and the expensive costs involved with rehabilitation and worse still the loss of topsoil due to erosion, soil structural vulnerability should be included in future land use planning. It is pertinent that soils that are most vulnerable to degradation under future scenario of climate change are to be excluded from intensive use such as cropping or are only to be used with preventive measures. In view of the possibility of changes in land use pattern for adaptation to climate change, e.g. development of bioenergy crops, and the suggestion of planting and using fast growing vegetation for biochar production using pyrolysis (Lehmann 2007), the more vulnerable soils under present cropping can be used for these permanent cropping systems. However, the obstacle to this approach is the current inadequate understanding of factors controlling resilience and the ranking of soil based on vulnerability (both stability and resilience). Further research is needed.

3.6.2 Adoption of Management Practices: Conservation Agriculture

In reality, due to scarcity of good soils for food production, some of the more marginal soils will have to be used for this purpose, and therefore the importance of protecting these soils under threat of structural degradation under climate change cannot be over-emphasised. However, because of the dominant influence of human activities on natural resources, management practices will play an important role in combating impact of climate change (Laddenbert 1993). As pointed out by Follett (1993), improved management technology on the farms was the reason for the significantly less adverse impact of the drought in the late 1980s in the USA, even though it was similar to that leading to the "dust bowl" in 1930s in severity and geographic expanse.

The principles of good management practices are encapsulated under the concept of conservation agriculture. The latter can be defined as minimal soil disturbance (no-till) and permanent soil cover (mulch) combined with rotations (Hobb 2007). Soil structural improvements due to these practices in terms of more stable soil aggregates, increased in water holding capacity have been documented and are closely associated with improved WUE (therefore higher crop yield, Table 3.2) and improved biological health of the soils, e.g. increases in biodiversity.

3.6.3 Soil Amendment and Amelioration

For many soils with inherent poor structure, e.g. the sodic soils, their improvement will require using ameliorants. For example, using gypsum in the case of sodic soils will greatly improve their structure (form, stability and may be even resilience) and will greatly assist to avoid soil degradation. Practices that increase soil organic carbon such as using organic amendments, e.g. compost, biosolids and other recycled organics will also help to improve soil structure (Chaps. 5 and 14).

However, efficacy of soil ameliorants such as gypsum under predicted climate change scenarios of drier soil conditions is not certain and requires research.

3.7 Conclusions

Soil structure through its controlling effects on soil functions serves important roles in agricultural production as well as environmental services. The review concludes that predicted climate change can adversely affect soil structure and soil functions with undesirable consequences on production and off-site impacts, but the magnitudes of changes will be dependent on the magnitude of climate changes and soil structural vulnerability. As a result, the actual impact will be location/soil specific. On the other hand, maintaining and improving soil structure through adoption of conservation practices can help to minimise potential impact and further help to adapt agricultural systems to climate change.

References

Alpert P, Ben-gai T, Baharad A, Benjamini Y, Yekutieli D, Colacino M, Diodato L, Ramis C, Homar V, Romero R, Michaelides S, Manes A (2002) The paradoxical increase of Mediteranean extreme daily rainfall in spite of decrease in total values. Geophys Res Lett 29:1–4

Aveyard JM (1993) Climate change for land degradation in NSW. In: Burgin S (ed) Climate change – implications for natural resources conservation. University of Western Sydney, Hawkesbury, pp 68–80

Baker GH, Whiby WA (2003) Soil pH preferences and the influences of soil type and temperature on the survival and growth of Aporrectodea long (Lumbricidae). Pedobiologia 47:745–753

Barros E, Grimaldi M, Sarrazin M, Chauvel A, Mitja D, Desjardins T, Levelle P (2004) Soil physical degradation and changes in macrofaunal communities in Central Amazon. Appl Soil Ecol 26:157–168

Bohl H, Roth CH (1993) A simple method to assess the susceptibility of soils to form surface seals under field conditions. Catena 20:247–256

Bridge BJ, Tunny J (1982) The effect of gypsum and irrigation frequency on irrigated clover pastures grown on three soils of the Riverine Plain of south-eastern Australia. Aust J Exp Agric 22:194–200

Briones MJI, Ineson P, Piearce TG (1997) Effect of climate change on soil fauna; responses of enchytraeids, Diptera larvae and tardigrades in a transplant experiment. Appl Soil Ecol 6:117–134

Bristow K, Cass A, Smettem KRJ, Ross PJ (1994) Water entry into sealing, crusting and hardsetting soils: a review and illustrative simulation study. Proceedings of second international symposium on "Sealing, Crusting, Hardsetting Soils: Productivity and Conservation". Brisbane, 7–11 February, 1994

Cerda A (2000) Aggregate stability against water forces under different climates on agriculture land and scrubland in southern Bolivia. Soil Tillage Res 57:159–166

Chan KY (2004) Impact of tillage practices and burrows of a native Australian anecic earthworm on soil hydrology. Appl Soil Ecol 27:89–96

Chan KY, Heenan DP (1993) Surface hydraulic properties of a red earth under continuous cropping with different management practices. Aust J Soil Res 31:13–24
Chan KY, Mullins CE (1994) Slaking characteristics of some Australian and British soils. Eur J Soil Sci 45:273–283
Chan KY, Heenan DP, Roberts WP (1992) Organic carbon levels and associated soil properties of a red earth after 10 years of rotation under different stubble and tillage practices. Aust J Soil Res 30:71–83
Chaney K, Swift RS (1984) The influence of organic matter on aggregate stability in some British soils. J Soil Sci 35:223–230
Chartres CJ, Helyar KR, Fitzpatrick RW, Williams J (1992) Land degradation as a result of European settlement of Australia and its influences on soil properties. In: Gifford RM, Barson MM (eds) Australia's renewable resources: sustainability and global change. Bureau of Rural Resources Proceedings No. 14, AGPS, Canberra
Colla G, Mitchell JP, Joyce BA, Huyet LM, Wallender WW, Temple SR, Hsiao TC, Poudel DD (2000) Soil physical properties and tomato yield and quality in alternative cropping systems. Agron J 92:924–932
Couteauz M, Bolger T (2000) Interactions between atmospheric CO2 enrichment and soil fauna. Plant Soil 224:123–134
CSIRO (2007) Technical report 2007– climate change in Australia. CSIRO, Australia
Dalal RC, Wang W, Robertson GP, Parton WJ (2003) nitrous oxide emission from Australian lands and mitigation options: a review. Aust J Soil Res 41:165–195
Dalal RC, Allen DE, Livesley SJ, Richards G (2008) Magnitude and biophysical regulators of methane emission and consumption in the Australian agricultural, forest, and submerged landscape: a review. Plant Soil 309:43–76
Daniel O (1991) Leaf litter consumption and assimilation by juveniles of *Lumbricus terrestris* L. (Oligochaeta, Lumbricidae) under different environmental conditions. Biol Fertil Soils 12:202–208
DeBano LF (1971) The effect of hydrophobic substances on water movement in soil during infiltration. Proc Soil Sci Soc Am 35:340–343
Dexter AR (1988) Advances in characterisation of soil stricture. Soil Tillage Res 11:199–238
Dore MHI (2005) Climate change and changes in global precipitation patterns: what do we know? Environ Int 31:1167–1181
Edwards CA, Bohlen PJ (1966) Biology and ecology of earthworms (Chapter 10). Chapman & Hall, London, pp 196–212
Emerson WW (1991) Structural decline of soils, assessment and prevention. Aust J Soil Res 29:905–921
Follett RF (1993) Global climate change, U.S. agriculture, and carbon dioxide. J Prod Agric 6:181–190
Garnaut R (2008) The Garnaut climate change review – final report. Cambridge University Press, Cambridge
Greacen EL (1983) Tillage research in a semi-arid environment. Soil Tillage Res 3:107–109
Greene RSB, Chartres CJ, Hodgkinson KC (1990) The effect of fire on the soil in a degraded semi-arid woodland. 1. Cryptogam cover and physical and micro-morphological properties. Aust J Soil Res 28:755–777
Harris RF, Chesters G, Allen ON (1966) Dynamics of soil aggregation. Adv Agron 18:107–168
Hobb PR (2007) Conservation agriculture: what is it and why is it important for future sustainable production? J Agric Sci 45:127–137
Howell M (1987) Gypsum use in the wheat belt. J Agric W Aust 28:40–43
Huang GB, Zhang RZ, Li GD, Li LL, Chan KY, Heenan DP, Bellotti WD, Chen W, Unkovich M, Robertson M, Cullis BR (2008) Productivity and sustainability of a spring wheat – field pea rotation in semi-arid environment. Field Crops Res 107:43–55
IPCC (2007) Climate change 2007: impact, adaptation and vulnerability. In: Parry ML, Canziani OF, Palutikof JP, van der Linden PJ, Hanson CE (eds) Contribution of working group II to

the fourth assessment report of the intergovernmental panel on climate change. Cambridge University Press, Cambridge

Kay BD (1990) Rates of change in soil structure under different cropping systems. Adv Soil Sci 12:1–52

Kay BD, Angers DA (2000) Soil structure. In: Sumner ME (ed) Handbook of soil science. CRC, Boca Raton, pp A229–A276

Kay BD, Dexter AR, Rasiah V, Grant CD (1994) Weather, cropping practices and sampling depth effects on tensile strength and aggregate stability. Soil Tillage Res 32:135–148

Kertesz A, Mika J (1999) Aridification – climate change in South- Eastern Europe. Phys Chem Earth 24:913–920

Laddenbert LL (1993) Climate change as a factor in soil degradation. In: Burgin S (ed) Climate change – implications for natural resources conservation. University of Western Sydney, Hawkesbury, pp 61–67

Lal R (1982) Effects of ten years of no-tillage and conventional plowing on maize yield and properties of a tropical soil. The 9th Conference of the International Soil & Tillage Research Organisation, Osijek, Yugoslavia

Lal R (2000) A modest proposal for the year 2001: we can control greenhouse gases and feed the world... with proper soil management. J Soil Water Conserv 55:429–433

Lavelle P, Bignell D, Lepage M, Wolters V, Roger P, Ineson P, Heal OW, Dhillion S (1997) Soil function in a changing world: the role of invertebrate ecosystem engineers. Eur J Soil Sci 33:159–193

Lee KE (1985) Earthworms – their ecology and relationships with soils and land use (Chapter 10). Academic, Sydney, pp 173–199

Lee KE, Foster RC (1991) Soil fauna and soil structure. Aust J Soil Res 29:745–775

Lehmann J (2007) A handful of carbon. Nature 447:143–144

Linn DM, Doran JW (1984) Effects of water-filled pore space on carbon dioxide and nitrous oxide production in tilled and nontilled soils. Soil Sci Soc Am J 48:1267–1272

McIntyre DS (1958) Permeability measurements of soil crusts formed by raindrop impact. Soil Sci 85:185–189

McKenzie DC, So HB (1989) Effect of gypsum on vertisols of the Gwydir Valley, New South Wales 1. Soil properties and wheat growth. Aust J Exp Agric 29:51–62

Mullins CE, MacLeod DA, Northcote KH, Tisdall JM, Young IM (1990) Hardsetting soils: behaviour, occurrence and management. In: Lal R, Stewart BA (eds) Soil degradation advances in soil science, vol 11. Springer, Berlin, pp 37–99

Nearing MA, Pruski FF, O'Neal MR (2004) Expected climate change impacts on soil erosion rates: a review. J Soil Water Conserv 59:43–50

Nishimura T, Nakano M, Miyazaki T (1993) Changes in soil physical properties due to surface crust formation under simulated rainfall. Exchange Processes at the Land Surface for a Range of Space and Time Scales, Proceedings of the Yokohama Symposium. IAHS Publication, no. 212

Oades JM (1984) Soil organic matter and structural stability: mechanisms and implications for management. Plant Soil 76:319–337

Quirk JP, Murray RS (1991) Towards a model for soil structural behaviour. Aust J Soil Res 29:829–867

Russell R (1977) Traditional and modern methods of tillage. In: Plant root systems: their function and interaction with the soil. McGraw-Hill, London, pp 241–249

Sanchis MPS, Torri D, Borselli L, Poesen J (2008) Climate effects on soil erodibility. Earth Surf Process Land 33:1082–1097

Sarah P (2005) Soil aggregation response to long- and short-term differences in rainfall amount under arid and Mediterranean climate conditions. Geomorphology 70:1–11

Schjonning P, Elmholt S, Christensen BT (2004) Soil quality management – concepts and terms. In: Schjonning P, Elmholt S, Christensen BT (eds) Managing soil quality – challenges in modern agriculture (Chapter 1). CABI Publishing, Wallingford, pp 1–15

Sedgley RH (1962) Effects of disruption and flocculation o pore-space changes in beds of clay aggregates. Soil Sci 94:357–365

Shakesby RA, Doerr SH (2006) Wildfire as a hydrological and geomorphological agent. Earth Sci Rev 74:269–307

So HB, Grabski A, Desborough P (2008) The impact of 14 years of conventional and no-till cultivation on the physical properties and crop yield of a loam soil at Grafton, NSW, Australia. Soil Tillage Res 104:180–184

Springett JA (1985) Effect of introducing Allolobophora longa Ude on root distribution and some soil properties. In: Fitter Ah (ed) Ecological interactions in soils, plants, microbes and animals. Blackwell, Oxford, pp 399–405

Teepe RT, Brumme R, Beese F, Ludwig B (2004) Nitrous oxide emission and methane consumption following compaction of forest soils. Soil Sci Soc Am J 68:605–611

Thonon I, Klok C (2007) Impact of a changed inundation regime caused by climate change and floodplain rehabilitation on population viability of earthworms in a lower River Rhine floodplain. Sci Total Environ 372:585–594

Tisdall JM, Oades JM (1982) Organic matter and water-stable aggregates in soils. J Soil Sci 33:141–163

Wani S, Rockstorm J, Oweis T (eds) (2007) Rainfed agriculture – unlocking the potential (Chapter 1). CABI, Wallingford, pp 1–35

Warburg MR, Linsentnair KE, Bercoviz K (1984) The effect of climate on the distribution and abundance of isopods. Symp Zool Soc Lond 53:339–3567

Zhang GS, Chan KY, Li GD, Huang GB (2008) Effect of plastic mulching and straw management under contrasting tillage practices on the physical quality of an erodible Loess soil. Soil Tillage Res 98:113–119

Chapter 4
Soil pH, Soil Health and Climate Change

Zdenko Rengel

4.1 General Introduction: Soil pH and Its Importance for Soil Health

Throughout the history, human activities and changes in climate have influenced physical and biogeochemical processes in soil (Grieve 2001), thus influencing surface water as well as groundwater and directly or indirectly all terrestrial and aquatic ecosystems. Current climate changes and variability are likely to have significant short- and long-term impacts on soil properties. One of the important measures of soil's capacity to fulfil environmental and economic functions is pH (Rengel 2002b, 2003). This chapter will assess the available knowledge about the role of temperature and rainfall changes caused by variable and changing climate on soil pH and some related properties.

4.1.1 Soil pH

The soil pH represents measure of the H^+ activity in the soil solution that is in a dynamic equilibrium with a negatively charged solid phase. The H^+ ions are strongly attracted to these negative sites and can replace other cations from them. The diffuse layer in a vicinity of a negatively charged surface has higher H^+ activity than the bulk soil solution (Rengel 2002c). However, the pH value most relevant to soil and plant biogeochemical processes is the pH of the soil solution. For example,

Z. Rengel (✉)
Soil Science and Plant Nutrition, M087, School of Earth and Environment, The University of Western Australia, 35 Stirling Highway, Crawley, WA 6009, Australia
e-mail: Zed.Rengel@uwa.edu.au

B.P. Singh et al. (eds.), *Soil Health and Climate Change*, Soil Biology 29,
DOI 10.1007/978-3-642-20256-8_4, © Springer-Verlag Berlin Heidelberg 2011

the pH of the soil solution is a better predictor of maize yields than the soil pH measured in soil/extractant mixtures (Moody et al. 1998).

Despite the obvious importance of soil solution pH, the pH of soil is usually measured in slurry of soil and extractant (water, $CaCl_2$ or KCl) because extraction of soil solution is a tedious process. The soil pH measured in the soil/extractant mixture, even though a useful and widely applied index, should be regarded as purely empirical.

Soil pH is a dynamic parameter, with significant spatial (Gottlein et al. 1999) and temporal differences (Gregory and Hinsinger 1999). Diurnal fluctuations of as much as one pH unit may occur, as well as season-to-season variations. During seasons with low to moderate rainfall (evapotranspiration exceeds precipitation), salts are not being removed by deep percolation; increased salts tend to reduce pH by forcing more of the exchangeable H^+ ions into the soil solution. Conversely, during wet seasons, salts are diluted/removed from the topsoil and pH increases (Rengel 2002c). This season-to-season fluctuation in the total salt content should not be confused with long-term effects occurring over decades and centuries when increased rainfall leaches basic cations from the soil causing acidification (Tang and Rengel 2003). Alterations of the rainfall patterns on a large scale occurring as a consequence of climate change and variability will have an effect on soil pH that has yet to be modelled and eventually quantified over the relevant time scales.

4.1.2 *Soil Health*

Even though the term "soil health" should be viewed broadly, comprising all soil properties and influencing soil productivity, it is worth mentioning that soil capacity to carry pathogens and cause plant diseases is strongly influenced by pH. Alteration of availability of micronutrients due to pH changes is case in point (Grewal et al. 1996; Huber and Graham 1999; Rengel 2001; Rengel et al. 1996; Streeter et al. 1999, 2001). For example, Mn fertilization and other agricultural practices that increase Mn availability decrease take-all (Huber and Graham 1999; Rengel 2001). Conversely, liming (which increases pH and decreases Mn availability) increases incidence of take-all (Mason et al. 1994).

There are other soil-borne diseases that are influenced by alterations of soil pH acting via changes in the structure and functioning of microbial communities. Cavity spot disease of carrots, caused by *Pythium coloratum*, was alleviated by liming that increased a number of non-streptomycete actinomycetes antagonistic to *P. coloratum* (El-Tarabily et al. 1996). Potato scab disease is increased in high-pH soils due to prevalence of Mn-oxidizing vs. Mn-reducing microbial populations (Rengel 2001; Rogers 1969).

4.2 The Role of pH in Soil Fertility and Productivity in Native and Managed Ecosystems

While both low or high soil pH alters soil properties and poses considerable problems for terrestrial ecosystems, greater attention in the literature has been devoted to acidic than alkaline soils. Hence, this chapter will also concentrate on acidic soils, giving just a brief overview of high-pH soils (calcareous and alkaline).

4.2.1 High-pH Soils

High-pH soils range from calcareous (around pH 7) to alkaline saline and sodic soils (pH 8 and higher) (Bennett et al. 2000; Marlet et al. 1998; Qadir et al. 2007, 2001; Rao et al. 1998; Rengasamy 2002; Vukadinovic and Rengel 2007). Such soils have major nutritional constraints (deficiency of cationic micronutrients and P) and/or toxicity of Na (and B in some soils, Brennan and Adcock 2004), coupled with excess HCO_3^-, water deficiency (such soils frequently occur in arid areas), mechanical impedance and poor aeration. Under extreme cases of sodic soils, soil mechanical and hydraulic properties would be severely affected (Lu et al. 2004; Quirk 2001).

The FAO database of degraded soils in the world does not include calcareous or alkaline soils as such, but lists saline and sodic soils as examples of major land degradation; currently, there are four million km^2 of saline (3% of total land area) and 4.3 million km^2 of sodic land (http://www.fao.org/ag/agl/agll/terrastat/; accessed 05 Aug 2009). About 0.76 million km^2 (76 million ha) of land worldwide are affected by secondary salinity (i.e. induced by human activities) (Bell 1999). Between 25% (see Barrett-Lennard 2003) and 33% of the irrigated land worldwide (Carter 1975) is affected by secondary salinity, and the problem is becoming increasingly severe in many areas. In Australia, secondary salinization affects about four million hectares currently (Niknam and McComb 2000), with projections that the area affected will increase to 17 million ha by the year 2050 (National Land and Water Resources Audit 2001). Hence, it is quite obvious that the large areas of salt-affected land linked to high soil pH represent a significant issue that is likely to be influenced by climate change. In areas with reduced rainfall, salts will accumulate, exacerbating salinity problems, whereas increased rainfall will result in enhanced leaching of not only salts (which will help alleviate salinity problem, Chhabra 2002) but also leaching of basic cations, which would contribute to soil acidification.

4.2.2 Acidic Soils

Close to four billion hectares (about 30% of the ice-free soils) in the world are acidic (Sumner and Noble 2003). The worst situation is in the south-east and pacific

Asia and South America, with well over 50% of the total land in these areas having low pH. Almost two-third of all acidic soils in the world belong to Ultisols, Entisols and Oxisols. In terms of use of acidic soils, they are mostly under forests (about 66%) and pastures (about 11%), with a relatively small proportion (5%) supporting crops (Sumner and Noble 2003). In Australia alone, 33 million ha of agric land are highly acidic and further 55 million ha are at risk. Soil acidity has been identified as the most serious land degradation issue for Australia, with an estimated $1 billion in lost production per year (CSIRO 2004).

4.2.3 Soil Fertility and Productivity

Cation exchange capacity (CEC) refers to the sum of negative sites on the soil solid phase that can bind cations (Bohn et al. 1985; Rengel 2002c; Tan 1998). Given that dissociation of -OH and -COOH groups on organic matter and clay minerals is pH-dependent, CEC increases with an increase in pH (see Rengel 2002c). Such an increase can be up to twofold (Aitken et al. 1998; Hunter et al. 1997). In contrast, increased acidity causes a decrease in CEC (Blake et al. 1999) because of a decrease in the capacity of solid soil particles to adsorb cations. As a consequence, large amounts of cations (especially basic cations such as Ca^{2+} and Mg^{2+}) are present in soil solution and are therefore prone to leaching (e.g. Blake et al. 1999). Liming of acidic soils results in an increase in CEC, with extra negatively charged sites neutralized by Ca^{2+} binding (Hochman et al. 1992; Rengel 2002a).

Chemical conditions in the rhizosphere (a thin cylinder of soil surrounding roots), including pH, are usually quite different from those in the bulk soil farther away from roots (Rengel 2002c). Root-induced changes in the rhizosphere pH are a result of the balance between H^+ and HCO_3^- (OH^-) excretion, the release of CO_2 by respiration, and the excretion and decomposition of various organic compounds collectively known as root exudates (Crowley and Rengel 1999; Hinsinger et al. 2003).

The balance between excretion of H^+ and HCO_3^- (OH^-) by plant roots depends on the cation/anion uptake ratio. A greater net excretion of H^+ that accompanies a greater absorption of cations than anions results in rhizosphere acidification; the reverse occurs when uptake of anions exceeds that of cations, and therefore net excretion of HCO_3^- exceeds that of H^+ (Haynes 1990; Rengel 2002c). The form of nitrogen is the most influential one in determining the cation/anion ratio: ammonium-fed plants take up more cations than anions, and they usually have the more acidic rhizosphere than the bulk soil, whereas nitrate-fed plants take up more anions than cations and show the opposite relationship between rhizosphere and the bulk soil pH (Tang et al. 2000; Tang and Rengel 2003). It should, however, be borne in mind that the effects on rhizosphere pH are also dependent on plant species as well as the genotype of the particular species (Rengel 1999; Rengel and Marschner 2005).

4.2.4 Ion Toxicity

Soil pH has a dominant effect on solubility and therefore availability and potential phytotoxicity of ions (nutrients as well as toxic ones). While low pH shift the equilibrium towards free metal cations and protonated anions, higher pH favours carbonate or hydroxyl complexes. Therefore, availability of the micronutrient and toxic ions present in soil solution as cations (e.g. Al^{3+} and Fe^{2+}) increases with increasing soil acidity (Khabaz-Saberi et al. 2010a, b), while availability of those present as anions [$MoO_4{}^{2-}$, $CrO_4{}^{2-}$, $SeO_4{}^{-}$, $SeO_3{}^{-}$ and $B(OH)_4{}^{-}$] increases with increasing pH (Rengel 2002c, 2003).

At pH $\leq$ 5.5, Mn oxides solubilize and release Mn^{2+} into soil solution (Khabaz-Saberi et al. 2010b). At pH $<$ 4.2, Al^{3+} ions may become dominant ions in the soil solution; at pH $<$ 3.8, Fe^{2+} becomes the dominant ion, with H^+, Fe and Al as the main exchangeable ions. At pH $<$ 3.2, H^+ and Fe^{2+} ions are the main exchangeable ions (Rengel 2002c).

Aluminium is the most yield-limiting factor in many acid soils throughout the world, possibly affecting about 40% and perhaps up to 70% of the world's arable land that is potentially useable for food and biomass production. Due to influences related to industrial and agricultural development, soil acidity and therefore toxicity of Al are expanding in time and space (Rengel 1996).

Solubility of Fe compounds in soils is relatively low. Consequently, toxicity of Fe may occur only in very acidic soils (pH $<$ 3.2) or soils contaminated with Fe-containing waste materials or under anaerobic conditions (Khabaz-Saberi et al. 2010a). Reduced conditions and lowering of pH favour conversion of non-available (non-toxic) Fe^{3+} into plant-available Fe^{2+} ions (Genon et al. 1994; Singh et al. 1992) that is toxic if present at high ionic activity (Khabaz-Saberi et al. 2010a; Rengel 2002c).

Manganese toxicity occurs when pH is $\leq$5.5 if sufficient total Mn is present in the soil, and also at higher pH in poorly drained soils where reducing conditions prevail (Bartlett and James 1994; Khabaz-Saberi et al. 2010b; Rengel 2000). In case of Mn toxicity, redox potential is more important than just pH. The expression of (pE + pH) (pE = $-$log of free-electron activity) is frequently used to characterize the soil with respect to solubilization of Mn compounds [values around 16 are associated with relatively high solubility of Mn^{2+}, the reverse applies to (pE + pH) values around 20] (Sumner et al. 1991).

In acidic soils, not only increased availability of Al^{3+} and Fe^{2+}, but also lower availability of essential basic cations Ca^{2+}, Mg^{2+} and K^+ may cause growth problems. In addition, phosphate availability in acid soils is low because of the formation of sparingly soluble Al and Fe phosphates. This low availability of P, Ca^{2+}, Mg^{2+} and K^+ are important constraints to productivity of weathered acidic soils (Rengel 2002c). For example, health decline of sugar maple (*Acer saccharum*) trees in temperate forests was caused by the foliar nutrient imbalances created by soil acidification (St Clair et al. 2008).

Availability of ammonium and conversion to ammonia, and thus possible ammonia toxicity to plants, are exacerbated by an increase in the soil pH. However, even in a slightly acidic soil, pH could increase in the vicinity of the fertilizer band (Zhang and Rengel 1999, 2000); such a pH increase could lead to high ammonia pressure in the soil.

4.3 Mechanisms of Soil Acidification

Soil acidification is a slow, continuous natural process that occurs during pedogenesis and results in acid soils being common in areas where soil development continued for long, geological periods of time and under climatic conditions where rainfall exceeds evapotranspiration. The process of soil acidification is aided by water leaching basic cations to lower (subsoil) horizons; acid soils are therefore common in wet tropics (Rengel 1996; Von Uexkull and Mutert 1995).

Soil acidification is linked with the reactions occurring at the surface of amorphous Al oxides. Protonation of amorphous Al hydroxides protects against further acidification; Al hydroxides can thus be regarded as a temporary storage of acid neutralizing capacity of the soil (Funakawa et al. 2008). In contrast, soil alkalinization results from deprotonation of oxide surfaces and/or dissociation of acidic functional groups of soil humus.

Soil acidification can be either accelerated or slowed down by farming practices. The causes of soil acidification in agricultural systems have mainly been attributed to imbalance in the carbon, nitrogen and sulphur cycles (Porter et al. 1995; Tang and Rengel 2003). The major processes leading to soil acidification include: (1) net H^+ excretion by plant roots due to excess uptake of cations over anions; (2) removal of alkalinity in farm products such as grain, hay, meat and wool; (3) accumulation of organic anions in the form of soil organic matter; (4) mineralization of organic matter, nitrification of ammonium and subsequent leaching of nitrate; and (5) input of acidifying substances such as NH_4^+-based fertilizers (Rengel 2004; Tang and Rengel 2003). Nitrogen-fixation by legumes (Peoples et al. 2004) and acidic depositions from the atmosphere (e.g. Forsius et al. 1997; Gbondo and Driscoll 2002; Welsch et al. 2006) also contribute to soil acidification.

4.3.1 Ameliorating Soil Acidity

Soil acidity is frequently ameliorated by adding Ca-containing materials to soils. The average Ca content in soils is around 10 g/kg, which is about five times lower than the Ca content in igneous and sedimentary rocks, reflecting ease of Ca leaching in the weathering process as well as in well-drained soils (see Rengel 2002a). Leaching of Ca from soils in wet climates can be up to around 300 kg Ca/ha per year. In areas where climate change is likely to result in increased rainfall, there will be increased Ca^{2+} leaching and therefore more intensive soil acidification.

Increased adoption of liming in Australia occurred in the late 1990s. The rate of soil acidification in the 0–20 cm depth in areas with the average annual rainfall of 525–625 mm was estimated to be 1.52 kmol H^+/ha per year. This would require 76 kg lime/ha per year to neutralize (Scott et al. 2007). In reality, larger amounts are applied but less frequently to neutralize accumulated acidity.

Addition of Ca-containing material to soil can effectively ameliorate topsoil acidity (e.g. Edmeades and Ridley 2003). However, the development of subsoil acidity (see Tang et al. 2003) is of particular concern because incorporation of lime into deeper horizons is technically difficult and economically unfeasible. Alternatively, the surface-applied liming material may in some light-textured, sandy soils allow sufficient leaching of Ca^{2+} (and alkalinity) down the profile to provide effective amelioration of subsoil acidity (Whitten et al. 2000); this approach requires lime rates high enough to neutralize topsoil acidity, with the surplus alkalinity then leaching into deeper horizons (Tang et al. 2003).

The primary pathway for dissolution of agricultural lime may result in retention of CO_2 in the form of bicarbonate (HCO_3^-) and/or carbonic acid (H_2CO_3) in soils. However, depending on the soil pH and the form of nitrogen fertilizer used, a fraction of agricultural lime may react with strong acids (such as nitric) and release CO_2 (West and McBride 2005). Given that soil acidification is a continuous process in most soils, re-acidification after lime application will result in release of CO_2 from bicarbonate. Intergovernmental Panel on Climate Change estimated that eventually all carbon in the liming materials is released into the atmosphere as CO_2, prompting USA Environmental Protection Agency to estimate that approximately nine million tonnes of CO_2 might have been released from 20 million tonnes of lime applied in 2001. However, subsequent revised estimates put the figure at 4.4–4.6 million tonnes of CO_2 released from 20 to 30 million tonnes of lime applied in 2001 (West and McBride 2005).

4.4 Acid Sulphate Soils

In parts of the world where climate change and variability are causing decreased rainfall and diminishing runoff, there is increased groundwater abstraction and drainage of wetlands for new urban developments. All these actions result in a drop in groundwater levels to expose overlying soil layers to oxygen. In soil layers with large content of reduced Fe and S minerals, such conditions would result in oxidation of these minerals to generate acidity (acid sulphate soils) (Hicks et al. 1999). Increased acidity of the soil and groundwater results in degradation of groundwater-dependent ecosystems, potable water supply and infrastructure. The pace of these processes and reactions (e.g. the Swan Coastal Plain around Perth, Western Australia) can be quite quick, with a decrease in groundwater pH from around 5 to around 3 occurring within weeks (Appleyard 2005).

Acid sulphate soils are widespread in Australia (e.g. Lin 1999). More than 40,000 km^2 of coastal soils in Australia contain significant amounts of iron sulphides (e.g. FeS and pyrite FeS_2), which are chemically stable in an anoxic environment but generate acids upon oxidation (Ahern et al. 2002). Human interventions (e.g. draining of wetlands and some other changes in land use, groundwater extraction beyond the replenishment capacity, prolonged drought in the context of recorded and predicted climate change) (Environmental Protection Authority 2007) result in the aeration of these soils and thus oxidation of iron sulphides, production of sulphate (SO_4^{2-}), ferrous iron (Fe^{2+}) and acidity (H^+) (Hicks et al. 1999), and hence formation of acid sulphate soils.

Sulphides are sinks for pollutants (acidity and trace metals/metalloids) and thus have a role in maintaining quality of water (e.g. Smith 2004). However, under oxidizing conditions, these pollutants get released into the environment, with potentially a dramatic impact. The acidity can mobilize toxic metals (Appleyard et al. 2004, 2006; Kjoller et al. 2004; Larsen and Postma 1997; Rengel 2004) as well as leach nutrients like Ca and Mg (Tang and Rengel 2003) and greatly reduce the biodiversity of wetlands and other aquatic ecosystems (e.g. Sommer and Horwitz 2001). Moreover, exposing soil layers to oxygen allows increased microbiological activity that oxidizes organic matter (Kirk 2004). Hence, large amounts of greenhouse gases can be released into the atmosphere, contributing to global warming.

Oxidation of peat layers in soil due to drought would result in the oxidation of sulphides into sulphates, resulting not just in a pH fall but also in a decrease in dissolved organic carbon (DOC) concentration (Clark et al. 2006), suggesting significant alteration of structure and function of microbial communities.

4.4.1 Environmental Risk Assessment of Acid Sulphate Soil Formation and Impact

In spite of obvious importance of capacity to assess the risk of acid sulphate soil formation and impact, there have been only sporadic published attempts to conduct environmental risk assessment of acid sulphate soils; even then, the emphasis was on the formation of these soils rather than their impact on the environment (Lin 2001). An environmental risk assessment (Abdul et al. 2006) incorporating weight-of-evidence (cf. Hagger et al. 2006) and probabilistic approach (cf. van Sprang et al. 2004) should be used in evaluating the potential risk of acid sulphate soils forming and having an adverse environmental impact. Such a risk assessment comprises: hazard identification, dose–response assessment, exposure assessment and risk characterization (see Oberg and Bergback 2005). The risk assessment must quantify the uncertainty in risk predictions (e.g. model and parameter uncertainty) arising from spatial and temporal variability by applying probabilistic methods (e.g. Monte Carlo simulations). Monte Carlo analysis can be used to mathematically integrate best-fitting exposure concentrations (e.g. trace metals mobilized as a function of the severity of acid sulphate soils), and plant and animal species-sensitivity distributions into a probabilistic risk quotient distribution (cf. van Sprang et al. 2004).

As groundwater levels drop in many parts of the world affected by drying climate, and given specific soil mineralogy, environmental risk assessment of acid sulphate soil formation and impact may become a crucial component of (1) planning and development in urban areas and (2) natural resource management in sensitive areas.

4.5 The Effects of Climate Change and Altered Rainfall Patterns on Soil pH and Soil Health

In different parts of the world, variable and changing climate will result in either increased or decreased rainfall accompanied by increased temperatures. As mentioned above, increased precipitation will result in enhanced leaching of basic cations, which can exacerbate soil acidification. Greater acidification will require larger amounts of lime for amelioration of soil acidity, but there are significant spatial, transportation and other financial issues associated with lime application, especially in extensive agricultural systems. Absence of adequate liming will result in soils acidifying to a point of becoming unproductive and, moreover, uneconomic to be brought back into production.

Greater soil acidification may also arise due to greater biomass production (caused by increased temperatures and increased CO_2 partial pressure in the air). It should be borne in mind that most plant material contains excess cations to balance negative charges on organic molecules (H^+ was exuded into the soil in exchange for nutrient cations). If such plant material gets decomposed in situ, alkalinity accumulated in the plant material is returned to soil, thus neutralizing soil acidity. However, in managed ecosystems that produce food, feed and fibre for economic benefit, a large proportion of the biomass produced is harvested and removed from the site (hence, biomass containing alkalinity is removed, leaving non-neutralized acidity in soil – this represents unbalanced C cycle). Regarding the N cycle, when plant material is produced and decomposed in situ (balanced N cycle), there are two equivalents of H^+ produced in soil (in nitrification) and two equivalents consumed (one during nitrate uptake and the other in converting ammonia to ammonium). If nitrate produced in decomposition of organic matter is leached, one equivalent of H^+ is not consumed in soil (i.e. absence of nitrate uptake), thus leaving non-neutralized acidity in soil (unbalanced N cycle). In scenarios of variable and changing climate that will result in increased biomass production in agricultural and forest systems due to increased temperature and/or rainfall, the harvest and removal of that increased biomass will result in increased removal of alkalinity from such systems as well as in unbalanced C and N cycles (Tang and Rengel 2003; see Chap. 6). On the other side, increased partial pressure of CO_2 in the soil air, though, is unlikely to have an effect on soil pH on its own (because H_2CO_3 is a weak acid).

In areas where climate is expected to become warmer and wetter, microbial activity may increase, resulting in increased soil air CO_2 concentrations (Reth et al. 2005) as well as higher leaching of basic cations due to increased rainfall. Enhanced cation leaching will initially increase stream alkalinity at the expense of reducing base saturation levels on the soil exchange sites. With time, however, such scenario is likely to result in increased soil acidity, reduced input of alkalinity into surface waters and eventually acidification of these waters (Welsch et al. 2006).

The results from acidification of surface waters (streams and lakes) in south-western parts of Czech Republic indicated that increasing temperatures as a consequence of climate change may decrease Al concentration in water. The inverse relationship between Al solubility and temperature caused lower Al mobilization in soil horizons and/or enhanced precipitation of Al in the lakes at higher temperature. Hence, global warming might enhance recovery of surface waters from acidification (Vesely et al. 2003).

A capacity for recovery of soils and surface waters from acidification needs to be assessed, given the likelihood of future climate change, and recent identification of links between climatic fluctuations and surface water trends on decadal time scales. Evans (2005) used the MAGIC dynamic model (Model of Acidification of Groundwater In Catchments) and current scenarios of rising temperature, and falling rainfall and acid deposition in the UK, to forecast effects of climate change on soil and water recovery from acidification. Increasing DOC and elevated organic acidity would be expected to lower soil pH and increase leaching of basic cations into the surface waters, bringing about recovery of these waters from acidity (Evans 2005).

Potential climate change resulting in significant changes in rainfall amounts, seasonal distribution of rainfall or evapotranspiration will change net soil water flux and, consequently, will affect the dynamics of the acidification response to continued sulphate loading (Huntington 1996). In Catalonian mountains (north-eastern Spain), soil weathering rates increased with increased temperature and increased precipitation, but were unaffected by increased temperature and decreased precipitation (Avila et al. 1996), emphasizing a complex nature of the relationships of climatic factors on terrestrial ecosystems.

4.5.1 Carbon and Nitrogen Cycling

Higher temperature may result in greater decomposition of soil organic matter. An increase in the average soil temperature by 2°C at the forest site in Hungary was accompanied by a 22% increase in the soil respiration (Kotroczo et al. 2008). This would suggest that global warming could be increased by a positive feedback mechanism via increasing soil respiration (see Chap. 7). Testing three meadow, two bare fallow and one forest sites, Reth et al. (2005) found that the spatial variation in microbial respiration (i.e. soil CO_2 emission) correlated positively with the soil pH and fine root mass, explaining respectively up to 24 and 31% of the variability.

The interactions driven by climate change and influencing terrestrial ecosystems (soils, water, plants, animals, microbes) are complex and interdependent. For example, climate change altered the species composition of forest (Kotroczo et al. 2008), and decreased the total leaf litter production, which would influence structure and function of microbial communities (a rise in average soil temperature by 2°C would result in an increase in soil respiration by 22%). Increased tree growth, enhanced soil acidification and a decrease in the C/N ratio in soils accompanied an upward shift of vegetation along the slopes in France due to an increase in CO_2 concentration and increased temperatures (Dupouey et al. 1998).

In many parts of the world (e.g. south-west of Western Australia), climate change is expected to result in the climate getting hotter and drier. Such changes may alter N and C cycling processes in soils and result in soil acidification and consequent changes in plant, animal and microbial community structures. However, Smith et al. (2002) used an elevation gradient as an analogue of climate change to analyse climatic influence on soil microbial activity and soil properties, concluding that a predicted increase in temperature and decrease in precipitation over the next 100 years would eventually cause the soil pH to increase and the soil electrical conductivity to decrease. This would be accompanied by a decreased nitrification potential and an increase in ammonium concentration. However, an increase in NO_3 leaching with consequent acidification of the surface water was predicted as a consequence of climate change affecting N cycling in soils in a small forested catchment in southern Finland (Forsius et al. 1997).

Volatilization of ammonia occurs at soil pH above neutral. Because addition of urea frequently results in an increase in soil pH, ammonia volatilization from urea-amended soil may be significant (Zia et al. 1999; see Chap. 10). Ammonia volatilization decreases the N content of manure slurries and can lead to soil acidification and eutrophication problems. Ammonia volatilization increased with increasing temperature (up to 35°C) (Van der Stelt et al. 2007). Hence, an increase in ammonia volatilization may be one of the consequences of climate change causing higher temperatures, which may exacerbate N pollution. Indeed, in most parts of the Netherlands, the N ceilings (the acceptable N inputs into a farming system) are limited by ammonia emissions, rather than limits for nitrate content in surface and groundwater. Given that N inputs in the Netherlands are up to about twofold higher than the N ceilings (de Vries et al. 2001), the contribution of ammonia volatilization to N pollution needs to be seriously considered.

Nitrous oxide (N_2O) is one of the main greenhouse gases, and accurate estimation of the N_2O flux from fertilized arable land is required (Barton et al. 2008). Acidic soils have a higher N_2O production activity than neutral arable soils because soil acidity increases the denitrification potential (McColl and Firestone 1987) and increases the ratio of N_2O produced vs. molecular nitrogen (Firestone et al. 1980). In addition to long-term heavy nitrogen fertilization, subsequent soil acidification would also enhance the N_2O flux in the field. For example, in tea-growing soils (which are as a rule acidic), the N_2O flux was proportional to temperature, increasing with an increase in soil temperature to reach peaks in summer and autumn and then decline with incoming winter (Tokuda and Hayatsu 2004). In contrast, soil drying decreased N_2O flux (Tokuda and Hayatsu 2004). Hence, the effects of climate change and variability regarding N_2O fluxes would differ in different soils, farming systems and locations.

4.6 Conclusions

There is no doubt that variable and changing climate will impact on soil pH and thus indirectly on a myriad of biogeochemical and physical processes occurring in the soil–water–microbe–plant continuum in native and managed terrestrial and aquatic ecosystems. However, such impacts would be dependent on soil type and environmental factors. Further work in discerning the effects of increasing air and soil temperatures as well as changes in rainfall distribution and intensity (getting either wetter or drier) on soil pH and other properties is urgently needed. Such knowledge will underpin management decisions required to protect soils as one of the most precious natural resources.

References

Abdul M, Hickson R, Barratt BI (2006) Principles of environmental risk assessment with emphasis on the New Zealand perspective. In: Bigler F, Babendreier D, Kuhlmann U (eds) Environmental impact of invertebrates for biological control of arthropods: methods and risk assessment. CABI Publishing, Wallingford, pp 241–253

Ahern C, Johnston P, Watling K (2002) Acid sulfate soils: an important coastal water quality issue. Natural Resources and Mines, Queensland

Aitken RL, Moody PW, Dickson T (1998) Field amelioration of acidic soils in south-east Queensland: I. Effect of amendments on soil properties. Aust J Agric Res 49:627–637

Appleyard S (2005) Potential geochemical and environmental changes caused by lowering of water tables. Waters and Rivers Commission, Perth

Appleyard S, Wong S, Willis-Jones B, Angeloni J, Watkins R (2004) Groundwater acidification caused by urban development in Perth, Western Australia: source, distribution, and implications for management. Aust J Soil Res 42:579–585

Appleyard SJ, Angeloni J, Watkins R (2006) Arsenic-rich ground water in an urban area experiencing drought and increasing population density, Perth, Australia. Appl Geochem 21:83–97

Avila A, Neal C, Terradas J (1996) Climate change implications for streamflow and streamwater chemistry in a Mediterranean catchment. J Hydrol 177:99–116

Barrett-Lennard EG (2003) The interaction between waterlogging and salinity in higher plants: causes, consequences and implications. Plant Soil 253:35–54

Bartlett RJ, James BR (1994) Redox chemistry of soils. Adv Agron 50:152–208

Barton L, Kiese R, Gatter D, Butterbach-Bahl K, Buck R, Hinz C, Murphy DV (2008) Nitrous oxide emissions from a cropped soil in a semi-arid climate. Glob Change Biol 14:177–192

Bell DT (1999) Australian trees for the rehabilitation of waterlogged and salinity-damaged landscapes. Aust J Bot 47:697–716

Bennett DR, Hecker FJ, Entz T, Greenlee GM (2000) Salinity and sodicity of irrigated solonetzic and chernozemic soils in east-central Alberta. Can J Soil Sci 80:117–125

Blake L, Goulding KWT, Mott CJB, Johnston AE (1999) Changes in soil chemistry accompanying acidification over more than 100 years under woodland and grass at Rothamsted Experimental Station, UK. Eur J Soil Sci 50:401–412

Bohn HL, McNeal BL, O'Connor GA (1985) Soil chemistry, 2nd edn. Wiley-Interscience, New York

Brennan RF, Adcock KG (2004) Incidence of boron toxicity in spring barley in southwestern Australia. J Plant Nutr 27:411–425

Carter DL (1975) Problems of salinity in agriculture. In: Poljakoff-Mayber A, Gale J (eds) Plants in saline environments. Springer, Berlin, pp 25–35

Chhabra R (2002) Salt-affected soils and their management for sustainable rice production – key management issues: a review. Agric Rev 23:110–126

Clark JM, Chapman PJ, Heathwaite AL, Adamson JK (2006) Suppression of dissolved organic carbon by sulfate induced acidification during simulated droughts. Environ Sci Technol 40:1776–1783

Crowley DE, Rengel Z (1999) Biology and chemistry of nutrient availability in the rhizosphere. In: Rengel Z (ed) Mineral nutrition of crops: fundamental mechanisms and implications. Food Products Press, New York, pp 1–40

CSIRO (2004) Acid soils – a ticking time bomb? CSIRO Plant Industry Communication Group, Clayton South

de Vries W, Kros H, Oenema O, Erisman JW (2001) Assessment of nitrogen ceilings for Dutch agricultural soils to avoid adverse environmental impacts. Sci World J 1:898–907

Dupouey J-L, Becker M, Bert D, Cadel G, Lefevre Y, Picard J-F, Thimonier A (1998) Recent changes of soil, vegetation and productivity in French Mountain Forests. Ecologie 29:341–349

Edmeades DC, Ridley AM (2003) Using lime to ameliorate topsoil and subsoil acidity. In: Rengel Z (ed) Handbook of soil acidity. Marcel Dekker, New York, pp 297–336

El-Tarabily KA, Hardy GESJ, Sivasithamparam K, Kurtböke ID (1996) Microbiological differences between limed and unlimed soils and their relationship with cavity spot disease of carrots (*Daucus carota* L.) caused by *Pythium coloratum* in Western Australia. Plant Soil 183:279–290

Environmental Protection Authority (2007) State of the environment report: Western Australia 2007. Department of Environment and Conservation, Perth

Evans CD (2005) Modelling the effects of climate change on an acidic upland stream. Biogeochem 74:21–46

Firestone MK, Firestone RB, Tiedje JM (1980) Nitrous oxide from soil denitrification: factors controlling its biological production. Science (Washington) 208:749–751

Forsius M, Johansson M, Posch M, Holmberg M, Kamari J, Lepisto A, Roos J, Syri S, Starr M (1997) Modelling the effects of climate change, acidic deposition and forest harvesting on the biogeochemistry of a boreal forested catchment in Finland. Boreal Environ Res 2:129–143

Funakawa S, Hirooka K, Yonebayashi K (2008) Temporary storage of soil organic matter and acid neutralizing capacity during the process of pedogenetic acidification of forest soils in Kinki District, Japan. Soil Sci Plant Nutr 54:434–448

Gbondo TSS, Driscoll CT (2002) Retrospective analysis of the response of soil and stream chemistry of a northern forest ecosystem to atmospheric emission controls from the 1970 and 1990 Amendments of the Clean Air Act. Environ Sci Technol 36:4714–4720

Genon JG, Hepcee ND, Duffy JE, Delvaux B, Hennebert PA (1994) Iron toxicity and other chemical soil constraints to rice in highland swamps of Burundi. Plant Soil 166:109–115

Gottlein A, Heim A, Matzner E (1999) Mobilization of aluminium in the rhizosphere soil solution of growing tree roots in an acidic soil. Plant Soil 211:41–49

Gregory PJ, Hinsinger P (1999) New approaches to studying chemical and physical changes in the rhizosphere: an overview. Plant Soil 211:1–9

Grewal HS, Graham RD, Rengel Z (1996) Genotypic variation in zinc efficiency and resistance to crown rot disease (*Fusarium graminearum* Schw. Group 1) in wheat. Plant Soil 186:219–226

Grieve IC (2001) Human impacts on soil properties and their implications for the sensitivity of soil systems in Scotland. Catena 42:361–374

Hagger JA, Jones MB, Leonard P, Owen R, Galloway TS (2006) Biomarkers and integrated environmental risk assessment: are there more questions than answers? Integr Environ Assess Manag 2:312–329

Haynes RJ (1990) Active ion uptake and maintenance of cation-anion balance: a critical examination of their role in regulating rhizosphere pH. Plant Soil 126:247–264

Hicks W, Bowman G, Fitzpatrick R (1999) East trinity acid sulfate soils. Part 1: environmental hazards. CSIRO Land and Water, Adelaide

Hinsinger P, Plassard C, Tang C, Jaillard B (2003) Origins of root-mediated pH changes in the rhizosphere and their responses to environmental constraints: a review. Plant Soil 248:43–59

Hochman Z, Edmeades DC, White E (1992) Changes in effective cation exchange capacity and exchangeable aluminium with soil pH in lime-amended field soils. Aust J Soil Res 30:177–187

Huber DM, Graham RD (1999) The role of nutrition in crop resistance and tolerance to diseases. In: Rengel Z (ed) Mineral nutrition of crops: fundamental mechanisms and implications. The Haworth Press, New York, pp 169–204

Hunter DJ, Yapa LGG, Hue NV (1997) Effects of green manure and coral lime on corn growth and chemical properties of an acid Oxisol in Western Samoa. Biol Fertil Soils 24:266–273

Huntington TG (1996) Predicting watershed acidification under alternate rainfall conditions. Water Air Soil Pollut 90:429–450

Khabaz-Saberi H, Rengel Z, Wilson R, Setter TL (2010a) Variation for tolerance to high concentration of ferrous iron (Fe^{2+}) in Australian hexaploid wheat. Euphytica 172:275–283

Khabaz-Saberi H, Rengel Z, Wilson R, Setter TL (2010b) Variation of tolerance to manganese toxicity in Australian hexaploid wheat. J Plant Nutr Soil Sci 173:103–112

Kirk G (2004) The biogeochemistry of submerged soils. Wiley, New York

Kjoller C, Postma D, Larsen F (2004) Groundwater acidification and the mobilization of trace metals in a sandy aquifer. Environ Sci Technol 38:2829–2835

Kotroczo Z, Fekete I, Toth JA, Tothmeresz B, Balazsy S (2008) Effect of leaf- and root-litter manipulation for carbon dioxide efflux in forest soil. Cer Res Commun 36:663–666

Larsen F, Postma D (1997) Nickel mobilization in a well field: release by pyrite oxidation and desorption by Mn oxides. Environ Sci Technol 31:2589–2595

Lin C (1999) Acid sulfate soils in Australia: characteristics, problems and management. Pedosphere 9:289–298

Lin C (2001) Analytical methods for environmental risk assessment of acid sulfate soils: a review. Pedosphere 11:301–310

Lu SG, Tang C, Rengel Z (2004) Combined effects of waterlogging and salinity on electrochemistry, water-soluble cations and water dispersible clay in soils with various salinity levels. Plant Soil 264:231–245

Marlet S, Barbiero L, Valles V (1998) Soil alkalinization and irrigation in the Sahelian Zone of Niger II: agronomic consequences of alkalinity and sodicity. Arid Soil Res Rehab 12:139–152

Mason MG, Porter WM, Cox WJ (1994) Effect of an acidifying nitrogen fertiliser and lime on soil pH and wheat yields. 2. Plant response. Aust J Exp Agric 34:247–253

McColl JG, Firestone MK (1987) Cumulative effects of simulated acid rain on soil chemical and microbial characteristics and conifer seedling growth. Soil Sci Soc Am J 51:794–800

Moody PW, Aitken RL, Dickson T (1998) Field amelioration of acidic soils in south-east Queensland: III. Relationships of maize yield response to lime and unamended soil properties. Aust J Agric Res 49:649–656

National Land and Water Resources Audit 2001 Australian Dryland Salinity Assessment (2000) Extent, impacts, processes, monitoring and management options. Land and Water Australia, Canberra

Niknam SR, McComb J (2000) Salt tolerance screening of selected Australian woody species – a review. For Ecol Manag 139:1–19

Oberg T, Bergback B (2005) A review of probabilistic risk assessment of contaminated land. J Soils Sedim 5:213–224

Peoples MB, Angus JF, Swan AD, Dear BS, Hauggaard-Nielsen H, Jensen ES, Ryan MH, Virgona JM (2004) Nitrogen dynamics in legume-based pasture systems. In: Mosier AR, Syers JK, Freney JR (eds) Agriculture and the nitrogen cycle: assessing the impacts of fertilizer use on food production and the environment. Island Press, Washington, pp 103–114

Porter WM, McLay CDA, Dolling PJ (1995) Rates and sources of acidification in agricultural systems of southern Australia. In: Plant soil interactions at low pH: principles and management. Proceedings of the third international symposium, Brisbane, Queensland. Kluwer Academic, Dordrecht, pp 75–83

Qadir M, Schubert S, Ghafoor A, Murtaza G (2001) Amelioration strategies for sodic soils: a review. Land Degrad Develop 12:357–386

Qadir M, Oster JD, Schubert S, Noble AD, Sahrawat KL (2007) Phytoremediation of sodic and saline-sodic soils. Adv Agron 96:197–247

Quirk JP (2001) The significance of the threshold and turbidity concentrations in relation to sodicity and microstructure. Aust J Soil Res 39:1185–1217

Rao BRM, Dwivedi RS, Sreenivas K, Khan QI, Ramana KV, Thammappa SS, Fyzee MA (1998) An inventory of salt-affected soils and waterlogged areas in the Nagarjunsagar Canal command area of southern India, using space-borne multispectral data. Land Degrad Develop 9:357–367

Rengasamy P (2002) Transient salinity and subsoil constraints to dryland farming in Australian sodic soils: an overview. Aust J Exp Agric 42:351–361

Rengel Z (1996) Tansley review No. 89: uptake of aluminium by plant cells. New Phytol 134:389–406

Rengel Z (1999) Physiological mechanisms underlying differential nutrient efficiency of crop genotypes. In: Rengel Z (ed) Mineral nutrition of crops: fundamental mechanisms and implications. The Haworth Press, New York, pp 227–265

Rengel Z (2000) Uptake and transport of manganese in plants. In: Sigel A, Sigel H (eds) Metal ions in biological systems. Marcel Dekker, New York, pp 57–87

Rengel Z (2001) The role of micronutrient fertilization in managing soil-borne plant diseases. In: Singh K, Mori S, Welch RM (eds) Perspectives on the micronutrient nutrition of crops. Scientific Publishers, Jodhpur, pp 53–69

Rengel Z (2002a) Calcium. In: Lal R (ed) Encyclopedia of soil science. Marcel Dekker, New York, pp 135–138

Rengel Z (ed) (2002b) Handbook of plant growth. pH as the master variable. Marcel Dekker, New York

Rengel Z (2002c) Role of pH in availability of ions in soil. In: Rengel Z (ed) Handbook of plant growth. pH as a master variable in plant growth. Marcel Dekker, New York, pp 323–350

Rengel Z (ed) (2003) Handbook of soil acidity. Marcel Dekker, New York

Rengel Z (2004) Aluminium cycling in the soil-plant-animal-human continuum. Biometals 17:669–689

Rengel Z, Marschner P (2005) Nutrient availability and management in the rhizosphere: exploiting genotypic differences. New Phytol 168:305–312

Rengel Z, Gutteridge R, Hirsch P, Hornby D (1996) Plant genotype, micronutrient fertilization and take-all infection influence bacterial populations in the rhizosphere of wheat. Plant Soil 183:269–277

Reth S, Reichstein M, Falge E (2005) The effect of soil water content, soil temperature, soil pH-value and the root mass on soil CO_2 efflux – a modified model. Plant Soil 268:21–33

Rogers P (1969) Organic manuring for potato scab control and its relations to soil manganese. Ann Appl Biol 63:371–378

Scott BJ, Fenton IG, Fanning AG, Schumann WG, Castleman LJC (2007) Surface soil acidity and fertility in the eastern riverina and western slopes of southern New South Wales. Aust J Exp Agric 47:949–964

Singh BP, Das M, Prasad RN, Ram M (1992) Characteristics of Fe-toxic soils and affected plants and their correction in acid Haplaquents of Meghalaya. Int Rice Res Newsl 17:18–19

Smith J (2004) Chemical changes during oxidation of iron monosulfide-rich sediments. Aust J Soil Res 42:659–666

Smith JL, Halvorson JJ, Bolton H Jr (2002) Soil properties and microbial activity across a 500 m elevation gradient in a semi-arid environment. Soil Biol Biochem 34:1749–1757

Sommer B, Horwitz P (2001) Water quality and macroinvertebrate response to acidification following intensified summer droughts in a Western Australian wetland. Marine Freshw Res 52:1015–1021

St Clair SB, Sharpe WE, Lynch JP (2008) Key interactions between nutrient limitation and climatic factors in temperate forests: a synthesis of the sugar maple literature. Can J For Res 38:401–414

Streeter TC, Graham RD, Lake AWH, Rengel Z (1999) Potential for breeding *Medicago spp.* for zinc efficiency to increase tolerance to *Rhizoctonia solani*. In: Gissel-Nielsen G, Jensen A (eds) Plant nutrition – molecular biology and genetics. Kluwer Academic, Dordrecht, pp 349–352

Streeter TC, Rengel Z, Neate SM, Graham RD (2001) Zinc fertilisation increases tolerance to *Rhizoctonia solani* (AG 8) in *Medicago truncatula*. Plant Soil 228:233–242

Sumner ME, Noble AD (2003) Soil acidification: the world story. In: Rengel Z (ed) Handbook of soil acidity. Marcel Dekker, New York, pp 1–28

Sumner ME, Fey MV, Noble AD (1991) Nutrient status and toxicity problems in acid soils. In: Ulrich B, Sumner ME (eds) Soil acidity. Springer, Berlin, pp 149–182

Tan KH (1998) Principles of soil chemistry. Marcel Dekker, New York

Tang C, Rengel Z (2003) Role of plant cation/anion uptake ratio in soil acidification. In: Rengel Z (ed) Handbook of soil acidity. Marcel Dekker, New York, pp 57–81

Tang C, Raphael C, Rengel Z, Bowden JW (2000) Understanding subsoil acidification: effect of nitrogen transformation and nitrate leaching. Aust J Soil Res 38:837–849

Tang C, Rengel Z, Diatloff E, Gazey C (2003) Responses of wheat and barley to liming on a sandy soil with subsoil acidity. Field Crops Res 80:235–244

Tokuda S-I, Hayatsu M (2004) Nitrous oxide flux from a tea field amended with a large amount of nitrogen fertilizer and soil environmental factors controlling the flux. Soil Sci Plant Nutr 50:365–374

Van der Stelt B, Temminghoff EJM, Van Vliet PCJ, Van Riemsdijk WH (2007) Volatilization of ammonia from manure as affected by manure additives, temperature and mixing. Bioresour Technol 98:3449–3455

van Sprang PA, Verdonck FA, Vanrolleghem PA, Vangheluwe ML, Janssen CR (2004) Probabilistic environmental risk assessment of zinc in Dutch surface waters. Environ Tox Chem 23:2993–3002

Vesely J, Majer V, Kopacek J, Norton SA (2003) Increasing temperature decreases aluminum concentrations in Central European lakes recovering from acidification. Limnol Oceanogr 48:2346–2354

Von Uexkull HR, Mutert E (1995) Global extent, development and economic impact of acid soils. Plant Soil 171:1–15

Vukadinovic V, Rengel Z (2007) Dynamics of sodium in saline and sodic soils. Commun Soil Sci Plant Anal 38:2077–2090

Welsch DL, Cosby BJ, Hornberger GM (2006) Simulation of future stream alkalinity under changing deposition and climate scenarios. Sci Total Environ 367:800–810

West TO, McBride AC (2005) The contribution of agricultural lime to carbon dioxide emissions in the United States: dissolution, transport, and net emissions. Agric Ecosyst Environ 108:145–154

Whitten MG, Wong MTF, Rate AW (2000) Amelioration of subsurface acidity in the south-west of Western Australia: downward movement and mass balance of surface-incorporated lime after 2–15 years. Aust J Soil Res 38:711–728

Zhang XK, Rengel Z (1999) Gradients of pH, ammonium, and phosphorus between the fertiliser band and wheat roots. Aust J Agric Res 50:365–373

Zhang XK, Rengel Z (2000) Role of soil pH, Ca supply, and banded P fertilisers in modulating ammonia toxicity to wheat. Aust J Agric Res 51:691–699

Zia MS, Aslam M, Rahmatullah AM, Ahmed T (1999) Ammonia volatilization from nitrogen fertilizers with and without gypsum. Soil Use Manag 15:133–135

Chapter 5
Soil Organic Matter, Soil Health and Climate Change

Ram C. Dalal, Diane E. Allen, K. Yin Chan, and Bhupinder Pal Singh

5.1 Introduction

Soil organic matter (SOM) is essential in maintaining physical, chemical and biological functions in soil, and hence soil health. In fact, SOM is the key indicator of soil health. It contains both living and non-living components. Living components include soil microbial biomass and living roots. Non-living SOM is a heterogeneous organic matter, variously described as labile, slow and recalcitrant SOM, light fraction (free or occluded) and heavy fraction, particulate (>53 μm) and non-particulate SOM. It is also described by its chemical constituents such as proteins, lipids, starch, carbohydrates, hemicelluloses, celluloses, lignins, polyphenols, pectins and tannins or by humic acid, fulvic acid and humins. Soil organic carbon (SOC) constitutes about 50% of SOM (Pribyl 2010) and contains labile, slow and recalcitrant C pools (Dalal and Chan 2001).

The SOC stocks are the resultant of the net balance between inputs and outputs of C in soil. While inputs are primarily from above-ground and below-ground phytobiomass or their consumed products such as manure and microbial biomass, the outputs are primarily CO_2 emission from soil, with minor contribution from leaching and runoff of dissolved and particulate organic C (POC), and in hydromorphic environments, significant contribution from CH_4 emission to the atmosphere. While both inputs and outputs of organic carbon are affected by changes to

R.C. Dalal (✉) • D.E. Allen
Environment and Resource Sciences, Department of Environment and Resource Management, 41 Boggo Road, Dutton Park, QLD 4102, Australia
e-mail: Ram.Dalal@derm.qld.gov.au

K.Y. Chan
NSW Department of Primary Industries, Locked Bag 4, Richmond, NSW 2753, Australia

B.P. Singh
Forest Science Centre, NSW Department of Primary Industries, PO Box 100, Beecroft, NSW 2119, Australia

B.P. Singh et al. (eds.), *Soil Health and Climate Change*, Soil Biology 29,
DOI 10.1007/978-3-642-20256-8_5, © Springer-Verlag Berlin Heidelberg 2011

global environment, we will specifically review the effects of changing climate on stocks of total SOC and SOC pools.

Global climate change scenarios considered are IPCC projection of global temperature increase of 1.6–6.4°C by 2100, atmospheric CO_2 concentration increase of up to 550 ppm and precipitation change by at least 20% (Denman et al. 2007). We have also considered the influence of atmospheric N deposition, an important component of global environmental change; the rates of N deposition have increased by threefold to fivefold over the past century (Janssens et al. 2010) and may continue to increase rapidly in densely populated areas. The increasing rates of atmospheric N deposition may play a major role in modulating climate change impacts.

We will first consider the impacts of these individual global change drivers (temperature, CO_2, atmospheric N deposition and precipitation) to understand the processes governing total SOC and SOC pools. Then we will consider the influence of interactions of these drivers on total SOC and SOC pools, with implications for soil health and climate change. Soil management practices that assist in C sequestration and/or reducing C loss from soil are also considered.

5.2 Effect of Temperature on SOC and SOC Pools

5.2.1 Soil Organic Carbon

Soil organic carbon (C) decomposition is an enzymatic process, and the temperature dependence of SOC decomposition follows the Arrhenius equation (5.1), that is:

$$k = a\exp^{-E_a/RT}, \tag{5.1}$$

where k is the reaction rate constant, a is frequency factor, E_a is the activation energy, R is the gas constant (8.314 J K^{-1} mol^{-1}) and T is the temperature in Kelvin (Arrhenius 1889).

The temperature dependence of SOC decomposition has been also described by Q_{10} based functions, which is the factor by which the rate of reaction increases with every 10° rise in temperature (see Davidson et al. 2006; Kirschbaum 1995, 2000). While in the Arrhenius (5.1), an activation energy (the minimum energy needed to create a reaction) parameter also goes along with temperature to describe relative reaction rates (SOC decomposition) as a function of temperature.

It follows from (5.1) that the Q_{10} value of a chemical reaction (SOC decomposition) with reactant's E_a around 50 kJ mol^{-1}, and at temperatures between 273 and 303°K, is usually close to 2, that is, the rate of reaction increases twofold for each 10° increase in temperature (Davidson and Janssens 2006). Furthermore, as E_a increases, that is, as less reactive and more recalcitrant pool increases, temperature sensitivity, or Q_{10} value may also increase (Glasstone et al. 1941; Bosatta and Agren 1999; Davidson and Janssens 2006). However, it must be noted that turnover

Table 5.1 Potential changes in organic C stocks in major pools in a simulated upland soil and a permafrost soil in response to global warming by 2100, and not accounting for the interactive effects of elevated [CO_2], atmospheric N deposition and changes in precipitation (modified from Davidson and Janssens 2006)

Organic C pools and temperature-sensitive ecosystems	Organic C stock (Gt C)	Turnover time (years)	Potential loss by 2100 (Gt C)	References
Upland soil (litter layer)	200	6	30	Jones et al. (2005)
Upland soil (1 m depth)				
Labile pool	20	6	3	Jones et al. (2005)
Slow pool	700	18	40	Jones et al. (2005)
Recalcitrant pool	100	1,000	0.1	Jones et al. (2005)
Permafrost (3 m depth)	400	4	100	Gruber et al. (2004)

rate of less reactive and more recalcitrant SOC pool is more than an order of magnitude slower than the reactive or labile pool (Table 5.1). Thus, although the temperature sensitivity of the recalcitrant SOC pool may be large than the labile SOC pool, the absolute rate of loss of the recalcitrant SOC pool is still likely to be smaller than the labile SOC pool as temperature increases (Table 5.1).

Provided substrate availability remains unchanged during the reaction, the Arrhenius-type equation describes the temperature dependence of the rates of biochemical reactions such as SOC decomposition satisfactorily. However, substrate availability changes during decomposition of SOC due to a number of factors, including spatial heterogeneity, physical inaccessibility (in aggregates), and chemical recalcitrance (due to inherent chemistry) and protection (upon reactions with mineral surfaces), and these factors could influence the observed (apparent) temperature dependence of SOC decomposition (Davidson and Janssens 2006). Since SOC decomposition is governed primarily by a series of enzymatic controlled processes, substrate availability and concentration at the active site of the enzyme are described by Michaelis–Menten (5.2), that is:

$$v = [S]\, V_{\max}/[S] + K_{\mathrm{m}}, \tag{5.2}$$

where $[S]$ is the substrate availability, v is the initial reaction rate, $V_{\max}$ is the maximum reaction rate at a given temperature and K_{m} is the Michaelis–Menten constant, and represents $[S]$ when $V_{\max}/2$ (Michaelis and Menten 1913). When substrate availability is unlimited, that is, $[S]>>>K_{\mathrm{m}}$, the temperature response of $V_{\max}$ determines the reaction rate in the range of optimum temperature, that is, when no enzyme denaturation occurs. Under such conditions, the temperature response of SOM decomposition follows the Arrhenius-type equation (Davidson and Janssens 2006; Gershenson et al. 2009). Under substrate-limited conditions, however, the K_{m} value of microbial enzymes becomes significant, which increases with increasing temperature, and counteracts the temperature response of $V_{\max}$. Thus, when substrate supply at the active enzyme site is limited, temperature dependence of SOC decomposition may be lower than modelled by (5.1).

Indeed, Liski et al. (1999), Giardina and Ryan (2000) and Kirschbaum (1995, 2006) argued that decomposition of SOC was mainly controlled by substrate supply, and consequently, the temperature response of SOC decomposition can vary considerably, e.g. from no response (Giardina and Ryan 2000; Luo et al. 2001; Melillo et al. 2002), slight response (Schimel 1994), to large response (Jenkinson et al. 1991; Trumbore et al. 1996). Therefore, interpretation of many studies on temperature dependence of SOC decomposition is probably complicated by the limitation of substrate supply during the experimental period (Gershenson et al. 2009). The substrate supply may also become limiting with increasing temperature, either due to increased temperature response of K_m or through interactive effects of temperature and soil water content on substrate release and diffusion to the active enzyme site (Davidson and Janssens 2006). It must be acknowledged that for heterogeneous soil environment, Arrhenius and Michaelis–Menten kinetics may not always describe the temperature sensitivity of SOC decomposition very well, because of co-variations in substrate supply and quality, substrate use efficiency by microbes, and composition and size of microbial populations as well as properties of enzymes they produce, commonly referred to as "temperature acclimation" (Allison et al. 2010; Davidson and Janssens 2006). Hence, to generalize the temperature response of SOC decomposition, the influences of factors (i.e. substrate limitation, microbial processes and acclimation of microbial response to temperature) need to be explicitly incorporated into C turnover models (Curiel Yuste et al. 2010; Davidson and Janssens 2006).

5.2.2 SOC Pools

Since SOM is heterogeneous in nature, various SOC pools such as labile pool [particulate organic matter (POM), light fraction organic matter], slow pool (SOM within aggregates, humus) or recalcitrant pool (organic-mineral complexes, charcoal C, phytolith C) may show different temperature sensitivity to SOC decomposition. Unfortunately, there is no consensus on relative temperature sensitivity of different SOM pools. For example, the temperature sensitivity of decomposition of recalcitrant SOM pools may be less than (Liski et al. 1999; Rey and Jarvis 2006), similar to (Fang et al. 2005; Reichstein et al. 2005; Conen et al. 2006; Plante et al. 2009), or greater than (Fierer et al. 2005; Conant et al. 2008) that of labile pools.

Again, these conflicting observations about the temperature sensitivity of SOC pools can primarily be explained by limitations in substrate supply (production and diffusion of substrate) or accessibility (physical and chemical protection) to microbes and exo-enzymes (Curiel Yuste et al. 2010). For example, as the substrate supply or accessibility becomes limiting, and K_m becomes equal to or greater than [S], the intrinsic temperature sensitivity or Q_{10} value of SOM decomposition decreases. This may occur more frequently for recalcitrant SOC pools than labile pools.

Curiel Yuste et al. (2010), however, showed that depletion of labile carbon decreased the overall temperature sensitivity of SOM decomposition and suggested that the supply and accessibility of labile carbon is more important in governing the temperature response of SOM decomposition than the intrinsic temperature sensitivity of SOM pools (Davidson and Janssens 2006). On the contrary, Conant et al. (2008) found that Q_{10} values of CO_2 efflux from the labile SOM pool, estimated from initial stages of CO_2 efflux, were consistently lower than the more resistant or recalcitrant SOM pool, estimated from later stages of CO_2 efflux from incubation of grassland and cultivated soils at different temperatures (Q_{10} labile pool, 2.1–2.3 vs. Q_{10} recalcitrant pool, 3.3–3.8). The findings of Conant et al. (2008) are consistent with other soil C and litter decomposition studies in which temperature sensitivity of relatively more resistant organic C is greater than that of relatively more labile organic C (Fierer et al. 2005; Leifeld and Fuhrer 2005). This also supports the hypothesis that in the presence of abundant or high substrate supply, that is, when $[S]>>>K_m$, Q_{10} value for labile pools (i.e. low E_a values) is usually close to 2, and for recalcitrant pools (high E_a values) is much more than 2, at temperatures between 273 and 303 K (Davidson and Janssens 2006; Hartley and Ineson 2008; Gershenson et al. 2009).

There is increasing evidence that large global stocks of "old" SOC in upland and permafrost ecosystems can be lost via decomposition as the temperature increases (Table 5.1). Apparently, the "old" SOC in permafrost contains large labile C pool, which is prone to decomposition on thawing due to its increased lability, increased diffusivity to decomposers and increased decomposer activity, although this C pool is relatively stable under permafrost conditions (Schuur et al. 2009). Since tundra soils contain large labile SOC stocks (Table 5.1), a net loss of organic C from these soils in coming years due to climate warming is a cause for concern.

5.3 Effect of Elevated CO_2 on SOC and SOC Pools

Carbon dioxide levels have increased from 280 ppm_v in 1750 to 387 ppm_v in 2009, an increase in CO_2 concentration by 40% since the industrial revolution. Experiments under elevated CO_2 concentrations have shown that most C_3 plants, which include all trees, many crops and vegetation from cold and temperate regions, respond with increased rates of photosynthesis, increased productivity and increased biomass (Ainsworth and Long 2005; Dalal and Allen 2008), especially under N inputs (Luo et al. 2006; Reich et al. 2006) and increased water availability (Housman et al. 2006). However, C_4 grasses rarely respond to elevated CO_2 concentrations even under high N inputs (van Kessel et al. 2006).

Hungate et al. (2009) compared four recent meta-analyses (Jastrow et al. 2005; de Graaff et al. 2006; Luo et al. 2006; van Groeningen et al. 2006) on the effect of elevated CO_2 on changes in soil C stocks under low N (<30 kg N ha^{-1} $year^{-1}$) and high N (>30 kg N ha^{-1} $year^{-1}$) supply. They concluded that elevated CO_2 increases net SOC accumulation with the addition of N fertilizer although at low N input there was no significant increase in SOC stocks (Fig. 5.1). This may be

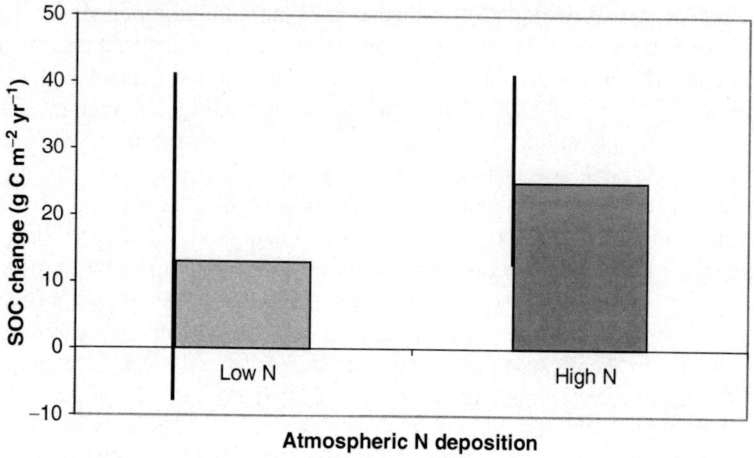

Fig. 5.1 Elevated CO_2 response to net change in soil organic C (SOC) under low N (<30 kg ha^{-1} year^{-1}) and high N (>30 kg ha^{-1} year^{-1}) from meta-analysis (n = 222 for low N and n = 187 for high N). Weighting function used for different datasets was the inverse of the variance. Bar height shows the 95% confidence interval (adapted from Hungate et al. 2009)

limited by the small SOC changes not detected by statistical technique under low N input, within a decade of experimental duration. It is likely, however, that these small SOC changes are biogeochemically important, and thus, there is a need for long-term experiments to verify the changes in SOC stocks under low N input over longer timescales, for example over 25–100 years (Conen et al. 2003). Furthermore, increased contribution from biological nitrogen fixation and increased N use efficiency of ecosystem N (Billings et al. 2002; de Graaff et al. 2006; Pepper et al. 2007), and supply of other limiting nutrients such as P, most likely will determine the medium-term (25–50 years) and long-term (50–100 years) responses of elevated CO_2 to net accumulation of SOC (van Groeningen et al. 2006).

Elevated CO_2 not only increases total plant biomass but may also increase total fine root production, especially under high N availability (Franklin et al. 2009), and hence leading to net SOC accumulation, especially in POC pool. Fine root biomass has been shown to be closely associated with POC or light fraction SOC (Dalal et al. 2005). Furthermore, elevated CO_2 may not alter the decomposition rate of new C inputs or labile and slow pools (Pendall and King 2007), and this may be because elevated CO_2 enhances protection of new root-derived C inputs into micro-aggregates (Jastrow et al. 2005). It is likely, however, that in the long term there may be a hierarchical saturation of soil C pools, with increasing amounts of SOC enter large macro-aggregates (Kool et al. 2007), which can be subjected to rapid turnover in soil. On the other hand, Heath et al. (2005) stated that under elevated CO_2 there may be only a small amount of C that is transferred from root-derived C to slow C pool, because overall SOC turnover was found to be enhanced (see Chap. 7).

5.4 Effect of Atmospheric N Deposition on SOC and SOC Pools

Atmospheric N deposition varies from 1 to >20 kg N ha^{-1} $year^{-1}$ (Galloway et al. 2004; Phoenix et al. 2006), with major deposition (>10 kg N ha^{-1} $year^{-1}$) occurring in populated areas of North America, Western Europe, South Asia and East Asia (Galloway et al. 2004). Several authors (Hyvonen et al. 2007; Magnani et al. 2007; Oren et al. 2001; Pepper et al. 2007; Reich et al. 2006) have suggested that additional N supplied through atmospheric N deposition can stimulate plant growth and thus increases C input as well as adds to soil N supply. The additional N supply then not only enhances the impact of elevated CO_2 concentration on SOC (Fig. 5.1), it may also slow the rate of organic matter decomposition in certain ecosystems (Swanston et al. 2004; Pepper et al. 2007; see Chaps. 6 and 7). Furthermore, C cycle-climate feedback models which do not consider the importance of N availability predict SOC decrease with increasing temperature, but those models which include N predict SOC increase or slower loss of SOC with increased N availability (Hungate et al. 2003; see Chap. 6); although as N saturation of the ecosystem approaches, the effect of increased N availability on SOC sequestration may diminish. For example, van Kessel et al. (2006) showed that whole soil profile C did not increase in N fertilized pastures as a response to prolonged elevated CO_2 concentration.

In general, increased atmospheric N deposition stabilize SOM, and SOM pools such as POC and heavy fractions, by reducing C mineralization, and microbial biomass and respiration, especially in the long term (Fogg 1988; Preston and Newman 1995; Scott et al. 1998; Swanston et al. 2004: Janssens et al. 2010). The stabilization of the POC fraction appears to be due to increased microbial efficiency and slow turnover of lignin and other relatively recalcitrant organic compounds in this fraction. On the other hand, stabilization of slow SOC pool due to increased soil N availability under elevated CO_2 concentration (Luo et al. 2006) may be attributed to enhanced formation of new aggregates, thereby reducing C accessibility within micro-aggregates (Jastrow et al. 2005).

5.5 Effect of Precipitation and Water Availability on SOC and SOC Pools

Similar to the N limitation, biomass response to CO_2 fertilization is limited by precipitation and hence water availability (Housman et al. 2006). Therefore, net SOC accumulation due to CO_2 fertilization will be constrained not only by N limitation but also by water availability (Garten et al. 2009). To some extent, limitation to water availability is overcome by increased water use efficiency under elevated CO_2 concentration and vegetation shift from C_3 to C_4 vegetation, the latter being more water use efficient than the former, and the increased water use efficiency may in turn accelerate any positive CO_2 fertilization response under water-limited conditions. However, the effect of improved water use efficiency on net SOC accumulation is ambiguous.

Generally, increased soil water availability or increased precipitation appears to be more important than temperature in regulating net SOC accumulation (Hungate et al. 2007; Liu et al. 2009), especially under water-limited ecosystems. For example, Garten et al. (2009) observed that soil moisture was more important than elevated CO_2 and temperature as a control on soil C stocks in a multi-factor (CO_2 $\times$ temperature $\times$ soil moisture) climate change experiment on a constructed old-field ecosystem: elevated CO_2 had no significant effect on total SOC, POC and clay + silt-associated SOC, whereas precipitation (1,300 mm year^{-1}) decreased SOC and POC stocks over 4 years (Fig. 5.2). The drought treatment (~20% lower soil water content or 92% lower precipitation) slowed soil C cycling in the POC pool, as compared to the watering (25 mm precipitation) treatment. It remains to be seen whether these observations hold in other ecosystems and thus emphasizes the need for comprehensive global climate-C cycle models which include the influence of multiple drivers of climate change and their feedback effects on total SOC and SOC pools.

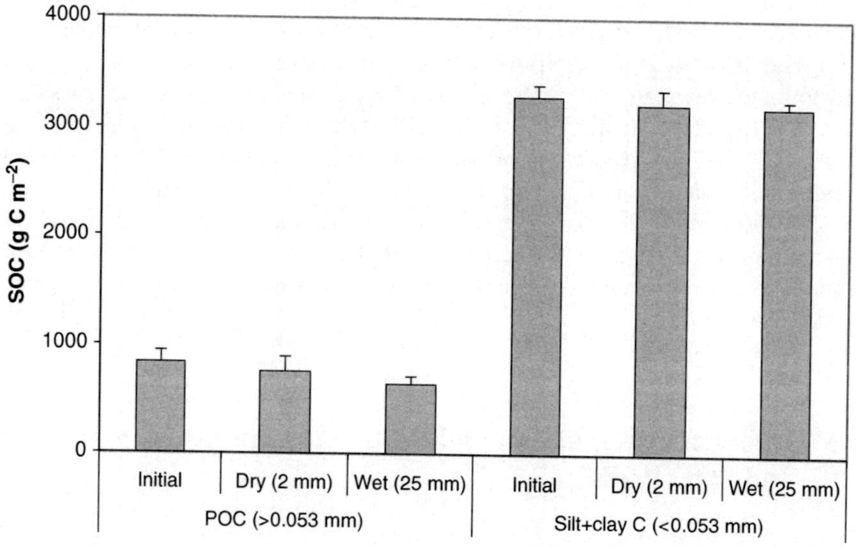

Fig. 5.2 Effect of precipitation on soil organic carbon (SOC) pools after 4 years in a multi-factor experiment (elevated CO_2, temperature and precipitation), growing two legumes and five non-legume plant species, on a silty Ultisol. The SOC pools are POC (>0.053 μm, POC) and silt + clay organic C (<0.053 μm). Dry treatment received 2 mm week^{-1} (drought) and wet treatment received 25 mm week^{-1} precipitation (equivalent to annual rainfall). Bar on the *top* of the histogram shows standard error of the mean (redrawn from Garten et al. 2009)

5.6 Effect of Fire on SOC and SOC Pools

Climate change is expected to produce less frequent but more intense fires. During forest and grassland fires, large amounts of CO_2 are released into the atmosphere from above-ground biomass and litter, burning and SOM charring. Above-ground biomass is assumed to incorporate CO_2 from the atmosphere in the subsequent growing vegetation within a short period and thus considered to have a net neutral C balance (Levine et al. 1995). However, the fire impact on SOC and SOC pools is much less understood.

Generally, only a small fraction of heat generated during the vegetation fire is radiated to soil. The degree of soil heating depends on temperature and extent of heat radiated to the soil, soil texture, soil moisture, soil porosity and temperature diffusivity (Gonzalez-Perez et al. 2004). Soil temperatures in the top 20–30 mm depths may reach up to 200°C in forest fires, but <100°C in grassland fires, however, may have little change below 50 mm depths.

The effect of fire on the total SOC is highly variable. While after a moderate fire an increase in SOC may be observed due to incorporation of partially charred biomass, an intense fire usually leads to reduction in total SOC for several months after the fire (Gonzalez-Perez et al. 2004). On the other hand, 10 years after burning, Johnson and Curtis (2001) observed lower SOC after prescribed burning and higher SOC after wildfire, the latter due to increased particulate charcoal (leading to the accumulation of altered lignin-like C, from O-aryl-C to aryl-C, and recalcitrant hydrophobic SOM, and increased occurrence of N_2-fixing vegetation.

It has been suggested that accumulation of charcoal C from biomass burning and/or wild fires may account for up to 30% of SOC in some Australian soils, and this may reduce a positive feedback to greenhouse gas emissions with elevated temperatures from climate change (Lehmann et al. 2008). However, overall impact of charcoal on soil health is unclear.

5.7 Climate Change and Changes in SOC Stocks

Long-term assessment of the impact of climate change on changes in SOC stocks has been examined only in few studies. Bellamy et al. (2005) found that SOC was lost from the soils of England and Wales from 1978 to 2003, at a mean rate of 0.6% $year^{-1}$, irrespective of land use and hence this loss was claimed to be due to climate change. However, Hopkins et al. (2009) measured no consistently significant changes in SOC stocks from the long-term experimental grassland plots (>100 years) in the UK, and found no evidence of the impact of climate change on SOC stocks.

These contradictory findings are due to differences in methodology used to estimate SOC stocks. In Bellamy et al. (2005) studies, soil bulk density data were derived from pedotransfer functions using clay, sand and organic C contents in regression equations, which may not be applicable across different soil types (Smith et al. 2007). In contrast, Hopkins et al. (2009) measured soil bulk density directly. Moreover, no account was taken of change in bulk density at two time periods by Bellamy et al. (2005), which could have resulted in assessing different soil masses on each occasion. Also, changes in management and C inputs such as manures during 1978 and 2003 could be responsible for changes in SOC stocks during this period (Smith et al. 2007). From further analysis of Bellamy et al. (2005) data, Kirk and Bellamy (2010) concurred that past changes in land use (conversion of grassland and natural vegetation to crops) and management (improvement in land drainage, greater use of mineral fertilizers and greater animal stocking rates) were more likely responsible for decrease in SOC stocks than climate change between 1978 and 2003. In fact, Janssens et al. (2003) found that the European soils, as a whole, may be current net CO_2 sinks, based on the estimates of absorption of 7–12% of European anthropogenic CO_2 emissions by terrestrial biosphere. The contradictory results of these three studies emphasize the need for application of similar assessment methodology, soil sampling and measurement protocols common to all SOC stock estimates, and there is also a need to account for any land use change and management differences across studies.

5.8 Impact of Climate Change on SOM Management and Soil Health

SOM is among the best indicator of soil health since it controls many soil properties (Waksman 1938; Dalal et al. 2004) and major biogeochemical cycles including C, N, S and P cycling, and is usually a strong indicator of soil fertility and land degradation (Dalal and Chan 2001; Manlay et al. 2007). It also provides the basis for ecosystem services such as through C sequestration, production of food and maintenance of water quality, by serving as an energy source for soil organisms, enhancing pesticide degradation, and by contributing to retention of nutrient ion in coarse-textured and low-activity clay soils. Furthermore, since SOC pool is three times larger than that in the atmosphere (Denman et al. 2007), any change in SOC pool due to climate change will have an effect on soil health. Hence, appropriate SOM management is essential for maintaining or improving soil health in the context of climate change.

5.9 Management Practices to Restore, Maintain and Increase SOC, or Reduce SOC Turnover

In agricultural systems, SOC is affected by management practices and this subject has been reviewed extensively (Lal et al. 2007; Smith et al. 2008). In the past, large quantity of SOC has been lost due to the clearing of natural vegetation for agriculture and the use of exploitative practices such as clean fallow and excessive tillage operations. It has been estimated that 78 Gt C has been lost via these human activities globally and as such they represent a significant CO_2 source to the atmosphere (Lal et al. 2007). Emissions of CO_2 could be mitigated by sequestering it in soil as SOC.

It has been well documented that SOC can increase under a range of improved management practices (e.g. Lal et al. 2007; Luo et al. 2010), and the rates of SOC sequestration under a range of land use and management practices have been documented. Table 5.2 summarizes the rates of C sequestration under a range of management practices and land use (see Chap. 1). These management practices can be divided into a number of categories, namely crop management including conservation tillage, pasture management, use of organic amendments and land use conversion.

It is worthwhile noting that for any specific practice, a range of SOC sequestration rates have been reported (Table 5.2). This is expected in view of the factors involved in controlling the production and decomposition of SOM.

As many of these management practices were developed for other purpose, such as for soil health improvement, a better understanding of the processes involved in controlling SOC fluxes under these management practices will help to quantify their soil C sequestration potential.

Table 5.2 Reported soil organic carbon (SOC) sequestration rates associated with management practices and land use practices that can increase SOC

Management category	Management practices	C sequestration rates (t C ha^{-1} $year^{-1}$)	References
Crop management	Soil fertility enhancement	0.05–0.15	
	Better rotation	0.10–0.30	
	Irrigation	0.05–0.15	Lal et al. (2003)
	Fallow elimination	0.10–0.30	
Conservation tillage	Stubble retention		
	Reduced tillage	0.24–0.40	Lal et al. (2003)
	No-tillage		
Pasture management	Fertilizer management	0.30	
	Grazing management	0.35	
	Earthworm introduction	2.35	
	Irrigation	0.11	Conant et al. (2001)
	Improved grass species	3.04	
	Introduction of legumes	0.75	
	Sown pasture	0.26–0.72	Gifford et al. (1992); Chan et al. (2010)
Organic amendments	Animal manure	0.1–0.6	Jarecki and Lal (2003)
	Biosolids	1.0	Brown and Leonard (2004)
Land conversion	Degraded cropland to pasture	0.8–1.1	Grogan and Matthews (2001)
	Bioenergy crop	0.98	

Since soil organic C change in a farming system is the balance of addition and removal of organic C, SOC can be increased by increasing C inputs to soil, through enhanced plant productivity, but often the increase in plant productivity is associated with increasing costs (e.g. increased fertilizer use) and greenhouse emissions, and biomass yield increases may not automatically lead to increases in SOC. For example, in the UK, wheat yield increased from 2.82 to 7.08 t ha^{-1} between 1948 and 2001, yet many cropping soils are still losing SOM, hence SOC (Rees et al. 2005). Therefore, in addition to increasing C inputs, strategies for reducing loss by reducing C turnover and hence increasing C residence time should also be important considerations for enhancing soil C sequestration (Jastrow et al. 2007).

5.9.1 No-Tillage

It is well documented that in no-tillage (NT) cropping, SOC is stored in a more protective form with slower turnover time compared to that under conventional tillage (CT) (Six et al. 2000). Physical protection of SOC occurs due to the influence of soil aggregation on SOC turnover. Soil structure in the form of pore- and aggregate size distribution acts as physical barriers between SOM and microbes and enzymes. There is much evidence indicating that SOC under pastures and NT cropping is protected within soil aggregates (Beare et al. 1994; Six et al. 2000; Chevallier et al. 2004; see Chap. 9). Six et al. (2000) showed that the increase in C storage under NT was in the form protected within stable micro-aggregates formed under the prevailing reduced rate of turnover of macro-aggregates compared to CT. They found that within macro-aggregates, the concentration of micro-aggregate POM-C was threefold greater in NT compared to CT, while the concentration of inter-aggregate POM-C was twofold greater in CT compared to NT. More than 90% of the difference in SOC between NT and CT was due to difference in micro-aggregate-associated C in all three soils studied. Using ^{13}C technique, Balesdent et al. (1990) also reported that the mean residence time (MRT) of C was greater under NT than conventionally tilled soil, e.g. for corn in France MRT was 127 years for NT vs. 55 years under CT.

5.9.2 Pastures/Agroforestry

Due to the long-term nature of forestry and permanent pasture and the minimal soil disturbance associated with these land conversion processes, the SOC sequestration process under these scenarios usually favours the formation of relatively stable forms of soil C, for example, protected C within soil aggregates resulting in higher soil C stocks. With the exception of pastureland conversion to pine plantations where a 12–15% reduction in SOC stocks was found, conversion of croplands to plantations resulted in 18% increase in soil C stocks (Guo and Gifford 2002).

5.9.3 Types of Carbon Amendment: Biochar

Returning of crop residues and other recycled organics to agricultural soils has been a time-honoured practice by farmers to maintain/increase soil organic C of agricultural soils. Recently, there is considerable interest in the application of biochar to soil as a way to sequester C to mitigate climate change (Lehmann et al. 2006). Biochars are relatively stable to decomposition in soil, but the stability of biochars is dependent on the conditions under which they are produced, as well as on biomass type. More research is needed to assess the soil C sequestration potential of biochars compared with other forms of recycled organics, as well as biochar's potential to provide various soil health benefits (see Chap. 15).

5.9.4 Subsoil C Sequestration

The potential of subsoil in increasing SOC sequestration has been raised by various researchers (e.g. Lorenz and Lal 2005; Chan 2008). Decomposition of SOC tends to be slower in the subsoil environment. While surface soils may be saturated with SOC (Six et al. 2002), there is evidence suggesting that many subsoils are unsaturated with the organo-mineral form (chemical protected) of SOC (Jastrow et al. 2007; Chan 2008). In the latter case, additional SOC can be stored in a more stabilized form with long turnover time. This can be achieved by promoting downward transport of organic C into the subsoil (Lorenz and Lal 2005; Wright et al. 2007).

The anecic earthworms are well known for their ability to transport organic matter into subsoil (Lee 1985). Tillage carried out appropriately according to soil types and environmental conditions can act in the same capacity as soil fauna, e.g. earthworms. Subsoil C storage can be increased by depositing crop residue deeper into the profile by tillage operations (see Chap. 9). Under Canadian and European conditions, there is evidence that ploughing can increase SOC storage by transferring organic matter into deeper soil layers (Angers et al. 1997; Nieder and Richter 2000). Angers et al. (1997) reported higher SOC in 20–40 and 40–60 cm depths under mouldboard ploughing (MP) than under NT and attributed this to the tillage action of incorporating crop residues to be in close contact with the mineral surfaces. More SOM was found associated with clay particles in the MP treatment than in the reduced tillage treatment in the subsoil layers. By increasing the tillage depth from <25 to >35 cm in West Germany, SOC increased by 16–26 t C ha^{-1} over a period of 28 years, which was equivalent to SOC sequestration rates of 0.6–0.9 t C ha^{-1} $year^{-1}$ (Nieder and Richter 2000). These rates are at the higher end of the SOC sequestration rates recorded for the NT practice (see Table 5.2). The higher SOC sequestration observed in the study was related to increased SOC in the 25–35 cm layer as a result of deeper tillage depth. This, therefore, highlights the potential of using tillage to redistribute the organic input within the soil profile to locations such as the subsoil, where they tend to have lower C turnover rates. However, specialized tillage techniques (implements) may be required to achieve the objective of transporting the organic materials to the desired locations within the soil profile but with minimal undesirable effects to soil health parameters.

5.9.5 Crop and Pasture Selection

More attention needs to be paid to the abilities of different plant species in sequestering C in the subsoil. Some grasses are particularly effective in increasing SOC in the subsoils partly due to their high root/shoot ratios (Fisher et al. 1994). By comparing three 20-year-old rotations, namely continuous soybean, wheat–soybean double crop and a sorghum–wheat–soybean under both CT and NT, Wright et al. (2007) reported significant differences in SOC not only due to tillage practices but also due to crop rotation. While SOC was 32% higher under NT than CT over 0–105 cm depth, significant differences in SOC at depth below 30 cm were detected only under NT and only for rotations involving 2–3 different species. For example, under NT, SOC at 30–55 cm depth under sorghum–wheat–soybean was more than twice that of continuous soybean (0.49 vs. 0.18%). However, no difference in SOC was detected at any depth among the three rotations under CT. These highlight the potential of crop and pasture selection for the purpose of increasing SOC in the subsoil and the interacting role played by tillage practice in such endeavour.

5.10 Summary and Conclusions

Soil organic C stocks and pools are the net result of organic C inputs and outputs; both of these will be affected by global change processes, such as, increasing temperature, elevated CO_2 concentration in the atmosphere, increasing atmospheric N deposition, and precipitation amounts and variability. However, the influence of these drivers of global environmental change on SOC stocks and dynamics is highly variable. For example, the differential temperature sensitivity of different SOC pools may largely be due to differences in substrate supply (production and diffusion of substrate to active enzyme sites), which indirectly depends on substrate quality, aggregation, water availability and available nutrients, especially N. As pointed out by Davidson and Janssens (2006), there is a need to quantify the multiple processes of environmental constraints, which govern substrate supply to the active sites of enzymes in soil within the context of climate change scenarios. Moreover, feedbacks provided by soil charcoal C, fire intensity and frequency, and enhanced soil C sinks in response to temperature, precipitation, CO_2 fertilization, atmospheric N deposition, and improved soil management, should be considered in estimating changes in soil C stocks due to climate change. Development of management practices that favour C sequestration or prevent further losses will help to reduce the potential SOC decline due to the climate change, and because of the importance of SOC to soil health and plant productivity, maintenance or restoration of SOC will help to reduce the threats of food security due to climate change.

Acknowledgments The authors appreciate helpful comments from an anonymous reviewer and Annette Cowie on an earlier version of this chapter.

References

Ainsworth EA, Long SP (2005) What have we learned from 15 years of free-air CO_2 enrichment (FACE)? A meta-analytic review of the responses of photosynthesis, canopy properties and plant production to rising CO_2. New Phytol 165:351–372

Allison SD, Wallenstein MD, Bradford MA (2010) Soil-carbon response to warming dependent on microbial physiology. Nat Geosci 3:336–340

Angers DA, Bolinder MA, Carter MR, Gregorich EG, Drury CF, Liang BC, Voroney RP, Simard RR, Donald RG, Beyaert RP, Martel J (1997) Impact of tillage practices on organic carbon and nitrogen storage in cool, humid soils of eastern Canada. Soil Tillage Res 41:191–201

Arrhenius S (1889) Uber die Reaktionsgesschwindigkeit bei der Inversion von Rohrzucker durch Sauren. Z fur Physik Chemie 4:226–248

Balesdent J, Boisgontier D, Mariotti A (1990) Effect of tillage on soil organic carbon mineralization estimated from ^{13}C abundance in maize fields. J Soil Sci 41:587–596

Beare MH, Cabrera ML, Hendrix PF, Coleman DC (1994) Aggregate-protected and unprotected organic matter pools in conventional- and no-tillage soils. Soil Sci Soc Am J 58:787–795

Bellamy PH, Loveland PJ, Bradley RI, Murray Lark R, Kirk G (2005) Carbon losses from all soils across England and Wales 1978–2003. Nature 437:245–248

Billings SA, Schaeffer SM, Zitzer S, Charlet T, Smith SD, Evans RD (2002) Alterations of nitrogen dynamics under elevated CO_2 in an intact Mojave Desert ecosystems: evidence from $\delta^{15}N$. Oecologia 131:462–467

Bosatta E, Agren G (1999) Soil organic matter quality interpreted thermodynamically. Soil Biol Biochem 10:215–221

Brown S, Leonard P (2004) Building carbon credits with biosolids recycling. BioCycle 45:25–29

Chan KY (2008) Potential role of tillage to increase soil carbon sequestration. CAB Rev Perspect Agric Vet Sci Nat Res 3(13):1–8

Chan KY, Oates A, Li GD, Conyers MK, Prangnell RJ, Poile G, Liu DL, Barchia IM (2010) Soil carbon stocks under different pastures and pasture management in the higher rainfall areas of south-eastern Australia. Aust J Soil Res 48:7–15

Chevallier T, Blanchart E, Albrecht A, Feller C (2004) The physical protection of soil organic carbon in aggregates: a mechanism of carbon storage in Vertisol under pasture and market gardening (Martinique, West Indies). Agric Ecosyst Environ 103:375–387

Conant RT, Paustian K, Elliott ET (2001) Grassland management and conversion into grassland: effects on soil carbon. Ecol Appl 11:343–355

Conant RT, Drijber RA, Haddix ML, Parton WJ, Paul EA, Plante AF, Six J, Steinweg JM (2008) Sensitivity of organic matter decomposition to warming varies with its quality. Glob Change Biol 14:868–877

Conen F, Yakutin MV, Sambuu AD (2003) Potential for detecting changes in soil organic carbon concentrations resulting from climate change. Glob Change Biol 9:1515–1520

Conen F, Leifeld J, Seth B, Alewell C (2006) Warming mineralises young and old soil carbon equally. Biogeosciences 3:515–519

Curiel Yuste J, Ma S, Baldocchi DD (2010) Plant-soil interactions and acclimation to temperature of microbial-mediated soil respiration may affect predictions of soil CO_2 efflux. Biogeochemistry 98:127–138

Dalal RC, Allen DE (2008) Greenhouse gas fluxes from natural ecosystems. Aust J Bot 56: 369–407

Dalal RC, Chan KY (2001) Soil organic matter in rainfed cropping systems of the Australian cereal belt. Aust J Soil Res 39:435–464

Dalal RC, Wang W, Mann S, Henry B (2004) Soil organic matter decline and restoration. Nat Resour Manage 7:2–15

Dalal RC, Harms B, Krull E, Wang W (2005) Total soil organic matter and its labile pools following Mulga (*Acacia aneura*) clearing for pasture development and cropping 1. Total and labile carbon. Aust J Soil Res 43:13–20

Davidson EA, Janssens IA (2006) Temperature sensitivity of soil carbon decomposition and feedbacks on climate change. Nature 440:165–173

Davidson EA, Janssens IA, Luo Y (2006) On the variability of respiration in terrestrial ecosystems: moving beyond Q_{10}. Glob Change Biol 12:154–164

De Graaff MA, van Groenigen KJ, Six J, Hungate BA, van Kessel C (2006) Interactions between plant growth and soil nutrient cycling under elevated CO_2: a meta-analysis. Glob Change Biol 12:2077–2091

Denman KL, Brasseur G, Chidthaisong A, Clais P, Cox PM, Dickinson RE, Hauglustaine D, Heinze C, Holland E, Jacob D, Lohmann U, Ramachandran S, da Silva Dias PL, Wofsy SC, Zhang X (2007) Coupling between changes in the climate system and biogeochemistry. In: Solomon S, Qin D, Chen Z, Marquis M, Averyt KB, Tignor M, Miller HL (eds) Climate change 2007: the physical science basis. Contribution of Working Group I to the Fourth Assessment Report of the Intergovernmental Panel on Climate Change. Cambridge University Press, Cambridge, pp 499–587

Fang CM, Smith P, Moncrieff JB, Smith JU (2005) Similar response of labile and resistant soil organic matter pools to changes in temperature. Nature 433:57–59

Fierer N, Craine JM, McLauchlan K, Schimel JP (2005) Litter quality and temperature sensitivity of decomposition. Ecology 86:320–326

Fisher MJ, Rao IM, Ayarza MA, Lascano CE, Sanz JI, Thomas RJ, Vera RR (1994) Carbon storage by introduced deep-rooted grasses in the South American savannas. Nature 371: 236–238

Fogg K (1988) The effect of added nitrogen on the rate of decomposition of organic matter. Biol Rev Camb Philos Soc 63:433–462

Franklin O, McMurtrie RE, Iversen CM, Crous KY, Finzi AC, Tissue DT, Ellsworth DS, Oren R, Norby RJ (2009) Forest fine-root production and nitrogen use under elevated CO_2: contrasting response in evergreen and deciduous trees explained by a common principle. Glob Change Biol 15:132–144

Galloway JN, Dentener FJ, Capone DG, Boyer EW, Howarth RW, Seitzinger SP, Asner GP, Cleveland CC, Green PA, Holland EA, Karl DM, Michaels AF, Porter JH, Townsend AR, Vorosmarty CJ (2004) Nitrogen cycles: past, present, and future. Biogeochemistry 70:153–226

Garten CT Jr, Classen AT, Norby RJ (2009) Soil moisture surpasses elevated CO_2 and temperature as a control on soil carbon dynamics in a multi-factor climate change experiment. Plant Soil 319:85–94

Gershenson A, Bader NE, Cheng W (2009) Effects of substrate availability on the temperature sensitivity of soil organic matter decomposition. Glob Change Biol 15:176–183

Giardina CP, Ryan MG (2000) Evidence that decomposition rates of organic carbon in mineral soil do not vary with temperature. Nature 404:858–861

Gifford RM, Cheney NP, Noble JC, Russel JS, Wellington AB, Zammit C (1992) Australian land use, primary production of vegetation and carbon pools in relation to atmospheric carbon dioxide concentration. In: Gifford RM, Barson MM (eds) Australia's renewable resources sustainability and global change. Bureaus of Rural Resources, CSIRO, Australia, pp 151–188

Glasstone S, Laidler M, Eyring H (1941) The theory of rate process. McGraw Hill, New York
Gonzalez-Perez JA, Gonzalez-Vila FJ, Almendros G, Knicker H (2004) The effect of fire on soil organic matter – a review. Environ Int 30:855–870
Grogan R, Matthews R (2001) Review of the potential for soil carbon sequestration under bioenergy crops in the UK. MAPF report on contract NF0418. Cranefield University, Silsoe
Gruber N, Friedlingstein P, Field CB, Valentini R, Heimann M, Richey JE, Romero-Lankao P, Schulze P, Chen C-T A (2004) The vulnerability of the carbon cycle in the 21st century: an assessment of carbon-climate-human interactions. In: Field C, Raupach M (eds) Toward CO_2 stabilisation: issues, strategies, and consequences. Island Press, Washington, pp 45–76
Guo LB, Gifford RM (2002) Soil carbon socks and land use change: a meta analysis. Glob Change Biol 8:345–360
Hartley I, Ineson P (2008) Substrate quality and the temperature sensitivity of soil organic matter decomposition. Soil Biol Biochem 40:1567–1574
Heath J, Ayres E, Possell M, Bardgett RD, Black HIJ, Grant H, Ineson P, Kerstiens G (2005) Rising atmospheric CO_2 reduces sequestration of root-derived soil carbon. Science 309: 1711–1713
Hopkins DW, Waite IS, McNicol JW, Poulton PR, MacDonald AJ, O'Donnell AG (2009) Soil organic carbon contents in long-term experimental grassland plots in the UK (Palace Leas and Park Grass) have not changed consistently in recent years. Glob Change Biol 15:1739–1754
Housman DC, Naumburg E, Huxman TE, Charlet TN, Nowak RS, Smith SD (2006) Increases in desert shrub productivity under elevated carbon dioxide vary with water availability. Ecosystems 9:374–385
Hungate BA, Dukes JS, Shaw MR, Luo Y, Field CB (2003) Nitrogen and climate change. Science 302:1512–1513
Hungate BA, Hart SC, Selmants PC, Boyle SI, Gehring CA (2007) Soil responses to management, increased precipitation, and added nitrogen in ponderosa pine forests. Ecol Appl 17: 1352–1365
Hungate BA, van Groenigen KJ, Six J, Jastrow JD, Luo Y, De Graaff MA, van Kessel C, Osenberg CW (2009) Assessing the effect of elevated carbon dioxide on soil carbon: a comparison of four meta-analyses. Glob Change Biol 15:2020–2034
Hyvonen R, Agren GI, Linder S, Persson T, Cotrufo F, Ekblad A, Freeman M, Grelle A, Janssens IA, Jarvis PG, Kellomaki S, Lindroth A, Loustau D, Lundmark T, Norby RJ, Oren R, Pilegaard K, Ryan MG, Sigurdsson BD, Stromgren M, van Oijen M, Wallin G (2007) The likely impact of elevated [CO_2], nitrogen deposition, increased temperature and management on carbon sequestration in temperate and boreal forest ecosystems: a literature review. New Phytol 173: 463–480
Janssens IA, Freibauer A, Ciais P, Smith P, Nabuurs G-J, Folberth G, Schlamadinger B, Hutjes RWA, Ceulemans R, Schulze E-D, Valentini R, Dolman AJ (2003) Europe's terrestrial biosphere absorbs 7–12% of European anthropogenic CO_2 emissions. Science 300: 1538–1542
Janssens IA, Dieleman W, Luyssaert S, Subke JA, Reichstein M, Ceulemans R, Ciais P, Dolman AJ, Grace J, Matteucci G, Papale D, Piao SL, Schulze ED, Tang J, Law BE (2010) Reduction of forest respiration in response to nitrogen deposition. Nat Geosci 3:315–322
Jarecki MK, Lal R (2003) Crop management for soil carbon sequestration. Crit Rev Plant Sci 22:471–502
Jastrow JD, Miller RM, Matamala R, Norby JR, Boutton TW, Rice CW, Owensby CE (2005) Elevated atmospheric carbon dioxide increases soil carbon. Glob Change Biol 11:2057–2064
Jastrow JD, Amonette JE, Bailey VL (2007) Mechanisms controlling soil carbon turnover and their potential application for enhancing carbon sequestration. Climate Change 80:5–23
Jenkinson DS, Adams DE, Wild A (1991) Model estimates of CO_2 emissions from soil in response to global warming. Nature 351:304–306
Johnson DW, Curtis PS (2001) Effects of forest management on soil C and N storage: meta analysis. For Ecol Manage 140:227–238

Jones C, McConnell C, Coleman K, Cox P, Falloon P, Jenkinson D, Powlson D (2005) Global climate change and soil carbon stocks: predictions from two contrasting models for the turnover of organic carbon in soil. Glob Change Biol 11:154–166

Kirk GJD, Bellamy PH (2010) Analysis of changes in organic carbon in mineral soils across England and Wales using a simple single-pool model. Eur J Soil Sci 61:406–411

Kirschbaum MUF (1995) The temperature dependence of soil organic matter decomposition and the effect of global warming on soil organic carbon storage. Soil Biol Biochem 27:753–760

Kirschbaum MUF (2000) Will changes in soil organic carbon act as a positive or negative feedback on global warming. Biogeochemistry 48:21–51

Kirschbaum MUF (2006) The temperature dependence of soil organic matter decomposition – still a topic of debate. Soil Biol Biochem 38:2510–2518

Kool DM, Chung H, Tate KR, Ross DJ, Newton PC, Six J (2007) Hierarchical saturation of soil carbon pools near a natural CO_2 spring. Glob Change Biol 13:1282–1293

Lal R, Follett RF, Kimble JM (2003) Achieving soil carbon sequestration in the United States: a challenge to the policy makers. Soil Sci 168:827–845

Lal R, Follett RF, Stewart BA, Kimble JM (2007) Soil carbon sequestration to mitigate climate change and advance food security. Soil Sci 172:943–956

Lee KE (1985) Earthworms and pedogenesis. In: Lee KE (ed) Earthworms – their ecology and relationships with soils and land use. Academic, Sydney, pp 244–251

Lehmann J, Gaunt J, Rondon M (2006) Bio-sequestration in terrestrial ecosystems – a review. Mitig Adapt Strategy Climate Change 11:403–427

Lehmann J, Skjemstad J, Sohi S, Carter J, Barson M, Falloon P, Coleman K, Woodbury P, Krull E (2008) Australian climate-carbon cycle feedback reduced by soil black carbon. Nat Geosci 1: 832–835

Leifeld J, Fuhrer J (2005) The temperature response of CO_2 production from bulk soils and soil fractions is related to soil organic matter quality. Biogeochemistry 75:433–453

Levine JS, Cofer WR, Cahoon DR, Winstead EL (1995) Biomass burning: a driver for global change. Environ Sci Technol 29:120A–125A

Liski J, Ilvesniemi H, Makela A, Westman CJ (1999) CO2 emissions from soil in response to climate warming are overestimated– the decomposition of old soil organic matter is tolerant of temperature. Ambio 28:171–174

Liu W, Zhang Z, Wan S (2009) Predominant role of water regulating soil and microbial respiration and their responses to climate change in a semiarid grassland. Glob Change Biol 15:184–195

Lorenz K, Lal R (2005) The depth distribution of soil organic carbon in relation to land use and management and the potential of carbon sequestration in subsoil horizons. Adv Agron 88: 35–66

Luo Y, Wan S, Hui D, Wallace LL (2001) Acclimatization of soil respiration to warming in a tall grass prairie. Nature 413:622–625

Luo Y, Hui D, Zhang D (2006) Elevated CO_2 simulates net accumulations of C and nitrogen in land ecosystems: a meta-analysis. Ecology 87:53–63

Luo Z, Wang E, Sun OJ (2010) Soil carbon change and its responses to agricultural practices in Australian agro-ecosystems: a review and synthesis. Geoderma 155:211–223

Magnani F, Mencuccini M, Borghetti M, Berbigier P, Berninger F, Delzon S, Grelle A, Hari P, Jarvis PG, Kolari P, Kowalski AS, Lankreijer H, Law BE, Lindroth A, Loustau D, Manca G, Moncrieff JB, Rayment M, Tedeschi V, Valentini R, Grace J (2007) The human footprint in the carbon cycle of temperate and boreal forests. Nature 447:848–850

Manlay RJ, Feller C, Swift MJ (2007) Historical evolution of soil organic matter concepts and their relationships with the fertility and sustainability of cropping systems. Agric Ecosyst Environ 119:217–233

Melillo JM, Steudler PA, Aber JD, Newkirk K, Lux H, Bowles FP, Catricala C, Magill A, Ahrens T, Morrisseau S (2002) Soil warming and carbon-cycle feedbacks to the climate system. Science 298:2173–2176

Michaelis L, Menten ML (1913) Die kinetic der invertin wirkung. Biochem Z 49:334–336

Nieder R, Richter J (2000) C and N accumulation in arable soils of West Germany and its influence on the environment – developments 1970–1998. J Plant Nutr Soil Sci 163:65–72

Oren R, Ellsworth DS, Johnsen KH, Nathan P, Ewers BE, Maler C, Schafer KVR, McCarthy H, Hendrey G, McNulty SG, Katul GG (2001) Soil fertility limits carbon sequestration by forest ecosystems in a CO_2-enriched atmosphere. Nature 411:469–472

Pendall E, King JY (2007) Soil organic matter dynamics in grassland soils under elevated CO_2: insights from long-term incubations and stable isotopes. Soil Biol Biochem 39:2628–2639

Pepper DA, Eliasson PE, McMurtrie RE, Corbeels M, Agren GI, Stromgren M, Linder S (2007) Simulated mechanisms of soil N feedback on the forest CO_2 response. Glob Change Biol 13: 1265–1281

Phoenix GK, Hicks WK, Cinderby S, Kuylenstierna JCI, Stock WD, Dentener FJ, Giller KE, Austin AT, Lefroy RDB, Gimeno BS, Ashmore MR, Ineson P (2006) Atmospheric nitrogen deposition in world biodiversity hotspots: the need for a greater global perspective in assessing N deposition impacts. Glob Change Biol 12:470–476

Plante AF, Six J, Paul EA, Conant RT (2009) Does physical protection of soil organic matter attenuate temperature sensitivity? Soil Sci Soc Am J 73:1168–1172

Preston CM, Newman RH (1995) A long-term effect of N fertilization on the 13C CPMAS NMR of de-ashed soil humin in a 2nd-growth Douglas-fir stand of coastal British Columbia. Geoderma 68:229–241

Pribyl DW (2010) A critical review of the conventional SOC to SOM conversion factor. Geoderma 156:75–83

Rees RM, Bingham IJ, Baddeley JA, Watson CA (2005) The role of plants and land management in sequestering soil carbon in temperate arable and grassland ecosystems. Geoderma 128: 130–154

Reich PB, Hobbie SE, Lee T, Ellsworth DS, West JB, Tilman D, Knops JMH, Naeem S, Trost J (2006) Nitrogen limitation constrains sustainability of ecosystem response to CO_2. Nature 440: 922–925

Reichstein M, Katterer T, Andren O, Ciais P, Schulze ED, Cramer W, Papale D, Valentini R (2005) Does the temperature sensitivity of decomposition vary with soil organic matter quality? Biogeosci Disc 2:737–747

Rey A, Jarvis P (2006) Modelling the effect of temperature on carbon mineralisation rates across a network of European forest sites (FORCAST). Glob Change Biol 12:1894–1908

Schimel DS (1994) Climatic, edaphic, and biotic controls over storage and turnover of carbon in soils. Glob Biogeochem Cycles 8:279–293

Schuur EAG, Vogel JG, Crummer KG, Lee H, TE Sickman O (2009) The effect of permafrost thaw on old carbon release and net carbon exchange from tundra. Nature 459:556–559

Scott NA, Parfitt RL, Ross DJ, Salt GJ (1998) Carbon and nitrogen transformations in New Zealand plantation forest soils from sites with different N status. Can J For Res 28:967–976

Six J, Elliott ET, Paustian K (2000) Soil macroaggregate turnover and microaggregate formation: a mechanism for C sequestration under no-tillage agriculture. Soil Biol Biochem 32: 2099–2103

Six J, Conant RT, Paul EA, Paustian K (2002) Stabilisation mechanisms of soil organic matter: implications for C-saturation of soils. Plant Soil 241:155–176

Smith P, Chapman SJ, Scott WA, Black HIJ, Wattenbach M, Milne R, Campbell CD, Lilly A, Ostle N, Levy PE, Lumsdon DG, Millard P, Towers W, Zaehle S, Smith JOU (2007) Climate change cannot be entirely responsible for soil carbon loss observed in England and Wales, 1978–2003. Glob Change Biol 13:2605–2609

Smith P, Martino D, Cai Z, Gwary D, Janzen HH, Kumar P, McCarl B, Ogle S, O'Mara F, Rice C, Scholes RJ, Sirotenko O, Howden M, McAllister T, Pan G, Romanenkov V, Schneider U, Towprayoon S, Wattenbach M, Smith J (2008) Greenhouse gas mitigation in agriculture. Phil Trans R Soc B Biol Sci 363:789–813

Swanston C, Homann PS, Caldwell BA, Myrold DD, Ganio L, Sollins P (2004) Long-term effects of elevated nitrogen on forest soil organic matter stability. Biogeochemistry 70:227–250

Trumbore SE, Chadwick OA, Amundson R (1996) Rapid exchange between soil carbon and atmospheric carbon dioxide driven by temperature change. Science 272:393–396

van Groeningen KJ, Six J, Hungate BA, van Kessel C, De Graaff MA, van Breemen N (2006) Element interactions limit soil carbon storage. Proc Nat Acad Sci USA 103:6571–6574

van Kessel C, Boots B, De Graaff MA, Harris D, Blum H, Six J (2006) Total soil C and N sequestration in a grassland following 10 years of free air CO_2 enrichment. Glob Change Biol 12:2187–2199

Waksman SA (1938) Humus, origin, chemical composition and importance in nature, 2nd edn. The Williams and Wilkins Company, Baltimore

Wright AL, Dou F, Hons FM (2007) Crop species and tillage effects on carbon sequestration in subsurface soil. Soil Sci 172:124–31

Chapter 6
The Nitrogen Cycle: Implications for Management, Soil Health, and Climate Change

Bijay-Singh

6.1 Introduction

Nitrogen (N) is the primary nutrient limiting plant growth throughout the world. While 78.08% of the earth's atmosphere is N, this is not readily available for use by the majority of living organisms, because the N_2 molecules do not easily enter into chemical reactions. The exception to this is biological N fixation, which until the preindustrial age was the only process where N_2 molecules were converted to reactive N forms, such as ammonium (NH_4^+) and nitrate (NO_3^-). Biological N fixation is a process undertaken by bacteria, living in soil or symbiotically within plant root nodules. However, in the postindustrial age, reactive N is increasingly being added through manufactured fertilizers, which accounts for more than half of the annual amount of N fixed by human activities (Schlesinger 2009). Burning of fossil fuels where the air becomes so hot that the N_2 molecules break apart constitutes another anthropogenic-driven N addition to terrestrial ecosystems. The N, which is lost from soils in gaseous forms to the atmosphere, can also redistribute across the landscape via wet and dry deposition.

A large amount of soil N exists in organic forms, which depending upon carbon (C)-to-N ratios and the chemical composition of the soil organic matter can be mineralized by microbes, transforming it into plant available forms, NH_4^+ and NO_3^- (mineralization) or immobilizing the mineral N forms to organic forms (immobilization). During microbially mediated processes of nitrification and denitrification, the reactive N is lost from soil as nitrous oxide (N_2O), a powerful greenhouse gas. The N cycle (Fig. 6.1) is completed through the process of denitrification, in which many soil microorganisms use NO_3^- as the electron acceptor and return N_2 molecules to the atmosphere, especially under highly anaerobic conditions (see Chap. 8). Losses of reactive N from the soil can also occur via leaching as NO_3^-, runoff, erosion, and ammonia volatilization (Fig. 6.1).

Bijay-Singh (✉)
Punjab Agricultural University, Ludhiana 141 004, Punjab, India
e-mail: bijaysingh20@hotmail.com

B.P. Singh et al. (eds.), *Soil Health and Climate Change*, Soil Biology 29,
DOI 10.1007/978-3-642-20256-8_6, © Springer-Verlag Berlin Heidelberg 2011

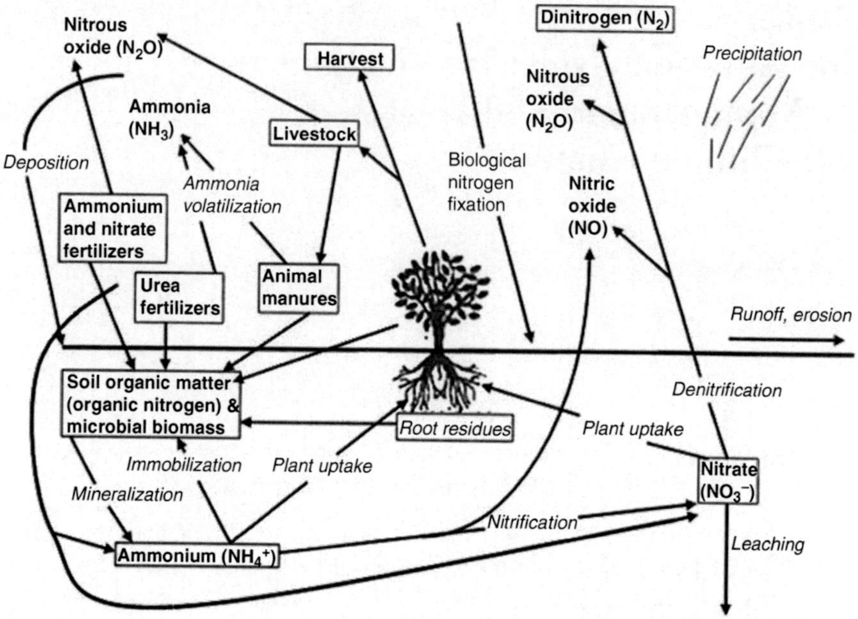

Fig. 6.1 The nitrogen cycle

The contribution to climate change by human-induced changes in the N cycle at regional and global scales is becoming more widely acknowledged. Concurrently, controlling the impact of agriculture on N cycling and vice versa is a growing challenge for sustainable development (Gruber and Galloway 2008). Increasing temperatures and atmospheric CO_2 levels, the two components of climate change, also influence a number of pathways in the terrestrial N cycle. For example, Rustad et al. (2000) reported that an increase in atmospheric temperature leads to significant overall increase in N mineralization and nitrification in terrestrial ecosystems. On the other hand, the direct influence of increasing atmospheric CO_2 is mainly limited to plant leaves (photosynthesis, stomatal aperture, and perhaps respiration), but there may be indirect effects on soil N transformations, mediated through the changes in the aboveground biomass and belowground C allocation (Koch and Mooney 1996). For example, Müller et al. (2009) observed that gross N mineralization and immobilization were increased in a temperate grassland soil under elevated CO_2 while nitrification was decreased. Thus, the effects of climate change on soil N transformations can be complex, and the long-term implications on N retention and N use efficiency are unclear.

The application of synthetic N fertilizers together with the development and introduction of rice and wheat cultivars has helped food production to keep pace with human population growth since the late 1960s. However, excessive use of N fertilizers can negatively influence soil properties, leading to decreased levels of exchangeable Ca, Mg, and K, reduced effective cation exchange capacity and acidification of soil, along with increased N leaching and production of N_2O. Management practices, such as the introduction of legumes into crop rotations, tillage practices, and stubble retention versus incorporation, also influence N availability and supply in agro-ecosystems, thereby providing significant implications for soil fertility and the environment.

In this chapter, an attempt has been made to evaluate the impacts of anthropogenic drivers (such as climate change, N deposition, and land management) on the terrestrial N cycle along with implications for soil health and climate change. Consequences of changes in soil health parameters such as N availability, C sequestration, and acidification in relation to managing fertilizer N use in agro-ecosystems with the aim of increasing N use efficiency and crop productivity but reducing greenhouse gas emissions have been discussed.

6.2 Nitrogen in the Biosphere

Globally, the creation of reactive N has increased nearly 50% between 1890 and 1990 (Galloway and Cowling 2002) and continues to increase every year. Although the creation of reactive N is dominated by agricultural activities, via the production and consumption of N fertilizers and N fixation by legumes, production of energy from fossil fuels also plays an important role (Table 6.1). The residence time of reactive N in vegetation and soils varies widely (ten to several hundred years) depending on the type of vegetation and its age. The residence time of reactive

Table 6.1 Budgets for nitrogen on the global land surface

	Preindustrial	Human-derived	Total
Inputs			
Biological nitrogen fixation	120	20[a]	140
Lightning	5	0	5
Industrial N fixation	0	125[b]	125
Fossil fuel combustion	0	25	25
Totals	125	170	295
Fates			
Biospheric increment	0	9	9
River flow	27	35	62
Groundwater	0	15	15
Denitrification	92[c]	17	109
Atmospheric transport to the ocean	6	48	54
Totals	125	124	249

All values are Tg N year^{-1}. Unless otherwise indicated, preindustrial values and human-derived inputs are for the mid-1990s from Galloway et al. (2004) and Duce et al. (2008). Fates of human-derived (anthropogenic) nitrogen are derived by Schlesinger (2009)

[a]Net of human activities

[b]US Geological Survey (2008) for 2007

[c]Calculated by difference to balance the N budget

N affects the permanence of the biospheric N sink. An adequate understanding of the fate of the reactive N in the biosphere is lacking; mass-balance studies of agricultural fields indicate that more than 50% of N in synthetic fertilizers escapes from the point of application (Deutsch et al. 2007). Globally, ~10% of the N applied as fertilizer is contained in food (Galloway and Cowling 2002); most of the remaining reactive N is lost to the environment during food production. Some of the N applied to agricultural land becomes unavailable to crop and pasture species through immobilization processes and can remain in the soil for long periods of time (Jagadamma et al. 2007).

Increasing emissions of nitrogen oxide (NO) from fossil fuel burning and NH_3 emissions associated with intensification of agriculture has led to a three- to five-fold increase in N emissions over the last century (Denman et al. 2007). Global NO and NH_3 emissions are mainly terrestrial in origin. In the year 2000, 52.1 Tg N year^{-1} of NO and 64.6 Tg N year^{-1} of NH_3 were emitted with between 30 and 50% of NO and around 40% of NH_3 deposited on the open ocean and coastal zones, while the rest is believed to have been distributed across terrestrial ecosystems (Dentener et al. 2006). Both NO and NH_3 emissions are predicted to increase further in many regions during the twenty-first century due to the growth of the global population leading to increased demand for food, particularly animal protein (Dentener et al. 2006). By 2100, N deposition over land may increase by a factor of 2.5 (Lamarque et al. 2005). Along with emissions of NO_x from fossil fuel combustion, gaseous forms of N emanating from agricultural soils produce an N enrichment of natural ecosystems, with documented effects on forests, deserts, grasslands, coastal ecosystems, and oceans (Boulart et al. 2006; Brooks 2003; Scavia and Bricker 2006).

It is estimated that 9 Tg N year^{-1} may be accumulating in the terrestrial biosphere in pools with residence times of ten to several hundred years (Schlesinger 2009). Atmospheric deposition of N on land has increased due to anthropogenic activities by ~46 Tg N year^{-1} (Galloway et al. 2004), with about one-third of the increase deposited in forests (18 Tg N year^{-1}) (Hudson et al. 1994), where trees provide a long-term sink for N in biomass. Studies indicate that ~25–30% of the N applied as fertilizer or through N deposition is retained in tree biomass (Nadelhoffer et al. 1999), and a similar amount is retained in forest soils (Schlesinger 2009). If ~50% of N deposition is stored within the ecosystem, then ~9 Tg N year^{-1} might be sequestered in terrestrial woody biomass (Schlesinger 2009). Some of this N is returned to the atmosphere as NO_x or N_2O when woody biomass is burned. As summarized by Schlesinger (2009), anthropogenic N applied to land as fertilizer and as N deposition is ~124 Tg N year^{-1}. The current estimate of reactive N produced by industry, biological nitrogen fixation, and from fossil fuel combustion is ~170 Tg N year^{-1} (Table 6.1). The budget for the terrestrial portion of the N cycle is thus not balanced and is thus linked with several serious environmental problems, such as increasing the greenhouse effect, reducing the protective ozone layer, adding to smog, contributing to acid rain, and contaminating drinking water.

6.3 Nitrogen Deposition, Soil Health, and Climate Change

The response of terrestrial ecosystems to N depositions can be interpreted to some extent from the conceptual "nitrogen saturation" model described by Aber et al. (1989). The N-limited systems initially retain the deposited N by using it for plant and microbial growth as well as via accumulation in plant biomass and soil organic matter. However, when inputs of N begin to exceed the biotic demands within the ecosystem, the excess N can be potentially lost via leaching and gaseous emissions (Matson et al. 2002). The nitrogen saturation model of Aber et al. (1989) predicts an eventual decrease in net primary production, occurring due to the loss of cations and nutrient imbalances. Plants may be less important than soils in retaining N when inputs are relatively low, especially in the range expected from current levels of atmospheric N depositions. However, chronic N additions may increase rates of soil N mineralization because of the incorporation of additional N into organic matter that could decrease its C:N ratio (Aber et al. 1998).

As soils in most agro-ecosystems already receive substantial inputs of N as fertilizer to achieve better crop yield, additional input through N deposition would not be expected to have a large effect on the cycling and sequestration of C in plant biomass or soil (Reay et al. 2008), unless N fertilization strategies are revised to account for additional N input through atmospheric deposition. Whereas, the unintended fertilization of forest ecosystems in the form of atmospheric N deposition has not only stimulated forest growth, but also affected soil microbial activity, and the cycling of C and nutrients in soils (Janssens et al. 2010).

Forest ecosystems contain a large part of the terrestrial C store, and also control the transfer of C between the atmosphere and the soil (Dixon et al. 1994). Since net primary production in both temperate and tropical forests is strongly N limited (LeBauer and Treseder 2008; Vitousek and Howarth 1991), even a small stimulation in tree growth rate in forest ecosystems due to an increase in the reactive N supply may cause a large change in the C sink capacity of forests. Nevertheless, a large amount of CO_2 that is taken up from the atmosphere by forest trees through photosynthesis are returned rapidly (back to the atmosphere) through respiration by roots and decomposer communities in soil (Bhupinderpal-Singh et al. 2003; Högberg et al. 2001; Högberg et al. 2008). Since C stocks in soil exceed those in vegetation by about 2:1 in northern temperate forests to over 5:1 in boreal forests (Dixon et al. 1994; Schlesinger 1997), changes in soil C stocks can be more important than changes in vegetation C stocks for a positive forest C budget (Medlyn et al. 2005).

The response of soil C to changing N deposition will be dictated by the balance between N-induced increases in C inputs, due to increased plant growth, and C outputs via soil organic C decomposition and microbial respiration (see Chaps. 5 and 7). However, only a few experiments have studied the effects of long-term additions of small amounts of N on soil organic matter decomposition and microbial respiration (Janssens et al. 2010). A meta-analysis by Knorr et al. (2005a) indicated that litter decomposition was stimulated with N deposition at sites containing high quality (low-lignin) litters, whereas it was decreased at sites containing low-quality (typically high-lignin) litters. Fog (1988) observed a positive relationship between C:N ratio and CO_2 evolution rate in the litter and humus layers in European forests, but no clear relationship was observed in the mineral soil (Persson et al. 2000). In two Swedish forest fertilization experiments, Persson et al. (2000) added N annually for up to 27 years. A 30% reduction in the mineralization rate was observed in the mor (litter) layer where N had been applied at a rate of 60 kg ha^{-1} $year^{-1}$ in comparison to unfertilized plots. Similarly, Franklin et al. (2003) showed that 100 years of fertilization with N applied at a rate of 30 kg ha^{-1} $year^{-1}$ could result in a doubling (1.3 kg C m^{-2}) of the amount of C stored in the mor layer. About 60% of this increase is estimated to be the result of decreased decomposition rate and the rest a result of increased litter production. A recent meta-analysis carried out by Janssens et al. (2010) also suggests that N deposition impedes organic matter decomposition and promotes C retention in temperate forest soils where N is not limiting microbial growth. According to Ågren et al. (2001), the retention of more C in N-fertilized forest soils may result from decreased growth rate of decomposers due to formation of recalcitrant compounds. The N-induced decrease in decomposition rate may also result from increased decomposer's C-use efficiency (production-to-assimilation ratio) (Ågren et al. 2001; Franklin et al. 2003).

Conversely, anthropogenic N deposition can have a detrimental impact on terrestrial ecosystems through soil acidification and a consequential reduction in plant biodiversity (Galloway et al. 2003). During the acidification process, soils release base cations, such as calcium and magnesium, neutralizing the increase in acidity. However, over time and with continued addition of N, the base cations can be depleted, at which time aluminum (Al^{3+}) is released from soil minerals, often reaching toxic levels. The acidification response may depend on the form of N added, the net balance between proton-producing and consuming processes, and the buffering capacity of the soil (Uehara and Gillman 1981). Soil acidification leads to reduced microbial N immobilization (Venterea et al. 2004). Most temperate-zone soils are buffered by base cations, which are replaced by Al^{3+} at pH ranges below pH 4.5, while many tropical soils are highly weathered, depleted in primary minerals and poorly buffered (Uehara and Gillman 1981), and hence these soils may be prone to acidification through N additions.

6.4 Interaction Between N and C Cycling in Relation to Climate Change

With human activities exerting so much impact on different systems of the planet, the interactions between the N cycle, the C cycle, and climate are expected to become an increasingly important determinant of the earth system (Gruber and Galloway 2008). Although understanding of the consequences of human alteration of the N cycle, and possible strategies to manage these, has increased over the last two decades, not enough emphasis has been placed on the study of the interactions of N cycling with that of C, and how these cycles interact with the climate system. In the recent past, the C cycle-climate change models generally assumed a strong CO_2 fertilization effect and did not consider N limitation of the terrestrial biosphere. Thus, these models may have overestimated the ability of the terrestrial biosphere to act as a CO_2 sink in the future (Thornton et al. 2009) and predict a slow rise in the atmospheric CO2 and a slow rate of climate change (Hungate et al. 2003). Furthermore, N limitation may become more pronounced in some ecosystems as atmospheric CO_2 concentration increases (Luo et al. 2004; Reich et al. 2006a). On the other hand, future increase in temperature may enhance soil N mineralization, thereby counteracting any adverse effects of elevated CO_2 on N availability (Hovenden et al. 2008). Additionally, interactions between C:N ratios in plants and soils, increased soil fertility, and microbial activity are also not very well understood and may need to be addressed in climate models. Thus the nature and importance of N–C-climate interactions are becoming increasingly pressing. The central question is: how will the availability of N affect the capacity of earth's biosphere to continue to absorb C from the atmosphere and hence mitigate climate change (Gruber and Galloway 2008)? However, increased supply of fertilizer N for food production may also have unintended negative environmental consequences such as elevated levels of N_2O production and increasing eutrophication of waterways. Thus, a tight coupling between N availability and plant uptake is desired and necessary to prevent detrimental environmental impacts.

The concentration of N_2O in the atmosphere is primarily determined by the magnitudes of nitrification and denitrification in soil – the two central microbial processes of the N cycle that result in gaseous emissions of N from soil. Over the past 60,000 years, close correspondence between atmospheric CO_2 levels, temperature, and atmospheric N_2O concentrations demonstrate that the N cycle is closely coupled to variations in the climate system and C cycle (Gruber and Galloway 2008). Nevertheless, anthropogenic alterations of the global C and N cycles appear to have led to the highest atmospheric concentrations of CO_2 and N_2O in 650,000 years (Forster et al. 2007). A significant correlation between the flux of CO_2 and N_2O from soils has been reported by Xu et al. (2008). The N_2O flux from soil may increase globally with rising atmospheric CO_2 and temperature, because these climate change drivers can alter N mineralization–immobilization turnover processes (Hoyle et al. 2006; Müller et al. 2009), and in turn nitrification and denitrification (Fig. 6.2).

Terrestrial C stocks have been altered by increasing atmospheric CO_2 concentration and N deposition, as well as by changing land use (Matson et al. 2002; see Chap. 1). In 2005, agriculture accounted for an estimated emission of 5.1–6.1 Gt CO_2-eq year^{-1}, which is 10–12% of total global anthropogenic emissions of greenhouse gases (Smith et al. 2007a). There are four main global sinks for these emissions: the atmosphere, the oceans, tropical vegetation, and temperate and boreal vegetation, mainly forests. It has long been recognized that N limitations often constrain C accumulations in mid- and high-latitude ecosystems, such as temperate and boreal forests (Tamm et al. 1982). Deposition of N to forests ranges

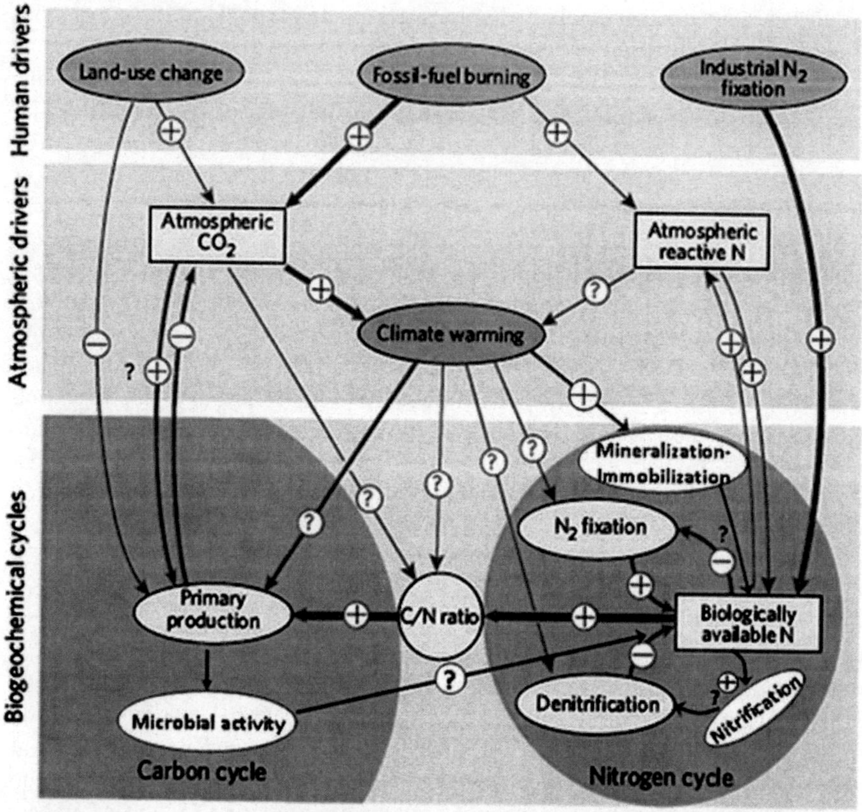

Fig. 6.2 Potential nitrogen–carbon-climate interactions. The main anthropogenic drivers of these interactions during the twenty-first century are shown. *Plus signs* indicate that the interaction increases the amount of the factor shown; *minus signs* indicate a decrease; *question marks* indicate an unknown impact (or, when next to a plus or minus sign, they indicate a high degree of uncertainty). *Arrow thickness* denotes strength of interaction. Only selected interactions are shown (adapted by permission from Macmillan Publishers Ltd: *Nature*, Gruber and Galloway (2008))

between 1 and 100 kg ha^{-1} $year^{-1}$; the smaller amounts generally occur in remote forests in rural areas at high latitudes and the large amounts in industrialized central Europe (Jarvis and Fowler 2001). However, there exists uncertainty as to the extent and for how long high annual rates of N additions will be able to stimulate the production of mature forests. Recent findings on plant responses to elevated CO_2 concentrations also confirm that low N availability can constrain C sequestration in terrestrial ecosystems (Luo et al. 2004; Reich et al. 2006b).

Thornton et al. (2007) and Sokolov et al. (2008) conducted modeling studies by introducing prognostic C and N cycle interactions in the stand-alone land surface component of an atmosphere-ocean general circulation model (AOGCM) and in a reduced-complexity climate model. They observed that the land-atmosphere components of the global climate-C cycle feedback are fundamentally influenced by C–N cycle interactions. Thornton et al. (2007) demonstrated that N limitation significantly reduced the stimulation of net C uptake on land associated with increased CO_2 concentration in the atmosphere. This should lead to predictions of higher CO_2 concentration in the atmosphere for a given level of fossil fuel consumption in a coupled climate-C cycle simulation. Studies carried out by Thornton et al. (2007) and Sokolov et al. (2008) predict a 53–78% reduction in the effect of elevated CO_2 on land C sink strength due to C–N coupling. Thornton et al. (2007) showed that C–N interaction fundamentally alters the terrestrial C cycle response to interannual variability in temperature and precipitation. Sokolov et al. (2008) found that the introduction of C–N coupling in a reduced-complexity climate model produced a change in the sign of the terrestrial C cycle response to warming, switching from a strong positive feedback in which warming leads to a net release of C from the terrestrial biosphere, to a weak negative feedback in which warming leads to a modest uptake of C aided by increased N mineralization. More recently, Thornton et al. (2009) demonstrated a weaker dependence of net terrestrial C flux on soil moisture changes and a stronger positive growth response to warming in tropical regions, than that predicted by a similar AOGCM model implemented without terrestrial C–N interactions.

6.5 Nitrogen in Agro-Ecosystems, Soil Health, and Climate Change

6.5.1 Nitrogen Cycling in Natural Versus Managed Ecosystems

An understanding of soil N turnover in natural or unmanaged ecosystems can provide clues about how to efficiently manage N fertilization while maintaining soil health and crop production. In natural ecosystems, atmospheric deposition and biological N_2 fixation constitute the external sources of N which then is returned to the soil either as plant litter or as residues from herbivore-based food chain. The accumulation of organic forms of N in soil is a characteristic feature of unmanaged terrestrial ecosystems. Mineral N released through mineralization–immobilization turnover can be taken up by plant roots or is lost from the system. Natural ecosystems often exhibit a high degree of temporal and spatial synchrony and synlocation between the release and uptake of N by mixed plant communities. This results in a minimal transfer across system boundaries relative to the extent of N cycling within the ecosystem, consequently leading to minimal losses via nitrate leaching or gaseous emissions (Christensen 2004). On the other hand, agro-ecosystems are often relatively open with respect to N cycling. In agro-ecosystems, besides inputs of N through deposition and biological N_2 fixation, N is also applied externally as chemical fertilizers or organic manures to compensate for the N removed in exported in harvested yield. In modern agro-ecosystems, the removal of as much as 300 kg N ha^{-1} in harvested aboveground portions of the crops each year necessitates substantial inputs of N to maintain productivity (Cassman et al. 2002). The intensive nature of agricultural cropping systems exhibits large N uptake during active but often relatively short growth phase. If the supply of N in synthetic fertilizers is not sufficient to meet crop demands in agro-ecosystems, soils can become N depleted. Reduced synlocation and synchrony in the N turnover and reduced return of organic matter to the soil adversely affect N use efficiency and soil health in managed agro-ecosystems. Furthermore, changes in hydrology through irrigation and drainage, and changes in soil structure through tillage, can further change the N dynamics within agro-ecosystems in comparison to those of natural systems.

6.5.2 Fertilizer N Use Effects on Soil Health in Agro-Ecosystems

One of the most important soil health parameters in relation to both climate change and N management is organic matter content of the soil. It is in fact a key indicator of soil health because it plays a role in a number of vital functions affecting soil fertility, productivity, and the environment. Managing soil organic matter for a maximum contribution to soil health and resilience presents a conundrum. Decomposition and mineralization of organic matter are required for functions such as providing energy and nutrients, whereas maintaining or increasing the quantity of organic matter can have positive effects on chemical, physical, and biological properties of soil (see Chaps. 2 and 5).

Maintenance of soil organic matter is an important goal in agriculture, both in terms of sustaining soil fertility and sequestering atmospheric CO_2. As C is sequestered in a soil, the availability of N can decrease as N binds with the organic matter and precluding it from participation in the N cycle. Thus, the role of C:N ratio in soil organic matter should also be considered in assessing fertilizer N use efficiency and designing long-term N management strategies. With growing dependence on chemical N fertilizer, the assertion has often been made that these inputs are a positive factor in maintaining or increasing soil organic matter as higher yields enhance the input of crop residues. Such a view is, however, at odds with the long-term changes in soil C reported for the Morrow Plots, the world's oldest experimental site under continuous corn (Khan et al. 2007). For example, after 40–50 years of applying chemical fertilizers that exceeded corn N removal by 60–190%, a net decline in soil C occurred despite the incorporation of residues, which is consistent with the data from numerous cropping experiments involving synthetic N fertilization (Khan et al. 2007). These data implicate fertilizer N in promoting the decomposition of crop residues and soil organic matter through the stimulation of microbial activity. Several contrary reports are also available in the literature, which show that N fertilization increases soil organic C sequestration in certain ecosystems (Khan et al. 2007). According to Khan et al. (2007), such contradictory observations may be due to the following reasons: (1) absence of baseline data in assessing soil organic C changes; (2) soil organic C data represent only a very limited depth of surface soil; or (3) the study period is inadequate for detecting the changes. Furthermore, fertilizer N applications may reduce microbial activity if acidity generated during NH_4^+ oxidation is not controlled (Vanotti et al. 1997), thereby leading to a decline in the rate of organic matter mineralization. Nitrogen additions in certain ecosystems have also been shown to adversely affect the activity of lignin-degrading microorganisms or enhance lignin-N complexes resistant to decomposition, thereby favoring organic matter accumulation in soil (Fog 1988).

To fully realize the potential benefits of N fertilizers, application rates must be adequate but not excessive, so as to maximize the economic profitability of crop production while minimizing microbial oxidation of residue C and native soil organic matter. This strategy merits serious consideration because soils hold more than twice as much C as the atmosphere, and even a minor change in terrestrial CO_2 balance could have a significant global impact (Powlson 2005). After reviewing data from long-term experiments worldwide to quantify the long-term impacts of N fertilizers on soil N cycling, Glendining and Powlson (1995) concluded that N applications increased N recycling from crop residues, root turnover, and exudates. Initially, mainly labile fractions of soil N were influenced from these inputs, but over years and decades the fractions of soil organic N that cycle more slowly were also affected. Soil productivity can be enhanced both through N fertilization and by growing legumes in rotation, although the two management strategies may have different influences on nutrient cycling and soil health (Liebig et al. 2002).

Logically, soil should gain N if fertilizer inputs exceed grain removal. However, Mulvaney et al. (2009) recorded a decline in total soil N similar to that recorded in soil organic C in the Morrow plots (Khan et al. 2007). This decline is in agreement with numerous long-term baseline data sets from chemical-based cropping systems involving a wide variety of soils, geographic regions, and tillage practices, as listed by Mulvaney et al. (2009). Furthermore, these trends from long-term experiments reveal that chemical fertilization is often ineffective in preventing soil N depletion, even in cases involving an ample input of N and greater production and incorporation of crop residues. As mentioned above, these effects are more likely due to the fact that mineral N, particularly in the form of ammoniacal fertilizers, can stimulate microbial carbon decomposition, thereby promoting the decay of added crop residues and also the indigenous organic matter that serves as the major reservoir of soil N.

When N used in the synthesis of plant biomass is returned to soil, it exists in equilibrium with a larger and more stable pool associated with humus. This equilibrium is shifted toward immobilization when the input of C is higher than N and toward mineralization when the input of N is higher than C. The ultimate effect in the latter case may be a net loss of applied organic N through crop uptake, leaching, or gaseous emissions of mineralized N. Dissolved organic N may also be lost through profile transport. As reviewed by van Kessel et al. (2009), dissolved organic N losses can range from a minimum of 0.3 kg N ha^{-1} $year^{-1}$ in a clover-based pasture to a maximum of 127 kg N ha^{-1} $year^{-1}$ in a grassland to which urine was applied. Mulvaney et al. (2009) have compiled a global data set to compare net mineralization with and without N fertilization and at different N application rates, which demonstrates the inherent potential of synthetic N to enhance microbial decomposition. This data set is remarkably consistent in documenting more rapid mineralization for fertilized than unfertilized soils, and in many cases identifying a positive effect from increasing the rate of N applied. Although Powlson et al. (2010) has criticized the way Mulvaney et al. (2009) interpreted the data from long-term experiments, they agree that the observation of significant soil C and N declines in subsoil layers deserves further consideration.

6.5.2.1 Fertilizer N Management in Agro-Ecosystems

In agro-ecosystems, crops do not always use N efficiently when it is applied through fertilizers, manures, and other sources (Galloway et al. 2003, Cassman et al. 2003). The unutilized N is susceptible to various loss pathways – leaching as nitrate, ammonia volatilization, denitrification, and runoff (Fig. 6.1). By reducing these losses, improved N use efficiency can also lead to reduced off-site N_2O emissions. Fertilizer N use efficiency can be improved by: (a) adjusting application rates based on precise estimation of crop needs; (b) using slow- or controlled-release fertilizer forms or nitrification inhibitors; (c) improved timing of N application; (d) placing the N more precisely into the soil to make it more accessible to crops roots; or (e) by avoiding N applications in excess of immediate plant requirements (Robertson 2004; Paustian et al. 2004; Bijay-Singh 2008; Monteny et al. 2006). Landholders apply N fertilizer on the assumption that in managed agro-ecosystems this is the sole source of plant available N. But considerable evidence obtained from ^{15}N-tracer investigations suggests that plant uptake is generally greater from native soil N than from N applied via synthetic fertilizers, even with excessive fertilization (Olson 1980; Kitur et al. 1984; Reddy and Reddy 1993; Stevens et al. 2005). Saito and Ishii (1987) showed that native soil N is an important source of N that is used by crops in managed agro-ecosystems. They investigated the uptake of N by maize from soil N and fertilizer N sources, from 12 soils, and showed that native soil N dictates the efficiency of applied fertilizer N as well as the quantity of N lost from the soil-plant system. Additionally, improvements in N use efficiency can indirectly lead to reduced greenhouse gas emissions from N fertilizer manufacture.

Given the fundamental coupling of microbial C and N cycling, the dominant occurrence of both elements in soil organic forms, and the close correlation between soil C and N mineralization (Dou et al. 2008), the practices that lead to loss of soil organic C will also have serious implications for the storage of N in soil. The loss of organic N decreases soil productivity and the agronomic efficiency of fertilizer N and has been implicated in reports of yield stagnation and the decline of grain production (Mulvaney et al. 2009). As already discussed, even with intensive fertilization, soil reserves usually supply the bulk of N uptake by nonleguminous crops. Thus, a decrease in soil N supply is inherently detrimental to productivity, although crop yields may be sustained or even increase because of introduction of improved varieties or due to higher fertilizer application rate despite the lower incremental return per unit of N applied. Eventually, however, soil degradation is likely to lead to a decline or stagnation in yield, an emerging concern for input-intensive agriculture.

6.5.3 *Nitrogen Availability in Relation to Non-CO_2 Emissions and Soil C Sequestration*

The basic principles of sound N management are well known and need to be promoted within the context of decreasing N losses from the soil-plant system including greenhouse gas emissions in the form of N_2O as well as for increasing economic profitability of crop production. It follows from the discussion in Sect. 6.5.2 that high N use efficiency can be achieved by avoiding excessive N applications and by synchronizing N supply with crop demand. The latter is more easily achieved when N is supplied in split doses using fertilizers compared to where organic N sources are used (both imported organic materials such as animal manures, and legumes or residues grown on site), where the release of nutrients is controlled by biological mineralization processes. In general, the release of N from organic sources continues beyond the period of crop production and when applied in amounts large enough to meet the nutrient requirement of agro-ecosystems can contribute to leaching losses and off-site pollution problems, including additional generation of N_2O (Duxbury 2006). Research has also shown that emissions of N_2O from cropland are higher when organic sources of N rather than inorganic fertilizers are used (Duxbury et al. 1982). As shown in Table 6.2, there exists a large global potential of N availability through N fixation by leguminous plants via intercropping and off-season cropping (Badgley et al. 2007) and from livestock

Table 6.2 Global nitrogen input in different agricultural sectors

Nitrogen derived from industrial production (by the Haber–Bosch process with fossil fuel combustion)	90–100 Mt N year^{-1}	Erisman et al. (2008); IFA (2009)
Potential nitrogen production by leguminous plants via intercropping and off-season cropping (without competing cash crops). This potential is underexploited by many conventional farmers	140 Mt N year^{-1}	Badgley et al. (2007)
Nitrogen from livestock feces of 18.3 billion farm animals (FAO, global figure). In specialized farming structure with strong segregation between crop and livestock production, nitrogen from manure and slurry is inefficiently used	160 Mt N year^{-1}	Niggli et al. (2009)

feces (Niggli et al. 2009). Nitrogen additions through these routes can also contribute to large N_2O emissions from soil (Rochette and Janzen 2005).

According to Smith et al. (2007a), agriculture accounted for 47 and 58% of total anthropogenic emissions of methane (CH_4) and N_2O during 2005. Methane and N_2O contributed 3.3 and 2.8 Gt CO_2-eq year^{-1}, respectively, to the atmosphere. Despite large annual exchanges of CO_2 between the atmosphere and agricultural lands, the fluxes are estimated to be approximately balanced with net CO_2 emissions ~0.04 Gt CO_2-C year^{-1} only. However, capturing CO_2 as soil organic matter can contribute to improved soil health and productivity (Drinkwater et al. 1998) and potentially mitigate climate change. In fact, soil C sequestration (enhanced sinks) is the mechanism responsible for most of the mitigation potential in the agriculture sector, with an estimated 89% contribution to the technical potential (per area estimate of mitigation potential for different greenhouse gases multiplied by the area for that practice in each region) (Smith et al. 2007a). In general, agronomic practices that increase crop yield and generate higher inputs of organic residue can lead to increased soil C storage, although reports do exist that N fertilizer use on soils to achieve higher crop yield may lead to the loss of organic C from soils (Khan et al. 2007; see Sect. 6.5.2). No-till agriculture is usually considered a useful agronomic technique to reduce the rate of loss of organic C and N from soils as opposed to conventional tillage practices (see Chap. 9). However, the overall rate of soil C sequestration under no-till agriculture may depend on local soil and climatic conditions, and hence the reported effects are inconsistent and not well-quantified globally (Smith and Conen 2004; Li et al. 2005; see Chap. 9). Emissions of N_2O per hectare can be reduced by adopting cropping systems with reduced reliance on fertilizers; for example, the use of rotations with legume crops (West and Post 2002). But, as already mentioned, legume-derived N can also serve as a source of N_2O. Furthermore, cultivated wetland rice soils emit significant quantities of CH_4 (Yan et al. 2003). This source of CH_4 emissions can be reduced by draining wetland rice during the growing season (Smith and Conen 2004; Yan et al. 2003). However, any reduction in CH_4 emissions achieved due to this strategy may be partly offset by increased N_2O emissions under aerobic conditions.

The oxidation of CH_4 by CH_4-oxidizing microorganisms (methanotrophs) in soil is important in defining the global CH_4 budget (see Chaps. 9 and 10). Oxic soils are a net sink while wetland soils are a net source of atmospheric CH_4. The consumption of CH_4 in upland as well as lowland systems may be inhibited by NH_4^+ produced in soil or added through ammoniacal fertilizers because NH_4^+ can competitively inhibit CH_4 oxidation by methanotrophs (Bedard and Knowles 1989). However, NH_4^+-based fertilization has also been demonstrated to stimulate CH_4 consumption in rice paddies, especially when methanotrophs are N limited (Bodelier and Laanbroek 2004). Available literature reveals that N limitation of CH_4 consumption occurs in a variety of lowland soils, upland soils, and sediments. Obviously, depriving CH_4-oxidizing bacteria of a suitable source of N hampers their growth and activity.

6.5.4 *Global Significance of N_2O Emissions from Terrestrial and Oceanic Sources*

Based on data pertaining to N_2O flux from soils, wetlands, and the sea, globally ~2.6–3.9% of the denitrification flux consists of N_2O (Denman et al. 2007; Seitzinger et al. 2006), which is increasing in earth's atmosphere by ~0.3% per year (Denman et al. 2007). Schlesinger (2009) has calculated recent changes in global denitrification by dividing the observed increase in N_2O in earth's atmosphere (~4 Tg N year^{-1}) by an estimate of the ratio of N_2O to the total $N_2 + N_2O$ produced by denitrifiers. For wetland, stream, and lake sediments, the values for this ratio are always low because the denitrification flux consists of greater proportion of N_2 than N_2O (Kralova et al. 1992; Riley and Matson 2000). For upland ecosystems, a significant percentage of the efflux due to denitrification occurs as N_2O. The mean ratios $N_2O/(N_2 + N_2O)$ for soils under natural or recovering vegetation, agricultural soils, and freshwater wetlands and flooded soils are 0.492 ± 0.066, 0.375 ± 0.035, and 0.082 ± 0.024, respectively (Schlesinger 2009). According to Schlesinger (2009), assuming that the rise in N_2O in the Earth's atmosphere is solely from denitrification in terrestrial ecosystems and if 53% of denitrification occurs in upland agricultural soils with a $N_2O/(N_2 + N_2O)$ ratio of 0.37, and 47% occurs in wetlands with a ratio of 0.08 (Bouwman et al. 2005), then the weighted mean ratio becomes 0.23 and the calculated total rate of denitrification is now 17 Tg N year^{-1} greater than in preindustrial times (Schlesinger 2009).

6.6 Nitrogen Management Under Future Climate Change Scenarios

As already discussed, sequestration of C in the soil is a mechanism that could potentially be responsible for significant mitigation potential for climate change. According to Smith et al. (2007a), reduction in CO_2, CH_4 and N_2O emissions from soils can contribute 89, 9, and 2%, respectively, to the total mitigation potential. Of course, the level of adoption of appropriate measures, effectiveness of these in enhancing C sinks, and persistence of mitigation, as influenced by future climatic trends, economic conditions, and social behavior will determine the exact contribution. The projected changes in climate in coming decades may influence emissions of greenhouse gases from agriculture and the effectiveness of practices adopted to minimize these emissions. For example, increasing atmospheric CO_2 concentrations may affect the functioning of terrestrial ecosystems through changes in plant growth rates, plant litter composition, drought tolerance, and N demands (van Groenigen et al. 2005; Jensen and Christensen 2004). Similarly, atmospheric N deposition and increasing temperature may affect the net C sequestration potential of forests and crop production systems. Increasing temperatures are likely to have a positive effect on forest and crop production in colder regions due to a longer growing season (Smith et al. 2007b), but increasing temperatures may also accelerate decomposition of soil organic matter, releasing stored soil C into the atmosphere (Knorr et al. 2005b; Smith et al. 2007b). Increasing mineralization rates will also elevate the rate with which N is released from soil organic matter and therefore its availability for uptake by plants. Similarly, higher rates of N deposition are expected to increase plant growth, but may alter C sequestration and soil N_2O emissions in unexpected directions. Furthermore, complex interactions between the climate change drivers and C and N cycles may produce high levels of uncertainties on the potential effects of climate change on soil N availability and CO_2 emissions (see Sects. 6.3 and 6.4).

With the prediction that the increase in atmospheric CO_2 levels by 2100 will result in an additional warming of up to several degrees Celsius (Forster et al. 2007), one can expect the intensification of N cycle in terrestrial ecosystems, in addition to direct anthropogenic inputs (Gruber and Galloway 2008). Thus, future management of the N cycle in terrestrial ecosystems is expected to be highly challenging. A meta-analysis of data generated in different soils and ecosystems carried out by Rustad et al. (2001) revealed that warming significantly increases N mineralization and nitrification. Furthermore, Groffman et al. (2009) found that N cycling processes are more sensitive to variations in soil moisture induced by global warming. It was observed that net N mineralization and nitrification were slower in warmer climate, but these processes were more driven by lower soil moisture content than higher temperature. In northeast China, Hu et al. (2009) observed that N mineralization rate in the soil under meadow steppes increased significantly due to a 2°C rise in soil surface temperature, although warming did not have a significant influence on total N content. In a semi-arid soil, Hoyle et al. (2006) observed that nitrification increased linearly with temperature and dominated over immobilization for available ammonium in soil incubated above 20°C, indicating that nitrification is often the principal process influencing NH_4^+ consumption. These findings suggest that the N mineralization–immobilization turnover at soil temperatures above 20°C is not tightly coupled, and there is a high potential for loss of N as nitrate.

Information about soil solution inorganic nitrogen ($NH_4^+ + NO_3^-$) is important from the point of fertilizer N management. Shaw and Harte (2001) recorded no increase in soil solution N despite increases in mineralization rate. This result was ascribed to increased plant uptake. While Ineson et al. (1998) showed a decrease, Peterjohn et al. (1994) observed no change in soil solution inorganic N concentration upon soil heating. According to Rustad et al. (2000), soil solution inorganic concentrations will increase only in N-saturated ecosystems, whereas most N will be taken up by vegetation in N-limited systems. Verburg (2005) found that by increasing air temperatures by 3°C in summer and 5°C in winter, soil inorganic pools increased indicating an enhanced mineralization which exceeded plant demand and led to leaching losses.

There is good reason for concern about sustaining world food production by supplying additional N because crop N uptake originates largely from the native soil source rather than fertilizer N (see Sect. 6.5.2.1). In view of the changing scenario with regard to N cycle process under a warmer climate, the immediate need is to use scientific and technological advances that can increase input efficiencies. The most important aspect of this strategy would be to more accurately match the input of ammoniacal N to crop N requirement by accounting for site-specific variations in soil N supplying capacity and by synchronizing application with plant N uptake. Across a wide range of crop-growing conditions, information to reduce uncertainty in synchronizing crop N demand and soil N supply can contribute substantially to reducing fertilizer application rates and the associated economic and environmental costs. In the long term, a transition toward agricultural diversification using legume-based crop rotations, which provide a valuable means to reduce the intensity of ammoniacal fertilization with the input of less reactive organic N, may be required. A number of management options which have limited potential now are also likely to have increased potential in the long term. Examples include better use of fertilizer through precision farming, wider use of slow and controlled-release fertilizers and adoption of nitrification inhibitors.

An analysis based on measurements and model studies as reported by Reay et al. (2008) indicates that the likelihood of greatly enhanced global CO_2 sequestration resulting from future changes in N deposition is low. A doubling of the N emission levels in the year 2000 may achieve approximately three billion tonnes of additional CO_2 sequestration in northern and tropical forests each year. However, using emission factor given by Crutzen et al. (2007), this would induce global annual emissions of 2.7 billion tonnes of CO_2 equivalent, in the form of N_2O, via increased nitrification and denitrification on land and in the oceans. Such pollution-swapping would greatly offset any climate change mitigation benefits of N deposition. Under such situations, protection of the existing terrestrial C sinks from deforestation and land use changes, rather than possible enhanced C sequestration through N deposition, will provide greater climate change mitigation benefits in coming decades (Reay et al. 2008).

6.7 Summary and Conclusions

The large generation of reactive N by industry, fossil fuel burning, and biological N_2 fixation has been able to maintain adequate availability of N for crop plants, thereby ensuring ample production of food and fiber to meet the increasing demand of a growing population. However, ~10% of the applied N remains in food and the remainder generally ends up in the environment, polluting the atmosphere and water bodies, and influencing the climate. The N cycle in natural ecosystems is highly conservative to loss pathways, but it is heavily disrupted in managed agro-ecosystems, leading to enhanced N_2O emissions and N leaching. Production of fertilizer N for application in managed agro-ecosystems is still accelerating, and this trend is not likely to change in the near future. Furthermore, many terrestrial and marine ecosystems will continue to receive an increasing supply of reactive N produced by industry and fossil fuel burning, and also indirectly via atmospheric deposition. Questions remain as to the likely consequences for terrestrial C sink capacity, specifically whether increasingly heavy inputs of reactive N will serve to significantly reduce rising atmospheric CO_2 concentrations.

The evidence for changes in soil C stocks under N deposition scenarios is contradictory, with some studies suggesting that soil C may decrease while others suggest no change or significant increases in soil C. It remains a challenge to obtain a clear understanding of the impacts of N additions through fertilizer application or atmospheric N deposition on soil health. Furthermore, changes in soil health due to excessive external N inputs such as acidification, cation exchange capacity, greenhouse gas emissions, and nutrient leaching are likely to have serious negative implications for plant productivity and the environment.

Climate change drivers such as elevated CO_2 and temperature can influence the terrestrial C and N cycling and alter soil N availability, which constrain the CO_2 sink capacity of earth's biosphere. As C–N interaction fundamentally alters the terrestrial C cycle response to interannual variability in temperature and precipitation (see Sect. 6.4), future atmospheric CO_2 concentration and associated anthropogenic climate change may be accurately predicted as climate change models introduce C–N interactions in their land components.

Effects of global warming on different processes and pathways of the N cycle are going to be very complex as several interactions will be involved and the overall effects are likely to be site specific. Management of N fertilizers in agro-ecosystems that improve fertilizer N use efficiency by avoiding excessive N applications and synchronizing N supply with crop demand should help produce enough food for growing populations with minimal contributions to climate change effects.

Acknowledgments I appreciate constructive comments from two anonymous reviewers as well as the editors, especially the senior editor, on earlier drafts.

References

Aber JD, Nadelhoffer KJ, Steudler P, Melillo JM (1989) Nitrogen saturation in northern forest ecosystems – hypothesis revisited. Bioscience 39:378–386

Aber JD, McDowell W, Nadelhoffer K, Magill A, Berntson G, Kamakea M, McNulty S, Currie W, Rustad L, Fernandez I (1998) Nitrogen saturation in temperate forest ecosystems. Bioscience 48:921–934

Ågren GI, Bosatta E, Magill AH (2001) Combining theory and experiment to understand effects of inorganic nitrogen on litter decomposition. Oecologia 128:94–98

Badgley C, Moghtader J, Quintero E, Zakem E, Chappell MJ, Avilés-Vàzquez K, Samulon A, Perfecto I (2007) Organic agriculture and the global food supply. Renewable Agric Food Syst 22:86–108

Bedard C, Knowles R (1989) Physiology, biochemistry, and specific inhibitors of CH_4, NH_4^+, and CO oxidation by methanotrophs and nitrifiers. Microbiol Rev 53:68–84

Bhupinderpal-Singh, Nordgren A, Lofvenius MO, Högberg MN, Mellander PE, Högberg P (2003) Tree root and soil heterotrophic respiration as revealed by girdling of boreal Scots pine forest: extending observations beyond the first year. Plant Cell Environ 26:1287–1296

Bijay-Singh (2008) Crop demand-driven site specific nitrogen applications in rice and wheat – some recent advances. Indian J Agron 53:157–166

Bodelier PLE, Laanbroek HJ (2004) Nitrogen as a regulatory factor of methane oxidation in soils and sediments. FEMS Microbiol Ecol 47:265–277

Boulart C, Flament P, Gentilhomme V, Deboudt K, Migon C, Lizon F, Schapira M, Lefebvre A (2006) Atmospherically-promoted photosynthetic activity in a well-mixed ecosystem: significance of wet deposition events of nitrogen compounds. Estuar Coast Shelf Sci 69:449–458

Bouwman AF, Van Drecht G, Knoop JM, Beusen AHW, Meinardi CR (2005) Exploring changes in river nitrogen export to the world's oceans. Global Biogeochem Cyc 19:GB1002. doi:10.1929/2004GB0002314

Brooks M (2003) Effects of increased soil nitrogen on the dominance of alien annual plants in the Mojave desert. J Appl Ecol 40:344–353

Cassman KG, Dobermann A, Walters D (2002) Agroecosystems, nitrogen-use efficiency, and nitrogen management. Ambio 31:132–140

Cassman KG, Dobermann A, Walters DT, Yang H (2003) Meeting cereal demand while protecting natural resources and improving environmental quality. Annu Rev Environ Res 28:315–358

Christensen BT (2004) Tightening the nitrogen cycle. In: Schjønning P, Elmholt S, Christensen BT (eds) Managing soil quality: challenges in modern agriculture. CAB International, Wallingford, pp 47–67

Crutzen PJ, Mosier AR, Smith KA, Winiwarter W (2007) N_2O release from agro-biofuel production negates global warming reduction by replacing fossil fuels. Atmos Chem Phys Discuss 7:11191–11205

Denman KL, Brasseur G, Chidthaisong A, Ciais P, Cox PM, Dickinson RE, Hauglustaine D, Heinze C, Holland E, Jacob D, Lohmann U, Ramachandran S, da Silva Dias PL, Wofsy SC, Zhang X (2007) Couplings between changes in the climate system and biogeochemistry. In: Solomon S, Qin D, Manning M, Chen Z, Marquis M, Averyt KB, Tignor M, Miller HL (eds) Climate change 2007: the physical science basis. Contribution of Working Group I to the Fourth assessment report of the Intergovernmental Panel on Climate Change. Cambridge University Press, Cambridge, pp 499–587

Dentener F et al. (2006) Nitrogen and sulfur deposition on regional and global scales: a multimodel evaluation. Global Biogeochem Cyc 20:GB4003. doi:10.1029/2005GB002672

Deutsch C, Sarmiento JL, Sigman DM, Gruber N, Dunne JP (2007) Spatial coupling of nitrogen inputs and losses in the ocean. Nature 445:163–167

Dixon RK, Brown S, Houghton RA, Solomon AM, Trexler MC, Wisniewski J (1994) Carbon pools and flux of global forest ecosystems. Science 263:185–190

Dou F, Wright AL, Hons FM (2008) Sensitivity of labile soil organic carbon to tillage in wheat-based cropping systems. Soil Sci Soc Am J 72:1445–1453

Drinkwater LE, Wagoner P, Sarrantonio M (1998) Legume-based cropping systems have reduced carbon and nitrogen losses. Nature 396:262–265

Duce RA, LaRoche J, Altieri K, Arrigo KR, Baker AR, Capone DG, Cornell S, Dentener F, Galloway J, Ganeshram RS, Geider RJ, Jickells T, Kuypers MM, Langlois R, Liss PS, Liu SM, Middelburg JJ, Moore CM, Nickovic S, Oschlies A, Pedersen T, Prospero J, Schlitzer R, Seitzinger S, Sorensen LL, Uematsu M, Ulloa O, Voss M, Ward B, Zamora L (2008) Impacts of atmospheric anthropogenic nitrogen on the open ocean. Science 320(5878):893–897

Duxbury JM (2006) Soil carbon sequestration and nitrogen management for greenhouse gas mitigation. In: Climate change and agriculture: promoting practical and profitable responses, pp IV-5–IV-7. www.climateandfarming.org/pdfs/FactSheets/IV.2Soil.pdf. Accessed 17 Aug 2009

Duxbury JM, Bouldin DR, Terry R, Tate RL (1982) Emissions of nitrous oxide from soils. Nature 298: 462–464

Erisman JW, Sutton MA, Galloway J, Klimont Z, Winiwarter W (2008) How a century of ammonia synthesis changed the world. Nat Geosci 1:636–639

Fog K (1988) The effect of added nitrogen on the rate of decomposition of organic matter. Biol Rev 63:433–462

Forster P, Ramaswamy V, Artaxo P, Berntsen T, Betts R, Fahey DW, Haywood J, Lean J, Lowe DC, Myhre G, Nganga J, Prinn R, Raga G, Schulz M, Van Dorland R (2007) Changes in atmospheric constituents and in radiative forcing. In: Solomon S, Qin D, Manning M, Chen Z, Marquis M, Averyt KB, Tignor M, Miller HL (eds) Climate change 2007: the physical science basis. Contribution of Working Group I to the Fourth assessment report of the Intergovernmental Panel on Climate Change. Cambridge University Press, Cambridge

Franklin O, Högberg P, Ekblad A, Ågren GI (2003) Pine forest floor carbon accumulation in response to N and PK additions – bomb ^{14}C modelling and respiration studies. Ecosystems 6:644–658

Galloway JN, Cowling EB (2002) Reactive nitrogen and the world: two hundred years of change. Ambio 31:64–71

Galloway JN, Aber JD, Erisman JW, Seitzinger SP, Howarth RW, Cowling EB, Cosby BJ (2003) The nitrogen cascade. Bioscience 53:341–356

Galloway JN, Dentener FJ, Capone DG, Boyer EW, Howarth RW, Seitzinger SP, Asner GP, Cleveland C, Green P, Holland E, Karl DM, Michaels AF, Porter JH, Townsend A, Vorosmarty C (2004) Nitrogen cycles: past, present and future. Biogeochemistry 70:153–226

Glendining MJ, Powlson DS (1995) The effects of long-continued applications of inorganic nitrogen fertilizer on soil organic nitrogen – a review. In: Lal R, Stewart BA (eds) Soil management: experimental basis for sustainability and environmental quality. Advances in Soil Science Series. Lewis, Boca Raton, pp 385–446

Groffman PJ, Hardy JP, Fisk MC, Fahey TJ, Driscoll CT (2009) Climate variation and soil carbon and nitrogen cycling processes in a northern hardwood forest. Ecosystems 12:927–943

Gruber N, Galloway JN (2008) An Earth-system perspective of the global nitrogen cycle. Nature 451:293–296

Högberg P, Nordgren A, Buchmann N, Taylor AFS, Ekblad A, Hogberg MN, Nyberg G, Ottosson-Lofvenius M, Read DJ (2001) Large-scale forest girdling shows that current photosynthesis drives soil respiration. Nature 411:789–792

Högberg P, Hogberg MN, Gottlicher SG, Betson NR, Keel SG, Metcalfe DB, Campbell C, Schindlbacher A, Hurry V, Lundmark T, Linder S, Nasholm T (2008) High temporal resolution tracing of photosynthate carbon from the tree canopy to forest soil microorganisms. New Phytol 177:220–228

Hovenden MJ, Newton PCD, Carran RA, Theobald P, Wills KE, Vander Schoor JK, Williams AL, Osanai Y (2008) Warming prevents the elevated CO_2-induced reduction in available soil nitrogen in a temperate, perennial grassland. Global Change Biol 14:1018–1024

Hoyle FC, Murphy DV, Fillery IRP (2006) Temperature and stubble management influence microbial CO_2–C evolution and gross N transformation rates. Soil Biol Biochem 38: 71–80

Hu L, Yang H, Wang W, Guo J (2009) Soil nutrient responses to one year of simulated global warming and nitrogen deposition on the Songnen meadow steppes, northeast China. http://www.paper.edu.cn/index.php/default/selfs/downpaper/huliangjun326400-self-200906-20. Accessed 27 July 2010

Hudson RJM, Gherini SA, Goldstein RA (1994) Modeling the global carbon cycle nitrogen-fertilization of the terrestrial biosphere and the missing CO_2 sink. Global Biogeochem Cyc 8:307–333

Hungate B, Dukes JS, Shaw MR, Luo Y, Field CB (2003) Nitrogen and climate. Science 302: 1512–1513

IFA (International Fertilizer Industry Association) (2009) Statistics. http://www.fertilizer.org/

Ineson P, Benham DG, Poskitt J, Harrison AF, Taylor K, Woods C (1998) Effects of climate change on nitrogen dynamics in upland soils. 1. A soil warming study. Global Change Biol 4:153–161

Jagadamma S, Lal R, Hoeft RG, Nafziger ED, Adee EA (2007) Nitrogen fertilization and cropping systems effects on soil organic carbon and total nitrogen pools under chisel-plow tillage in Illinois. Soil Tillage Res 95:348–356

Janssens IA, Dieleman W, Luyssaert S, Subke JA, Reichstein M, Ceulemans R, Ciais P, Dolman AJ, Grace J, Matteucci G, Papale D, Piao SL, Schulze ED, Tang J, Law BE (2010) Reduction of forest soil respiration in response to nitrogen deposition. Nat Geosci 3:315–322

Jarvis PG, Fowler DG (2001) Forests and the atmosphere. In: Evans J (ed) The forests handbook, vol 1. Blackwell Science, Oxford, pp 229–281

Jensen B, Christensen BT (2004) Interactions between elevated CO_2 and added N: effects on water use, biomass, and soil ^{15}N uptake in wheat. Acta Agric Scand 54B:175–184

Khan SA, Mulvaney RL, Ellsworth TR, Boast CW (2007) The myth of nitrogen fertilization for soil carbon sequestration. J Environ Qual 36:1821–1832

Kitur BK, Smith MS, Blevins RL, Frye WW (1984) Fate of 15N-depleted ammonium nitrate applied to no-tillage and conventional tillage corn. Agron J 76:240–242

Knorr M, Frey SD, Curtis PS (2005a) Nitrogen additions and litter decomposition: a meta-analysis. Ecology 86:3252–3257

Knorr W, Prentice IC, House JI, Holland EA (2005b) Long-term sensitivity of soil carbon turnover to warming. Nature 433:298–301

Koch GW, Mooney HA (eds) (1996) Carbon dioxide and terrestrial ecosystems. Academic, San Diego

Kralova M, Masscheleyn PH, Lindau CW, Patrick WH (1992) Production of dinitrogen and nitrous oxide in soil suspensions as affected by redox potential. Water Air Soil Pollut 61:37–45

Lamarque J-F, Kiehl J, Brasseur G, Butler T, Cameron-Smith P, Collins WJ, Granier C, Hauglustaine D, Hess P, Holland E, Horowitz L, Lawrence M, Mckenna D, Merilees P, Prather M, Rasch P, Rotman D, Shindell D, Thornton P (2005) Assessing future nitrogen deposition and carbon cycle feedback using a multimodel approach: Analysis of nitrogen deposition. J Geophys Res 110:D19303. doi: 10.1029/2005JD005825

LeBauer DS, Treseder KK (2008) Nitrogen limitation of net primary productivity in terrestrial ecosystems is globally distributed. Ecology 89:371–379

Li C, Frolking S, Butterbach-Bahl K (2005) Carbon sequestration in arable soils is likely to increase nitrous oxide emissions, offsetting reductions in climate radiative forcing. Clim Change 72:321–338

Liebig MA, Varvel GE, Doran JW, Wienhold BJ (2002) Crop sequence and nitrogen fertilization effects on soil properties in the Western Corn Belt. Soil Sci Soc Am J 66:596–601

Luo Y, Su B, Currie WS, Dukes JS, Finzi AC, Hartwig U, Hungate BA, McMurtrie RE, Oren R, Parton WJ, Pataki DE, Shaw MR, Zak DR, Field CB (2004) Progressive nitrogen limitation of ecosystem responses to rising atmospheric carbon dioxide. Bioscience 54:731–739

Matson P, Lohse KA, Hall SJ (2002) The globalization of nitrogen deposition: consequences for terrestrial ecosystems. Ambio 31:113–119

Medlyn BE, Berbigier P, Clement R, Grelle A, Loustau D, Linder S, Wingate L, Jarvis PG, Sigurdsson BD, McMurtrie RE (2005) The carbon balance of coniferous forests growing in contrasting climatic conditions: a model-based analysis. Agric Forest Meteorol 131:97–124

Monteny G-J, Bannink A, Chadwick D (2006) Greenhouse gas abatement strategies for animal husbandry. Agric Ecosyst Environ 112:163–170

Müller C, Rütting T, Abbasi MK, Laughlin RJ, Kammann C, Clough TJ, Sherlock RR, Kattge J, Jäger H-J, Watson CJ, Stevens RJ (2009) Effect of elevated CO_2 on soil N dynamics in a temperate grassland soil. Soil Biol Biochem 41:1996–2001

Mulvaney RL, Khan SA, Ellsworth TR (2009) Synthetic nitrogen fertilizers deplete soil nitrogen: a global dilemma for sustainable cereal production. J Environ Qual 38:2295–2314

Nadelhoffer KJ, Emmett BA, Gundersen P, Kjønaas OJ, Koopmans CJ, Schleppi P, Tietema A, Wright RF (1999) Nitrogen deposition makes a minor contribution to carbon sequestration in temperate forests. Nature 398:145–148

Niggli U, Fließbach A, Hepperly P, Scialabba N (2009) Low greenhouse gas agriculture: mitigation and adaptation potential of sustainable farming systems. Rev. 2 – 2009. Food and Agriculture Organization of the United Nations, Rome, 21 pp

Olson RV (1980) Fate of tagged nitrogen fertilizer applied to irrigated corn. Soil Sci Soc Am J 44: 514–517

Paustian K, Babcock BA, Hatfield J, Lal R, McCarl BA, McLaughlin S, Mosier A, Rice C, Robertson GP, Rosenberg NJ, Rosenzweig C, Schlesinger WH, Zilberman D (2004) Agricultural mitigation of greenhouse gases: science and policy options, CAST (Council on Agricultural Science and Technology) Report, R141 2004. ISBN 1-887383-26-3, 120 pp

Persson T, Karlsson PS, Seyferth U, Sjöberg RM, Rudebeck A (2000) Carbon mineralisation in European forest soils. In: Schulze ED (ed) Carbon and nitrogen cycling in European forest ecosystems, Ecological Studies 142. Springer, Berlin, pp 257–275

Peterjohn WT, Melillo JM, Steudler PA, Newkirk KM, Bowles FP, Aber JD (1994) Responses of trace gas fluxes and N availability to experimentally elevated soil temperatures. Ecol Appl 4:617–625

Powlson D (2005) Will soil amplify climate change? Nature 433:204–205

Powlson DS, Jenkinson DS, Johnston AE, Poulton PR, Glendining MJ, Goulding KWT (2010) Comments on "Synthetic nitrogen fertilizers deplete soil nitrogen: a global dilemma for sustainable cereal production," by R.L. Mulvaney, S.A. Khan, and T.R. Ellsworth in the Journal of Environmental Quality 2009 38:2295–2314. J Environ Qual 39:749–752

Reay DS, Dentener F, Smith P, Grace J, Feely RA (2008) Global nitrogen deposition and carbon sinks. Nat Geosci 1:430–437

Reddy GB, Reddy KR (1993) Fate of nitrogen-15 enriched ammonium nitrate applied to corn. Soil Sci Soc Am J 57:111–115

Reich PB, Hobbie SE, Lee T, Ellsworth DS, West JB, Tilman D, Knops JMH, Naeem S, Trost J (2006a) Nitrogen limitation constrains sustainability of ecosystem response to CO_2. Nature 440:922–925

Reich PB, Hungate BA, Luo Y (2006b) Carbon-nitrogen interactions in terrestrial ecosystems in response to rising atmospheric carbon dioxide. Ann Rev Ecol Evol Syst 37:611–636

Riley WJ, Matson PA (2000) NLOSS: a mechanistic model of denitrified N_2O and N_2 evolution from soil. Soil Sci 165:237–249

Robertson GP (2004) Abatement of nitrous oxide, methane and other non-CO_2 greenhouse gases: the need for a systems approach. In: Field CB, Raupach MR (eds) The global carbon cycle. Integrating humans, climate, and the natural world, SCOPE 62. Island Press, Washington, pp 493–506

Rochette P, Janzen HH (2005) Towards a revised coefficient for estimating N_2O emissions from legumes. Nutr Cycl Agroecosyst 73:171–179

Rustad LE, Melillo JM, Mitchell MJ, Fernandez IJ, Steudler PA, McHale PJ (2000) Effects of soil warming on C and N cycling in northern U.S. forest soils. In: Mickler R, Birdsey R, Hom J (eds) Responses of northern U.S. forests to environmental change. Springer, Berlin, pp 357–381

Rustad LE, Campbell J, Marion GM, Norby RJ, Mitchell MJ, Hartley AE, Cornelissen JHC, Gurevitch J (2001) A meta-analysis of the response of soil respiration, net N mineralization, and aboveground plant growth in experimental ecosystem warming. Oecologia 126:543–562

Saito M, Ishii K (1987) Estimation of soil nitrogen mineralization in corn-grown fields based on mineralization parameters. Soil Sci Plant Nutr 33:555–566

Scavia D, Bricker SB (2006) Coastal eutrophication assessment in the United States. Biogeochemistry 79:187–208

Schlesinger WH (1997) Biogeochemistry, an analysis of global climate change. Academic, San Diego

Schlesinger WH (2009) On the fate of anthropogenic nitrogen. Proc Natl Acad Sci USA 106: 203–208

Seitzinger S, Harrison JA, Böhlke JK, Bouwman AF, Lowrance R, Peterson B, Tobias C, van Drecht G (2006) Denitrification across landscapes and waterscapes: a synthesis. Ecol Appl 16:2064–2090

Shaw MR, Harte J (2001) Response of nitrogen cycling to simulated climate change: differential responses along a subalpine ecotone. Global Change Biol 7:193–210

Smith KA, Conen F (2004) Impacts of land management on fluxes of trace greenhouse gases. Soil Use Manage 20:255–263

Smith P, Martino D, Cai Z, Gwary D, Janzen H, Kumar P, McCarl B, Ogle S, O'Mara F, Rice C, Scholes B, Sirotenko O (2007a) Agriculture. In: Metz B, Davidson OR, Bosch PR, Dave R, Meyer LA (eds) Climate change 2007: mitigation: contribution of Working Group III to the Fourth Assessment Report of the Intergovernmental Panel on Climate Change. Cambridge University Press, Cambridge, pp 498–540. http://www.ipcc.ch/ipccreports/ar4-wg3.htm

Smith P, Martino D, Cai Z, Gwary D, Janzen HH, Kumar P, McCarl BA, Ogle SM, O'Mara F, Rice C, Scholes RJ, Sirotenko O, Howden M, McAllister T, Pan G, Romanenkov V, Schneider UA, Towprayoon S (2007b) Policy and technological constraints to implementation of greenhouse gas mitigation options in agriculture. Agric Ecosyst Environ 118:6–28

Sokolov AP, Kicklighter DW, Melillo JM, Felzer BS, Schlosser CA, Cronin TW (2008) Consequences of considering carbon-nitrogen interactions on the feedbacks between climate and the terrestrial carbon cycle. J Clim 21:3776–3796

Stevens WB, Hoeft RG, Mulvaney RL (2005) Fate of nitrogen-15 in a long-term nitrogen rate study: II. Nitrogen uptake efficiency. Agron J 97:1046–1053

Tamm CO, Lake JV, Miller HG (1982) Nitrogen cycling in undisturbed and manipulated boreal forest. Phil Trans R Soc Lond 296(1082):419–425

Thornton PE, Lamarque J-F, Rosenbloom NA, Mahowald NM (2007) Influence of carbon-nitrogen cycle coupling on land model response to CO_2 fertilization and climate variability. Global Biogeochem Cyc 21:GB4018. doi:10.1029/2006GB002868

Thornton PE, Doney SC, Lindsay K, Moore JK, Mahowald N, Randerson JT, Fung I, Lamarque J-F, Feddema JJ, Lee Y-H (2009) Carbon-nitrogen interactions regulate climate-carbon cycle feedbacks: results from an atmosphere-ocean general circulation model. Biogeosci Discuss 6:3303–3354

Uehara G, Gillman G (1981) The mineralogy, chemistry, and physics of tropical soils with variable charge clays. Westview, Boulder, 160pp

US Geological Survey (2008) Nitrogen (fixed) – ammonia. In: Mineral commodity summaries 2008. U.S. Department of the Interior, U.S. Geological Survey, pp 118–119

van Groenigen KJ, Gorissen A, Six J, Harris D, Kuikman PJ, van Groenigen JW, van Kessel C (2005) Decomposition of ^{14}C-labeled roots in a pasture soil exposed to 10 years of elevated CO_2. Soil Biol Biochem 37:497–506

van Kessel C, Clough T, van Groenigen JW (2009) Dissolved organic nitrogen: an overlooked pathway of nitrogen loss from agricultural systems? J Environ Qual 38:393–401

Vanotti MB, Bundy LG, Peterson AE (1997) Nitrogen fertilizer and legume-cereal rotation effects on soil productivity and organic matter dynamics in Wisconsin. In: Paul EA et al (eds) Soil organic matter in temperate agroecosystems: long-term experiments in North America. CRC, Boca Raton, pp 105–119

Venterea RT, Groffman PM, Verchot LV, Magill AH, Aber JD (2004) Gross nitrogen process rates in temperate forest soils exhibiting symptoms of nitrogen saturation. For Ecol Manage 196:129–142

Verburg PSJ (2005) Soil solution and extractable soil nitrogen response to climate change in two boreal forest ecosystems. Biol Fertil Soils 41:257–261

Vitousek PM, Howarth RW (1991) Nitrogen limitation on land and in the sea – how can it occur? Biogeochemistry 13:87–115

West TO, Post WM (2002) Soil organic carbon sequestration rates by tillage and crop rotation: a global data analysis. Soil Sci Soc Am J 66:1930–1946

Xu X, Tian H, Hui D (2008) Convergence in the relationship of CO_2 and N_2O exchanges between soil and atmosphere within terrestrial ecosystems. Global Change Biol 14:1651–1660

Yan X, Ohara T, Akimoto H (2003) Development of region-specific emission factors and estimation of methane emission from rice field in East, Southeast and South Asian countries. Global Change Biol 9:237–254

Chapter 7
Soil Respiration in Future Global Change Scenarios

Bhupinder Pal Singh, Vivien de Rémy de Courcelles, and Mark A. Adams

7.1 Soil Respiration

Soil respiration is a critical ecosystem process that regulates C cycling and climate in the earth system (Luo and Zhou 2006). Here, we define soil respiration as carbon dioxide (CO_2) released from soil to the atmosphere via the combined activity of (1) roots (root respiration) and associated micro-organisms, mainly respiring the recently assimilated C by plants (rhizomicrobial respiration), and (2) micro- and macro-organisms decomposing litter and organic matter (humus) in soil, referred to as "true" heterotrophic respiration (Fig. 7.1; Sect. 7.3; Högberg et al. 2005). Soil respiration is usually monitored in situ using classical dynamic or static chamber methodology to measure CO_2 efflux at the soil surface (Luo and Zhou 2006; Ohlsson et al. 2005). After gross primary production (total C fixed by plant photosynthesis), soil respiration is the second largest C flux in most terrestrial ecosystems and may account for ~70% of total ecosystem respiration on an annual basis (Yuste et al. 2005). Globally, soil respiration emits 98 ± 12 Pg CO_2-C year^{-1} to the atmosphere (Bond-Lamberty and Thomson 2010), which is approximately an order of magnitude larger than the current annual anthropogenic CO_2-C emissions from fossil fuel combustion (Boden et al. 2010). However, despite its global significance, we have only a limited understanding of the magnitude and responses of soil respiration, and especially of its components, to abiotic (temperature, moisture, soil fertility) and biotic (photosynthesis, seasonality of belowground C allocation patterns and root growth, quality and quantity of above and belowground litter) controls. Furthermore, soil respiration is generally greater (~20%) in

B.P. Singh (✉)
NSW Department of Primary Industries, Forest Science Centre, PO Box 100, Beecroft 2119, NSW, Australia
e-mail: bp.singh@sf.nsw.gov.au

V. de Rémy de Courcelles • M.A. Adams
Faculty of Agriculture, Food and Natural Resources, The University of Sydney, Sydney, NSW 2006, Australia

B.P. Singh et al. (eds.), *Soil Health and Climate Change*, Soil Biology 29,
DOI 10.1007/978-3-642-20256-8_7,

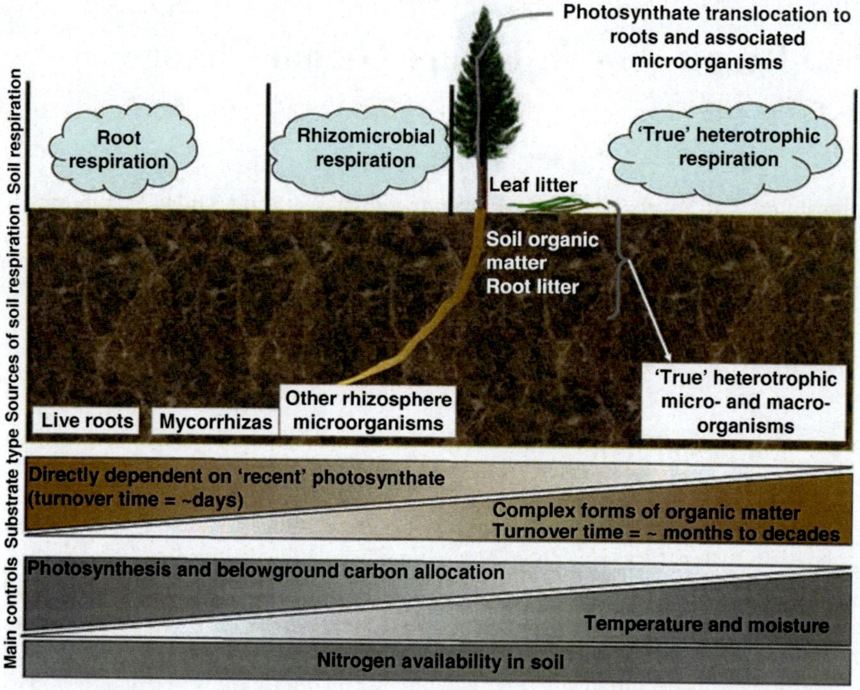

Fig. 7.1 Soil respiratory components, autotrophic (root, rhizomicrobial) and "true" heterotrophic respiration, arranged according to their dependency on substrate type (with their turnover times ranging from days for recent photosynthates to months to decades for complex forms of organic matter in soil), belowground carbon allocation supply and environmental controls (soil temperature and water availability). Nitrogen availability in soil may equally influence both autotrophic and heterotrophic components of soil respiration

grasslands than forest stands, and lower (~10%) in coniferous forests than adjacent broad-leaved forests, under similar edaphic and climatic conditions, demonstrating that vegetation type may modulate the influences of abiotic and biotic controls on soil respiration (Raich and Tufekcioglu 2000).

7.2 Soil Respiration in Relation to Climate Change and Soil Health

Globally, organic C reserves within the surface soil horizons (to the depth of 1 m) of terrestrial ecosystems are huge (1,576 Pg) and exceed the combined mass of C in the aboveground biomass (550 Pg) and the atmosphere (750 Pg) (Eswaran et al. 1993; Schlesinger and Andrews 2000). Increased storage of C in world soils could help offset further anthropogenic emissions of CO_2. In contrast, even a small change in the release of CO_2 from soil via soil respiratory processes would have profound impact on the atmospheric CO_2 (Rustad et al. 2000). Increases in root respiration may not directly affect soil C storage because root respiration is closely linked to C uptake by photosynthesis and, obviously, to the proportion of recently fixed C that is distributed to the belowground parts of plants (see Sect. 7.3 below); whereas increases in heterotrophic respiration may lead to significant reductions in soil C stocks. The rate at which C is "allocated" belowground and root exudation can also indirectly affect soil C turnover, through stimulation or the so-called priming of microbial respiration (Kuzyakov 2002; Kuzyakov and Cheng 2001). In short, any processes, climate perturbations or management practices that decrease organic C inputs to or enhance outputs from soil are likely to have adverse implications for soil health and climate change (Bond-Lamberty and Thomson 2010; see Chap. 5).

Soil respiration is one of the important measures of soil health, because it reflects the capacity of soil to support life (micro- and macro-organisms and plant roots) and is directly related to other functions, such as organic matter decomposition, nutrient mineralisation–immobilisation and microbial activity in general. Fast rates of soil respiration are indicative of intense biological activity in soil with consequences for plant growth and the environment, e.g. through increased rate of nutrient cycling in soil. Slow rates may indicate little or suppressed biological activity, which may be due to management-induced stresses and/or climate perturbations, or limitations of the resources (such as substrates, nutrients, O_2) required for such biological activity. Soil respiration usually increases following cultivation of lands as a result of increased accessibility of previously protected organic matter within soil aggregates to soil micro-organisms, and this may adversely affect soil C balance (Aslam et al. 2000; Schlesinger and Andrews 2000; Shi et al. 2010; Chaps. 5 and 9). On the other hand, land use and management practices such as conservation tillage, manure and crop residue application, and perennial and deep-rooted crops, which increase organic matter input to soil (Chap. 5), may also increase the rate of soil respiration, especially under non-limiting environmental (soil moisture and temperature) conditions. In anaerobic environments, incomplete turnover of soil organic matter may reduce CO_2 emissions, but increase emissions of non-CO_2 greenhouse gases, such as nitrous oxide and methane. Clearly, there are instances where soil respiration may not be a good indicator of changes in net greenhouse gas emissions from soil, and even overall soil health.

7.3 Sources of Soil Respiration and Their Quantification

Soil respiration is usually separated into autotrophic and heterotrophic respiration. Because root respiration and rhizomicrobial respiration are extremely difficult to measure separately, they are often grouped as autotrophic respiration (Cisneros-Dozal et al. 2006; Högberg et al. 2001; Kuzyakov 2006). Nevertheless, rhizomicrobial respiration results mainly from the activity of mycorrhizal fungi and other rhizosphere micro-organisms growing on plant-assimilated C, and should, taxonomically, be classified as heterotrophic respiration. In recognition of the difficulty of separating autotrophic and heterotrophic respiration under natural (field) conditions, Högberg et al. (2005) suggested considering an autotroph–heterotroph continuum. According to this autotroph–heterotroph continuum, autotrophic respiration includes the respiration of roots and associated micro-organisms that are directly dependent on recent plant-assimilated C for their activity (see Fig. 7.1). The "true" heterotrophic respiration includes respiration from free-living micro- and macro-organisms oxidising plant litter, and inherent soil organic matter, containing more complex organic molecules than recent plant assimilates (Cisneros-Dozal et al. 2006).

Critical elements in understanding mechanisms responsible for belowground C turnover and whether a soil will become a future CO_2 source or sink are: (1) quantification of individual soil respiratory components, and (2) assessment of their responses to different environmental and plant factors [e.g. temperature, moisture, nitrogen (N) availability, seasonality of C allocation]. Each component returns C to the atmosphere on different time scales, with rapid C cycling associated with root/rhizosphere respiration (days to months), followed by litter C decomposition (years), and then inherent (native) soil organic matter decomposition (decades to centuries) (Cisneros-Dozal et al. 2006; Kuzyakov 2006). The relatively slow turnover rate of soil organic matter makes heterotrophic respiration an important component of soil respiration with profound implications for long-term storage of organic C in soil. Quantifying and isolating heterotrophic respiration from autotrophic respiration remains an important objective for making accurate predictions of future C sequestration in terrestrial ecosystems and devising policies for appropriately managing the terrestrial C cycle in response to land use and climate changes.

Detailed reviews of the methods used to partition soil respiratory components have been presented in Hanson et al. (2000); Högberg et al. (2005) and Kuzyakov (2006). In summary, the methods used can be grouped into four broad categories: (1) integration of physically disintegrated respiratory components (root, leaf litter, soil organic matter, etc.); (2) exclusion of live roots from soil monoliths via trenching, with the assumption that severed fine roots disappear by decomposition within a short time and decomposition of coarse roots is very slow; (3) the use of non-invasive stable or radioactive isotopes that rely on different "isotopic signatures" of CO_2 derived from living roots and soil organic matter; and (4) stem girdling in forest ecosystems [involving instantaneous termination of photosynthetic C flow to roots and associated micro-organisms without affecting micro-climate, at least initially (Högberg et al. 2005)]. More recently, mycorrhizal mesh-collar chambers have been used in situ for isolation of the contributions of roots, extraradical ectomycorrhizal, and soil heterotrophic respiration to total soil respiration in a young Lodgepole pine forest (Heinemeyer et al. 2007). This technique has been only tested in a limited range of ecosystems.

Estimates of the contribution of autotrophic or heterotrophic respiration to total soil respiration vary widely depending on the type of ecosystem studied and the methods used to partition the fluxes (Hanson et al. 2000; Subke et al. 2006).

The relative contributions of soil respiratory components are also influenced by the seasonality of plant C allocation as well as by climatic controls on microbial and/or root activities (Högberg et al. 2005; Subke et al. 2006). In recent years, despite variations in biomes and ecosystems studied, the development and careful application of some of the more novel techniques, such as tree-girdling, grass clipping ± shading, trenching, and non-invasive C-isotope-based approaches have helped refine estimates (Table 7.1). For example, using a tree-girdling approach autotrophic respiration in a boreal forest was estimated to contribute between 50 and 65% of total soil respiration (Bhupinderpal-Singh et al. 2003; Högberg et al. 2001). Högberg et al. (2009) reported a maximum mean contribution of autotrophic respiration of ca. 56% for unfertilised boreal forests and ca. 44% for temperate forests. Contributions of autotrophic respiration in a subtropical eucalyptus forest were much smaller (24%) than those reported from boreal and temperate forests (Högberg et al. 2009) because of overestimation of heterotrophic respiration caused by continuous respiration of large root starch reserves for up to several months following tree girdling (Binkley et al. 2006). In the same way as tree girdling, a grass clipping ± shading approach has also reasonably separated soil respiratory components, such as root plus rhizomicrobial (30%), aboveground litter (14%) and belowground litter plus soil organic matter (56%), in a tall grass prairie (Wan and Luo 2003). Novel trenching approaches that eliminate or account for methodological artefacts associated with classical long-term trenching experiments seemingly allow robust partitioning of autotrophic and heterotrophic components of soil respiration (Diaz-Pines et al. 2010; Jassal and Black 2006; Sayer and Tanner 2010). Furthermore, non-invasive C-isotope methods involve fewer methodological artefacts than most other partitioning methods (Högberg et al. 2005; Subke et al. 2006), and have consistently reduced the variability attributed to estimates of the contributions of soil respiratory components to total soil CO_2 efflux (Paterson et al. 2009).

7.4 Regulation of Soil Respiration

7.4.1 Photosynthesis, Carbon Allocation and Substrate Supply

While girdling, clipping and shading are designed to instantaneously terminate C supply from aboveground plant tissues to roots (Bhupinderpal-Singh et al. 2003; Högberg et al. 2001; Kuzyakov and Cheng 2001; Wan and Luo 2003), pulse-labelling is used to trace the flux of recently fixed C through the plant–microbe–soil continuum (Högberg et al. 2008). Both approaches have shown a direct and rapid link between the supply of photosynthates and the autotrophic component of soil respiration. These studies clearly show that photosynthesis and patterns of C allocation are the major controls on root respiration. Globally, soil respiration is positively related to aboveground net primary production and consequently to

Table 7.1 Estimates of the contribution of autotrophic respiration to total soil respiration based on some novel partitioning methods

Biome	Vegetation type (age in years)	Approach	Ra/Rs[a]	References
Boreal	Scots pine forest (45–55)	Girdling	0.54	Högberg et al. (2001)
	Scots pine forest (45–55)	Girdling	0.65 (2nd year)	Bhupinderpal-Singh et al. (2003)
	Norway spruce forest (40)	Girdling	0.40/0.60[b]	Olsson et al. (2005)
	Norway spruce-dominated forest (120)	Girdling	0.53	Högberg et al. (2009)
	Scots pine-dominated forest (150)	^{13}C of respired CO_2 in response to seasonal variations in air relative humidity	0.65	Ekblad and Högberg (2001)
Temperate	Norway spruce forest (35)	Girdling	0.50	Subke et al. (2004)
	European chestnut (57)	Girdling	0.36	Frey et al. (2006)
	Mixed deciduous forest (~60)	Radiocarbon	0.66	Gaudinski et al. (2000)
	Mixed deciduous forest (~60)	Radiocarbon	0.71/0.80[c]	Borken et al. (2006)
	Tall grass prairie (na)	Clipping ± shading	0.30	Wan and Luo (2003)
	Montane spruce-dominated forest (130)	Trenching	0.44	Díaz-Pinés et al. (2010)
	Douglas-fir forest (55)	Trenching	0.39	Jassal and Black (2006)
	Loblolly pine forest (15)	^{13}C-depleted CO_2 (in free-air carbon dioxide enrichment plots)	0.55	Andrews et al. (1999)
Tropical	Eucalyptus plantation (2)	^{13}C (from C_4 to C_3 vegetation change)	0.68/0.73[b]	Giardina et al. (2004)
	Eucalyptus plantation (6.5)	Girdling	0.24	Binkley et al. (2006)
	Old moist lowland forest (na)	Root trenching	0.38	Sayer and Tanner (2010)
	Semi-arid perennial grasses (na)	Radiocarbon	0.74	Carbone et al. (2008)
	Semi-arid shrubs (na)	Radiocarbon	0.81	Carbone et al. (2008)
	Savannah (na)	^{13}C (from C_4 to C_3 vegetation gradients)	0.46	Millard et al. (2008)

[a]Maximal contributions of root respiration (Ra) to total soil respiration (Rs), either on day, seasonal or annual time steps, as reported by the authors are included here. *na* not available

[b]From fertilised/non-fertilised treatments, respectively

[c]From control/through fall exclusion treatments, respectively

litter production (Raich and Tufekcioglu 2000). For heterotrophic soil respiration, substrates are supplied via root exudates, leaf litter, root litter and native soil organic matter. Root exudates are mostly highly labile, whereas leaf and root litter are rather less labile and native soil organic matter is usually the least labile. Litterfall manipulation studies suggest a clear relationship between soil respiration and litterfall (Raich and Tufekcioglu 2000; Sayer et al. 2007); also, litterfall phenology helps determine patterns on an annual basis (DeForest et al. 2006). Factors that limit photosynthesis or the supply of photosynthates to roots (e.g. irradiation, water availability, soil fertility and aboveground herbivory) or the accessibility and availability to microbial enzymes of soil organic matter (e.g. aggregation, water availability, substrate quality) serve as confounding factors and may mask the effect of substrate supply on soil respiration. It is important to note, but difficult to isolate, the influence of the supply of substrate from such confounding factors. Any time lag between photosynthesis and root and root-derived respiration must also be considered when interpreting controls of autotrophic respiration (Kuzyakov and Gavrichkova 2010).

7.4.2 Temperature

In global models, temperature is widely considered the most important abiotic control of soil respiration (Cox et al. 2000; Fang and Moncrieff 2001; Xiao et al. 1998). In these models, the relationship of soil respiration to temperature is often described by Q_{10}-based or similar expressions (Davidson et al. 2006; Lloyd and Taylor 1994; Van't Hoff 1898) and applied at local, regional and even global scales. Often Q_{10} is assumed to be equal to ~2, i.e. the respiration rate doubles for every 10°C increase in temperature, and rather too rarely do authors acknowledge differences in short- and long-term responses and acclimation of soil respiration to changes in temperature (e.g. Atkin et al. 2000; Larigauderie and Korner 1995). Since Q_{10} of soil respiration is often derived in response to its seasonal variation in temperature, it is difficult to separate the temperature effect from the seasonality of C allocation to roots, as well as from other seasonal co-variables such as solar radiation and soil moisture. Some studies have suggested different seasonally derived Q_{10} values for root and heterotrophic respiration, with root respiration being twice as sensitive (Q_{10} ~4) as heterotrophic respiration (Q_{10} ~2) (Boone et al. 1998; Epron et al. 2001). However, this suggestion has been challenged because of the lack of consideration given to confounding variables (e.g. seasonality of belowground C allocation and root growth) in increasing Q_{10} of root respiration (Bhupinderpal-Singh et al. 2003; Högberg et al. 2001; Högberg 2010; Ruehr and Buchmann 2010; Schindlbacher et al. 2009). Several studies have now shown a lack of response of root and/or mycorrhizal respiration to temperature, especially during periods when plant C allocation remained more-or-less constant (Bhupinderpal-Singh et al. 2003; Heinemeyer et al. 2007; Olsson et al. 2005; Ruehr and Buchmann 2010).

Although there may be a positive correlation between soil respiration and soil and/or air temperatures (Högberg et al. 2009), temperature cannot automatically be considered the most important determinant of root respiration (Schindlbacher et al. 2009). Similarly, while temperature is a major regulator of the decomposition of litter and soil organic matter (Schindlbacher et al. 2009), the accessibility and supply of organic substrates to microbial enzymes in soil play key roles in limiting the temperature sensitivity of heterotrophic respiration (Davidson and Janssens 2006; Davidson et al. 2006; Schindlbacher et al. 2009). Soil moisture availability may also modulate the relationship between soil respiration and soil temperature. For example, the temperature sensitivity of soil respiration has been reported to decrease under very low soil moisture conditions (Almagro et al. 2009; Jassal et al. 2008).

7.4.3 Soil Moisture

The availability of moisture can limit many biological processes in soils, including respiration (Ruehr et al. 2009). The activity of both root and heterotrophic respiratory components is usually reduced when soils are either very dry or very wet. When very dry, the activity of heterotrophs can be limited by the rate of diffusion of extracellular enzymes and soluble C substrates. This can potentially lead to dormancy and/or death of micro-organisms as well as a reduction in microbial mobility (Orchard and Cook 1983). Soil fungi are generally better adapted to water stress than bacteria or microfauna in soil (Swift et al. 1979). Furthermore, the ability of deep-rooted plants to access water from deeper moist soil layers may confer significant advantage over shallow-rooted plants to maintain respiratory activity over a longer period of time under dry soil conditions (Almagro et al. 2009; Borken et al. 2006). On the other hand, gas exchange and soil oxygen concentrations are usually reduced in very wet soils, including at sites of microbial and root activity.

In field studies, seasonal covariance of temperature, photosynthetic rate, litterfall, precipitation and root and microbial activity can make interpretation of the effect of soil moisture on soil respiration highly difficult (Jassal et al. 2008). Nevertheless, across a range of ecosystems and when temperatures are moderate to high (e.g. >16°C, Almagro et al. 2009), soil moisture emerges as a main driver of soil respiration, especially in the range <20% volumetric soil moisture content (Almagro et al. 2009; Jassal et al. 2008).

7.4.4 Nitrogen Availability

Soil N availability indirectly affects soil respiration and its components by influencing plant growth and productivity, belowground C allocation, fine root growth, litter quantity and quality, microbial enzyme activity, composition of decomposer community and decomposability of recalcitrant organic matter in soil

(Janssens et al. 2010). Productive, N-rich soils usually support faster rates of plant growth and growth respiration is increased commensurately. Increased plant growth usually results in greater annual litterfall, thereby providing greater availability of substrates for heterotrophic activity. On the other hand, rates of fine root production, including mycorrhizal colonisation of roots, are often less under high-N conditions and both root and heterotrophic respiration may also be reduced (Janssens et al. 2010). The reverse trend might be expected in N-poor soil – greater investment by plants of fixed energy (photosynthate) into the growth of fine roots and mycorrhizas may well lead to increased rates of respiration.

The quality of litter and soil C, and N availability and its interaction with organic C, help regulate decomposition of organic matter in soil (Bending et al. 1998; Wang et al. 2004). There may be only weak responses of heterotrophic respiration to the application of N where organic matter with wide C-to-N (>30) and lignin-to-N ratios is present, e.g. as found in many forest ecosystems (Janssens et al. 2010; Knorr et al. 2005). In a major review, Fog (1988) hypothesised the reasons for slower decomposition of plant materials under optimal N availability, such as: (a) formation of additional recalcitrant compounds (such as polyphenol-organic N complexes), and/or (b) inhibition of appropriate growth conditions for certain fungi (white rot) with lignolytic enzyme synthesis capabilities (see Chaps. 5 and 6).

7.5 Responses of Soil Respiration to Future Global Change Scenarios

Current atmospheric CO_2 concentrations are ~110 ppm above preindustrial concentrations (280 ppm) and are predicted to increase further to >550 ppm by 2100 (Forster et al. 2007). The rising levels of CO_2 as well as those of other greenhouse gases in the Earth's troposphere are contributing to global warming and possibly to changes in the patterns of precipitation (Solomon et al. 2007). It has been estimated that global mean surface temperature will rise by 1.1–6.4°C by 2100 (Solomon et al. 2007), while the frequency of more intense rainfall events may increase by 5–10% per degree Celsius of warming, similar to the rate of increase in atmospheric water vapour content (Allison et al. 2009). Intensification of hydrological cycles seems likely to be accompanied by increased precipitation extremes, including heavy precipitation in already-wet areas and increased drought in already-dry areas (Allison et al. 2009). For example, subpolar and polar regions are likely to experience increases in precipitation, whereas an opposite trend (e.g. longer dry spells followed by heavier but shorter precipitation events) is expected in (sub)-tropical regions (Solomon et al. 2007). Concurrent with the above changes, combustion of fossil fuel and production of synthetic N fertiliser has also increased threefold to fivefold over the past century, thereby leading to greater atmospheric concentrations of gaseous forms of N and greater rates of N deposition (Janssens et al. 2010; see Chap. 6). These too are expected to increase further in many parts of

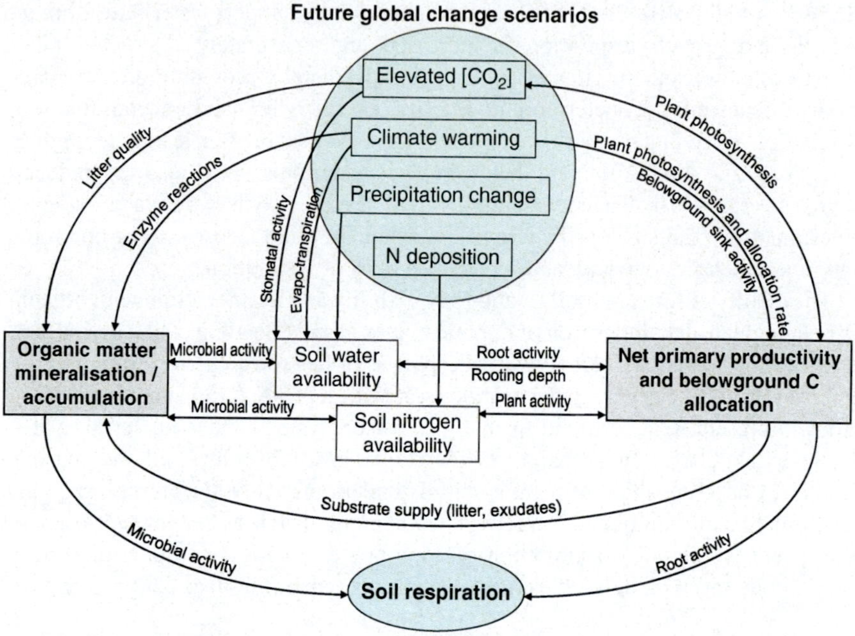

Fig. 7.2 Schematic of direct or indirect influences of future global change scenarios on soil respiration via a series of closely interrelated processes and factors affecting the activities of roots and microbes in soil

the world owing to rapid expansion in industrial and agricultural activities (Galloway et al. 2004). Such global changes have consequences for the functioning of terrestrial ecosystems, including soil respiration (Fig. 7.2), and a greater understanding of their interactive effects is required to accurately estimate uncertainties in global climate change projections and predict ecosystem feedbacks to atmospheric CO_2 levels (Rustad 2008).

7.5.1 Elevated Atmospheric [CO_2]

Elevated atmospheric [CO_2] generally increases rates of photosynthesis and consequently net primary productivity (Norby et al. 1999). The average increase in plant growth rate for a number of studies was 51% for grassy species and 42% for woody species (Zak et al. 2000). Numerous CO_2 experiments also report a general enhancement of C cycling through the belowground compartment and consequently increased rates of soil respiration (Bernhardt et al. 2006; Deng et al. 2010; Janssens et al. 1998; Trueman and Gonzalez-Meler 2005; Wan et al. 2007). Thus, although more C may be sequestered in plant biomass under elevated atmospheric [CO_2] than under ambient atmospheric [CO_2], there may also be a simultaneous and greater output of either "newly added" or "old" C through respiration (Heath et al. 2005; Trueman and Gonzalez-Meler 2005), and this may limit the potential for increased C sequestration in terrestrial ecosystems exposed to elevated atmospheric [CO_2]. Nevertheless, a meta-analysis by Jastrow et al. (2005) indicated that exposure of terrestrial ecosystems to elevated atmospheric [CO_2] could result in potential additional storage of C (~5.6%) in soils due to relatively large generation and protection of root-derived C into micro-aggregates.

Fine root biomass, the total number of roots and root length generally increase with increasing atmospheric [CO_2] (Matamala and Schlesinger 2000; Pritchard et al. 2008; Tingey et al. 2000), but specific root respiration (respiration per unit root biomass) is not affected to an extent that root and rhizosphere respiration are increased (Edwards and Norby 1999; Matamala and Schlesinger 2000). This suggests that root and rhizosphere respiration may be more closely related to photosynthetic activity and belowground C allocation under rising atmospheric [CO_2], rather than fine root growth per se.

Increased inputs of aboveground litter as well as increased root production, turnover and mortality under elevated atmospheric [CO_2] contribute to increased soil CO_2 efflux by increasing the amounts of organic substrate available for heterotrophs (Deng et al. 2010; Zak et al. 2000). Heath et al. (2005) showed that the increase in heterotrophic soil respiration with rising atmospheric [CO_2] could be attributed mainly to enhanced production and turnover of root exudates. Furthermore, Trueman and Gonzalez-Meler (2005) found that the "priming" effects of increased supply of root exudates enhanced the decomposition of "old" soil C. In contrast, litter produced by plants grown under elevated atmospheric [CO_2] may be richer in lignin and poorer in N than plants grown under ambient [CO_2] (Henry et al. 2005; Knops et al. 2007), and this may adversely affect soil heterotrophic respiration. There is some evidence that concentrations of N in foliage in deciduous forest decline over time under elevated atmospheric [CO_2] relative to ambient atmospheric [CO_2] (Norby et al. 2009). However, in other studies specific rates of litter decomposition were unchanged by CO_2 treatment (Henry et al. 2005; Knops et al. 2007). Elevated atmospheric [CO_2] may also have an indirect effect on organic matter decomposition through a shift in plant community composition. For example, there were more dicots, particularly legumes, in a grassland plot exposed to elevated atmospheric [CO_2] than in plots under ambient [CO_2]; this could enhance soil N availability and alleviate N limitations to litter decomposition (Allard et al. 2004).

In water-limited systems, CO_2-induced reductions in leaf stomatal conductance and plant transpiration may help increasing soil water content and thus soil respiration (Deng et al. 2010; Field et al. 1995; Pendall et al. 2003). The reverse may also be true in already-wet ecosystems – CO_2-induced reductions in stomatal conductance may decrease instantaneous soil respiration rates due to increased diffusional constraints on soil–atmosphere gas exchange (Bader and Körner 2010). Clearly, many of the potential CO_2 effects on soil respiration depend on phenology (e.g. greater effects in developing forest stands than mature stands), the length of exposure to elevated [CO_2] (e.g. less stimulation after long-term exposure) and soil conditions (Deng et al. 2010; Bader and Körner 2010). Tingey et al. (2006)

found that when a covariance model was used to remove the influence of temperature, soil moisture and elapsed time from planting, elevated atmospheric [CO_2] had a significant negative effect on mean soil respiration over a 3-year period of growth of ponderosa pine seedlings. An overall decline in the response of soil respiration over time may be attributed, at least in part, to declining soil N availability under conditions of elevated atmospheric [CO_2] (Bernhardt et al. 2006).

7.5.2 *Climate Warming*

Climate warming can potentially convert terrestrial ecosystems from C sinks to C sources. This prediction is supported by a range of studies which show that soil warming increases soil respiration by 20% overall, with forest ecosystems being more responsive to climate warming treatments than other ecosystems (Luo and Zhou 2006; Rustad et al. 2001). In a recent warming experiment, where soil temperature was increased by 4°C above ambient, Schindlbacher et al. (2009) noted continuous increases in soil respiration by up to ~45–47% for 2 years, with nearly similar responses of autotrophic and heterotrophic components. Similarly, in a soil-only warming experiment, Bronson et al. (2008) recorded a 24% increase in soil respiration in the first year and an 11% increase in the second, following a 5°C increase in soil temperature relative to the control treatment. Warming-induced prolongation of a plant growing season, and enhanced photosynthesis may contribute to increases in soil respiration (Niu and Wan 2008; Wan et al. 2005). Warming may also increase soil respiration by enhancing mycorrhizal fungal biomass and throughput of recent photosynthate through roots and mycorrhizal fungi (Hawkes et al. 2008). By increasing the rate of turnover of organic matter, climate warming may enhance organic N mineralisation, and this may counteract any N limitations to heterotrophic respiration induced by elevated [CO_2] (Dijkstra et al. 2010; Hovenden et al. 2008; see Sect. 7.5.1).

Several studies, however, suggest initial positive responses of soil respiration to warming may dissipate with time as a result of any of the following: acclimation of respiring organisms including roots and microbes (Atkin et al. 2000; Bradford et al. 2008; Luo et al. 2001; Melillo et al. 2002); rapid depletion of readily available soil C pool (Hartley et al. 2007); decreases in fine root biomass (Bronson et al. 2008) or microbial biomass (Allison et al. 2010); and reductions in soil moisture content (Almagro et al. 2009; Jassal et al. 2008). Microbial acclimation to climate warming can include a shift in microbial community composition with increasing abundance of temperature-tolerant and substrate-efficient fungi and possible reductions in the relative abundance or activity of bacteria (Zhang et al. 2005). Other researchers have discounted such acclimation as a major factor in reducing soil respiration in the long term following climate warming, focusing instead on the argument of warming-induced limitations of labile substrate supply (Hartley et al. 2008; Kirschbaum 2004) or reductions of microbial C-use efficiency with concomitant decreases in microbial biomass and their degradative enzymes (Allison et al. 2010).

Some warming experiments (e.g. mature *Picea abies* forests, Comstedt et al. 2006; grasslands, Luo et al. 2001; Wan et al. 2007) have shown no effects of increased soil temperature on soil respiration, while others have shown *reduced* soil respiration, by as much as 23–31%, e.g. in combined soil-plus-air warming treatments (Bronson et al. 2008). Clearly, more mechanistic studies are required to understand the effects of climate warming on soil respiration.

7.5.3 Precipitation Extremes

Experiments simulating future patterns of rainfall emphasise the importance of precipitation extremes (heavy rainfall to prolonged intervals between rainfall events) in altering the magnitude of soil respiration (Almagro et al. 2009; Borken et al. 2006; Davidson et al. 2008; Harper et al. 2005; Jarvis et al. 2007). In general, soil respiration rapidly increases following precipitation on soils previously subjected to prolonged drought, a phenomenon known as the "Birch effect", named after H.F. Birch, one of the first to describe this phenomenon (Davidson et al. 2008; Jarvis et al. 2007). The Birch effect is most commonly observed in Mediterranean and semi-(arid) ecosystems that experience long droughts followed by heavy precipitation events. Four reasons are commonly listed to explain this effect: (a) successive drying and wetting of soils disrupt soil aggregates and make the labile organic matter available for microbial decomposition; (b) dry spells kill soil micro-organisms that are subsequently decomposed when soils re-wet; (c) wetting of dry soils triggers a sudden growth and turnover of microbial biomass causing enhanced mineralisation of intercellular compounds in response to increased soil water potential; (d) microbes respond to osmotic shock during re-wetting by releasing labile C-rich solutes that accumulated in their cytoplasm during the dry periods (Almagro et al. 2009; Jarvis et al. 2007). Furthermore, displacement of CO_2 in soil pores following infiltration of rainwater may also contribute to rapid increases in soil CO_2 efflux, i.e. usually called soil respiration (Huxman et al. 2004).

Wetting of soil by heavy rainfall after a prolonged dry period may also lead to a short-term decrease in soil respiration when CO_2 diffusivity decreases due to soil saturation; however, soil respiration recovers quickly following a heavy rain event presumably due to increased heterotrophic and autotrophic activity at high soil water content (Jassal et al. 2005; Smart and Penuelas 2005; Vargas and Allen 2008; Heinemeyer et al. 2007). Heinemeyer et al. (2007) found a strong positive response of extraradical ectomycorrhizal respiration, as opposed to total soil respiration, to changes in soil moisture caused by prolonged drought and precipitation events, suggesting that autotrophic respiration may be more responsive than heterotrophic respiration to variations in soil moisture. Similarly, Vargas and Allen (2008) found that a significant fraction of the pulse of soil respiration following precipitation events can be associated with an increase in fine root and rhizomorph metabolic activity.

Rapid pulses of soil respiration accompanying re-wetting following an extended dry period may adversely impact annual net C gains in Mediterranean and other seasonally dry ecosystems (Jarvis et al. 2007; Sponseller 2007). The time taken for the stimulated soil respiration to return to pre-precipitation, basal rates may range from a few hours to 30 days (Jarvis et al. 2007; Sponseller 2007). The observation that a "re-wetting index" (amount of precipitation over time elapsed between a rainfall event and soil respiration measurement), rather than soil water content, best described CO_2 efflux during summer drought emphasises the need to consider precipitation timing and rates in any assessment of overall C balance, especially in xeric ecosystems (Almagro et al. 2009; Sponseller 2007).

However, there are also examples of precipitation having no effect on soil respiration (Muhr et al. 2008), or of short-lived CO_2 pulses producing a net decrease in soil respiration over the growing season (Harper et al. 2005). These inconsistent results have been attributed to the overriding effects of substrate availability, e.g. as observed in a cold desert ecosystem in the USA (Fernandez et al. 2006), or extreme water stress for plants or microbes, thereby limiting the response of soil respiration to precipitation events (Harper et al. 2005; Sponseller 2007).

Soil hydrophobicity can potentially increase with drought; this may delay the movement of water through the soil and thereby the response of soil respiration to precipitation (Muhr et al. 2008). Furthermore, irregular soil hydrophobicity and patchy vegetation cover in xeric ecosystems may cause uneven distribution and storage of soil moisture across the landscape, and this may increase spatial variation in soil respiration (Muhr et al. 2008; Sponseller 2007).

Extreme precipitation events, particularly prolonged drought, may also adversely affect autotrophic respiration by limiting net primary productivity (Harper et al. 2005). Furthermore, vegetation type may modulate the response of soil respiration to drought events (Almagro et al. 2009); for example, some deep-rooted plants may effect hydraulic lifting of lower soil layer water to upper soil layers, thus supporting the activity of roots and soil micro-organisms in the dry upper layers during drought (Caldwell et al. 1998).

Absolute rooting depth is strongly and positively related to mean annual precipitation in many vegetation types (Schenk and Jackson 2002). Therefore, any increase in new plant C input to deeper soil layers due to enhanced net primary productivity and absolute rooting depth under high precipitation scenarios may positively prime old soil C decomposition (Heimann and Reichstein 2008); this may have adverse implications for net C storage in soil.

7.5.4 Nitrogen Deposition

The mechanisms presented in Sect. 7.4.4 (above) are clearly applicable to N deposition considerations. Experiments simulating N deposition through N-fertiliser addition have shown highly variable responses of soil respiration, depending on ecosystem type and age, the rate of N deposition and initial N status of soil

(Ambus and Robertson 2006; Bowden et al. 2004; Ding et al. 2007; Janssens et al. 2010; Mo et al. 2008).

A recent meta-analysis suggests that N deposition or N-fertiliser addition decreases heterotrophic respiration in temperate forest ecosystems by an average of 15%, with responses ranging from +57% to −63%; the strong negative effect was observed in those forests where soil N was either not a limiting factor for microbial growth, or where the forest sites received >50 kg N ha^{-1} $year^{-1}$ (Janssens et al. 2010). The impact of N deposition on total soil respiration followed the same negative trend (averaging −10%), indicating that root respiration may be also suppressed in temperate forests, e.g. due to altered belowground plant C allocation patterns. The negative effect of N addition on soil respiration is often immediate and long lasting (Janssens et al. 2010). However, soil respiration under young (<4 years) or CO_2-fumigated temperate forest stands was positively affected by N deposition, probably through an increase in autotrophic respiration supported by rapid canopy development of young trees or sustained stimulatory effect of elevated [CO_2] on photosynthesis (Janssens et al. 2010).

Recent studies in China suggest soil respiration in young (<5 years) tropical forests, responds quickly and strongly to N deposition (>100 kg N ha^{-1} $year^{-1}$), likely due to quick growth of young seedlings, including root biomass. However, the stimulatory response declined over time and was attributed to reduced demand of plants and microbes for N and reduced N assimilation costs under sufficient N supply condition (Deng et al. 2010). In a N-saturated, mature tropical forest in China, soil respiration showed no response at low (50 kg N ha^{-1} $year^{-1}$) or medium (100 kg N ha^{-1} $year^{-1}$) rates of N addition and a 14% decline at high rates of 150 kg N ha^{-1} $year^{-1}$ (Mo et al. 2008). The reduction in soil respiration at high rates of N addition was associated with a significant reduction in fine root biomass and microbial biomass – indications of reduced autotrophic and heterotrophic soil respiration (Mo et al. 2008). High rates of N deposition may also induce soil acidification, and this may adversely affect soil heterotrophic respiration (Janssens et al. 2010).

In a 40-year-old boreal Norway spruce forest, fertiliser additions significantly decreased both autotrophic and heterotrophic soil respiration, as revealed by large-scale girdling, while aboveground production was three times higher in fertilised than in control plots (Olsson et al. 2005). Similarly, in a maize cropping system, N addition reduced soil respiration by ~10% (Ding et al. 2007), whereas N added at three different concentrations had no effect on soil respiration in successional grasslands established on abandoned farmland (Ambus and Robertson 2006). Measurements in root exclusion plots showed that fertilisation of a temperate Douglas-fir stand at 200 kg urea-N ha^{-1} resulted in a significant increase in soil respiration during the first 3–4 months due to increase in autotrophic soil respiration, and this was followed by a small decrease in heterotrophic soil respiration (Jassal et al. 2010). The variable responses of soil respiration to N deposition or additions observed in various studies highlight the need for understanding the relative importance of different processes in a wide range of climate, ecosystem-type and site-specific conditions.

7.6 Soil Respiration in a Changing World: Multi-Factor Interactions

Terrestrial ecosystems are expected to experience multiple, concurrent and interacting changes in global climate, and these global changes will directly or indirectly influence soil respiration via a series of closely interrelated processes and factors affecting the activities of roots and microbes in soil (see Fig. 7.2).

There are still very few studies that have included multi-factor analysis of soil respiration (e.g. Deng et al. 2010; Norby and Luo 2004; Wan et al. 2007; Zhou et al. 2006). The available information shows that responses of soil respiration to multiple factors can be non-interactive or interactive and usually non-additive (Luo et al. 2008; Rustad 2008). For example, interactive effects of multiple global change factors were non-significant for soil respiration between warming and increased precipitation (Zhou et al. 2006), elevated atmospheric [CO_2] and warming (Edwards and Norby 1999; Niinistö et al. 2004), elevated atmospheric [CO_2] and N addition (Butnor et al. 2003), and elevated [CO_2], air warming and water availability (Garten et al. 2009; Wan et al. 2007). On the other hand, Deng et al. (2010) found a strong interactive effect of elevated atmospheric [CO_2] and added N on soil respiration in young, subtropical forests, and the combined effect of this two-factor treatment on soil respiration was greater (increased by 50%) than either of the treatments alone (29% increase by elevated [CO_2] and 8% increase by N addition). Similarly, belowground C turnover was affected more by warming plus elevated [CO_2] than by elevated [CO_2] alone, and the interaction was strongly mediated by N supply (Loiseau and Soussana 1999).

The highly variable, non-additive and complex responses of soil respiration to combinations of global change drivers have constrained our ability to predict feedback effects on the global C cycle. This underlines the need for further multi-factor experiments in a range of ecosystems and over longer time scales than used to date (Rustad 2008).

7.7 Conclusions and Future Directions

Soil respiration, comprising root and microbial respiration, is an important indicator for assessing soil health. The large annual emissions of CO_2 through soil respiration, 98 ± 12 Pg CO_2-C year^{-1}, signify the importance of soil respiration in the global C cycle. Soil respiration in terrestrial ecosystems is regulated by multiple abiotic (environmental) and biotic (plant-related) factors, and a better understanding of their impacts is fundamental to predicting responses of soil respiration to global change. Novel partitioning approaches to quantifying the responses of soil respiratory components can further assist in improving predictions of future C sequestration in terrestrial ecosystems. The global change drivers (such as rising atmospheric CO_2 concentration, elevated temperatures, precipitation extremes, rising N deposition) variably influence soil respiration with potential to further alter the global C cycle through feedback effects. While responses of soil respiration to individual regulatory drivers have been studied widely, there is still no clear consensus as to those responses under scenarios of multiple changes in global environment. A lack of consensus and inconsistencies among studies are mainly due to variations in the type and age of ecosystem studied, the timescale of measured responses, artefacts induced by methodologies used (especially when seeking to partition soil respiratory components) or to the size of experimental units (e.g. glasshouse, whole-plant chambers, ecosystem-scale as in free-air carbon dioxide enrichment experiments), and lack of consideration of the confounding influences of other factors not included in experimental designs. There is a clear need for studies considering multi-factor approaches, together with appropriate partitioning techniques and modelling efforts, to disentangle complex responses of soil respiration to global change scenarios and assist in improving climate change projections (Luo et al. 2008; Rustad 2008).

References

Allard V, Newton PCD, Lieffering M, Soussana JF, Grieu P, Matthew C (2004) Elevated CO_2 effects on decomposition processes in a grazed grassland. Global Change Biol 10:1553–1564

Allison I, Bindoff NL, Bindschadler RA, Cox PM, de Noblet N, England MH, Francis JE, Gruber N, Haywood AM, Karoly DJ, Kaser G, Le Quéré C, Lenton TM, Mann ME, McNeil BI, Pitman AJ, Rahmstorf S, Rignot E, Schellnhuber HJ, Schneider SH, Sherwood SC, Somerville RCJ, Steffen K, Steig EJ, Visbeck M, Weaver AJ (2009) The Copenhagen diagnosis, 2009: updating the world on the latest climate science. The University of New South Wales Climate Change Research Centre (CCRC), Sydney, p 60

Allison SD, Wallenstein MD, Bradford MA (2010) Soil-carbon response to warming dependent on microbial physiology. Nat Geosci 3:336–340

Almagro M, Lopez J, Querejeta JI, Martinez-Mena M (2009) Temperature dependence of soil CO_2 efflux is strongly modulated by seasonal patterns of moisture availability in a Mediterranean ecosystem. Soil Biol Biochem 41:594–605

Ambus P, Robertson GP (2006) The effect of increased N deposition on nitrous oxide, methane and carbon dioxide fluxes from unmanaged forest and grassland communities in Michigan. Biogeochemistry 79:315–337

Andrews JA, Harrison KG, Matamala R, Schlesinger WH (1999) Separation of root respiration from total soil respiration using carbon-13 labeling during free-air carbon dioxide enrichment (FACE). Soil Sci Soc Am J 63:1429–1435

Aslam T, Choudhary MA, Saggar S (2000) Influence of land-use management on CO_2 emissions from a silt loam soil in New Zealand. Agric Ecosyst Environ 77:257–262

Atkin OK, Edwards EJ, Loveys BR (2000) Response of root respiration to changes in temperature and its relevance to global warming. New Phytol 147:141–154

Bader MK-F, Körner C (2010) No overall stimulation of soil respiration under mature deciduous forest trees after 7 years of CO_2 enrichment. Global Change Biol 16:2830–2843

Bending GD, Turner MK, Burns IG (1998) Fate of nitrogen from crop residues as affected by biochemical quality and the microbial biomass. Soil Biol Biochem 30:2055–2065

Bernhardt ES, Barber JJ, Pippen JS, Taneva L, Andrews JA, Schlesinger WH (2006) Long-term effects of free air CO_2 enrichment (FACE) on soil respiration. Biogeochemistry 77: 91–116

Bhupinderpal-Singh, Nordgren A, Ottosson-Löfvenius M, Högberg MN, Mellander P-E, Högberg P (2003) Tree root and soil heterotrophic respiration as revealed by girdling of boreal Scots pine forest: extending observations beyond the first year. Plant Cell Environ 26:1287–1296

Binkley D, Stape JL, Takahashi EN, Ryan MG (2006) Tree-girdling to separate root and heterotrophic respiration in two Eucalyptus stands in Brazil. Oecologia 148:447–454

Boden TA, Marland G, Andres RJ (2010) Global, regional, and national fossil-fuel CO_2 emissions. Carbon Dioxide Information Analysis Center, Oak Ridge National Laboratory, U.S. Department of Energy, Oak Ridge, TN, USA. doi 10.3334/CDIAC/00001_V2010

Bond-Lamberty B, Thomson A (2010) Temperature-associated increases in the global soil respiration record. Nature 464:579–582

Boone RD, Nadelhoffer KJ, Canary JD, Kaye JP (1998) Roots exert a strong influence on the temperature sensitivity of soil respiration. Nature 396:570–572

Borken W, Savage K, Davidson EA, Trumbore SE (2006) Effects of experimental drought on soil respiration and radiocarbon efflux from a temperate forest soil. Global Change Biol 12:177–193

Bowden RD, Davidson E, Savage K, Arabia C, Steudler P (2004) Chronic nitrogen additions reduce total soil respiration and microbial respiration in temperate forest soils at the Harvard Forest. For Ecol Manage 196:43–56

Bradford MA, Davies CA, Frey SD, Maddox TR, Melillo JM, Mohan JE, Reynolds JF, Treseder KK, Wallenstein MD (2008) Thermal adaptation of soil microbial respiration to elevated temperature. Ecol Lett 11:1316–1327

Bronson DR, Gower ST, Tanner M, Linder S, Van Herk I (2008) Response of soil surface CO_2 flux in a boreal forest to ecosystem warming. Global Change Biol 14:856–867

Butnor JR, Johnsen KH, Oren R, Katul GG (2003) Reduction of forest floor respiration by fertilization on both carbon dioxide-enriched and reference 17-year-old loblolly pine stands. Global Change Biol 9:849–861

Caldwell MM, Dawson TE, Richards JH (1998) Hydraulic lift: consequences of water efflux from the roots of plants. Oecologia 113:151–161

Carbone MS, Winston GC, Trumbore SE (2008) Soil respiration in perennial grass and shrub ecosystems: linking environmental controls with plant and microbial sources on seasonal and diel timescales. J Geophys Res 113:G02022. doi:10.1029/2007JG000611

Cisneros-Dozal LM, Trumbore S, Hanson PJ (2006) Partitioning sources of soil-respired CO_2 and their seasonal variation using a unique radiocarbon tracer. Global Change Biol 12:194–204

Comstedt D, Bostrom B, Marshall JD, Holm A, Slaney M, Linder S, Ekblad A (2006) Effects of elevated atmospheric carbon dioxide and temperature on soil respiration in a boreal forest using delta C-13 as a labeling tool. Ecosystems 9:1266–1277

Cox PM, Betts RA, Jones CD, Spall SA, Totterdell IJ (2000) Acceleration of global warming due to carbon-cycle feedbacks in a coupled climate model. Nature 408:184–187

Davidson EA, Janssens IA (2006) Temperature sensitivity of soil carbon decomposition and feedbacks to climate change. Nature 440:165–173

Davidson EA, Janssens IA, Luo YQ (2006) On the variability of respiration in terrestrial ecosystems: moving beyond Q_{10}. Global Change Biol 12:154–164

Davidson EA, Nepstad DC, Ishida FY, Brando PM (2008) Effects of an experimental drought and recovery on soil emissions of carbon dioxide, methane, nitrous oxide, and nitric oxide in a moist tropical forest. Global Change Biol 14:2582–2590

DeForest JL, Noormets A, McNulty SG, Sun G, Tenney G, Chen JQ (2006) Phenophases alter the soil respiration-temperature relationship in an oak-dominated forest. Int J Biometeorol 51:135–144

Deng Q, Zhou G, Liu J, Liu S, Duan H, Zhang D (2010) Responses of soil respiration to elevated carbon dioxide and nitrogen addition in young subtropical forest ecosystems in China. Biogeosciences 7:315–328

Diaz-Pines E, Schindlbacher A, Pfeffer M, Jandl R, Zechmeister-Boltenstern S, Rubio A (2010) Root trenching: a useful tool to estimate autotrophic soil respiration? A case study in an Austrian mountain forest. Eur J For Res 129:101–109

Dijkstra FA, Blumenthal D, Morgan JA, Pendall E, Carrillo Y, Follett RF (2010) Contrasting effects of elevated CO_2 and warming on nitrogen cycling in a semiarid grassland. New Phytol 187:426–437

Ding WX, Cai Y, Cai ZC, Yagi K, Zheng XH (2007) Soil respiration under maize crops: effects of water, temperature, and nitrogen fertilization. Soil Sci Soc Am J 71:944–951

Edwards NT, Norby RJ (1999) Below-ground respiratory responses of sugar maple and red maple saplings to atmospheric CO_2 enrichment and elevated air temperature. Plant Soil 206:85–97

Ekblad A, Högberg P (2001) Natural abundance of ^{13}C in CO_2 respired from forest soils reveals speed of link between tree photosynthesis and root respiration. Oecologia 127:305–308

Epron D, Le Dantec V, Dufrene E, Granier A (2001) Seasonal dynamics of soil carbon dioxide efflux and simulated rhizosphere respiration in a beech forest. Tree Physiol 21:145–152

Eswaran H, Van Den Berg E, Reich P (1993) Organic carbon in soils of the world. Soil Sci Soc Am J 57:192–194

Fang C, Moncrieff JB (2001) The dependence of soil CO_2 efflux on temperature. Soil Biol Biochem 33:155–165

Fernandez DP, Neff JC, Belnap J, Reynolds RL (2006) Soil respiration in the cold desert environment of the Colorado Plateau (USA): abiotic regulators and thresholds. Biogeochemistry 78:247–265

Field CB, Jackson RB, Mooney HA (1995) Stomatal responses to increased CO_2 – implications from the plant to the global-scale. Plant Cell Environ 18:1214–1225

Fog K (1988) The effect of added nitrogen on the rate of decomposition of organic-matter. Biol Rev Cambr Philos Soc 63:433–462

Forster P, Ramaswamy V, Artaxo P, Berntsen T, Betts R, Fahey DW, Haywood J, Lean J, Lowe DC, Myhre G, Nganga J, Prinn R, Raga G, Schulz M, Van Dorland R (2007) Changes in atmospheric constituents and in radiative forcing. In: Solomon S, Qin D, Manning M, Chen Z, Marquis M, Averyt KB, Tignor M, Miller HL (eds) Climate change 2007: the physical science basis contribution of Working Group I to the Fourth Assessment Report of the Intergovernmental Panel on Climate Change. Cambridge University Press, Cambridge

Frey B, Hagedorn F, Giudici F (2006) Effect of girdling on soil respiration and root composition in a sweet chestnut forest. For Ecol Manage 225:271–277

Galloway JN, Dentener FJ, Capone DG, Boyer EW, Howarth RW, Seitzinger SP, Asner GP, Cleveland CC, Green PA, Holland EA, Karl DM, Michaels AF, Porter JH, Townsend AR, Vorosmarty CJ (2004) Nitrogen cycles: past, present, and future. Biogeochemistry 70: 153–226

Garten CT, Classen AT, Norby RJ (2009) Soil moisture surpasses elevated CO_2 and temperature as a control on soil carbon dynamics in a multi-factor climate change experiment. Plant Soil 319:85–94

Gaudinski JB, Trumbore SE, Davidson EA, Zheng SH (2000) Soil carbon cycling in a temperate forest: radiocarbon-based estimates of residence times, sequestration rates and partitioning of fluxes. Biogeochemistry 51:33–69

Giardina CP, Binkley D, Ryan MG, Fownes JH, Senock RS (2004) Belowground carbon cycling in a humid tropical forest decreases with fertilization. Oecologia 139:545–550

Hanson PJ, Edwards NT, Garten CT, Andrews JA (2000) Separating root and soil microbial contributions to soil respiration: a review of methods and observations. Biogeochemistry 48:115–146

Harper CW, Blair JM, Fay PA, Knapp AK, Carlisle JD (2005) Increased rainfall variability and reduced rainfall amount decreases soil CO_2 flux in a grassland ecosystem. Global Change Biol 11:322–334

Hartley IP, Heinemeyer A, Ineson P (2007) Effects of three years of soil warming and shading on the rate of soil respiration: substrate availability and not thermal acclimation mediates observed response. Global Change Biol 13:1761–1770

Hartley IP, Hopkins DW, Garnett MH, Sommerkorn M, Wookey PA (2008) Soil microbial respiration in Arctic soil does not acclimate to temperature. Ecol Lett 11:1092–1100

Hawkes CV, Hartley IP, Ineson P, Fitter AH (2008) Soil temperature affects carbon allocation within arbuscular mycorrhizal networks and carbon transport from plant to fungus. Global Change Biol 14:1181–1190

Heath J, Ayres E, Possell M, Bardgett RD, Black HIJ, Grant H, Ineson P, Kerstiens G (2005) Rising atmospheric CO_2 reduces sequestration of root-derived soil carbon. Science 309: 1711–1713

Heimann M, Reichstein M (2008) Terrestrial ecosystem carbon dynamics and climate feedbacks. Nature 451:289–292

Heinemeyer A, Hartley IP, Evans SP, Carreira De La Fuente JA, Ineson P (2007) Forest soil CO_2 flux: uncovering the contribution and environmental responses of ectomycorrhizas. Global Change Biol 13:1786–1797

Henry HAL, Cleland EE, Field CB, Vitousek PM (2005) Interactive effects of elevated CO_2, N deposition and climate change on plant litter quality in a California annual grassland. Oecologia 142:465–473

Högberg P (2010) Is tree root respiration more sensitive than heterotrophic respiration to changes in soil temperature? New Phytol 188(1):9–10. doi:10.1111/j.1469-8137.2010.03366.x

Högberg P, Nordgren A, Buchmann N, Taylor AFS, Ekblad A, Högberg MN, Nyberg G, Ottosson-Lofvenius M, Read DJ (2001) Large-scale forest girdling shows that current photosynthesis drives soil respiration. Nature 411:789–792

Högberg P, Nordgren A, Högberg MN, Ottosson-Löfvenius M, Bhupinderpal-Singh, Olsson P, Linder S (2005) Fractional contributions by autotrophic and heterotrophic respiration to soil-surface CO_2 efflux in boreal forests. In: Griffiths H, Jarvis PG (eds) The carbon balance of forest biomes. Taylor & Francis, Oxford, pp 251–267

Högberg P, Högberg MN, Gottlicher SG, Betson NR, Keel SG, Metcalfe DB, Campbell C, Schindlbacher A, Hurry V, Lundmark T, Linder S, Nasholm T (2008) High temporal resolution tracing of photosynthate carbon from the tree canopy to forest soil microorganisms. New Phytol 177:220–228

Högberg P, Bhupinderpal-Singh, Löfvenius MO, Nordgren A (2009) Partitioning of soil respiration into its autotrophic and heterotrophic components by means of tree-girdling in old boreal spruce forest. For Ecol Manage 257:1764–1767

Hovenden MJ, Newton PCD, Carran RA, Theobald P, Wills KE, Vander Schoor JK, Williams AL, Osanai Y (2008) Warming prevents the elevated CO_2-induced reduction in available soil nitrogen in a temperate, perennial grassland. Global Change Biol 14:1018–1024

Huxman TE, Snyder KA, Tissue D, Leffler AJ, Ogle K, Pockman WT, Sandquist DR, Potts DL, Schwinning S (2004) Precipitation pulses and carbon fluxes in semiarid and arid ecosystems. Oecologia 141:254–268

Janssens IA, Crookshanks M, Taylor G, Ceulemans R (1998) Elevated atmospheric CO_2 increases fine root production, respiration, rhizosphere respiration and soil CO_2 efflux in Scots pine seedlings. Global Change Biol 4:871–878

Janssens IA, Dieleman W, Luyssaert S, Subke JA, Reichstein M, Ceulemans R, Ciais P, Dolman AJ, Grace J, Matteucci G, Papale D, Piao SL, Schulze ED, Tang J, Law BE (2010) Reduction of forest soil respiration in response to nitrogen deposition. Nat Geosci 3:315–322

Jarvis P, Rey A, Petsikos C, Wingate L, Rayment M, Pereira J, Banza J, David J, Miglietta F, Borghetti M, Manca G, Valentini R (2007) Drying and wetting of Mediterranean soils stimulates decomposition and carbon dioxide emission: the "Birch effect". Tree Physiol 27:929–940

Jassal RS, Black TA (2006) Estimating heterotrophic and autotrophic soil respiration using small-area trenched plot technique: theory and practice. Agric For Meteorol 140:193–202

Jassal RS, Black TA, Novak MD, Morgenstern K, Nesic Z, Gaumont-Guay D (2005) Relationship between soil CO_2 concentrations and forest-floor CO_2 effluxes. Agric For Meteorol 130: 176–192

Jassal RS, Black TA, Novak MD, Gaumont-Guay D, Nesic Z (2008) Effect of soil water stress on soil respiration and its temperature sensitivity in an 18-year-old temperate Douglas-fir stand. Global Change Biol 14:1305–1318

Jassal RS, Black TA, Trofymow JA, Roy R, Nesic Z (2010) Soil CO_2 and N_2O flux dynamics in a nitrogen-fertilized Pacific Northwest Douglas-fir stand. Geoderma 157:118–125, doi:10.1016/j.geoderma.2010.04.002

Jastrow JD, Michael Miller R, Matamala R, Norby RJ, Boutton TW, Rice CW, Owensby CE (2005) Elevated atmospheric carbon dioxide increases soil carbon. Global Change Biol 11:2057–2064

Kirschbaum MUF (2004) Soil respiration under prolonged soil warming: are rate reductions caused by acclimation or substrate loss? Global Change Biol 10:1870–1877

Knops JMH, Naeemw S, Reich PB (2007) The impact of elevated CO_2, increased nitrogen availability and biodiversity on plant tissue quality and decomposition. Global Change Biol 13:1960–1971

Knorr M, Frey SD, Curtis PS (2005) Nitrogen additions and litter decomposition: a meta-analysis. Ecology 86:3252–3257

Kuzyakov Y (2002) Review: factors affecting rhizosphere priming effects. J Plant Nutr Soil Sci Z Pflanzenernahr Bodenkd 165:382–396

Kuzyakov Y (2006) Sources of CO_2 efflux from soil and review of partitioning methods. Soil Biol Biochem 38:425–448

Kuzyakov Y, Cheng W (2001) Photosynthesis controls of rhizosphere respiration and organic matter decomposition. Soil Biol Biochem 33:1915–1925

Kuzyakov Y, Gavrichkova O (2010) Time lag between photosynthesis and carbon dioxide efflux from soil: a review of mechanisms and controls. Global Change Biol 16(12):3386–3406

Larigauderie A, Korner C (1995) Acclimation of leaf dark respiration to temperature in alpine and lowland plant-species. Ann Bot 76:245–252

Lloyd J, Taylor JA (1994) On the temperature dependence of soil respiration. Funct Ecol 8:315–323

Loiseau P, Soussana JF (1999) Elevated [CO_2], temperature increase and N supply effects on the turnover of below-ground carbon in a temperate grassland ecosystem. Plant Soil 210:233–247

Luo Y, Zhou X (2006) Soil respiration and the environment. Elsevier, London

Luo YQ, Wan SQ, Hui DF, Wallace LL (2001) Acclimatization of soil respiration to warming in a tall grass prairie. Nature 413:622–625

Luo Y, Gerten D, Le Maire G, Parton WJ, Keough C, Weng E, Beier C, Ciais P, Cramer W, Dukes JS, Hanson PJ, Knapp AAK, Linder S, Nepstad D, Rustad L, Sowerby A (2008) Modelled effects of precipitation on ecosystem carbon and water dynamics in different climatic zones. Global Change Biol 14:2365–2379

Matamala R, Schlesinger WH (2000) Effects of elevated atmospheric CO_2 on fine root production and activity in an intact temperate forest ecosystem. Global Change Biol 6:967–979

Melillo JM, Steudler PA, Aber JD, Newkirk K, Lux H, Bowles FP, Catricala C, Magill A, Ahrens T, Morrisseau S (2002) Soil warming and carbon-cycle feedbacks to the climate system. Science 298:2173–2176

Millard P, Midwood AJ, Hunt JE, Whitehead D, Boutton TW (2008) Partitioning soil surface CO_2 efflux into autotrophic and heterotrophic components, using natural gradients in soil delta C-13 in an undisturbed Savannah soil. Soil Biol Biochem 40:1575–1582

Mo J, Zhang W, Zhu W, Gundersen P, Fang Y, Li D, Wang H (2008) Nitrogen addition reduces soil respiration in a mature tropical forest in southern China. Global Change Biol 14:403–412

Muhr J, Goldberg SD, Borken W, Gebauer G (2008) Repeated drying-rewetting cycles and their effects on the emission of CO_2, N_2O, NO, and CH_4 in a forest soil. Wiley, New York

Niinistö SM, Silvola J, Kellomaki S (2004) Soil CO_2 efflux in a boreal pine forest under atmospheric CO_2 enrichment and air warming. Global Change Biol 10:1363–1376

Niu SL, Wan SQ (2008) Warming changes plant competitive hierarchy in a temperate steppe in northern China. J Plant Ecol 1:103–110

Norby RJ, Luo YQ (2004) Evaluating ecosystem responses to rising atmospheric CO_2 and global warming in a multi-factor world. New Phytol 162:281–293

Norby RJ, Wullschleger SD, Gunderson CA, Johnson DW, Ceulemans R (1999) Tree responses to rising CO_2 in field experiments: implications for the future forest. Plant Cell Environ 22:683–714

Norby R, Warren J, Iversen C, Garten C, Medlyn B, McMurtrie R (2009) CO_2 enhancement of forest productivity constrained by limited nitrogen availability. Available via Nature Precedings. http://hdl.handle.net/10101/npre.2009.3747.1. Cited 24 June 2010

Ohlsson KEA, Bhupinderpal-Singh, Holm S, Nordgren A, Lövdahl L, Högberg P (2005) Uncertainties in static closed chamber measurements of the carbon isotopic ratio of soil-respired CO_2. Soil Biol Biochem 37:2273-2276

Olsson P, Linder S, Giesler R, Högberg P (2005) Fertilization of boreal forest reduces both autotrophic and heterotrophic soil respiration. Global Change Biol 11:1745–1753

Orchard VA, Cook FJ (1983) Relationship between soil respiration and soil moisture. Soil Biol Biochem 15:447–453

Paterson E, Midwood AJ, Millard P (2009) Through the eye of the needle: a review of isotope approaches to quantify microbial processes mediating soil carbon balance. New Phytol 184:19–33

Pendall E, Del Grosso S, King JY, LeCain DR, Milchunas DG, Morgan JA, Mosier AR, Ojima DS, Parton WA, Tans PP, White JWC (2003) Elevated atmospheric CO_2 effects and soil water feedbacks on soil respiration components in a Colorado grassland. Global Biogeochem Cycle 17:13

Pritchard SG, Strand AE, McCormack ML, Davis MA, Finzi AC, Jackson RB, Matamala R, Rogers HH, Oren R (2008) Fine root dynamics in a loblolly pine forest are influenced free-air-CO_2-enrichment: a six-year-minirhizotron study. Global Change Biol 14:588–602

Raich JW, Tufekcioglu A (2000) Vegetation and soil respiration: correlations and controls. Biogeochemistry 48:71–90

Ruehr NK, Buchmann N (2010) Soil respiration fluxes in a temperate mixed forest: seasonality and temperature sensitivities differ among microbial and root-rhizosphere respiration. Tree Physiol 30:165–176

Ruehr NK, Offermann CA, Gessler A, Winkler JB, Ferrio JP, Buchmann N, Barnard RL (2009) Drought effects on allocation of recent carbon: from beech leaves to soil CO_2 efflux. New Phytol 184:950–961

Rustad LE (2008) The response of terrestrial ecosystems to global climate change: towards an integrated approach. Sci Total Environ 404:222–235

Rustad LE, Huntington TG, Boone RD (2000) Controls on soil respiration: implications for climate change. Biogeochemistry 48:1–6

Rustad LE, Campbell JL, Marion GM, Norby RJ, Mitchell MJ, Hartley AE, Cornelissen JHC, Gurevitch J (2001) A meta-analysis of the response of soil respiration, net nitrogen mineralization, and aboveground plant growth to experimental ecosystem warming. Oecologia 126:543–562

Sayer EJ, Tanner EVJ (2010) A new approach to trenching experiments for measuring root-rhizosphere respiration in a lowland tropical forest. Soil Biol Biochem 42:347–352

Sayer EJ, Powers JS, Tanner EVJ (2007) Increased litterfall in tropical forests boosts the transfer of soil CO_2 to the atmosphere. PLoS ONE 2:e1299. doi:10.1371/journal.pone.0001299

Schenk HJ, Jackson RB (2002) Rooting depths, lateral root spreads and below-ground/above-ground allometries of plants in water-limited ecosystems. J Ecol 90:480–494

Schindlbacher A, Zechmeister-Boltenstern S, Jandl R (2009) Carbon losses due to soil warming: do autotrophic and heterotrophic soil respiration respond equally? Global Change Biol 15:901–913

Schlesinger WH, Andrews JA (2000) Soil respiration and the global carbon cycle. Biogeochemistry 48:7–20

Shi XM, Li XG, Long RJ, Singh BP, Li ZT, Li FM (2010) Dynamics of soil organic carbon and nitrogen associated with physically separated fractions in a grassland-cultivation sequence in the Qinghai-Tibetan plateau. Biol Fertil Soils 46:103–111

Smart DR, Penuelas J (2005) Short-term CO_2 emissions from planted soil subject to elevated CO_2 and simulated precipitation. Appl Soil Ecol 28:247–257

Solomon S, Qin D, Manning M, Alley RB, Berntsen T, Bindoff NL, Chen Z, Chidthaisong A, Gregory JM, Hegerl GC, Heimann M, Hewitson B, Hoskins BJ, Joos F, Jouzel J, Kattsov V, Lohmann U, Matsuno T, Molina M, Nicholls N, Overpeck J, Raga G, Ramaswamy V, Ren J, Rusticucci M, Somerville R, Stocker TF, Whetton P, Wood RA, Wratt D (2007) 2007: Technical summary. In: Solomon S, Qin D, Manning M, Chen Z, Marquis M, Averyt KB, Tignor M, Miller HL (eds) Climate change 2007: the physical science basis contribution of Working Group I to the Fourth Assessment Report of the Intergovernmental Panel on Climate Change. Cambridge University Press, Cambridge

Sponseller RA (2007) Precipitation pulses and soil CO_2 flux in a Sonoran Desert ecosystem. Global Change Biol 13:426–436

Subke JA, Hahn V, Battipaglia G, Linder S, Buchmann N, Cotrufo MF (2004) Feedback interactions between needle litter decomposition and rhizosphere activity. Oecologia 139:551–559

Subke JA, Inglima I, Cotrufo MF (2006) Trends and methodological impacts in soil CO_2 efflux partitioning: a metaanalytical review. Global Change Biol 12:921–943

Swift MJ, Heal OW, Anderson JM (1979) Decomposition in terrestrial ecosystem. Blackwell Scientific, Oxford

Tingey DT, Phillips DL, Johnson MG (2000) Elevated CO_2 and conifer roots: effects on growth, life span and turnover. New Phytol 147:87–103

Tingey DT, Johnson MG, Lee EH, Wise C, Waschmann R, Olszyk DM, Watrud LS, Donegan KK (2006) Effects of elevated CO_2 and O_3 on soil respiration under ponderosa pine. Soil Biol Biochem 38:1764–1778

Trueman RJ, Gonzalez-Meler MA (2005) Accelerated belowground C cycling in a managed agriforest ecosystem exposed to elevated carbon dioxide concentrations. Global Change Biol 11:1258–1271

Van't Hoff JH (1898) Lectures on theoretical and physical chemistry. Part 1. Chemical dynamics. Edward Arnold, London

Vargas R, Allen MF (2008) Environmental controls and the influence of vegetation type, fine roots and rhizomorphs on diel and seasonal variation in soil respiration. New Phytol 179:460–471

Wan SQ, Luo YQ (2003) Substrate regulation of soil respiration in a tallgrass prairie: results of a clipping and shading experiment. Global Biogeochem Cycle 17:12

Wan SQ, Hui DF, Wallace L, Luo YQ (2005) Direct and indirect effects of experimental warming on ecosystem carbon processes in a tallgrass prairie. Global Biogeochem Cycle 19:13

Wan S, Norby RJ, Ledford J, Weltzin JF (2007) Responses of soil respiration to elevated CO_2, air warming, and changing soil water availability in a model old-field grassland. Global Change Biol 13:2411–2424

Wang WJ, Baldock JA, Dalal RC, Moody PW (2004) Decomposition dynamics of plant materials in relation to nitrogen availability and biochemistry determined by NMR and wet-chemical analysis. Soil Biol Biochem 36:2045–2058

Xiao X, Melillo JM, Kicklighter DW, McGuire AD, Prinn RG, Wang C, Stone PH, Sokolov A (1998) Transient climate change and net ecosystem production of the terrestrial biosphere. Global Biogeochem Cycle 12:345–360

Yuste JC, Nagy M, Janssens IA, Carrara A, Ceulemans R (2005) Soil respiration in a mixed temperate forest and its contribution to total ecosystem respiration. Tree Physiol 25:609–619

Zak DR, Pregitzer KS, King JS, Holmes WE (2000) Elevated atmospheric CO_2, fine roots and the response of soil microorganisms: a review and hypothesis. New Phytol 147:201–222

Zhang W, Parker KM, Luo Y, Wan S, Wallace LL, Hu S (2005) Soil microbial responses to experimental warming and clipping in a tallgrass prairie. Global Change Biol 11:266–277

Zhou XH, Sherry RA, An Y, Wallace LL, Luo YQ (2006) Main and interactive effects of warming, clipping, and doubled precipitation on soil CO_2 efflux in a grassland ecosystem. Global Biogeochem Cycle 20:12

Chapter 8
Soil Biota, Soil Health and Global Change

P.M. Mele

8.1 Soil Health and Soil Biota

The concept of soil health is intuitively simple because of the obvious connections with human health. Scotland's Macaulay Institute have taken up this anthropomorphic approach in their soil health knowledge and outreach programme titled "dirt doctors" in which they ascribe human health and behavioural traits to the major soil types in Scotland (www.macaulay.ac.uk/news/dirtdoctors). The term "soil health" has, however, created some challenges when applied in contexts other than for communication and knowledge transfer. From a technical perspective, soil health is difficult to measure and therefore hard to manage; from a political perspective, it is hard to ascribe a value to, particularly with respect to public good benefits. From an economic viewpoint, it is ascribed significance only when plant production systems decline or fail. The "soil health as a useful concept" debate will therefore continue to generate new or slightly modified definitions reflecting the objectives of the particular interest group. For example, the policy context defines soil health as "A goal for land owners and managers that embraces an accountable system of soil management for maintaining and improving soil functions and productivity for a range of purposes across generations. Healthy soil is a soil that is managed as an ecosystem on a 'fit-for-purpose' basis and within its capabilities" (*A policy framework for investment by the Department of Primary Industries, Victoria, Australia June 2007*). A popular and enduring technical definition is provided by Doran et al (1996) as "the capacity of a living soil to function, within natural or managed ecosystem boundaries, to sustain plant and animal productivity, maintain or enhance water and air quality, and promote plant and animal health". A more contemporary definition embraces a broader ecosystems context; a soil that is capable of supporting the production of food and fibre, to a level and with a quality

P.M. Mele (✉)
Biosciences Research Division, Department of Primary Industries Victoria, 1 Park Drive, Bundoora, VIC 3083, Australia
e-mail: pauline.mele@dpi.vic.gov.au

B.P. Singh et al. (eds.), *Soil Health and Climate Change*, Soil Biology 29,
DOI 10.1007/978-3-642-20256-8_8, © Springer-Verlag Berlin Heidelberg 2011

sufficient to meet human requirements, together with continued delivery of other ecosystem services that are essential for maintenance of the quality of life for humans and the conservation of biodiversity (Falkowski et al. 2008; Kibblewhite et al. 2008; also see Chap. 1 and Preface of the book).

Collectively, the challenges associated with the soil health concept reflect the reality; soils *are* a complex ecosystem and the "health status" of that ecosystem depends not only on the prevailing farming system (e.g. a healthy soil for viticulture may be different from a healthy soil for grain production) but also on many interacting physical, chemical and biological factors operating on temporal and spatial scales that may lack relevance for human needs and activities. Understanding the complexity of functions and interactions within the soil ecosystem therefore is not without challenges. Soils are, afterall, *are* the most diverse ecosystem on the planet (Torsvik et al. 1990; Torsvik and Øvreås 2002; Venter et al. 2004), and although the knowledge of these systems is growing exponentially (see Sects. 8.3 and 8.4), the community composition of the soil biota, particularly the micro-organisms, remains largely unclassified (Chen et al. 2008; Keller and Zengler 2004), with the underpinning ecological concepts, rudimentary and sometimes misattributed (Bardgett et al. 2005).

Notwithstanding these challenges, the requirement for greater acknowledgement and understanding of the soil as an ecosystem that provides essential services for society is, however, critical. The already widespread degradation of soils (EU 2002) and the growing population pressures for food will require an estimated 63% increase in average cereal yields by 2050, placing even more pressure on land already under agriculture (Lal 2009). Global change will add to this pressure thereby further highlighting the need for more robust knowledge to underpin both political and landholder risk management strategies for soil-based food production industries.

The soil biota is linked to the soil health concept through its role in the mediation of processes that provide agricultural goods (e.g. nutrients and disease control for food and fibre production) and ecosystem services (water quality and supply, erosion control, atmospheric composition and climate regulation, pollutant attenuation and degradation, non-agricultural pest and disease control and biodiversity conservation) (Kibblewhite et al. 2008). It is therefore imperative when addressing issues of global change impacts on soil health to consider the trilogy of interrelated properties: the physical, chemical *and* the biological (see Chaps. 1 and 2).

8.2 Soil Biota: Multiple Roles and Responses to Under Global Change

Soil biota has many and varied roles in all global change. It has a direct causal role, it can also contribute to mitigation and can be changed (either adversely of beneficially) by global change. Both the causal and mitigation roles relate to the capacity of microbes such as bacteria, archaea and fungi to produce *and* consume all the greenhouse gases, NO, N_2O, CH_4 and CO_2, respectively (Fig. 8.1). In other words, these gases are both starting substrates and by-products of energy-generating pathways

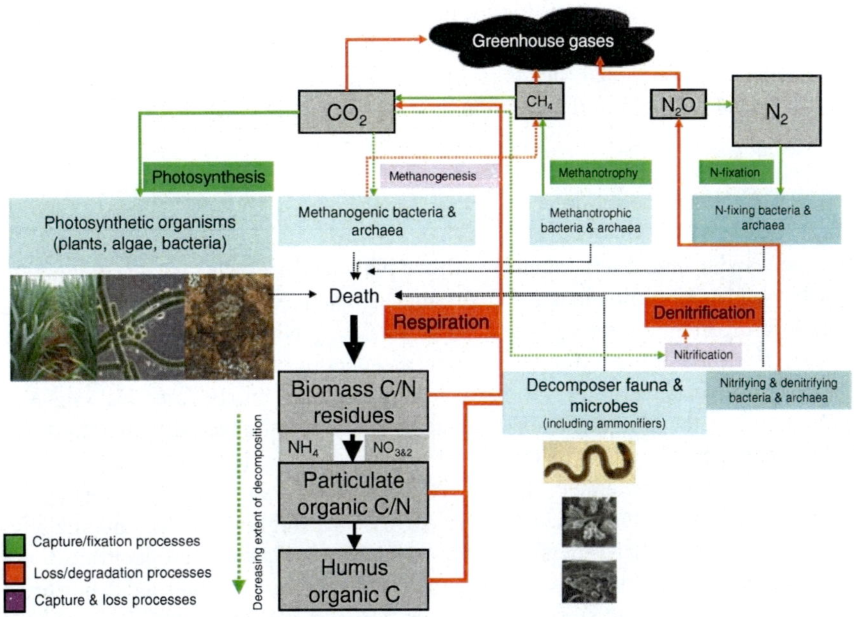

Fig. 8.1 The microbial carbon and nitrogen cycles illustrating the multiple microbial capture and loss pathways associated with the generation of greenhouse gases

necessary for microbial growth and integral to fundamental biogeochemical cycles and a whole range of ecosystem goods and services (Falkowski et al. 2008; Kibblewhite et al. 2008). These processes are cyclic, dynamic and adaptive and are regulated by temperature (ambient and soil), moisture availability, soil and plant management. The soil biota is likely to be affected directly (e.g. physiological stress and adaptation responses) and indirectly (e.g. through habitat modification) by global change scenarios. These scenarios include elevated temperature (projected increase from 1990 of 1.4–5.8°C by 2100), elevated atmospheric CO_2 (projected increase from 368 ppm in 2000, to between 540 and 970 ppm by 2100), elevated atmospheric N (projected increase of N_2O from 316 ppb in 2000) and fluctuating CH_4 concentrations (e.g. a rapid rise from about 700 ppb in 1,750 to about 1,775 ppb in 2005, followed by a projected decline), and precipitation changes by an average of 20% on current levels (Solomon et al. 2007). Soil biota can also provide solutions for mitigation and adaptation to global change. An example of this is where sensitive species can be used to predict impending biogeochemical changes or where new species can displace existing species to reduce loss of greenhouse gases to the atmosphere. The following sections will provide further detail of some of the key roles that soil biota play in global change.

8.2.1 Soil Microbiota and Greenhouse Gas Production

Anthropogenic disturbance of the biogeochemical cycles is perhaps today's greatest environmental challenge, and C and N cycling are probably most profoundly affected. These cycles, while highly interdependent, differ in the microbes that mediate the underlying processes and specifically those involved in greenhouse gas emissions. The functional and taxonomic differentiation of these microbes will inform greenhouse gas mitigation options.

8.2.1.1 N-Gas Production

Nitrogen is a key nutrient for plant growth and plays a critical role in plant community structure and composition in many environments. There are nine forms of nitrogen in soil, differing in their degree of oxidation (Paul 2007). Nitrous oxide (N_2O) and to a lesser extent, nitric oxide (NO) are important greenhouse gases that are generated from plant and microbial (organic) N forms which are converted into N-gases, N_2O and N_2, through NO as the obligatory intermediate. It has been estimated that two-thirds of N_2O originate from soil and that of the 12 sources identified in the USA, agricultural soil management generates almost 70% of emissions (http://www.epa.gov/nitrousoxide/sources.html; cited 18 Dec 2010). Two highly coupled, microbially mediated processes that generate N_2O and NO are nitrification and denitrification. Nitrous oxide is emitted from soil during the processes of nitrification and denitrification. Nitrification is the microbial oxidation of NH_4^+ (to NH_3^+) to NO to NO_2^- to NO_3^- and of NH_4^+ (to NH_3^+) to gaseous forms, NO and N_2O. Although nitrification represents a key step in the conversion of ammonia nitrogen into its gaseous forms, this process is less relevant to N_2O emissions. Furthermore, a commonly held view is that nitrifier numbers are typically low in soils. Denitrification is the reduction of NO_3^- to NO, N_2O, and N_2 in anaerobic soil conditions (water filled pore space >60%). The identification of the microbial species associated with nitrification and denitrification is ongoing and being rapidly facilitated by microbial genome sequencing technologies (Cavicchioli et al. 2007; Green et al. 2010).

Nitrification is the most complex of the two processes being carried out most energy efficiently by autotrophic nitrifiers that derive C-energy from CO_2 or carbonate and less efficiently by heterotrophic nitrifiers that derive C-energy from organic matter. The autotrophic nitrifiers are divided into two discrete taxonomic groups, the ammonia oxidisers (to NH_3^+ to NO_2^-) and the nitrite oxidisers (NO_2^- to NO_3^-). The ammonia oxidisers have been classified as bacteria belonging exclusively to the Betaproteobacteria and specifically to genera, *Nitrosomonas*, *Nitrosolobus*, *Nitrospira* and *Nitrosovibrio* (Paul 2007; Prosser 1986). The nitrite oxidisers are more broadly distributed in the Proteobacteria with *Nitrobacter* in the Alphaproteobacteria, *Nitrospina* and *Nitrospira* in Deltaproteobacteria and *Nitrosococcus* in the Gammaproteobacteria. The fungi are considered to be the most numerous and efficient heterotrophic nitrifiers and include the common soil fungi *Fusarium*, *Aspergillus* and *Penicillium* (Lin et al. 2008; Prosser 1986; Zhang et al. 2002); however, it is known that the production of N_2O in heterotrophic nitrification is not energy generating and thought to be a result of the release of oxidases and peroxidases during cell lysis and lignin degradation. More recently, a whole third domain discovered in the late 1970's, the Archaea, has been implicated in nitrification. While largely taxonomically unresolved, the ammonia oxidisers are described as archaeal ammonia oxidisers (AOA) and the nitrite oxidisers as Nitrospira-like and Nitrotoga-like nitrite oxidisers. Of significance is the assertion that these archaeal nitrifiers are dominant in many natural and agricultural soil systems (Cavicchioli et al. 2007; Wagner 2009).

Denitrification by contrast is relatively straightforward and the denitrifiers are the most important producers of the major greenhouse gases, and are also the most physiologically diverse encompassing the organotrophs, chemo- and photolithotrophs, N_2-fixers, thermophiles, halophiles and various pathogens. These facultatively anaerobic bacteria are represented by more than 60 genera of bacteria of culturable species, particularly *Pseudomonas* and *Algaligenes*, *Bacillus*, *Agribacterium* and *Flavibacterium* (Paul 2007) and unculturable phyla Acidobacteria and Thermomicrobia (Morales et al. 2010).

Edaphic factors including soil water content, temperature, pH, salinity, plant species and availability of organic carbon influence the rate of N-gas generation. The literature contains many examples demonstrating the degree to which these factors influence actual and potential nitrification and denitrification either by direct measurement of gas efflux (Chen et al. 2010; Davidson et al. 2008; Falloon et al. 2009; Gu et al. 2009; Ma et al. 2010; Reth et al. 2005) or by the measurement of the abundance of specific enzymes associated with each process using quantitative-PCR (Q-PCR)-based approaches (Colloff et al. 2008; Liu et al. 2010; Morales et al. 2010; Okano et al. 2004; Rotthauwe et al. 1997; Wakelin et al. 2007). Knowledge of how these edaphic factors selectively regulate the nitrifier and denitrifier communities, particularly those that use NO_3^- as the electron acceptor, will provide soil management strategies for reducing N loss via denitrification (Faulwetter et al. 2009; Siciliano et al. 2009; Zhang et al. 2009). Modification of soil moisture and oxygen content and C:N input ratios (quality and quantity) by altering soil structure and residue input regimes respectively is likely to have a significant impact on the magnitude and scale of biological N-loss. Currently, however, N_2O efflux reduction strategies focus mainly on the use of nitrification inhibitors, an old strategy for reducing N-fertiliser loss (McGuinn 1924). In the contemporary context of global change, there has been a renaissance in the types of inhibitors and modes of application. These are either nitrification or urease inhibitors, the former reducing the availability of NO_3^- for denitrification, and the latter slowing the hydrolysis of urea to ammonium, resulting in less availability of substrate for nitrification (Malla et al. 2005; see Chap. 10). The efficacy and environmental impacts of nitrification inhibitors are highly dependent on soil texture, water and temperature with largely benign effects on physical– chemical and biological soil properties reported (Cuttle 2008; Edmeades 2004; Granli and Bockman 1994; O'Callaghan et al. 2010).

8.2.1.2 C-Gas Production

Carbon, like nitrogen, is a key nutrient and has many forms differing in their degree of oxidation. The global soil organic carbon pool is estimated to be ~1,395 $\times$ 10^{15} g

(Post et al. 1982), and accumulation is controlled principally by net primary productivity while losses are mainly a function of the biological stability of the various chemical forms. These forms are lignin, cellulose, tannins, starches, sugars and the gases, carbon dioxide (CO_2) and methane (CH_4). The two principal C-loss pathways are through microbial respiration (see Chap. 7) and methanogenesis.

The degradation of organic C in soil and the generation of CO_2 in the process of aerobic respiration are carried out by the heterotrophic microbial community with the assistance of the micro-, meso- and macrofauna. It is widely estimated that the vast majority of the microbial community have a predominantly heterotrophic existence, and the soil biota is collectively regarded as the "waste management crew" in that they balance intake of CO_2 by photosynthetic processes and release it back into the environment by respiration processes (Paul 2007). The age and quality of the organic matter can influence the rate of release of CO_2 back into the atmosphere, with the particulate organic matter being the most readily decompose and the humus being more slowly decomposed (Hernández and Hobbie 2010; Skjemstad et al. 2004). The addition of labile sources can stimulate the decomposition of more recalcitrant C via a priming effect on the microbial community (Blagodatskaya et al. 2007; Blagodatsky et al. 2010; Kuzyakov 2010). Furthermore, the activity of specific groups of soil biota such as enchytraeids (Annelida: Oligochaeta) can also facilitate release of CO_2 from older C pools (Briones et al. 2010).

The generation of CH_4 in the relatively more complex microbial process of methanogenesis is also a respiration process that uses CO_2 and other forms of C instead of oxygen. Methane is roughly 25 times more effective as a greenhouse gas than carbon dioxide (IPCC 2007) and thus even at lower atmospheric concentrations contributes about half the radiative climate forcing of CO_2 (Beerling et al. 2009). Large quantities of CH_4 are annually produced (methanogens) and consumed (methanotrophs) by interdependent microbial communities ("consortia") living in the soil and ocean sediment. Methane can be produced by both acetotrophic and hydrogenotrophic methanogens, and differences in environmental conditions affect the composition of these communities (Demirel and Scherer 2008). Cultured methanogens are grouped into the orders Methanobacteriales, Methanococcales, Methanomicrobiales, Methanosarcinales (Whitman et al. 2001) and uncultured methanogens are obligate archaeal anaerobes that may represent new orders with complex interactions with other anaerobes and the physical and chemical environment (Liu and Whitman 2008). Like methanogens, methanotrophs have many uncultivated members that are either aerobic or more most recently, anaerobic CH_4 oxidising archaea (ANME) (Alperin and Hoehler 2009). The aerobic methanotrophs belong to two groups; type I (Methylococcaceae) or type II (Methylocystaceae) with molecular studies indicating that many are still uncultivated (Chen et al. 2009). Peatlands, considered to be highly significant carbon reservoirs, have active methanogenesis and methanotrophy with a dominance of the Methylocystis-related species of methanotrophs (Chen et al. 2008). The degree to which methanogenesis is coupled with methanotrophy ultimately determines how much CH_4 is released into the atmosphere. The extent to which each group contributes to CH_4 emissions and the regulatory role of soil physical and chemical conditions is of major interest (Andert et al. 2009).

8.2.2 Soil Biota and Greenhouse Gas Capture

Just as there are biological processes responsible for the generation of greenhouse gases, so are there biological processes that capture or sequester greenhouse gases, or their precursors, thereby countering emissions. The preceding section provides a good example of how methanotrophs use the CH_4 generated by the methanogens and convert it into a more environmentally benign C-form (see Chap. 10). Biological processes that capture or sequester greenhouse gases are therefore of great relevance. For example, soil biota can theoretically capture soil C through processes such as deep residue burial (e.g. dung beetles and earthworms), microbial photosynthesis (light carbon fixation), microbial chemosynthesis (dark carbon fixation), microbial methanotrophy (see Sect. 8.2.1.2) and humification (the diversion of litter into humus through microbial and chemical reactions) (Albrecht et al. 2010; Novis et al. 2007; Prescott 2010; Vargas-García et al. 2010). The extent to which these processes occur in various soil ecosystems and the assemblages of soil microbes that mediate these specific C-capture processes requires a systems approach and a highly integrated research effort.

8.2.3 Novel Applications for Soil Microbes in Global Change Mitigation Strategies

An emergent role for microbes is their potential as sensitive indicators of adverse effects of mitigation strategies. For example, geosequestration, the process of deep ocean subsurface deposition of CO_2, is an important strategy for reducing atmospheric CO_2. The environmental impact of CO_2 leakage from these deep reservoirs into near-surface terrestrial environments is largely unknown. A CO_2GeoNet project team (http://www.co2geonet.com, cited 18 Dec 2010) is using the abundance and diversity of "environmentally important" bacteria living in deep ocean sediment to provide sensitive indicators for the leakage of CO_2 from deep reservoirs into near-surface terrestrial ecosystems (Krüger et al. 2009). This study suggests that the different CO_2 gradients select for distinctive and measurable microbial communities that shift towards anaerobic and acidophilic species under elevated CO_2 concentrations. It is proposed that these communities can provide a sensitive and reliable fingerprint of the extent of leakage. Another study by Cunningham and co-workers (2009) is using engineered microbial biofilms which are capable of precipitating crystalline calcium carbonate to form a plug that may result in the long-term sealing of preferential leakage pathways.

8.2.4 The Impact of Global Change on Soil Biota

The impact of global change and specifically elevated temperature and carbon dioxide on soil processes has largely focused on the measurement and predictive modelling of greenhouse gas emission trends over space and time. A better mechanistic understanding of many global change processes and the opportunities for adaptation and mitigation could be obtained from a greater focus on microbial physiology and metabolism, the factors controlling microbial assemblages, and of their spatial and temporal distribution (Scow and Louise 1997). A limited number of multidimensional, multinational programmes have been established to map trends in greenhouse gas evolution (e.g. NitroEurope; http://www.nitroeurope.eu and the Austrian research network, MICDIF; http://www.nfn.oberwalder.info/NFN). These programmes incorporate the measurement of microbial community diversity and function as important components of understanding greenhouse gas efflux patterns that are linked to global change scenarios.

There are two investigative approaches for assessing the likely impacts of global change on soil microbial diversity. The first is the use of free air carbon dioxide enrichment experiments (usually abbreviated as FACE) which are field-based facilities of defined area containing a representative plant ecosystem that is exposed to ambient or elevated CO_2 with or without elevated ambient temperature. There are numerous examples of FACE experiments (e.g. BioCON, University of Minnesota; http://www.biocon.umn.edu, TasFACE; University of Tasmania; http://www.utas.edu.au). An alternative approach is to focus on ecosystems that are particularly vulnerable such as the high altitude forests (N-enrichment threats) and Polar Regions. The microbial communities in both these regions can be viewed as sentinels and as amplifiers of global change factors and polar regions in particular are regarded as ideal model ecosystems for exploring microbial changes associated with elevated temperature and CO_2 (Vincent 2010). Other reasons cited for the focus on polar regions are that soils are poorly developed with a simplified soil food-web structure, and that significant warming has already occurred with greatest global warming rates being recorded here (Rinnan et al. 2009; Robinson 2009). There are many more studies that examine the effects of temperature and CO_2 separately, and for this reason they are treated separately below.

8.2.4.1 Elevated Temperature

Global surface temperatures have increased 0.76°C from the average calculated from 1850 to 1899, to the current average calculated from 2001 to 2005. Eleven of the last 12 years have been the warmest experienced since instrumental records began in 1850 (Houghton et al. 2007). Several studies have demonstrated the impacts of elevated temperature (eT) on gross soil biological processes such as respiration and decomposition with a clear trend for enhanced rates of decomposition with increasing temperature in medium- to long-term studies. For example, organic matter decomposition rates in Maritime Antarctic terrestrial ecosystems (Antarctic Peninsula regions of Anchorage and Signy Islands) were contrasted with those at the cool temperate Falkland Islands. Higher temperatures applied in field studies led to small increases in organic matter breakdown in both ecosystems with effects more strongly influenced by local substratum characteristics (especially soil N availability) and plant functional type than by large-scale temperature differences. It was concluded that substantial warming (4–8°C) is required before significant effects can be detected (Robinson 2009). Studies have also indicated that warming effects are influenced by the plant rhizosphere and season. For example, an increase in temperature 3°C above ambient did not effect root rhizosphere respiration as much as bulk soil respiration. Furthermore, there was a significant stimulation of belowground respiration during the coldest part of the year when plant productivity was low. These findings have clear consequences for the modelling of belowground respiration and emphasise the need for separating root and bulk soil respiration fluxes and for accounting for seasonal effects (see Chap. 7). Most importantly, they suggest that the warming-induced increase in decomposition will reduce C storage in the soils of temperate ecosystems (Hartley et al. 2007; Kirschbaum 1995; Zhang et al. 2005).

A limited number of studies have examined the impact of eT on soil biological communities. For soil fauna, global warming and changes in the quality and quantity of litter are the main factors that are expected to increase mesofaunal biomass and diversity and alter life history features (Coûteaux and Bolger 2000). Microbial communities are more adaptable to eT and responses can vary depending on the historical exposure of the community to a range of soil temperatures (Waldrop and Firestone 2006). In an Antarctic climate gradient experiment, where eT was applied in open top chambers, bacterial communities adapted to the mean annual temperature of their environment. It was also estimated that the predicted increase of 2.6°C for the Antarctic Peninsula would increase the minimum temperature for bacterial growth by 0.6–1°C. Interestingly, the temperature sensitivity of bacterial communities increased with mean annual soil temperatures suggesting that communities from colder regions were less temperature sensitive than those from warmer regions (Rinnan et al. 2009). These adaptive responses have also been demonstrated in laboratory-based experiments for both fungal and bacterial communities over a wide range of temperatures (5–50°C) and particularly above the optimum for microbial growth (about 30°C). Increasing soil temperature above this optimum resulted in an increase in the optimum for bacterial and fungal growth with minor effects on growth below this optimum, although lower temperatures selected for communities growing better at the lowest temperature (Barcenas-Moreno et al. 2009). This adaptive response to eT has been correlated with changes in community structure with distinct microbial heterotrophic communities observed at different temperatures (Andrews et al. 2000) and declines in soil fungi and Gram-negative bacteria relative to Gram-positive bacteria (Feng and Simpson 2009). These changes in community structure have also been associated with decreases in microbial biomass, enzyme activities and microbial respiration (Waldrop and Firestone 2006). Feng and Simpson (2009), however, cautioned about the confounding effects of substrate utilisation in elevated temperature response studies concluding that faster exhaustion of available nutrient sources at higher temperatures rather than temperature per se primarily regulated microbial biomass, and that SOM quality intimately controls microbial responses to global warming.

It is clear that soil fauna and microbes respond directly and to varying degrees to eT, but other factors such as substrate availability and quality also play a role in this response (see Chap. 5). The thermal acclimation potential of soil communities in various global ecosystems needs to be resolved using high resolution approaches to accurately predict the likely outcome of prolonged temperature increases. Furthermore, a focus on eT impacts on key soil functional guilds associated with the provision of ecosystem goods and services and specifically greenhouse gas loss and capture mechanisms is also required. Technologies that resolve microbial community structure and function will provide unprecedented insight into eT in global ecosystems (see Sect. 8.4).

8.2.4.2 Elevated CO_2

The effects of elevated CO_2 (eCO_2) are mainly considered to be beneficial for plants, and hence by association, for soil biota too, since plant ecosystems respond to rising CO_2 concentrations with increased photosynthesis, growth and resource allocation. This in turn is thought to affect all soil dwelling species as a result of quantitative and qualitative alterations in C and other nutrients, and in water availability (Coûteaux and Bolger 2000; Drigo et al. 2007; Hu et al. 1999; Kanerva et al. 2008; Moscatelli et al. 2005; Yuan et al. 2006). Given the complexities and highly interactive associations in soil, the response to eCO_2, as with eT, is likely to be variable and unlike eT, mainly indirect.

Comprehensive reviews have highlighted the major impacts of eCO_2 on soil fauna (Coûteaux and Bolger 2000) and microbial communities (Waldrop and Firestone 2006; Zak et al. 2000). In contrast to the effects of eT, the effects of eCO_2 are largely indirect and influenced by litter quality and quantity, seasonality of substrate supply and global warming (Coûteaux and Bolger 2000). Zak and colleagues (2000) summarised data from 47 published reports on soil microbial C and N cycling under eCO_2 in an attempt to generalise whether process rates increase, decrease or remain unchanged. This synthesis focussed on changes in soil respiration, microbial respiration, microbial biomass, gross N mineralisation, microbial immobilisation and net N mineralisation, representing important control points for the belowground flow of C and N. With few exceptions, and as with eT, the rates of soil and microbial respiration were more rapid under eCO_2, indicating that (1) greater plant growth under eCO_2 enhanced the amount of C entering the soil, and (2) additional substrate was being metabolised by soil micro-organisms (see Chap. 7). In contrast, microbial biomass, gross N mineralisation, microbial immobilisation and net N mineralisation under eCO_2 were highly variable, and it was concluded that there were insufficient data to predict how microbial activity and rates of soil C and N cycling will change as the atmospheric CO_2 concentration continues to rise (Zak et al. 2000). There has, however, been significant research activity in the last 20 years and not surprisingly the whole spectrum of soil biological responses to eCO_2 (e.g. stimulatory, inhibitory, no effect, transient and disparate) have been reported and mainly from experiments conducted at FACE facilities globally. For example, inhibitory responses to eCO_2 have been recorded for methanotrophy from a coniferous forest free-air CO_2 enrichment (FACE) site with net CH_4 consumption (soil methanotrophic activity), being 47% lower despite similar moisture, temperature, NO_3^- and NH_4^+ contents. It was concluded that a reduction in the strength of the CH_4 soil sink may ultimately affect the atmospheric CH_4 budget and, consequently, future climates (Phillips et al. 2001). Other studies have demonstrated little or no changes in microbial activity in response to eCO_2. This has been reported for N_2 fixation (Billings and Ziegler 2005), for rhizosphere populations of bacteria, NH_4^+-oxidising bacteria and *Rhizobium leguminosarum bv. trifolii* (Drigo et al. 2007; Schortemeyer et al. 1996), for microbial biomass nitrogen, rates of nitrogen mineralisation and nitrification, bacterial substrate utilisation and extracellular enzyme activities (Sinsabaugh et al. 2003), for total PLFA biomass, profiles and specific subgroups for 16S rRNA gene clone libraries (Kanerva et al. 2008), extracellular enzyme activity and potential soil N mineralisation and nitrification (Austin et al. 2009), and for arbuscular mycorrhizal fungi (AMF) measurements (hyphal length, glomalin-related soil protein) (Clark et al. 2009). Many more studies have demonstrated either variable or transient shifts in soil microbial communities. The wide range of responses is commonly attributed to the confounding influence of the plant response to eCO_2 in terms of compositional change in litter, in seasonality and in competition with the microbial community for nutrients. For example, microbial enzyme activities associated with N or P mineralisation only increased when plant nutrient limitation was strongly exerted under eCO_2 (Kang et al. 2005) and soil microbial population shifted to a more fungal dominated community with reduced plant residue N quality (Drigo et al. 2007; Kandeler et al. 2008). Seasonality related to plant biomass quantity and fertilisation were important overriding factors in determining changes in the nematode abundance and diversity (Li et al. 2009). Furthermore, increased competition between the plant and the microbial community for nutrients and particularly nitrogen has also been demonstrated under eCO_2 (Freeman et al. 1998; Insam et al. 1999).

The impacts of eCO_2 are also complicated by the multiple interactions within microbial communities. For example, a study of a grassland microcosm soil by Barnard and co-workers demonstrated how soil nitrifying enzyme activity had a tendency to increase while denitrifying enzyme activity decreased under eCO_2, but a possible large immobilisation of N by soil micro-organisms concurrently may have reduced soil nitrate concentrations under eCO_2, leading potentially to decreased nitrate leaching and denitrification (Barnard et al. 2005). Free-living bacterial-feeding protozoa increased or decreased under eCO_2 depending on whether mycorrhiza was present or absent (Rønn et al. 2002). A similar disparate response to eCO_2 was observed for rhizosphere bacterial and fungal communities depending on the presence of mycorrhiza (Drigo et al. 2007, 2009). Other examples of highly interactive effects of eCO_2 are related to water availability and faunal lifecycles (Coûteaux and Bolger 2000), soil drying and the capacity of mycorrhizal fungi and plant growth promoting bacteria to contribute to aggregate stability (Kohler et al. 2009), and soil type on rhizosphere nematode and mycorrhizal fungi (Drigo et al. 2007).

8.3 Specific Challenges for Resolving the Role of Soil Biota in Global Change

The soil biota remains the most elusive and challenging component of soil health in terms of precise measurement of who is present and active and meaningful interpretation of global change impacts. Several challenges will be highlighted related to: (1) community structure (who is there) to function (what they are doing), (2) microhabitat diversity and scale and (3) applying ecological rules.

8.3.1 Soil Communities: Structure and Function

Describing the soil biota according to a range of parameters such as size, abundance, and taxonomic and functional guilds is the first step in elucidating the relative impact of physicochemical factors related to climate, geomorphic soil class, major land-use and soil management. The cataloguing of assemblages according to their taxonomy is a work-in-progress with estimates of only 1% of bacterial and viral, 5% of fungal, 12% of collembolan, 15% of mite and 50% of earthworm species having been described (Gatson and Spicer 2004; Hawkesworth 2001; Lavelle 1996; Williamson et al. 2005). The capacity to identify single species (or genomes) and entire microbial communities (or metagenomes) through DNA-based sequencing technologies is increasing knowledge of taxonomy of soil biota at an unprecedented rate (Sect. 8.4). But is taxonomy enough? Of all the technical definitions of soil health, biological function rather than taxonomy is highlighted. There are several levels of functional classification of soil biota from the low resolution or food-web-based categories of ecosystem engineers, litter transformers, phytophages and parasites, micro-predators and microflora (Lavelle 1996) to categories based on gene abundance linked to enzyme targets and to phylogenetically linked gene expression (Tringe and Hugenholtz 2008). While the food-web-based functional categories take into account the potential top–down regulatory controls of larger organisms (e.g. the ecosystem engineers) over smaller ones, a higher resolution classification is required to more precisely attribute soil health functions and the organisms involved in agroecosystem goods and services (Kibblewhite et al. 2008b; Van Der Heijden et al. 2008). The linkage of soil biological community structure to function and then to ecosystem goods and services is the ultimate goal. Not only will this assist in creating a societal value for specific functions associated with soil health, it will also enable more targeted management of these soil functions to improve soil health to achieve multiple outcomes in a global change context.

8.3.2 Microhabitat Diversity and Scale

The extreme heterogeneity of the soil microhabitat arises because of inherent geomorphological and hydrogeological soil features, as well as the non-uniform effect of external conditions (e.g. climate and nutrient inputs). One consequence of a large diversity of microhabitats is that the predicted impact of a disturbance (e.g. eCO_2 and eT) on a given species may not be the same even over a small area (Bardgett et al. 2005). This can be problematic when applying the "sentinel or indicator" species concept because the relative sensitivities to say, eT for example, may differ over a small spatial scale due to the large habitat diversity, and the identified sensitive species may therefore not be present in a closely located soil ecosystem (Bünemann et al. 2006). Examples of important mechanistic questions like "how will global change factors such as elevated CO_2 and temperature influence soil N and C turnover?" will require more targeted examination at smaller scales where intracellular and intercellular interactions occur. Clearly, soil biological processes occur at a much lower scale (micrometre to metre), and the compositing of samples across an area can mask effects (Manter et al. 2010). Important agronomic issues that are likely to be impacted by global change such as disease spread and suppression and nitrogen mineralisation and fixation and are mediated by specific rhizosphere biological processes will also require a much finer scale of investigation (from micrometre to metre) (Watt et al. 2006). Another important consideration relates to the way some microbes transcend scale in their exploitation of soil niches. A good example is where the prey, a fungus, extends hyphal networks across a larger scale (metres) than their predators (springtails and mites), which may extend only a few centimetres (Wall et al. 2005). This suggests that in some cases, several scales must be considered simultaneously to explain soil N and C turnover.

The impact of global change on these functions must therefore apply sampling strategies and statistical approaches that take into account heterogeneity of microhabitat (Baker et al. 2009) and interpolates across scale to account for the principal that the sum of soil functions at the micrometre scale, controlled by spatial organisation (of soil), affects metre scale function (Herrmann et al. 2007).

8.3.3 Applying Ecological Rules

Soil is a complex ecosystem and the soil biological community performs multiple, highly interactive functions that collectively reflect soil health status. These are nutrient cycling, pesticide degradation, disease control and plant growth promotion. Soil ecosystems behave differently to aboveground or plant-based ecosystems, so that the rules that are generally applied to ecosystem processes do not always hold when applied belowground (Wall et al. 2005). These "conceptual anomalies" must be considered in monitoring strategies and in the interpretation of data to assess the impacts of human activities and global change. One feature of belowground communities is "functional interdependency". This concept was highlighted in a review co-chaired by Handelsman et al. (2007), who described the ability of microbes to eat rock, to breathe metals, transform the inorganic to the organic and crack the toughest chemicals as a microbial bucket brigade. "Each microbe performs its own task, and its end product becomes the starting fuel for its neighbor". This concept is more specifically illustrated as syntrophy (or literally, "feeding together"), which is where two or more species with contrasting nutritional characteristics require each other to function (Paul 2007). In syntrophy, one species may produce a product or environment that is essential for a second species to function. In most cases, the nature of syntrophic reactions involves H_2 being produced by one partner and consumed by another. This "interspecies H transfer" allows anaerobic microbes to grow in aerobic environments. There are many examples of syntrophic systems including the nitrification–denitrification coupling (see Sect. 8.2.1.1) and the degradation of ethanol to CH_4 and CO_2 (*Desulfovibrio* spp. and a methanogen, *Methanobacterium* spp.) (Fenchel and Finlay 1995).

Another concept worthy of note in the context of global change is functional redundancy, where one function can be carried out by a range of different microorganisms. It is generally used to explain the different levels of stability in soil ecosystems. The underlying principle is that microbial communities with functionally redundant species may display similar levels of indicator enzyme activities across a range of environmental conditions (Marschner et al. 2003; Perez-Piqueres et al. 2006). Loss or reduction in the population size of a particular species following normal disturbance events (e.g. eCO_2 and eT) thus does not impair key functional characteristics of the soil such as biodegradation, nitrogen transformations, disease suppression and plant growth promotion.

Community ecological concepts related to interdependency, including symbioses or syntrophies and redundancy, must therefore guide experimental design, measurement and interpretation when elucidating roles of soil biota in global change. The proceeding section describes how emergent biotechnological and biostatistical approaches are providing unprecedented access to the soil microbial communities with all the complexities associated with temporal-spatial scale and functional interdependencies.

8.4 Enabling Technologies and Applications in Global Change

The unprecedented growth in knowledge of soil biota and particularly of the microbial component can be attributed to the arrival of sequencing and scanning technologies and greatly enhanced computational power to generate and process large, multiparametric datasets. Many of the emergent biotechnologies will provide "warp speed" advancement in knowledge of soil biological systems by virtue of the following features: (1) they are rapid throughput, allowing for parallel detection of a range of features of interest (species, functions), (2) they integrate large data sets and reveal relationships between these datasets in a visual output format and (3) they provide high level visualisation of mechanistic information.

8.4.1 Soil "Omics" Technologies

The suffix "ome" is derived from the Greek word for "all" or "every". The original term was "genomics" referring to the study of the genetic material contained in an organism or cell. Many technologies are now described using the "omics" suffix, and several are beginning to be applied to the study of soil biological systems. For example, metagenomics studies the DNA of communities of soil organisms (answers who is in the community), and transcriptomics studies the RNA (and answers simultaneously who is there and what is functioning). Proteomics is the study of proteins in an organism. Metabolomics is the study of the entire range of metabolites, the precursors and products of enzymatic activity and can provide an overview of the metabolic status of a biological system (Yu Lin et al. 2006).

8.4.1.1 Soil Metagenomics

The soil metagenome is defined as the collective genomes (bacterial, archaeal, fungal and viral) in or recovered from soil and starts with the extraction of both DNA and RNA (Daniel 2004, 2005; Handelsman et al. 1998). Soil metagenomics is by far the most widely embraced of the new technologies to apply to soil addressing a serious shortcoming of conventional soil microbiology, the inability to identify the vast majority (99%) of soil organisms and the functions they perform (Chen et al. 2008; Keller and Zengler 2004; Xu 2006). It is therefore providing an unprecedented view of the taxonomic diversity, metabolic potential and ecological role of soil microbial communities enabling powerful resolution of the multiple ways in which soil microbes can benefit society. The importance of this technology lies within the emergent ecological concept that vital ecosystem services (e.g. mitigation of CO_2 plant growth promotion and disease suppression, nutrient cycling and bioremediation) are undertaken by a whole community of microbes working together in a self-organised pattern. Understanding the dynamic role of microbial communities is currently an unmet challenge. Metagenomics technologies will have a central role in answering some big questions that have been eloquently posed by Handelsman et al. (2007) and subcommittee members. These are:

- How resilient are microbial communities in the face of rapid global change?
- Can microbial communities, versatile as they are, help to buffer and mediate key elemental cycles now undergoing rapid shifts?

- Can changes in microbial communities serve as sensors and early-alarm systems of environmental perturbations?
- To what extent can we manage microbial communities to modulate the effects of human activities on natural elemental cycles sensibly and deliberately?

These questions will be answered neither by culturing microbes nor by sequencing single genomes or conserved genes (e.g. 16S and 18S rRNA). Considerable momentum is now building for global coordination of soil metagenomics research efforts to ensure consistency in (1) sampling strategies to address spatial and temporal variability, (2) DNA extraction procedures to maximise representation and (3) interpretation of output sequence data (http://www.terragenome.org).

While metagenomics provides an overarching community approach for understanding anthropogenic impacts, there are some specific and elegant tools that will allow the linkage of community structure with function and specifically the construction of gene-to-metabolite networks related to functions such as greenhouse gas production and C sequestration. There are several emergent enabling technologies that will greatly assist in this aim. The two receiving most attention and therefore worthy of mention are metatranscriptomics and stable isotope probing (SIP) or SIP-coupled techniques.

8.4.1.2 Soil Metatranscriptomics

Transcriptomics refers to the measurement of the transcriptome or the set of all messenger RNA (mRNA) molecules, or transcripts, produced by a cell. "Meta"transcriptomics is therefore the transcriptome of functions derived from a population of cells that would occur in soil. Because RNA is extracted in this analysis, the active or expressed subset of genes that are being transcribed under certain soil environmental conditions, rather than those simply present (as with DNA extraction) are measured. Expression profiling can be used to assess active groups of soil microbes present under certain environmental conditions (Urich et al. 2008) and to detect biocatalysts and other resource material associated with a range of applications including alternative energy production (Ferrer et al. 2009; Warnecke and Matthias 2009).

8.4.1.3 Stable Isotope Probing

SIP particularly when combined with other molecular technologies is a powerful approach for linking community function and structure. SIP is a technique that is used to identify the micro-organisms in environmental samples that use a particular growth substrate. The method relies on the incorporation of a substrate, including bacteria, that is highly enriched in a stable isotope, such as ^{13}C or ^{15}N and the identification of active micro-organisms by the selective recovery and analysis of isotope-enriched cellular components from the microbial community (Dumont and Murrell 2005; Dumont et al. 2006; Friedrich 2006; Manefield et al. 2002; Radajewski et al. 2000). Some examples of SIP-coupled technologies that are relevant to global change studies are SIP of phospholipid fatty acids (PLFA-SIP) and rRNA (RNA/DNA-SIP) to analyse the metabolically active methanotrophic community in soil and water (Boschker et al. 1998; Cebron et al. 2007; Qiu et al. 2008), to detect benzoate-utilising denitrifying bacteria (Gallagher et al. 2005) to assess the primary production functions in a dark environment (Chen et al. 2009) and to determine the role of earthworms in soil CH_4 emissions (Hery et al. 2007). Finally, and most recently, the coupling of SIP and metagenomics has enabled a genome-wide analysis of methylotrophs and has been used to target a novel methylotroph, *Methylotenera mobilis*, and to reconstruct its metabolism (Chen et al. 2009; Kalyuzhnaya and Lapidus 2008).

8.5 Future Needs

Soil biota is a critical element in the search for new knowledge leading to improved prediction and management of adverse impacts of global change. There are outstanding needs and challenges that must take account of three key facts (Chapin et al. 2009): (1) soil is the most diverse ecosystem on the planet, (2) only 1% of the soils microbial diversity is catalogued and (3) ecological concepts that apply to aboveground plant communities do not always apply belowground. These highlight that knowledge of soil biota per se and as mediators of global change processes is relatively rudimentary. The timely arrival of "omics" technologies has provided high resolution, purpose built tools for providing unprecedented access to the hidden majority and has spawned a new science discipline in bioinformatics. It will generate new and/or modified ecological rules and hence a much needed framework for understanding a whole range of anthropogenic impacts. Also critical to a renewed focus on soil biota is the need for greater intellectual organisation, and strong leadership and advocacy from the global Research and Development community. Once achieved, research efforts can be coordinated to generate shared outcomes such as: (1) new knowledge of the underlying soil biological processes and the multiple regulators that control responses to eCO_2 and eT *in a relevant time frame and spatial context*, (2) deep exploration of the mainly "hidden" microbial community for alternative bioenergy applications, and (3) data for enhanced predictive modelling capability to ensure that the current anthropogenic impacts can be managed efficiently and cost-effectively.

References

Albrecht R, Périssol C, Ruaudel F, Petit JL, Terrom G (2010) Functional changes in culturable microbial communities during a co-composting process: carbon source utilization and co-metabolism. Waste Manage 30:764–770

Alperin MJ, Hoehler TM (2009) Anaerobic methane oxidation by archaea/sulfate-reducing bacteria aggregates: 1. Thermodynamic and physical constraints. Am J Sci 309:869–957

Andert J, Borjennsen G, Hallin S (2009) Response of microbial communities involved in the emission of greenhouse gases from peat soil to season and environment. In: Janssen J (ed)

BAGECO-10, Bacterial genetics and ecology-coexisting on a changing planet. Uppsala Universitet

Andrews JA, Matamala R, Westover KM, Schlesinger WH (2000) Temperature effects on the diversity of soil heterotrophs and the [delta]^{13}C of soil-respired CO_2. Soil Biol Biochem 32: 699–706

Austin EE, Castro HF, Sides KE, Schadt CW, Classen AT (2009) Assessment of 10 years of CO_2 fumigation on soil microbial communities and function in a sweetgum plantation. Soil Biol Biochem 41:514–520

Baker K, Langenheder S, Nicol G, Ricketts D, Killham K, Campbell C, Prosser J (2009) Environmental and spatial characterisation of bacterial community composition in soil to inform sampling strategies. Soil Biol Biochem 41:2292–2298

Barcenas-Moreno G, Gomez-Brandon M, Rousk J, Baath E (2009) Adaptation of soil microbial communities to temperature: comparison of fungi and bacteria in a laboratory experiment. Global Change Biol 15:2950–2957

Bardgett RD, Usher MB, Hopkins DW (2005) Biological Diversity and Function in Soils. Cambridge University Press, Cambridge

Barnard R, Leadley PW, Lensi R, Barthes L (2005) Plant, soil microbial and soil inorganic nitrogen responses to elevated CO_2: a study in microcosms of *Holcus lanatus*. Acta Oecologica 27:171–178

Beerling D, Berner RA, Mackenzie FT, Harfoot MB, Pyle JA (2009) Methane and the CH_4 related greenhouse effect over the past 400 million years. Am J Sci 309:97–113

Billings SA, Ziegler SE (2005) Linking microbial activity and soil organic matter transformations in forest soils under elevated CO_2. Global Change Biol 11:203–212

Blagodatskaya EV, Blagodatsky SA, Anderson TH, Kuzyakov Y (2007) Priming effects in Chernozem induced by glucose and N in relation to microbial growth strategies. Appl Soil Ecol 37:95–105

Blagodatsky S, Blagodatskaya E, Yuyukina T, Kuzyakov Y (2010) Model of apparent and real priming effects: linking microbial activity with soil organic matter decomposition. Soil Biol Biochem 42(8):1275–1283

Boschker HTS, Nold SC, Wellsbury P, Bos D, de Graaf W, Pel R, Parkes RJ, Cappenberg TE (1998) Direct linking of microbial populations to specific biogeochemical processes by ^{13}C-labelling of biomarkers. Nature 392:801–805

Briones MJI, Garnett MH, Ineson P (2010) Soil biology and warming play a key role in the release of old C from organic soils. Soil Biol Biochem 42:960–967

Bünemann EK, Schwenke GD, Van Zwieten L (2006) Impact of agricultural inputs on soil organisms – a review. Aust J Soil Res 44:379–406

Cavicchioli R, DeMaere M, Thomas T (2007) Metagenomic studies reveal the critical and wide-ranging ecological importance of uncultivated archaea: the role of ammonia oxidizers. Bioessays 29:11–14

Cebron A, Bodrossy L, Stralis-Pavese N, Singer AC, Thompson IP, Prosser JI, Murrell JC (2007) Nutrient amendments in soil DNA stable isotope probing experiments reduce the observed methanotroph diversity. Appl Environ Microbiol 73:798–807

Chapin SF III, McFarland J, McGuire AD, Euskirchen ES, Ruess RW, Kielland K (2009) The changing global carbon cycle: linking plant and soil carbon dynamics to global consequences. J Ecol 97:840–850

Chen Y, Dumont MG, Neufeld JD, Bodrossy L, Stralis-Pavese N, McNamara NP, Ostle N, Briones MJI, Murrell JC (2008) Revealing the uncultivated majority: combining DNA stable-isotope probing, multiple displacement amplification and metagenomic analyses of uncultivated Methylocystis in acidic peatlands. Environ Microbiol 10:2609–2622

Chen Y, Boden R, Hillebrand D, Moussard H, Baciu M, Lu Y, Murrell JC (2009) Life without light: microbial diversity and evidence of sulfur- and ammonium-based chemolithotrophy in Movile Cave. ISME J 3:1093–1104

Chen D, Suter HC, Islam A, Edis R (2010) Influence of nitrification inhibitors on nitrification and nitrous oxide (N_2O) emission from a clay loam soil fertilized with urea. Soil Biol Biochem 42: 660–664

Clark NM, Rillig MC, Nowak RS (2009) Arbuscular mycorrhizal fungal abundance in the Mojave Desert: seasonal dynamics and impacts of elevated CO_2. J Arid Environ 73:834–843

Colloff MJ, Wakelin SA, Gomez D, Rogers SL (2008) Detection of nitrogen cycle genes in soils for measuring the effects of changes in land use and management. Soil Biol Biochem 40: 1637–1645

Coûteaux M-M, Bolger T (2000) Interactions between atmospheric CO_2 enrichment and soil fauna. Plant Soil 224:123–134

Cunningham AB, Gerlach R, Spangler L, Mitchell AC (2009) Microbially enhanced geologic containment of sequestered supercritical CO_2. Energy Procedia 1:3245–3252

Cuttle SP (2008) Impacts of pastoral grazing on soil quality. In: McDowell RW (ed) Environmental Impacts of Pasture-based Farming'. CABI, Wallingford

Daniel R (2004) The soil metagenome – a rich resource for the discovery of novel natural products. Curr Opin Biotechnol 15:199–204

Daniel R (2005) The metagenomics of soil. Nat Rev Microbiol 3:470–478

Davidson E, Nepstad D, Yoko Ishida F, Brando P (2008) Effects of an experimental drought and recovery on soil emissions of carbon dioxide, methane, nitrous oxide, and nitric oxide in a moist tropical forest. Global Change Biol 14:2582–2590

Demirel B, Scherer P (2008) The roles of acetotrophic and hydrogenotrophic methanogens during anaerobic conversion of biomass to methane: a review. Rev Environ Sci Biotechnol 7:173–190

Doran JW, Sarrantonio M, Liebig M (ed) (1996) *Soil health and sustainability*. Academic Press: San Diego, 1–54 pp

Drigo B, Kowalchuk GA, Yergeau E, Martijn Bezemer T, Boschker HTS, Van Veen JA (2007) Impact of elevated carbon dioxide on the rhizosphere communities of *Carex arenaria* and *Festuca rubra*. Global Change Biol 13:1365–2486

Drigo B, van Veen JA, Kowalchuk GA (2009) Specific rhizosphere bacterial and fungal groups respond differently to elevated atmospheric CO_2. ISME J 3:1204–1217

Dumont MG, Murrell JC (2005) Stable isotope probing linking microbial identify to function. Nat Rev Micro 3:499–504

Dumont MG, Neufeld JD, Murrell JC (2006) Isotopes as tools for microbial ecologists. Curr Opin Biotechnol 17:57–58

Edmeades DF (2004) Nitrification and urease inhibitors: a review of the national and international literature on their effects on nitrate leaching, greenhouse gas emissions and ammonia volatilisation from temperate legume-based systems. Environment Waikato Technical Report 2004/22. Environment Waikato, Hamilton. ISSN: 1172–4005

EU (2002) Towards a thematic strategy for soil protection. The Council, the European Parliament, the Economic and Social Committee and the Committee of the Regions, Brussels

Falkowski P, Fenchel T, deLong E (2008) The microbial engines that drive earth's biogeochemical cycles. Science 320:1034–1039

Falloon P, Smith P, Betts R, Jones C, Smith J, Hemming D, Challinor A (2009) Carbon Sequestration and Greenhouse Gas Fluxes from Cropland Soils- Climate Opportunities and Threats. Springer, Berlin

Faulwetter JL, Gagnon V, Sundberg C, Chazarenc F, Burr MD, Brisson J, Camper AK, Stein OR (2009) Microbial processes influencing performance of treatment wetlands: a review. Ecol Eng 35:987–1004

Fenchel T, Finlay BJ (1995) Ecology and evolution in anoxic worlds. Oxford University Press, Oxford

Feng X, Simpson MJ (2009) Temperature and substrate controls on microbial phospholipid fatty acid composition during incubation of grassland soils contrasting in organic matter quality. Soil Biol Biochem 41:804–812

Ferrer M, Beloqui A, Timmis KN, Golyshin PN (2009) Metagenomics for mining new genetic resources of microbial communities. J Mol Microbiol Biotechnol 35:265–274

Freeman C, Baxter R, Farrar JF, Jones SE, Plum S, Ashendon TW, Stirling C (1998) Could competition between plants and microbes regulate plant nutrition and atmospheric CO_2 concentrations? Sci Total Environ 220:181–184

Friedrich MW (2006) Stable-isotope probing of DNA: insights into the function of uncultivated microorganisms from isotopically labeled metagenomes. Curr Opin Biotechnol 17:59–66

Gallagher E, McGuinness L, Phelps C, Young LY, Kerkhof LJ (2005) ^{13}C-carrier DNA shortens the incubation time needed to detect benzoate-utilizing denitrifying bacteria by stable-isotope probing. Appl Environ Microbiol 71:5192–5196

Gatson KJ, Spicer JI (2004) Biodiversity- An Introduction. Blackwell Publishing, Cornwall

Granli T, Bockman OC (1994) Nitrous oxide from agriculture. Norw J Agric Sci 12:7–127

Green SJ, Prakash O et al (2010) Denitrifying bacteria isolated from terrestrial subsurface sediments exposed to mixed-waste contamination. Appl Environ Microbiol 76:3244–3254

Gu C, Maggi F et al (2009) Aqueous and gaseous nitrogen losses induced by fertilizer application. J Geophys Res 114:G01006

Handelsman J, Rondon MR, Brady SF, Clardy J, Gooman RM (1998) Molecular biological access to the chemistry of unknown soil microbes: a new frontier for science. Chem Biol 5:245–249

Handelsman J, Tiedje L et al (2007) The New Science of Metagenomics. Revealing the Secrets of Our Microbial Planet. National Academy of Sciences, Washington

Hartley IP, Heinemeyer A, Evans SP, Ineson P (2007) The effect of soil warming on bulk soil vs. rhizosphere respiration. Global Change Biol 13:2654–2667

Hawkesworth DL (2001) The magnitude of fungal diversity: the 1·5 million species estimate revisited. Mycol Res 105:1422–1432

Hernández D, Hobbie S (2010) The effects of substrate composition, quantity, and diversity on microbial activity. Plant and Soil 335:397–397

Herrmann AM, Ritz K, Nunan N, Clode PL, Pett-Ridge J, Kilburn MR, Murphy DV, O'Donnell AG, Stockdale EA (2007) Nano-scale secondary ion mass spectrometry – a new analytical tool in biogeochemistry and soil ecology: a review article. Soil Biol Biochem 39:1835–1850

Hery M, Singer A, Kumaresan D, Bodrossey L, Stralis-Pavese N, Prosser JI, Thompsen IP, Murrell JC (2007) Effect of earthworms on the community structure of active methanotrophic bacteria in a landfill cover soil. ISME J 2:92–104

Houghton JT, Ding Y, Griggs DJ, Noguer M, van der Linden PJ, Dai X, Maskell K, Johnson CA (eds) (2007) Climate change: the scientific basis. Cambridge University Press, Cambridge

Hu S, Firestone MK, Chapin FS (1999) Soil microbial feedbacks to atmospheric CO_2 enrichment. Trends Ecol Evol 14:433–437

Insam H, Bååth E, Berreck M, Frostegård Å, Gerzabek MH, Kraft A, Schinner F, Schweiger P, Tschuggnall G (1999) Responses of the soil microbiota to elevated CO_2 in an artificial tropical ecosystem. J Microbiol Meth 36:45–54

IPCC (2007) Climate change 2007: the physical science basis. In: IPoC Change (ed). Cambridge University Press, Cambridge

Kalyuzhnaya MG, Lapidus A (2008) High-resolution metagenomics targets specific functional types in complex microbial communities. Nat Biotech 26:1029–1034

Kandeler E, Mosier AR, Morgan JA, Milchunas DG, King JY, Rudolph S, Tscherko D (2008) Transient elevation of carbon dioxide modifies the microbial community composition in a semi-arid grassland. Soil Biol Biochem 40:162–171

Kanerva T, Palojärvi A, Rämö K, Manninen S (2008) Changes in soil microbial community structure under elevated tropospheric O_3 and CO_2. Soil Biol Biochem 40:2502–2510

Kang H, Kim S-Y, Fenner N, Freeman C (2005) Shifts of soil enzyme activities in wetlands exposed to elevated CO_2. Sci Total Environ 337:207–212

Keller M, Zengler K (2004) Tapping into microbial diversity. Nat Rev Micro 2:141–150

Kibblewhite M, Ritz K, Swift M (2008) Soil health in agricultural systems. Philos Trans R Soc Lond B Biol Sci 362:685–701

Kirschbaum MUF (1995) The temperature dependence of soil organic matter decomposition, and the effect of global warming on soil organic C storage. Soil Biol Biochem 27:753–760

Kohler J, Caravaca F, del Mar AM, Roldán A (2009) Elevated CO_2 increases the effect of an arbuscular mycorrhizal fungus and a plant-growth-promoting rhizobacterium on structural stability of a semiarid agricultural soil under drought conditions. Soil Biol Biochem 41: 1710–1716

Krüger M, West J et al (2009) Ecosystem effects of elevated CO_2 concentrations on microbial populations at a terrestrial CO_2 vent at Laacher See, Germany. Energy Procedia 1:1933–1939

Kuzyakov Y (2010) Priming effects: interactions between living and dead organic matter. Soil Biol Biochem 42:1363–1371

Lal R (2009) Soils and world food security. Soil Till Res 102:1–4

Lavelle P (1996) Diversity of soil fauna and ecosystem function. Biol Int 33:3–16

Li Q, Xu C, Liang W, Zhong S, Zheng X, Zhu J (2009) Residue incorporation and N fertilization affect the response of soil nematodes to the elevated CO_2 in a Chinese wheat field. Soil Biol Biochem 41:1497–1503

Lin Y, Kong HN, Wang RY, Li CJ, Yan L, He YL (2008) Characteristic and prospects of heterotrophic nitrification. Huanjing Kexue/Environ Sci 29:3291–3296

Liu Y, Whitman WB (2008) Metabolic, phylogenetic, and ecological diversity of the methanogenic archaea. Ann NY Acad Sci 1125:171–189

Liu B, Mørkved PT, Frostegård Å, Bakken LR (2010) Denitrification gene pools, transcription and kinetics of NO, N_2O and N_2 production as affected by soil pH. FEMS Microbiol Ecol 72: 407–417

Ma BL, Wu TY, Tremblay N, Deen W, Morrison MJ, Mclaughlin NB, Gregorich EG, Stewart G (2010) Nitrous oxide fluxes from corn fields: on-farm assessment of the amount and timing of nitrogen fertilizer. Global Change Biol 16:156–170

Malla G, Bhatia A, Pathak H, Prasad S, Jain N, Singh J (2005) Mitigating nitrous oxide and methane emissions from soil in rice-wheat system of the Indo-Gangetic plain with nitrification and urease inhibitors. Chemosphere 58:141–147

Manefield M, Whiteley AS, Griffiths RI, Bailey MJ (2002) RNA stable isotope probing, a novel means of linking microbial community function to phylogeny. Appl Environ Microbiol 68: 5367–5373

Manter DK, Weir TL, Vivanco JM (2010) Sample pooling masks PCR-based estimates of soil microbial richness and community structure. Appl Environ Microbiol 76:2086–2090

Marschner P, Kandeler E, Marschner B (2003) Structure and function of the soil microbial community in a long-term fertilizer experiment. Soil Biol Biochem 35:453–461

McGuinn AF (1924) The action of dicyandiamide and guanyl urea sulfate on plant growth. Soil Sci Soc Am J 17:487–500

Morales SE, Cosart T, Holben WE (2010) Bacterial gene abundances as indicators of greenhouse gas emission in soils. ISME J 4:799–808

Moscatelli MC, Lagomarsino A, Angelis PD, Grego S (2005) Seasonality of soil biological properties in a poplar plantation growing under elevated atmospheric CO_2. Appl Soil Ecol 30: 162–173

Novis P, Whitehead D, Gregorich EG, Hunt A, Sparrow A, Hopkins D, Elberling B, Greenfield L (2007) Annual carbon fixation in terrestrial populations of Nostoc commune (Cyanobacteria) from an Antarctic dry valley is driven by temperature regime. Global Change Biol 13: 1224–1237

O'Callaghan M, Gerard EM, Carter PE, Lardner R, Sarathchandra U, Burch G, Ghani A, Bell N (2010) Effect of the nitrification inhibitor dicyandiamide (DCD) on microbial communities in a pasture soil amended with bovine urine. Soil Biol Biochem 42(9):1425–1436

Okano Y, Hristova KR, Leutenegger CM, Jackson LE, Denison RF, Gebreyesus B, Lebauer D, Scow KM (2004) Application of real-time PCR to study effects of ammonium on population size of ammonia-oxidizing bacteria in soil. Appl Environ Microbiol 70:1008–1016

Paul EA (ed) (2007) Soil microbiology, ecology and biochemistry. Elsevier, Burlington

Perez-Piqueres A, Edel-Hermann V, Alabouvette C, Steinberg C (2006) Response of soil microbial communities to compost amendments. Soil Biol Biochem 38:460–470
Phillips RL, Whalen SC, Schlesinger WH (2001) Response of soil methanotrophic activity to carbon dioxide enrichment in a North Carolina coniferous forest. Soil Biol Biochem 33: 793–800
Post WM, Emanuel WR, Zinke PJ, Stangenberger AG (1982) Soil carbon pools and world life zones. Nature 298:156–159
Prescott C (2010) Litter decomposition: what controls it and how can we alter it to sequester more carbon in forest soils? Biogeochemistry 70:227–250
Prosser JI (1986) Nitrification. IRL Press, Oxford
Qiu Q, Noll M, Abraham W-R, Lu Y, Conrad R (2008) Applying stable isotope probing of phospholipid fatty acids and rRNA in a Chinese rice field to study activity and composition of the methanotrophic bacterial communities in situ. ISME J 2:602–614
Radajewski S, Ineson P, Parekh NR, Murrell JC (2000) Stable-isotope probing as a tool in microbial ecology. Nature 403:646–649
Reth S, Hentschel K, Drösler M, Falge E (2005) DenNit – experimental analysis and modelling of soil N_2O efflux in response on changes of soil water content, soil temperature, soil pH, nutrient availability and the time after rain event. Plant Soil 272:349–363
Rinnan R, Johannes R, Tergeau T, Kowalchuk GA, Baath E (2009) Temperature adaptation of soil bacterial communities along an Antarctic climate gradient: predicting responses to climate warming. Global Change Biol 15:2615–2625
Robinson SA (2009) Introduction: climate change biology at the ends of the Earth-International Polar year special issue. Global Change Biol 15:1615–1617
Rønn R, Gavito M, Larsen J, Jakobsen I, Frederiksen H, Christensen S (2002) Response of free-living soil protozoa and microorganisms to elevated atmospheric CO_2 and presence of mycorrhiza. Soil Biol Biochem 34:923–932
Rotthauwe JH, Witzel KP, Liesack W (1997) The ammonia monooxygenase structural gene amoA as a functional marker: molecular fine-scale analysis of natural ammonia-oxidizing populations. Appl Environ Microbiol 63:4704–4712
Schortemeyer M, Hartwig UA, Hendrey GR, Sadowsky MJ (1996) Microbial community changes in the rhizospheres of white clover and perennial ryegrass exposed to free air carbon dioxide enrichment (FACE). Soil Biol Biochem 28:1717–1724
Scow KM, Louise EJ (1997) Soil microbial communities and carbon flow in agroecosystems. In: Ecology in agriculture. Academic, San Diego, pp 367–413
Siciliano SD, Ma WK, Ferguson S, Farrell RE (2009) Nitrifier dominance of Arctic soil nitrous oxide emissions arises due to fungal competition with denitrifiers for nitrate. Soil Biol Biochem 41:1104–1110
Sinsabaugh RL, Saiya-Cork K, Long T, Osgood MP, Neher DA, Zak DR, Norby RJ (2003) Soil microbial activity in a Liquidambar plantation unresponsive to CO_2-driven increases in primary production. Appl Soil Ecol 24:263–271
Skjemstad JO, Spouncer LR, Cowie B, Swift RS (2004) Calibration of the Rothamsted organic carbon turnover model (RothC ver. 26.3), using measurable soil organic carbon pools. Aust J Soil Res 42:79–88
Torsvik V, Øvreås L (2002) Microbial diversity and function in soil: from genes to ecosystems. Curr Opin Microbiol 5:240–245
Torsvik V, Goksoyr J, Daae F (1990) High diversity in DNA of soil bacteria. Appl Environ Microbiol 56:782–787
Tringe SG, Hugenholtz P (2008) A renaissance for the pioneering 16s rRNA gene. Curr Opin Microbiol 11:442–446
Urich T, Lanza A, Qi J, Huson DH, Schleper C, Schuster SC (2008) Simultaneous assessment of soil microbial community structure and function through analysis of the meta-transcriptome. PLoS ONE 3:2527

Van Der Heijden MGA, Bardgett RD, Van Straalen NM (2008) The unseen majority: soil microbes as drivers of plant diversity and productivity in terrestrial ecosystems. Ecol Lett 11:651

Vargas-García MC, Suárez-Estrella F, López MJ, Moreno J (2010) Microbial population dynamics and enzyme activities in composting processes with different starting materials. Waste Manage 30:771–778

Venter J, Remington K, Heidelberg J, Halpern AR, Rusch D, Eisen J, Wu D, Smith H (2004) Environmental genome shotgun sequencing of the Sargasso Sea. Science 304:66–74

Vincent WF (2010) Microbial ecosystem responses to rapid climate change in the Arctic. ISME J 4(9):1087–1090

Wagner M (2009) Unexplored key players in nitrification. In: Janssen J (ed) BAGECO-10, Bacterial genetics and ecology-coexisting on a changing planet. Uppsala Universitet

Wakelin SA, Colloff MJ, Harvey PR, Marschner P, Gregg AL, Rogers SL (2007) The effects of stubble retention and nitrogen application on soil microbial community structure and functional gene abundance under irrigated maize. FEMS Microbiol Ecol 59:661–670

Waldrop MP, Firestone MK (2006) Response of microbial community composition and function to climate change. Microb Ecol 52:716–724

Wall DH, Fitter AH, Paul EA (eds) (2005) Biological diversity and function in soils. Developing new perspectives from advances in soil biodiversity research. Cambridge University Press, Dorset

Warnecke F, Matthias H (2009) A perspective: metatranscriptomics as a tool for the discovery of novel biocatalysts. J Biotechnol 142:91–95

Watt MJ, Kirkegaard JA, Passioura JB (2006) Rhizosphere biology and crop productivity – a review. Aust J Soil Res 44:299–317

Whitman W, Boone DR, Koga Y, JT K (2001) Taxonomy of methanogenic Archaea. Springer, Berlin

Williamson KE, Radosevich M, Wommack KE (2005) Abundance and diversity of viruses in six Delaware soils. Appl Environ Microbiol 71:3119–3125

Xu J (2006) Microbial ecologyin the age of genomics and metagenomics:concepts, tools and recent advances. Mol Ecol 15:1713–1731

Yu Lin C, Viant MR, Tjeerdema RS (2006) Metabolomics: methodologies and applications in the environmental sciences. J Pesticide Sci 31:241–251

Yuan X, Lin X, Chu H, Yin R, Zhang H, Hu J, Zhu J (2006) Effects of elevated atmospheric CO_2 on soil enzyme activities at different nitrogen application treatments. Acta Ecol Sin 26:48–53

Zak DR, Pregitzer KS, King JS, Holmes WE (2000) Elevated atmospheric CO_2, fine roots and the response of soil microorganisms: a review and hypothesis. New Phytol 147:201–222

Zhang L, Kudo T, Takaya N, Shoun H (2002) Distribution, structure and function of fungal nitric oxide reductase P450nor – recent advances. Int Congr Ser 1233:197–202

Zhang W, Parker K, Luo Y, Wan S, Wallace LL, HU S (2005) Soil microbial responses to experimental warming and clipping in a tallgrass prairie. Global Change Biol 11:266–277

Zhang H-H, He P-J, Shao L-M, Yuan L (2009) Minimisation of N_2O emissions from a plant-soil system under landfill leachate irrigation. Waste Manage 29:1012–1017

Part III
Conventional Land use Systems

Chapter 9
Tillage and Crop Stubble Management and Soil Health in a Changing Climate

Brian W. Murphy, Ian J. Packer, Annette L. Cowie, and Bhupinder Pal Singh

9.1 Introduction

Climate change is predicted to seriously impact many of the world's major cropping areas. The majority of the world's food and fibre is produced in cropping systems, and increasing food production to meet the needs of the increasing world population in an environment of uncertainties about climate change is going to be a major challenge facing communities (Wentworth Group 2009). Maintaining crop productivity is a key element in meeting these future challenges, and this will rely largely on maintaining soil health.

Some of the expected impacts of climate change on cropping areas include (IPCC 2007):

- Reduced total effective rainfall, which will change the potential amount of biomass produced and the amount of ground cover;
- Changing rainfall patterns with increased rainfall intensity and erosivity;
- Higher temperatures and evapotranspiration rates.

In response to these predictions, the adaptive land management strategies being investigated include varying planting time, sowing rates, nitrogen application,

B.W. Murphy (✉)
Organization NSW Office of Environment and Heritage, PO Box 445, Cowra, NSW 2794, Australia
e-mail: Brian.Murphy@environment.nsw.gov.au

I.J. Packer
Lachlan Catchment Management Authority, PO Box 510, Cowra, NSW 2794, Australia

A.L. Cowie
National Centre for Rural Greenhouse Gas Research, University of New England, Armidale, NSW 2351, Australia

B.P. Singh
Forest Science Centre, NSW Department of Primary Industries, PO Box 100, Beecroft, NSW 2119, Australia

B.P. Singh et al. (eds.), *Soil Health and Climate Change*, Soil Biology 29,
DOI 10.1007/978-3-642-20256-8_9, © Springer-Verlag Berlin Heidelberg 2011

cover and crop varieties, residue management, tillage type and depth, and length of fallow. Improvement in soil health through the adoption of best management strategies will have a profound positive impact, providing resilience and flexibility to meet the challenges of climate change and variability. For example, adopting conservation cropping practices with less soil disturbance and maintaining greater levels of crop residues has the potential to maintain and improve soil health, through positive impacts on key soil properties or processes (see Sects. 9.2 and 9.3) and consequently increasing agriculture sector's ability to adapt to climate change.

Furthermore, agriculture is responsible for significant release of carbon dioxide (CO_2), nitrous oxide (N_2O) and methane (CH_4) from soil (Smith et al. 2008; Yadvinder-Singh et al. 2005). Thus, changes in management practices on cropping lands that reduce or avoid greenhouse gas (GHG) emissions from soil, and/or enhance carbon sequestration, have great potential to mitigate climate change impacts of agriculture (Smith et al. 2008).

In this chapter, we review the current knowledge of the effect of tillage and crop stubble management on soil health and GHG emissions. Particular emphasis is given to describing factors that increase or maintain soil carbon levels, as this is seen as the basis for improving and maintaining soil health (Bhupinderpal-Singh and Rengel 2007; Bolliger et al. 2006; Dalal and Chan 2001; Lawrie et al. 2007; Luo et al. 2010a; Yadvinder-Singh et al. 2005) in the context of changing climate. We also address some of the apparently contradictory evidence on the effects of tillage and stubble management on soil carbon levels and GHG emissions, which may result from differences in soil type, climate and tillage/stubble management techniques. We emphasise the need for a better understanding of the role of conservation tillage practices in improving soil carbon stocks, maintaining soil health and reducing soil GHG emissions under diverse climate-crop type combinations, to enhance the contribution of croplands to climate change mitigation and adaptation.

9.2 Tillage/Crop Stubble Management: Impact on Soil Carbon Levels

9.2.1 General Principles and Defining Practices

Cropping involves a range of land management practices that can affect soil health and levels of soil organic carbon. Individual tillage operations and stubble management can have specific impacts on soil health, but it is the combination of management practices that ultimately determines the organic carbon levels in soils. Table 9.1 presents a gradation of different combinations of tillage and stubble management and indicates how these management combinations, based on the practices outlined in Lawrie et al. (2007) and Murphy et al. (2008), may affect soil carbon. This gradation of the effect of tillage/stubble management practices on soil carbon is based on the following general, well-recognised principles for maintaining soil health:

Table 9.1 Rating of cropping practices based on tillage, stubble management and sowing operations

Practice	Cropping practice	Median or typical tillage and stubble management operations
A	Long fallow	Stubble burning 250 days before sowing First tillage 200 days before sowing 2 Tillage operations with two-way disc 3 Tillage operations with one-way disc 3 Tillage operations with tyned implement 100% surface disturbance at sowing Ground traffic/3 years coverage by machinery is 200% of ground surface
B	Traditional tillage TTb	Stubble burning 100 days before sowing First tillage 100 days before sowing 1 Tillage operations with two-way disc 2 Tillage operations with one-way disc 2 Tillage operations with tyned implement 100% surface disturbance at sowing Ground traffic/3 years coverage by machinery is 100% of ground surface
C	Best traditional tillage TTa	Stubble grazed Stubble burning 45 days before sowing First tillage 40 days before sowing 1 Tillage operations with one-way disc 2 Tillage operations with tyned implement 100% surface disturbance at sowing Ground traffic/3 years coverage by machinery is 75% of ground surface
D	Reduced tillage/ minimum tillage DDb	Stubble normally grazed Stubble may or may not be burnt – if burnt done after break rains 30–50 days before sowing First tillage 30 days before sowing 1 Tillage operations with tyned implement 100% surface disturbance at sowing Ground traffic/3 years coverage by machinery is 50% of ground surface
E	Best direct drill DDa	Stubble grazed Stubble burnt 10–30 days before sowing No-tillage before sowing Full 100% surface disturbance at sowing Ground traffic/3 years is coverage by machinery 20–30% of ground surface
F	No-tillage	Stubble mainly retained – may burn Stubble not grazed No-tillage before sowing <30% surface disturbance at sowing, often using specialised tynes or <10% disturbance surface if using disc seeders Ground traffic/3 years is coverage by machinery 20–30% of ground surface
G	Zero-tillage and controlled traffic	Stubble retained Stubble not grazed No-tillage before sowing <10% surface disturbance if using disc seeders. Limited amount of zero-tillage with tyne seeders Ground traffic/3 years is coverage by machinery <5% of ground surface
H	Pasture cropping	Crops are sown into perennial pasture Cropping inputs – fertiliser and herbicides Improved grazing management that maintains groundcover throughout the year. Often based on perennial pastures – native or improved. Carbon status can often depend on initial carbon levels as this practice is often implemented on degraded soils

Arrow shows expected increasing soil carbon levels based on long-term equilibrium, where A is the set of cropping practices that is likely to result in the worst soil health, and G and H the best soil health. Coverage by machinery refers to the amount of the ground that is driven on and compacted by the passage of machinery in the course of cropping operations

1. Maximise ground cover and biomass;
2. Minimise soil disturbance;
3. Minimise soil compaction;
4. Maintain adequate plant nutrient levels;
5. Maintain soil biodiversity to enhance nutrient cycling and physical attributes.

This suggested that classification of management practices can be used as a general guide to identifying "better" practices that can improve soil health through increased soil carbon. However, the impacts of these practices on soil carbon may not be universal across all climates, soil types and crops. For this reason, a more detailed understanding of the effects of specific operations in particular circumstances is required.

9.2.2 Factors That Affect Soil Carbon Levels

9.2.2.1 Climate

Soil carbon stocks reflect the balance between inputs of organic matter and its decomposition. Climatic regimes that favour plant growth – warm, moist climates – have the highest rate of input to the soil, but these same environments also favour rapid decomposition of soil organic matter (see Chap. 5). Generally, soil carbon stocks are highest in cool temperate environments where organic carbon output via decomposition is inhibited by low temperature. On the other hand, soil carbon stocks are lower in the wet tropics where organic matter is turned over rapidly in the warm, moist climate, and lowest in dry environments where plant growth is limited by water availability. Slattery and Surapaneni (2002) reported impacts of rainfall on soil carbon in Australian cropping soils, with carbon concentration in the top 10 cm being 2.66% in wetter areas (~600 mm annual rainfall) and less than 1.15% in drier areas (<400 mm). Valzano et al. (2005) identified critical mean annual temperatures affecting soil carbon levels in south eastern Australia: <12.8°C tends to result in higher levels of soil carbon, and >17.4°C tends to result in lower values of soil carbon in cropping systems.

Climate interacts with management factors; so the effects of stubble management and tillage vary between climatic regimes (see Sect. 9.2.2.9).

9.2.2.2 Soil Type

The soil type determines how much organic matter can be stored in the soil: clayey soils are able to store more organic matter than sandier soils, as the clays adsorb organic molecules, aid in formation of aggregates and protect organic matter from microbial decay (Carter et al. 2003; Jenkinson 1977; Ladd et al. 1996). Liang et al. (2002) have estimated that there is a direct linear relationship between clay content and the potential to increase soil carbon by crop management practices. The impact of cropping practices may also depend on the soil type (Bolliger et al. 2006; Lawrie et al. 2007; Oades 1984). Soils with high levels of swelling clays (smectites) that are self-mulching will be more resilient to physical impacts of cultivation than soils dominated by less-active clays with low cation exchange capacity (illites and kaolinite) and lighter textured soils (sandy loams to loams or clay loams) with lower clay contents. Soils with high levels of free iron, the Ferrosols and Ultisols, are also likely to be more resilient to the effects of soil structural breakdown than the more fragile lighter textures soils. The lighter textured, lower activity or low cation exchange capacity soils are more dependent on soil carbon to maintain soil health.

9.2.2.3 Maintenance of Biomass Production: Agronomic and Nutrient Management

The maintenance or increase of soil organic carbon levels requires the input of organic matter. Plants growing in situ will often provide the major input of carbon into the soil, but additions through mulches, composts and manures can be important in some systems. Adequate soil moisture levels, plant nutrition and use of fertilisers, and improved agronomy based on selection of varieties, sowing times, weed control and appropriate crop–crop and crop–pasture rotations are required to promote plant growth and therefore biomass input to soil.

In a steady state, generally the carbon inputs from plant residues determine the organic carbon levels in soils (Dalal and Chan 2001). He and Lui (1992) estimated that about 3.8 t ha^{-1} $year^{-1}$ of crop residues were required to maintain soil organic carbon levels in a barley – early rice – late rice cropping sequence. In the tropical areas of Brazil with annual average temperatures above 20°C, residues of 8–10 t ha^{-1} are estimated to be required to maintain soil carbon levels because of high decomposition rates (Bolliger et al. 2006). At broader regional and inter-regional scales, the precipitation/evaporation ratio is considered more useful to predict the likely biomass production and therefore soil carbon inputs into the soil (Parton et al. 1996). As a general rule, about 2–5% of biomass produced by the plant is converted to soil carbon (Dalal and Chan 2001).

Nitrogen also has an important role to play in the accumulation of carbon in soils (Snyder et al. 2007), not only by increasing dry matter production (Paustian et al. 1997), but also by aiding the formation of soil organic matter or humus (Amado et al. 2006; De Maria et al. 1999). Furthermore, Chan et al. (2010) reported that additions of P were necessary to significantly increase soil carbon levels under improved pastures.

9.2.2.4 Crop–Pasture Rotations

Crop–pasture rotations are an important component of cropping systems involving a suite of management regimes including tillage and stubble management.

A pasture phase in crop rotations may modify tillage/stubble management impacts and consequently play an important role in rebuilding soil carbon levels (Dalal and Chan 2001; Grace et al. 1995; Russell and Williams 1982). Impact of the pasture phase will depend on:

- Length of the pasture phase;
- Adequate plant nutrition (including N and P) during the pasture phase;
- Intensity and duration of grazing, which determine the amount of biomass removed and the impact of stocking on soil structure;
- Management of soil acidification and sodicity where these are a limitation;
- The species balance (legume/grass and perennial/annual), which affect the quality and quantity of organic inputs.

9.2.2.5 Clearing of Native Vegetation for Cropping

It is well documented that major loss of soil carbon has resulted from clearing native vegetation for cropping (Bell et al. 1995; Dalal and Chan 2001; Harte 1984; Hermle et al. 2008; Luo et al. 2010a; Shi et al. 2010; West and Post 2002). An example is shown in Fig. 9.1 where the soil carbon level fell sharply after converting native vegetation (Bimble Box, *Eucalyptus populnea*) to cropping lands (Murphy et al. 2003). As indicated in Dalal and Chan (2001), Luo et al. (2010a) and Murphy et al. (2003), soil organic carbon commonly declines by about 50% within 10 years after the introduction of cropping. The loss of soil carbon on conversion to cropping is sometimes attributed solely to increased decomposition due to the physical disturbance of cultivation, but it is more likely a result of the combined effect of soil physical disturbance and changes in microclimatic conditions stimulating native organic matter decomposition and low levels of organic matter inputs during fallow and early stages of crop growth. While gross changes in land use have major impacts on soil carbon, more subtle management changes, such as those attributed to tillage practices, can be smaller and often more difficult to verify.

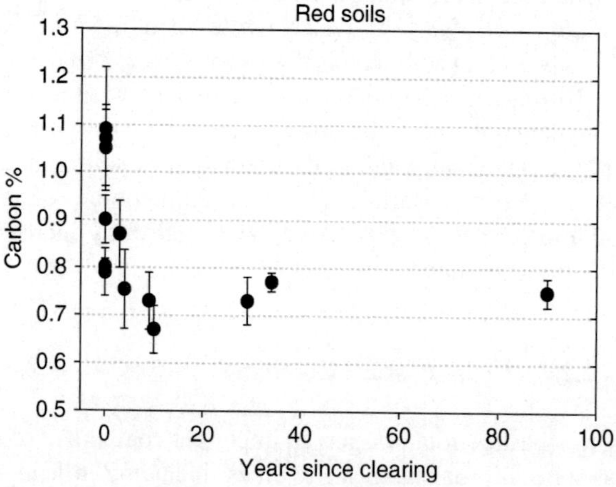

Fig. 9.1 Soil carbon levels in the red earth soils (Red Luvisols) of the Cobar Peneplain in central western New South Wales after clearing for cropping (after Murphy et al. 2003)

9.2.2.6 Tillage

As mentioned above (see Sect. 9.2.2.5), regularly cultivated soils have lower soil carbon stocks than natural vegetation. The effect of tillage management on soil carbon levels depends on the number, type, depth and intensity of tillage operations before sowing; these management variations affect the extent of organic matter input and output and its distribution in soil (Luo et al. 2010b). Intense tillage operations such as shattering (by rotary hoe) and inversion (by disc plough) will disrupt soil aggregates and expose organic matter for microbial attack, change soil physical and microclimatic conditions, and bury crop stubble in soil to a greater extent, thereby leading to reduced soil carbon compared with less aggressive conservation tillage operations. The length of time from the first tillage before sowing also affects soil carbon levels. Early tillage operations a short time after harvest expose organic matter to greater microbial attack and thus encourage rapid stubble breakdown. This usually results in a bare soil surface for extended periods before sowing, which may increase soil temperature thus further stimulating organic matter decomposition, as well as enhance risk of soil carbon removal through water and wind erosion (see Sect. 9.4).

Six et al. (2002) in a broad review of literature on soil organic matter dynamics in temperate and tropical soils concluded that after 6–8 years, there was a relative increase in soil organic carbon under zero-till systems in comparison to tilled systems, averaging 0.325 (±0.113) t C ha^{-1} $year^{-1}$. A recent review of results from Australian studies found similar impacts of tillage on soil carbon: no-tillage treatments stored 0.34 t C ha^{-1} $year^{-1}$ relative to traditional tillage (Sanderman et al. 2010). Similarly, Mishra et al. (2010) in northeastern USA showed that the amount of soil organic carbon in the 0–40 cm depth was greater under long-term no-tillage (>20–40 years) compared to traditional tillage practices; on average, soils under no-tillage treatment accumulated 0.31–0.82 t C ha^{-1} $year^{-1}$ more than traditional tillage across a diverse soil and crop type and management regimes. However, no response to tillage treatment was observed where the practices were implemented for a shorter duration (~6–20 years). It was suggested that differences in cropping systems and management intensity and depth of measurements could have masked consistent responses of soil carbon to tillage practices (Mishra et al. 2010).

9.2.2.7 Controlled Traffic

Controlled traffic involves confining machinery traffic to a narrow specified set of wheel tracks. This avoids extensive soil compaction and when combined with no-tillage or zero-tillage, controlled traffic ensures that the soil physical and biological conditions between the machinery tracks are improved/maintained (Chamen et al. 1992; Li et al. 2001; Tullberg et al. 2001; Wang et al. 2008).

9.2.2.8 Stubble Management

The following factors determine the influence of stubble management on build up of soil carbon.

- *The amount of stubble retained* – this can vary from near 0 for an early hot burn after harvest to near 100% for stubble retained with no-grazing. The quantity of stubble retained varies, even at paddock scale, through time as a result of variations in farming operations, grazing activities and decomposition/burning gradients. Between the extremes of 0 and 100% retention are: baling (~20% retention), grazed (~60% retention if lightly stocked and ~20% if heavily stocked) and late burnt just before sowing (~35% retention) (Lawrie et al. 2007). Stubble burning will also result in the loss of nutrients from stubble. Sharma and Mishra (2001) have estimated that burning wheat stubble results in the loss of 87% of nitrogen, 2% of phosphorus, 18% of potassium and 25% of sulphur.
- *Decomposition rate* – decomposition of straw is most rapid when the soil moisture is between 30 and 100 kPa tension (Doel et al. 1990). Burying stubble by tillage to shallow depths (less than ~20 cm) may accelerate decomposition as it increases soil contact and breaks stubble into smaller pieces, but this may not be the case in a cooler climate (Hermle et al. 2008). Stubble treatments such as breaking up the stubble into smaller sizes, rolling and cracking the stubble to increase water penetration or even adding cellulose digestion agents can increase the rate of decomposition (Lawrie et al. 2007). Smaller particles of stubble (<1 mm size) may decompose much more rapidly than larger pieces of stubble (50–100 mm) (Angers and Racous 1997); however, Bhupinderpal-Singh et al. (2006) found no effect of particle size on residue decomposition rate in a nutrient-poor sandy soil, where nutrient limitation had an overriding effect, restricting decomposition rate irrespective of particle size.
- *Stubble chemical characteristics* – the C/N ratio and lignin content determine the decomposition rate and the amount of stubble that is converted into soil organic matter (Campbell et al. 2002; Yadvinder-Singh et al. 2005). For example, legume stubbles with a low C/N ratio in combination with a low content of lignin breakdown more rapidly than cereal stubbles (Bhupinderpal-Singh and Rengel 2007).

9.2.2.9 Tillage and Crop Stubble Management, Soil Type and Climate Interactions

There are conflicting reports as to whether the traditional (e.g. mouldboard plough, disc plough, stubble burning/incorporation) or conservation farming (direct drill, chisel till, disc/tyne seeders, limited soil disturbance/stubble retention) practices increase or decrease soil carbon levels (Bhupinderpal-Singh and Rengel 2007; Yadvinder-Singh et al. 2005). As discussed above, the extent of the impact of tillage and stubble management on soil organic carbon depends on the intensity and type of tillage operations and stubble management regimes, the condition of the soil at the time of tillage, and duration of management practice, and also on soil and crop type, and climate (Balesdent et al. 2000; Chan and Heenan 2005; Mishra et al. 2010; Luo et al. 2010a). Furthermore, depth of sampling can influence the apparent effects on soil carbon of tillage and stubble management (Luo et al. 2010b).

The combination of early, hot burns of stubble and multiple tillage operations before sowing has been shown to result in the lowest soil carbon levels in cropping systems in southeastern Australia (Bell et al. 1995; Chan and Mead 1988; Chan et al. 1992; Harte 1984; Packer et al. 1992; Slattery and Surapaneni 2002). These studies have often compared practice extremes as defined in Table 9.1.

In a long-term tillage trial (19 years) at Wagga Wagga in the wheat belt of south eastern Australia, on a fine sandy loam Luvisol, tillage has had a greater negative effect than stubble burning on soil carbon levels (Chan and Heenan 2005). Soil carbon was about 0.6% greater in no-tillage than traditional tillage treatments in the 0–5 cm layer ($\approx$4.5 t ha^{-1}), while it was 0.19% ($\approx$1.425 t ha^{-1}) lower in stubble-burned treatments than where stubble was retained. The lowest soil organic carbon levels occurred in the traditional tillage–stubble burn treatment. The carbon change was restricted to the surface 5 cm depth.

From trials across southeastern Australia, Valzano et al. (2005) estimated that soil carbon stock to 30 cm depth was 5–10 t ha^{-1} higher on average, with some differences up to 14 t ha^{-1} higher, under conservation tillage compared with traditional tillage, but the magnitude of increase was dependent on rainfall (Fig. 9.2): where rainfall was ~660 mm, direct drill treatments had 57 t C ha^{-1} (30 cm depth) and traditional tillage had 42.5 t C ha^{-1}. For areas where annual rainfall was <660 mm, rainfall, rather than cropping practice, was a more important determinant of soil carbon level. While a difference between tillage treatments was found in the 450–570 mm rainfall class, at average rainfall <450 mm, no differences between the tillage treatments were evident (Valzano et al. 2005). Based on a review of Australian studies, Luo et al. (2010a) reported that adoption of conservation tillage (cf. traditional tillage) increased soil carbon in the surface soil (0–10 cm) across diverse rainfall and temperature regimes; the greatest increase occurred in regions with 500–600 mm mean annual rainfall or at mean annual temperatures <16°C, whereas in regions with annual rainfall less than 300–400 mm or greater than 600 mm, or where mean annual temperatures fall between 18 and 19°C, the increases in soil carbon were relatively small. Furthermore, Franzluebbers and Steiner (2002) found that the capacity for increased soil carbon storage under no-tillage was greatest in the mesic-subhumid regions of North America where the mean annual rainfall to evaporation ratio was 1.1–1.4 mm mm^{-1}, whereas more extreme environments on the drier Great Plains and the cold, humid eastern provinces of Canada had much lower capacity for increased storage of soil carbon under no-tillage. Luo et al. (2010a) hypothesised that although there may be much higher C input into soil from high productivity in high temperature and high rainfall regions, soil carbon output may also be higher under these scenarios, thus suggesting that the interactive effects of rainfall and temperature on soil carbon accumulation in different

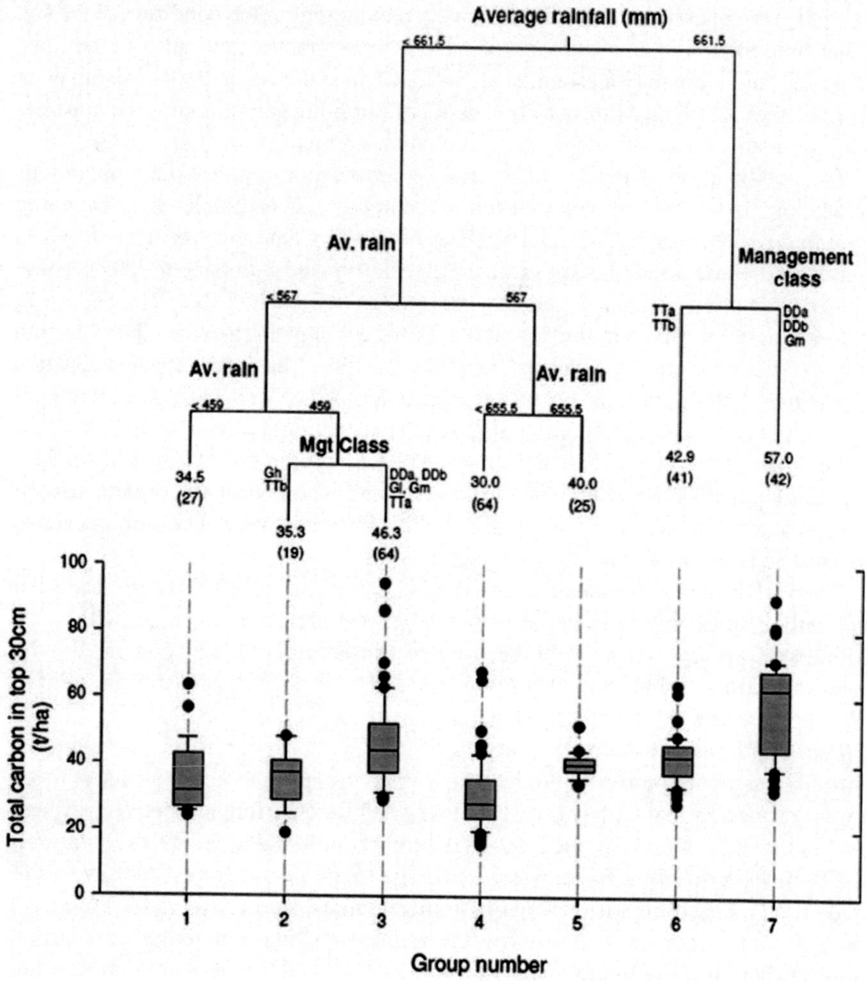

Fig. 9.2 Tree regression of a set of soil carbon densities (t ha to 30 cm) for tillage trials in eastern Australia. It demonstrates that there is an interaction between climate and the tillage, stubble management and land management systems to determine the soil carbon levels in the soils (after Valzano et al. 2005). The *symbols* are: *TTb* is worst traditional tillage; *TTa* is the best traditional tillage; *DDb* is the less-ideal direct drill/no-till–zero-till and includes reduced tillage; *DDa* is the best direct drill/no-till–zero-till; and *Gh*, *Gm* and *Gl* are heavy, moderate and light grazing. Values at the *foot* of each arm are soil carbon stores to 30 cm in t ha^{-1}, and the *brackets* show the number of data points in each set. Values on the *X*-axis are group numbers from the tree regression

crop production and tillage management systems need a systematic analysis in future studies.

One effect that can complicate comparisons of soil carbon levels between cropping systems is that the depth distribution of soil carbon differs between less-intensive tillage (conservation tillage) and the more intensive tillage (traditional tillage). It is clear that in some cases, reported impacts of tillage management on soil carbon result from a change in the depth distribution of soil carbon and not necessarily any change in the total amount of soil carbon (Baker et al. 2007; Hermle et al. 2008; Poirier et al. 2009; VandenBygaart et al. 2003). For example, using meta-analysis of global data from 69 paired experiments, where soil sampling was extended to >40 cm depth, Luo et al. (2010b) showed that conversion from traditional tillage to no-tillage significantly altered the vertical distribution of soil carbon, and increased soil carbon by 3.15 ± 2.42 t ha^{-1} in the surface (0–10 cm), but decreased soil carbon by 3.30 ± 1.61 t ha^{-1} in the deeper (20–40 cm) soil layers. Thus, there was no overall increase in soil carbon stocks under no-tillage (cf. conventional tillage) to 40 cm depth, except in double cropping systems (two crops per year). Furthermore, Hermle et al. (2008) could not show any changes in soil carbon between tillage regimes in Switzerland over a 19-year period. We feel it is significant that these observations were made in wet, cold climates and where mouldboard ploughs were used to invert stubbles to 20–30 cm depth. The study site of Hermle et al. (2008) had an annual average temperature of 8.4°C and annual average rainfall of 1,183 mm. We caution against generalising these results; in cool climates and where tillage practices involve mouldboard ploughs which bury relatively large amounts of stubble to 20–30 cm depths, the decomposition rate of stubble carbon is likely to be slowed by tillage. While soil carbon stocks may be maintained by deep incorporation of residues in cool climates, this may not give the best outcomes for soil health: maintaining high soil carbon levels higher in the surface soil is probably preferable for minimising soil erosion and maximising water infiltration.

The potential for soil carbon differences between tillage treatments to be confounded by different depth distributions of soil carbon is probably less relevant in the semi-arid, hotter environments such as the Australian wheat belt, parts of Brazil and other areas with similar climates and less-intensive tillage practices. In the Australian wheat belt, tillage operations usually affect the surface 5–10 cm and usually avoid inversion. In fact, a range of Australian studies has shown that differences in soil carbon levels between tillage treatments are restricted to the top 5–30 cm (Chan and Heenan 2005; Murphy et al. 2003; Wilson et al. 2008). In Brazilian studies, Bolliger et al. (2006) noted that zero-till soils have highly stratified soil carbon distributions in the absence of soil inversion and mixing, with most of the soil carbon being concentrated near the soil surface. They concluded that soil carbon measurements to at least 20 cm are required to compare soil carbon levels between tillage systems. In some cases, they suggested that it may be necessary to measure soil carbon to 100 cm. However, some critical conclusions from the review of Bolliger et al. (2006) were first that the zero-till soils had higher levels of labile soil carbon or particulate organic matter (POM), and this indicates the soil condition would be expected to be better. Second, a large body of work in Brazil confirms the conclusion that higher levels of soil carbon accumulate under zero-till than under tilled land. They also note the importance of appropriate liming and fertiliser regimes to achieve these increases in soil carbon. More recently, Roper et al. (2010) reported increases in total soil carbon under no-tillage/conservation tillage treatments of 2.6–4.4 t ha^{-1} (to 10 cm depth) during a 7-year tillage experiment on a deep sandy soil in Western Australia. The evidence from the various published studies is that whether conservation tillage regimes have higher soil carbon levels than traditional tillage regimes will depend on climate (temperature and rainfall), soil and crop type, agronomic management, nutrient management, and the type, intensity and duration of tillage practiced, as summarised in Table 9.2. Several reviews have reached similar conclusions, including

Table 9.2 Suggested checklist to assess the potential for an increase in soil carbon levels under no-tillage/zero-tillage cropping systems in comparison to traditional tillage systems

Factor	Details to consider
Biomass production	Has the biomass productivity been restricted by a. Nutrient deficiencies b. Agronomic management factors (poor emergence, wrong variety for district, etc.) c. Soil physical condition – are the surface soils sodic? d. Soil chemical imbalance – soil acidity
Annual average temperature	As a general indication a. <12°C high biomass production and low decomposition rates – build up of soil carbon is likely b. >17°C high decomposition – high decomposition rates – build up of soil carbon less likely
Average annual rainfall	As a general indication a. <450 mm – less likely for build up of soil carbon under no-tillage b. 550–700 mm – possible conditions for highest build up of soil carbon under no-tillage c. >1,000 mm – may not build soil carbon under no-tillage
Tillage type	a. Inversion deep (>20 cm) – likely to result in large amounts of organic matter at depth – unlikely to have build up of soil carbon under no-tillage – interaction with climate is also expected b. Inversion shallow (<20 cm) – soil carbon maintained largely near surface – likely to have soil carbon build up under no-tillage c. Tyne deep (>20 cm) – likely to result in moderate amounts of organic matter at depth – likely to have build up of soil carbon under no-tillage d. Tyne shallow (<20 cm) – soil carbon maintained largely near surface – likely to have soil carbon build up under no-tillage
Stubble management	a. Stubble burnt in traditional tillage treatment (early burn) – likely to have soil carbon build up under no-tillage b. Stubble burnt in traditional tillage treatment (late burn) likely to have small soil carbon build up under no-tillage c. Stubble incorporated in traditional tillage treatment – effect will depend on tillage type and climate
Initial soil carbon levels	If initial soil carbon levels are very high, modified cropping practices may not increase soil carbon levels. For example: a. It may be a case of implementing cropping practices that reduce the rate of soil carbon loss b. It is not reasonable to expect a conservation tillage cropping practice to increase soil carbon levels after a well-managed pasture phase

These values and guidelines are tentative and are largely indicative rather than being definite values

Bhupinderpal-Singh and Rengel (2007), Bolliger et al. (2006), Franzluebbers and Steiner (2002), Luo et al. (2010a, b) and Valzano et al. (2005).

Despite the mixture of results in the literature, it is clear that over a wide range of soils and under a range of climatic conditions in general, conservation tillage practices including stubble retention, direct drill, no-tillage and zero-tillage can generally increase surface soil carbon levels relative to the more traditional tillage practices, but whether conservation tillage practices increase soil carbon in deeper soil layers, and under what cropping systems and climatic conditions, needs further research (Luo et al. 2010a, b).

9.3 Tillage/Crop Stubble Management: Impacts on Soil Physical, Chemical and Biological Health

In addition to effects on soil carbon, tillage and stubble management can directly or indirectly affect important soil physical, chemical and biological parameters with implications for soil health. This section presents a brief overview of such impacts; detailed reviews can be seen in Kumah and Goh (2000), Bhupinderpal-Singh and Rengel (2007), and Yadvinder-Singh et al. (2005).

9.3.1 Soil Physical Health

The important soil physical changes impacted by tillage and stubble management are as follows.

9.3.1.1 Aggregate Stability

Organic matter from crop stubbles can increase the binding of soil particles into aggregates. Soil aggregation is a two-tiered process with smaller aggregates (<250 μm) being held together by the more resistant humic fraction of the organic matter, and the larger aggregates (>250 μm) being held together by the less stable (labile) fraction (Tisdall and Oades 1979). Loss of soil carbon, such as may occur through tillage, usually results in lower aggregate stability (Hamblin 1980).

9.3.1.2 Water-Holding Capacity

The potentially higher soil organic matter under conservation tillage could contribute to increase in water-holding capacity by increasing the volume of macro- and micro-pores in soils (Manrique and Jones 1991; Kaur et al. 2002; Tranter et al. 2007).

9.3.1.3 Infiltration/Permeability

Stubble retained on the soil surface protects the soil from sealing through raindrop impact, thereby enhancing water infiltration. Improved aggregate stability associated with higher levels of soil organic matter, which may result under conservation tillage, enhances soil porosity, which also increases infiltration and permeability. Higher levels of soil organic matter are also associated with increased root growth and soil faunal activity which add to macropore development (Chan and Mead 1988; Murphy et al. 1993; Packer et al. 1992).

9.3.1.4 Bulk Density and Air Porosity

Improved aggregate stability and porosity and potentially higher soil organic matter under conservation tillage may result in lower bulk density and improved soil aeration. In contrast, reduction in bulk density and increase in soil porosity resulting from tillage are temporary effects that diminish quickly as soil compacts due to raindrop impact (Murphy 1998; Onstad et al. 1984).

9.3.1.5 Friability and Compactability

Macks et al. (1996) observed increased friability with increasing soil carbon levels, and a severe loss of friability in soils after extended periods of frequent tillage and stubble burning.

Plastic limit, or the water content when soil aggregates begin to break down, is higher when the organic matter increases (Cotching et al. 2001). Soils with higher organic carbon do not compact as readily due to lower bulk densities and better structure (Hamblin 1987).

9.3.2 Soil Chemical Health

Tillage and stubble management can also affect soil chemical health.

9.3.2.1 Soil pH

Traditional tillage may decrease soil pH by enhancing carbon and nitrogen cycling rate in soil (Bolan and Hedley 2003) or by bringing deep acidic layers to the surface during tillage operations (Chan et al. 1992). Stubble burning in situ can increase soil pH (Brennan et al. 2004). Conservation tillage involving retention of crop stubbles may increase soil pH through a series of acid-neutralising processes including:

(a) decarboxylation of organic anions, (b) ligand exchange of OH^- by specific adsorption of organic anions on aluminium and iron hydroxides and (c) addition of excess base cations in crop stubbles (Bhupinderpal-Singh and Rengel 2007).

9.3.2.2 Soil Nutrient Availability

Tillage and stubble management can affect soil nutrient availability: (a) by altering soil organic matter content and quality, (b) by modifying the rate of cycling of nutrients in soil and (c) by directly supplying nutrients from crop stubble following their burning and retention of ash in situ (Bhupinderpal-Singh and Rengel 2007). Although non-volatile nutrients are rapidly released following stubble burning, hot burns lead to substantial mean losses of nitrogen (85%) and sulphur (74%) via volatilization (Angus et al. 1998; Heard et al. 2006). While if a crop stubble is retained, significant quantities of nutrients are returned to soil (Scott et al. 2010). The proportion of nutrients that will be available to subsequent crops will depend on the rate and extent of decomposition, losses through erosion and leaching, and immobilisation in microbial biomass.

Repeated and variable retention of crop stubble on the soil surface can cause a significant gradient in horizontal and vertical distribution of nutrients in soil, generally leading to greater accumulation of nutrients on the soil surface than at greater depths, whereas incorporation of crop stubble may reduce the nutrient distribution gradient with depth (Du Preez et al. 2001; Salinas-Garcia et al. 2001). Retention or incorporation of low carbon-to-nutrient (e.g. carbon-to-nitrogen ratio) residues may lead to rapid release of nutrients (e.g. nitrogen) through mineralisation. In contrast, when, for example, a high carbon-to-nitrogen ratio stubble is retained, microbial immobilisation may cause a temporary deficiency in nitrogen (Rice 2002), which needs to be compensated by supplying extra nitrogen from other organic or inorganic sources. However, temporary immobilisation of nitrogen or other nutrients in microbial biomass is a less significant issue than the absolute loss of nutrients that would occur if the stubble was burned (Scott et al. 2010).

9.3.3 Soil Biological Health

Improved soil biological health, including increased biodiversity, is important for maintaining soil's ability to perform important soil functions such as nutrient cycling, soil aggregation and porosity development, stubble and crop residue breakdown and the biological control of crop pathogens. Broadly, a biologically healthy soil has a balance of micro-organisms from the bacteria, fungi, protozoan and nematode groups. It also has an active population of meso- and macro-fauna such as earthworms, springtails and other arthropods (Pankhurst and Lynch 1994). As Pankhurst and Lynch suggest:

> A major aim of soil biotic management is to manipulate the processes of residue decomposition, nutrient mobilisation, and mineralisation so that nutrient release is synchronised with plant growth (p. 7).

The enhancement of soil biological activity is seen as one of the benefits of tillage and stubble management practices that minimise soil disturbance and compaction and increase organic matter content. Soil microbial biomass tends to increase with reduced tillage and stubble retention (Carter and Mele 1992; Dalal and Chan 2001), whereas stubble removal or burning in situ may decrease soil microbial biomass (Hoyle et al. 2006). Roper et al. (2010) showed higher microbial C and N quotients and lower metabolic quotient under conservation tillage because of less soil disturbance than under traditional tillage, thus suggesting that these indices may serve as sensitive indicators of the impact of tillage management on soil health.

9.4 Tillage/Crop Stubble Management: Impacts on Soil Erosion

Wind and water erosion can result in significant loss of soil organic matter and nutrients, which in turn results in a decline in soil carbon stocks and in soil health. Both wind erosion and water erosion are increased by a reduction in ground cover and by an increase in number and intensity of tillage operations (Leys 2007; Leys and McTainsh 1994; Marston 1978; Packer et al. 1992). The greatest risk of erosion occurs when soil is in a loose tilled condition and has no surface cover (Renard et al. 1997; Rosewell and Edwards 1988). Therefore, it is likely that soil erosion has occurred on many of the soils where native vegetation has been cleared and cropping implemented. Long-term annual average rates of sheet and rill erosion for the cropping area of New South Wales in eastern Australia are often up to 5–10 t ha^{-1} $year^{-1}$, depending on rainfall erosivity, degree of slope, length of slope and the cropping practices (Edwards and Zierholz 2007). Given that the topsoil has the highest soil organic carbon levels, erosion is estimated to result in a loss of at least 0.05–0.1 t ha^{-1} of soil organic carbon per year, assuming a soil organic carbon concentration of 1%. Hairsine et al. (1993) estimated that about 2.8 t ha^{-1} of soil organic carbon (28 mm topsoil) was lost from the surface soil in one major rainfall event when 81 mm of rain fell in 45 min.

In Nigeria, Lal (1976) demonstrated dramatic effects of stubble management and tillage on soil losses: the annual rates of soil erosion on a 15% slope were 230 t ha^{-1} in bare fallow, 41 t ha^{-1} under conventional (traditional) cultivation, 0.1 t ha^{-1} in no-tillage and 0.0 t ha^{-1} with mulched surface. These data demonstrate the additive effects of tillage and stubble management with the most erodible being the cultivated soil with no surface cover that occurs in bare fallow. In the USA, similar results are recorded with the typical annual soil loss being 48.5 t ha^{-1} under conventional (traditional) cultivation and 2.0 t ha^{-1} under no-tillage wheat (Angima 2002).

Soil losses associated with wind erosion are generally of a similar order of magnitude to those of water erosion; some of the critical control mechanisms are similar to those of water erosion, with maintenance of cover and reductions in tillage being critical to the control of wind erosion (Leys 2007).

The stubble of legumes, having low C:N ratio, decomposes more rapidly than stubble with relatively high C:N ratio, leading to bare soil that is susceptible to wind and water erosion.

9.5 Tillage/Crop Stubble Management: Impacts on GHG Emissions

The importance of CO_2 to climate change has always been emphasised; however, emissions of non-CO_2 GHGs (such as CH_4, N_2O) also need to be taken into account when considering the net climate change impacts of land management actions.

9.5.1 Carbon Dioxide

Clearing of land for cultivation and traditional cropping practices has been a major source of CO_2 emissions through decomposition of plant biomass and soil organic matter. Now there is the chance to reverse this trend by sequestering soil carbon through changes in land management such as through adopting conservation tillage/stubble management practices. Slattery and Surapaneni (2002) estimated that if soil carbon levels were to be increased by 1% (0–10 cm soil layer) in the cropping soils of Australia over 30 years, this would result in the sequestration of 11 Mt C year^{-1} (40 Mt CO_2 -e), offsetting up to 8–10% of Australia's emissions. Luo et al. (2010a) have reported that there is the potential to sequester up to 325 million tonnes of carbon annually worldwide by the adoption of conservation agriculture practices. This is equivalent to 1,200 million tonnes of CO_2 per year, or 4% of global emissions from fossil fuels; these numbers are consistent with those reported by Smith et al. (2008), i.e. ~1,300 million tonnes of CO_2 -e global mitigation potential through conservation management of cropland.

9.5.2 Methane

Globally, aerobic, well-drained soils are a net sink for CH_4 (CH_4), an important GHG with global warming potential 25 times larger than that of CO_2 (Denman et al. 2007). These soils consume (oxidise) about 30 Tg CH_4 year^{-1} during 2000–2004, which is equivalent to 5% of the annual load of CH_4 to the atmosphere (Denman et al. 2007). The major sources of CH_4 emissions from cropping systems are: enteric fermentation associated with domestic stock in cropping systems, flooded rice fields, anaerobic decomposition of plant residues and organic matter, and burning of plants and plant residues (Muller and Bartsch 1999). The impacts of conservation vs. traditional tillage practices on CH_4 oxidation are variable. For example, tillage may adversely affect CH_4 oxidation due to direct or indirect impacts of mechanical disturbance on soil nitrogen availability, physical soil properties (e.g. porosity, soil structure) and consequently soil's capacity to oxidise atmospheric CH_4 (Suwanwaree and Robertson 2005; Ussiri et al. 2009), but a lack of impact of tillage on CH_4 oxidation has also been reported (Suwanwaree and Robertson 2005). In contrast, no-tillage practice may enhance soil CH_4 oxidation by improving soil macroporosity and pore continuity, thereby enhancing CH_4 diffusivity and uptake by methanotrophs (Ussiri et al. 2009). On the other hand, CH_4 oxidation may be lower under no-tillage and stubble retention (cf. traditional tillage) in irrigated cropping systems, because higher water-filled pore space coupled with presence of easily decomposable crop stubble in no-tillage systems may create ideal conditions for greater CH_4 production than oxidation in soil (Alluvione et al. 2009). The net effects of tillage/crop stubble management on soil CH_4 flux can be quite complex and need further research.

9.5.3 *Nitrous Oxide*

Nitrous oxide (N_2O) is a powerful GHG having a global warming potential 300 times that of CO_2. Agriculture is the main (about 80%) source of N_2O emissions (Smith et al. 2008). The main sources of N_2O in agriculture are:

- N fertiliser use: it is estimated that for every 100 kg of fertiliser N that is applied, on an average 1 kg of N will be emitted directly as N_2O, and up to 0.325 kg indirectly from N lost to volatilisation and leaching (IPCC 2006);
- Soil disturbance and the expansion of legume-based pastures;
- Livestock urine and manure;
- Grassland and savannah burning.

Nitrous oxide production is favoured in low-oxygen environments, such as waterlogged soils, or anoxic microsites in moist soils, and in soils with low aeration due to compaction or surface sealing. The highest rates of N_2O release occur at 60–80% water-filled pore space (Dalal et al. 2003a).

Management practices that reduce N_2O emissions include the following (see Dalal et al. 2003a, b):

- Matching the addition of N fertilisers with crop and pasture needs;
- Applying N fertiliser in granulated or slow release forms;
- Maintaining stubble at the surface and reducing cultivation rather than incorporating stubble, to minimise the mineralisation of nitrogen during the fallow period;
- Improving soil structure so that soils are well-drained and compaction zones minimised. Ball et al. (1999) showed that N_2O emissions were increased under compacted soils compared to uncompacted soils;

- Using nitrification inhibitors (Bhatia et al. 2010);
- Incorporating crop residues with high C/N ratios tends to reduce N_2O emissions;
- Managing soil moisture in irrigation systems to ensure that soils are not at 60–80% water-filled pore space when N fertilisers are added.

There is conflicting evidence about whether no-tillage systems result in reduced N_2O emissions in comparison to chisel till (conservation tillage) or traditional till systems. Bhatia et al. (2010) showed that no-tillage resulted in higher N_2O emissions compared to traditional tillage, largely as a result of increased water-filled pore space under no-tillage. On the other hand, Jacinte and Dick (1997) and Ussiri et al. (2009) showed that no-tillage systems had lower N_2O emissions. In these cases, it appears that the no-till had decreased bulk densities and improved soil aeration. Rochette (2008) and Rochette et al. (2008) found that N_2O emissions from no-till vs. tilled soils were similar in a well-aerated loamy soil, but were higher in a poorly aerated heavy clay soil because N_2O emission via denitrification was favoured.

In general, the mitigation of N_2O emissions requires matching the addition of N fertilisers to the needs of the growing crops and pastures, as well as maintaining soils in a good physical condition to avoid compaction zones. It seems that soil properties and tillage practices that favour better soil aeration are likely to mitigate soil N_2O emissions.

9.6 Role of Tillage and Stubble Management in Climate Change Mitigation and Adaptation

Climate change can be expected to impose stresses on soils and land resources. Increases in temperature, higher drought events associated with reductions in annual rainfall and higher storm frequencies with higher intensity rainfall are likely impacts of climate change (Bates et al. 2008). A recent investigation confirms that increases in temperature and reduction in annual rainfall are likely for southern Australia (DECCW 2010). These climatic changes have the potential to significantly impact on soil carbon and soil health (Luo et al. 2010a; see Chaps. 1 and 5). For example, the increased temperatures are likely to increase the decomposition rates of soil carbon, and in combination with lower annual rainfall can adversely impact plant productivity and reduce soil carbon levels (see Chap. 5). Furthermore, extreme rainfall events have the potential to increase the rate of soil carbon cycling (see Chap. 7).

Likewise, higher potential water erosion rates can be expected from increased storm frequencies and rainfall erosivity under future climate change scenarios, and this may adversely affect plant productivity and consequently ground cover. Reduced ground cover and lower rainfall events are also likely to increase the wind erosion hazard. These potential climate change effects will place more emphasis on the need to maintain or implement better land management practices to avoid soil degradation, increase soil carbon and maintain plant productivity. In a cropping enterprise, stubble retention combined with no or reduced tillage will play a major role in protecting the soil surface from increased rainfall erosivity, wind erosion and higher surface soil temperatures. Furthermore, in areas expected to suffer reduced annual rainfall, tillage and crop stubble management practices that maintain or enhance soil carbon, and therefore water-holding capacity, will be particularly important to increase water use efficiency and maintain crop yields. Thus, an improved understanding of the role of tillage and stubble management practices in increasing soil carbon stocks and reducing GHG emissions under future climate change scenarios (increased temperatures, altered rainfall patterns) will assist in mitigating and adapting to adverse impacts of climate change on soil health (Luo et al. 2010a). The growing use of crops for biofuel production will increase pressure on soils and land resources, adding to the need for implementation of better cropping practices.

9.7 Conclusions and Implications

Increasing food and biofuel production to meet the needs of the growing world population in an environment of uncertainties about climate change is going to be a major challenge for communities. Maintaining and improving soil health is a key element in meeting these demands for food, fibre and fuel for present and future generations. In a cropping enterprise, minimal tillage and stubble retention can improve soil health by directly and indirectly altering soil physical, chemical and biological properties. They will also provide protection against degradation by water and wind erosion and losses of soil organic matter and nutrients. There is potential for reduced tillage and stubble retention to reduce emissions of CH_4 and N_2O GHGs, provided appropriate considerations are given to crop and soil type, climatic conditions and fertiliser regimes while implementing conservative tillage/stubble management for improving soil health. Furthermore, these practices are likely to improve adaptation to climate change, through their positive effects on soil health and system resilience due to improved rainfall infiltration, moisture holding capacity, soil friability and better erosion management. However, there still remain some unresolved issues, and further research is needed to identify and quantify these effects in different soil types, environments and climate change scenarios.

The magnitude of impact of minimal tillage and stubble retention on soil carbon in particular, and soil health in general, is highly dependent on site-specific factors such as climate and soil type. It is also dependent on the specific tillage/farming operations, land management history, duration of implementation of conservation farming practices and associated land management practices such as crop rotations and inclusion of a pasture phase.

Some caution is needed in taking a set of research results for one environment and soil type and applying them beyond the climate, soil type and tillage conditions in which they were produced. There is a need for the development of a climate by soil type by cropping practice database or framework in which to place available research results so that the significance of available research can be better understood.

Acknowledgement We gratefully acknowledge helpful comments from Yin Chan on an earlier draft of this chapter.

References

Alluvione F, Halvorson AD, Del Grosso SJ (2009) Nitrogen, tillage and crop rotation effects on CO_2 and CH_4 fluxes for irrigated crops. J Environ Qual 38:2023–2033

Amado TJC, Bayer C, Conceic¸a˜o PC, Spagnollo E, Campos BC, da Veiga M (2006) Potential of carbon accumulation in zero tillage soils with intensive use and cover crops in Southern Brazil. J Environ Qual 35:1599–1607

Angers A, Racous S (1997) Decomposition of wheat and rye residue as affected by particle size. Plant Soil 189:197–203

Angima SD (2002) Vegetative techniques for erosion and sedimentation control. In: Lal R (ed) Encyclopaedia of soil science. Marcel Dekker, New York

Angus J, Poss R, Kirkegaard J (1998) Long term benefits of stubble. Aust Grain 7:29–30

Baker JM, Achnsner TE, Venterea RT, Griffis TJ (2007) Tillage and soil carbon sequestration – what do we really know? Agric Ecosyst Environ 118:1–5

Balesdent J, Chenu C, Balabane M (2000) Relationship of soil organic matter dynamics to physical protection and tillage. Soil Tillage Res 53:215–230

Ball BC, Parker JP, Scott A (1999) Soil and residue management effects on cropping conditions and N_2O flux under controlled traffic in Scotland. 2. Nitrous oxide, soil nitrogen status and weather. Soil Tillage Res 52:191–201

Bates BC, Kundzewicz ZW, Wu S, Palutikof JP (eds) (2008) Climate change and water. IPCC technical paper VI. Intergovernmental panel on climate change. IPCC Secretariat, Geneva, 210 pp

Bell MJ, Harch GR, Bridge BJ (1995) Effects of continuous cultivation on Ferrosols in subtropical southeast Queensland. I. Site characterisation, crop yields and soil chemical status. Aust J Agric Res 46:237–253

Bhatia A, Sasma S, Jain N, Pathak H, Kumar R, Singh A (2010) Mitigating N_2O emission from soil under conventional and no-tillage in wheat using nitrification inhibitors. Agric Ecosyst Environ 136:247–253

Bhupinderpal-Singh, Rengel Z, Bowden JW (2006) Carbon, nitrogen and sulphur cycling following incorporation of canola residue of different sizes into a nutrient-poor sandy soil. Soil Biol Biochem 38:32–42

Bhupinderpal-Singh, Rengel Z (2007) The role of crop residues in improving soil fertility. In: Marschner P, Rengel Z (eds) Soil biology, vol 10. Nutrient cycling in terrestrial ecosystems. Springer, Heidelberg

Bolan NS, Hedley MJ (2003) Role of carbon, nitrogen, and sulfur cycles in soil acidification. In: Rengel Z (ed) Handbook of soil acidity. Marcel Dekker, New York, pp 29–56

Bolliger A, Magid J, Amado TJC, Neto FS, de Fatima dos Santos Ribeiro M, Calegari A, Ralisch R, de Neergaard A (2006) Taking stock of the Brazilian "Zero-till revolution": a review of landmark research and farmers practice. Adv Agron 91:41–110

Brennan RF, Bolland MDA, Bowden JW (2004) Potassium deficiency, and molybdenum deficiency and aluminium toxicity due to soil acidification, have become problems for cropping sandy soils in south-western Australia. Aust J Exp Agric 44:1031–1039

Campbell CA, McConkey BG, Gameda S, Izaauralde RC, Liang BC Zenter RP, Sabourin D (2002) Efficiencies of conversion of residue C to soil C. In: Kimble JM, Lal R, Follett RF (eds) Agricultural practices and policies for soil carbon sequestration in soil. CRC, New York

Carter MR, Mele PM (1992) Changes in microbial biomass and structural stability at the surface of a duplex soil under direct drilling and stubble retention in north-eastern Victoria. Aust J Soil Res 30:493–503

Carter MR, Angers DA, Gregorich EG, Bolinder MA (2003) Characterizing organic matter retention for surface soils in eastern Canada using density and particle size fractions. Can J Soil Sci 83:11–23

Chamen W, Vermeulen GD, Campbell DJ, Sommer C (1992) Reduction of traffic-induced soil compaction: a synthesis. Soil Tillage Res 24:303–318

Chan KY, Heenan DP (2005) The effects of stubble burning and tillage on soil carbon sequestration and crop productivity in south eastern Australia. Soil Use Manage 21:427–431

Chan KY, Mead JA (1988) Surface physical properties of a sandy loam soil under different tillage practices. Aust J Soil Res 26:549–559

Chan KY, Roberts WP, Heenan DP (1992) Organic carbon and associated soil properties of a red earth after 10 years of rotation under different stubble and tillage practices. Aust J Soil Res 30:71–83

Chan KY, Oates A, Li GD, Conyers MK, Prangnel RJ, Poile G, Liu DL, Barchia IM (2010) Soil carbon stocks under different pastures and pasture management in the higher rainfall areas of south-eastern Australia. Aust J Soil Res 48:7–15

Cotching WE, Cooper J, Sparrow LA, McCorkell BE, Rowley W (2001) Effects of agricultural management on Sodosols in northern Tasmania. Aust J Soil Res 39:711–735

Dalal RC, Chan KY (2001) Soil organic matter in rainfed cropping systems of the Australian cereal belt. Aust J Soil Res 39:435–464

Dalal R, Wang W, Robertson GP, Parton WJ, Myer CM, Raison RJ (2003a) Emission sources of N_2O from Australian agricultural and forest lands and mitigation options. Technical report no 35, National carbon accounting system, Australian Greenhouse Office, Australian Government, Canberra

Dalal RC, Wang W, Robertson GP, Parton WJ (2003b) Nitrous oxide emission from Australian agricultural lands and mitigation options: a review. Aust J Soil Res 41:165–195

De Maria IC, Nabude PC, Castro OM (1999) Long-term tillage and crop rotation effects on soil chemical properties of a Rhodic Ferralsol in southern Brazil. Soil Tillage Res 51:71–79

DECCW (2010) The impacts of climate change on the biophysical environment of New South Wales. Department of Environment, Climate Change and Water NSW, Sydney. ISBN 978 1 74232 259 9

Denman KL, Brasseur G, Chidthaisong A, Ciais P, Cox PM, Dickinson RE, Hauglustaine D, Heinze C, Holland E, Jacob D, Lohmann U, Ramachandran S, da Silva Dias PL, Wofsy SC, Zhang X (2007) Couplings between changes in the climate system and biogeochemistry. In: Solomon S, Qin D, Manning M, Chen Z, Marquis M, Averyt KB, Tignor M, Miller HL (eds) Climate change 2007: the physical science basis. Contribution of working group I to the fourth assessment report of the intergovernmental panel on climate change. Cambridge University Press, Cambridge

Doel DS, Honeycutt CW, Halteman WA (1990) Soil water effects on the use of heat units to predict crop residue carbon and nitrogen mineralisation. Biol Fertil Soils 10:102–106

Du Preez CC, Steyn JT, Kotze E (2001) Long-term effects of wheat residue management on some fertility indicators of a semi-arid Plinthosol. Soil Tillage Res 63:25–33

Edwards K, Zierholz C (2007) Soil formation and erosion rates. In: Charman PEV, Murphy BW (eds) Soils: their properties and management. Oxford University Press, Melbourne

Franzluebbers AJ, Steiner JL (2002) Climatic influences on soil organic carbon storage with no tillage. In: Kimble JM, Lal R, Follett RF (eds) Agricultural practices and policies for carbon sequestration in soils. CRC, New York

Grace PR, Oades JM, Keith H, Hancock TW (1995) Trends in wheat yields and soil organic carbon in the permanent rotation trial at the Waite Agricultural Research Institute, South Australia. Aust J Exp Agric 35:857–864

Hairsine P, Murphy BW, Packer IP, Rosewell CJ (1993) Profile of erosion from a major storm in the south east cropping zone. Aust J Soil Water Conserv 6:50–55

Hamblin AP (1980) Changes in aggregate stability and associated organic matter properties after direct drilling and ploughing on some Australian soils. Aust J Soil Res 18:27–36

Hamblin AP (1987) The effect of tillage on soil physical conditions. In: Cornish PS, Pratley JE (eds) Tillage – new directions in Australian agriculture. Inkata Press, Melbourne

Harte AJ (1984) The effects of tillage on the stability of three red soils of the northern wheat belt. J Soil Conserv Serv New S Wales 40:94–101

He N, Lui Z (1992) Effect of organic and chemical fertilisers on the grain yields and soil properties. In: Proceedings of the international symposium on nutrient management for sustained productivity, vol II. Punjab Agricultural University, Ludhiana, pp 130–132

Heard J, Cavers C, Adrian G (2006) Up in smoke – nutrient loss with straw burning. Better Crops Plant Food 90:10–11

Hermle S, Anken T, Leifeld J, Weisskopf P (2008) The effect of the tillage system on soil organic carbon content under moist cold-temperate conditions. Soil Tillage Res 98:94–105

Hoyle FC, Murphy DV, Fillery IRP (2006) Temperature and stubble management influences microbial CO_2-C evolution and gross N transformation rates. Soil Biol Biochem 38:71–80

IPCC (2006) IPCC guidelines for national greenhouse gas inventories, prepared by the National Greenhouse Gas Inventories Programme. In: Eggleston HS, Buendia L, Miwa K, Ngara T, Tanabe K (eds). IGES, Japan

IPCC (2007) Synthesis report. Climate change 2007, An assessment of the intergovernmental panel on climate change, Intergovernmental panel on climate change, Spain, 2007

Jacinte P, Dick WA (1997) Soil management and N_2O emissions from cultivated fields in southern Ohio. Soil Tillage Res 41:221–235

Jenkinson DS (1977) Studies on decomposition of plant material in soil. V. The effects of plant cover and soil type on the loss of soil carbon from ^{14}C labelled ryegrass decomposing under field conditions. J Soil Sci 28:424–434

Kaur R, Kumar S, Gurung HP (2002) A pedotransfer function (PTF) to estimate soil bulk density and its comparison with existing PTFs. Aust J Soil Res 40:847–857

Kumah K, Goh KM (2000) Crop residues and management practices: effects on soil quality, nitrogen dynamics, crop yields and nitrogen recovery. Adv Agron 68:197–319

Ladd JN, van Gestell M, Monrozier LJ, Amato M (1996) Distribution of organic ^{14}C and ^{15}N in particle size fractions in of soils incubated with ^{14}C, ^{15}N-labelled glucose, NH_4, legume and wheat straw residues. Soil Biol Biochem 28:893–905

Lal R (1976) Soil erosion on Alfisols in Western Nigeria. Geoderma 16:363–375

Lawrie JW, Murphy BW, Packer IJ, Harte AJ (2007) Soils and sustainable farming systems. In: Charman PEV, Murphy BW (eds) Soils: their properties and management. Oxford University Press, Melbourne

Leys JF (2007) Wind erosion. In: Charman PEV, Murphy BW (eds) Soils: their properties and management. Oxford University Press, Melbourne

Leys JF, McTainsh G (1994) Soil loss and nutrient declines by wind erosion – cause for concern. Aust J Soil Water Conserv 7:30–35

Li YX, Tullberg JN, Freebairn DM (2001) Traffic and residue cover effects on infiltration. Aust J Soil Res 39:239–247

Liang BC, McConkey BG, Campbell CA, Johnston AM, Moulin AP (2002) Short term crop rotation and tillage effects on soil organic carbon on the Canadian Prairies. In: Kimble JM, Lal R, Follett RF (eds) Agricultural practices and policies for soil carbon sequestration in soil. CRC, New York

Luo Z, Wang E, Sun OJ (2010a) Soil carbon change and its response to agricultural practices in Australian agro-ecosystems: a review and synthesis. Geoderma 155:211–223

Luo Z, Wang E, Sun OJ (2010b) Can no-tillage stimulate carbon sequestration in agricultural soils? A meta-analysis of paired experiments. Agric Ecosyst Environ 139:224–231

Macks SP, Murphy BW, Creswell HP, Koen TB (1996) Soil friability in relation to management history and suitability for direct drilling. Aust J Soil Res 34:343–360

Manrique LA, Jones CA (1991) Bulk density of soil in relation to soil physical and chemical properties. Soil Sci Soc Am J 55:476–481

Marston D (1978) Conventional tillage systems as they affect erosion. J Soil Conserv Serv New S Wales 34:194–198

Mishra U, Ussiri DAN, Lal R (2010) Tillage effects on soil organic carbon storage and dynamics in Corn Belt of Ohio, USA. Soil Tillage Res 107:88–96

Muller B, Bartsch U (1999) The modelling of anthropogenic CH_4 emissions: methodology and estimates. EV29. Oxford Institute for Energy Studies. ISBN 1 901795 14 4

Murphy BW (1998) Mechanisms that cause changes in soil structure: emphasis on the wheat belt of south eastern Australia. PhD Thesis, University of New South Wales, Sydney

Murphy BW, Koen TB, Jones BA, Huxedurp LM (1993) Temporal variation of hydraulic properties of some soils with fragile structure. Aust J Soil Res 31:179–197

Murphy BW, Rawson A, Ravenscroft L, Rankin M, Millard R (2003) Paired site sampling for soil carbon estimation – NSW. Australian Greenhouse Office Technical Report No. 34, Canberra

Murphy B, Cauchi J, Lawrie J, Taylor S, Welch A (2008) Land and soil capability – how we safely manage our soil. Land classifications of the Central West – 2008. Central West Catchment Management Authority, Wellington

Oades JM (1984) Soil organic matter and structural stability: mechanisms and implications for management. Plant Soil 76:319–367

Onstad CA, Wolfe ML, Larson CL, Slack DC (1984) Tilled soil subsidence during repeated wetting. Trans Am Soc Agric Eng 27:733–736

Packer IJ, Hamilton GJ, Koen TB (1992) Runoff, soil loss and soil physical property changes of light textured surface soils from long term tillage treatments. Aust J Soil Res 30:789–806

Pankhurst CE, Lynch JM (1994) The role of soil biota in sustainable agriculture. In: Pankhurst CE, Doube BM, Gupta VVSR, Grace PR (eds) Soil biota – management in sustainable farming systems. CSIRO, Melbourne. ISBN 0 643 05598 3

Parton WJ, Ojima DS, Schimel DS (1996) Models to evaluate soil organic matter storage and dynamics. In: Carter MR, Stewart BA (eds) Structure and organic matter storage in agricultural soils, Advances in soil science. CRC, New York, pp 421–466

Paustian K, Collins HP, Paul EA (1997) Management controls on soil carbon, Chap. 2. In: Paul EA, Paustian K, Elliot ET, Cole CV (eds) Soil organic matter in temperate agroecosystems. CRC, New York, pp 39–41

Poirier V, Angers DA, Rochette P, Chantigny MH, Ziada N, Tremblay G, Fortin J (2009) Interactive effects of tillage and mineral fertilisation on soil carbon profiles. Soil Sci Soc Am J 73:255–261

Renard KG, Foster GR, Weesies GA, McCool DK, Yoder DC (1997) Predicting soil erosion by water: a guide to conservation planning with the revised universal soil loss equation (RUSLE), Agriculture Handbook No 703. US Department of Agriculture, Washington

Rice CW (2002) Organic matter and nutrient dynamics. In: Lal R (ed) Encyclopaedia of soil science. Marcel Dekker, New York

Rochette P (2008) No-till only increases N_2O emissions in poorly-aerated soils. Soil Tillage Res 101:97–100

Rochette P, Angers DA, Chantigny MH, Bertrand N (2008) N_2O emissions respond differently to no-till in a loam and a heavy clay soil. Soil Sci Soc Am J 72:1363–1369

Roper MM, Gupta VVSR, Murphy DV (2010) Tillage practices altered labile soil organic carbon and microbial function without affecting crop yields. Aust J Soil Res 48:274–285

Rosewell CJ, Edwards K (1988) SOILOSS. A program to assist in the selection of management practices to reduce erosion. Technical Handbook No 11, Soil Conservation Service of NSW, Sydney

Russell JS, Williams CH (1982) Biogeochemical interactions of carbon, nitrogen, sulphur and phosphorus in Australian ecosystems. In: Galbally LE, Freney JR (eds) The cycling of carbon, nitrogen, sulphur and phosphorus in terrestrial and aquatic ecosystems. Australian Academy of Sciences, Canberra, pp 61–75

Salinas-Garcia JR, Báez-González AD, Tiscareño-López M, Rosales-Robles E (2001) Residue removal and tillage interaction effects on soil properties under rain-fed corn production in Central Mexico. Soil Tillage Res 59:67–79

Sanderman J, Farquharson R, Baldock J (2010) Soil carbon sequestration potential: a review for Australian agriculture. http://www.csiro.au/resources/Soil-Carbon-Sequestration-Potential-Report.html (http://www.csiro.au/files/files/pwiv.pdf). Accessed 3 Dec 2010

Scott BJ, Eberbach PL, Evans J, Wade LJ (2010) EH Graham Centre Monograph No. 1: stubble retention in cropping systems in southern Australia: benefits and challenges. In: Clayton EH, Burns HM (eds) Industry & investment NSW, Orange. http://www.csu.edu.au/research/grahamcentre/. Accessed 17 Dec 2010

Sharma PK, Mishra B (2001) Effect of burning rice and wheat crop residues: loss of N, P, K and S from soil and changes in nutrient availability. J Ind Soc Soil Sci 49:425–429

Shi XM, Li XG, Long RJ, Singh BP, Li ZT, Li FM (2010) Dynamics of soil organic carbon and nitrogen associated with physically separated fractions in a grassland-cultivation sequence in the Qinghai-Tibetan plateau. Biol Fertil Soils 46:103–111

Six J, Feller C, Denef K, Ogle SM, Sa´ JCM, Albrecht A (2002) Soil organic matter, biota and aggregation in temperate and tropical soils – effects of no-tillage. Agronomie 22:755–775

Slattery WJ, Surapaneni A (2002) Effect of management practices on the sequestration of carbon in duplex soils in south eastern Australia. In: Kimble JM, Lal R, Follett RF (eds) Agricultural practices and policies for carbon sequestration in soils. CRC, New York

Smith P, Martino D, Cai Z, Gwary D, Janzen H, Kumar P, McCarl B, Ogle S, O'Mara F, Rice C, Scholes B, Sirotenko O, Howden M, McAllister T, Pan G, Romanenkov V, Schneider U, Towprayoon S, Wattenbach M, Smith J (2008) Greenhouse gas mitigation in agriculture. Philos Trans R Soc B 363:789–813

Snyder CS, Bruulsema TW, Jensen TL (2007) Greenhouse gas emissions from cropping systems and the influence of fertilizer management – a literature review. International Plant Nutrition Institute, Norcross

Suwanwaree P, Robertson GP (2005) Methane oxidation in forest, successional, and no-till agricultural ecosystems: effects of nitrogen and soil disturbance. Soil Sci Soc Am J 69:1722–1729

Tisdall JM, Oades JM (1979) Stabilisation of soil aggregates by the root systems of ryegrass. Aust J Soil Res 17:429–441

Tranter G, Minasny B, McBratney AB, Murphy B, McKenzie NJ, Grundy M, Brough D (2007) Building and testing conceptual and empirical models for predicting soil bulk density. Soil Use Manage 23:437–443

Tullberg JN, Ziebarth PJ, Li YX (2001) Tillage and traffic effects on runoff. Aust J Soil Res 39:249–257

Ussiri DAN, Lal R, Jarecki MK (2009) Nitrous oxide and CH_4 emissions from long term tillage under a continuous corn cropping system. Soil Tillage Res 104:247–255

Valzano F, Murphy BW, Koen T (2005) The impact of tillage on changes in soil carbon density with special emphasis on Australian conditions. Technical report no 43, National carbon Accounting System, Australian Greenhouse Office, Canberra

VandenBygaart AJ, Gregorich EG, Angers DA (2003) Influence of agricultural management on soil organic carbon: a compendium and assessment of Canadian studies. Can J Soil Sci 83:363–380

Wang X, Gao H, Tullberg JN, Li H, Kuhn N, McHugh AD, Li Y (2008) Traffic and tillage effects on runoff and soil loss on the Loess Plateau of northern China. Aust J Soil Res 46:667–675

Wentworth Group (2009) Optimum carbon in the Australian landscape. How to guide the carbon terrestrial market to deliver multiple economic and environmental benefits. Wentworth Group of Concerned Scientists, Canberra, October 2009

West TO, Post WM (2002) Soil organic carbon sequestration rates by tillage and crop rotation: a global data analysis. Soil Sci Soc Am J 66:1930–1946

Wilson BR, Growns I, Lemon J (2008) Land-use effects on soil properties on the north western slopes of New South Wales: implications for soil condition assessment. Aust J Soil Res 46:359–367

Yadvinder-Singh, Bijay-Singh, Timsina J (2005) Crop residue management for nutrient cycling and improving soil productivity in rice-based cropping systems in the tropics. Advances in agronomy, vol 85. Elsevier, Amsterdam

Chapter 10
Intensification in Pastoral Farming: Impacts on Soil Attributes and Gaseous Emissions

Surinder Saggar, Jiafa Luo, Dong-Gill Kim, and Neha Jha

10.1 Introduction

Grasslands worldwide cover about 25% of the earth's surface, occupy 117 million km^2 of vegetated land and provide forage for over 1,800 million livestock units (Saggar et al. 2009a). They are also one of the key contributors to potent non-carbon dioxide (CO_2) greenhouse gases (GHGs) (Clark et al. 2005). Methane (CH_4) produced by the fermentation of organic matter in an anaerobic environment has a global warming potential (GWP) of at least ~25 times greater than CO_2 (Shindell et al. 2009). The GWP of nitrous oxide (N_2O), which is produced in pastoral soils from mineral nitrogen (N) originating from dung, urine, biologically fixed dinitrogen (N_2) (BFN), applied fertiliser and mineralisation of soil organic matter, is even greater (310). Livestock production is responsible for 18% of global GHG emissions from all human activities measured on a CO_2-equivalent basis (Steinfeld et al. 2006). Intensification of managed pastoral soils affects GHGs emissions and modifies soil properties that have wider environmental impacts on water and air quality. The emissions per unit of milk production are highest in developing regions and least in North America and Europe, and are higher in grazing systems than mixed systems (FAO 2008). Although intensification can produce less GHGs per

S. Saggar (✉) • D.-G. Kim
Global Change Processes Group, Landcare Research, Private Bag 11052, Palmerston North 4442, New Zealand
e-mail: Saggars@landcareresearch.co.nz

J. Luo
Climate, Land and Environment Section, AgResearch, Ruakura Research Centre, Private Bag 3123, Hamilton 3240, New Zealand

N. Jha
Soil and Earth Sciences, Institute of Natural Resources, Massey University, Private Bag 11222, Palmerston North 4442, New Zealand

B.P. Singh et al. (eds.), *Soil Health and Climate Change*, Soil Biology 29,
DOI 10.1007/978-3-642-20256-8_10,

unit of output, intensification can impact negatively on soil quality, biodiversity and eutrophication of water. Moreover, “Compassion in World Farming” – the farm animal welfare charity, considers that intensification is a deeply flawed strategy from the point of view of halting climate change and from environmental and animal welfare considerations (FAO 2008).

Pastoral landscapes are generally managed at the individual farm level, but the impact occurs at catchment level. Farmers have several goals – economic, environmental and lifestyle – that will affect how they manage their land and resources. In countries such as New Zealand, where the economy relies heavily on export income from pastoral agriculture, farmers face pressure to increase efficiency of production to maintain their financial and economic viability, as well as their position as low-cost producers in an internationally competitive market. One way of achieving this is to increase an efficient use of inputs to stimulate a larger increase in returns, and hence an increased output–input ratio. Present pastoral farming trends in New Zealand show that the sector is growing and using higher inputs, including fertiliser, energy, water for irrigation and capital to produce more output from the same area of land (PCE 2004). Reviews by Saggar et al. (2004b, 2009a) and Bolan et al. (2009) provide some data on major changes in New Zealand dairy industry. We can also assume that intensification will continue to occur to meet ever-increasing demands for food, even within landscapes already farmed relatively intensively.

Pastoral agriculture has traditionally focussed on outputs of products, i.e. meat, fibre and milk. However, farming systems have other outputs/effects, e.g. loss of nutrients to water and other impacts on soil properties, and GHG emissions. Pastoral farms can be regarded as forage supply platforms, so more forage produced per unit area is a primary aim. The environmental costs associated with intensive livestock farming (including confined livestock operations) are the disposal of waste products that may cause soil, air and water pollution, increased disease risks to animals and humans, the reduction in biodiversity and increased GHG emissions.

In this chapter, we describe intensification of pastoral agriculture and address its impacts on physical and chemical soil attributes of intensively managed pasture land, with particular emphasis on temperate-grazed pasture systems. We investigate how increased chemical inputs can affect the productivity and environmental quality of these pastoral soils. We describe how intensive management of the pastoral system influences emissions of GHGs. We also explore options for reducing the negative impacts of intensification, and identify current gaps and limitations for developing future sustainable pastoral systems management strategies while maintaining productivity, profitability and the environment. As only a brief description of intensification impacts on pastoral farming is given in this chapter, the reader is referred to appropriate reviews that provide in-depth coverage of relevant topics (Bilotta et al. 2007; Bolan et al. 2004; Carroll et al. 2004; Cuttle 2008; DeKlein and Eckard 2008; Drewry 2006; Kemp and Michalk 2005; Kurtz et al. 2006; Ledgard and Luo 2008; Luo et al. 2010; Oliver et al. 2005; Saggar et al. 2004b, 2009a).

10.2 Intensification of the Pastoral Land Use

Pastoral intensification is a broad term and includes increases in the level of inputs such as fertiliser, stocking rate, irrigation, chemicals, plant and animal germplasm, machinery, labour, and biotechnologies. Intensification here refers to any practice that increases productivity per unit land area at some cost in labour or capital inputs. Intensification of pastoral land use may be considered as a system to feed the world while avoiding the Malthusian outcome. But this unprecedented success has come at a large cost both to the environment and to human health. Responses to different inputs of intensification can have different consequences, e.g. fertiliser application may increase, decrease or have no effect on soil carbon (C) content, while increases in stocking rate may decrease soil C or may have some positive effect on soil C.

In more humid regions including Australia, New Zealand, Europe and parts of North and South America, most pastoral land is intensively managed with substantial inputs of N fertiliser. While pasture production commonly increases with increasing rate of N application, N use efficiency decreases (McKenzie et al. 2006). Pasture generally responds linearly to N application rates up to 200–400 kg N ha^{-1} $year^{-1}$ (Whitehead 1995; Sun et al. 2008). Higher inputs of N fertiliser can result in a large N surplus (i.e. N inputs–N outputs in products). For example, there have been N surplus of 150–250 kg N ha^{-1} $year^{-1}$ in highly productive dairy farm systems in the Netherlands and northern Germany (Rotz et al. 2005), mainly from excessive application of animal manure/excretal deposition. N surpluses in these intensively managed pasture systems are likely to exacerbate N losses to waterways and the atmosphere. The environmental effects of NO_3^- leached to groundwater and other waterways and the potential damage to soils are a major concern to the farming industry, the scientific community and the society. In New Zealand, the declining water quality of Lake Taupo (Vant and Huser 2000), the Rotorua Lakes and algal blooms occurring in Lake Rotoiti has been linked to the land use within the catchment. Environment Waikato data suggest the quality of about 10% of the groundwater in the livestock farming area of the region is below World Health Organisation drinking water standards (Annon 2005). Thus, excessive N additions can contaminate pastoral ecosystems and alter both their ecological functioning and the living communities they support. Another example here is the large dead zone in the Gulf of Mexico, which is a direct result of nutrients and agrochemical run-off from intensively managed agricultural land in the USA via the Mississippi river (McIsaac et al. 2001).

Recent trends in intensive pastoral land use in New Zealand include higher stocking rates and stocking densities, increased use of fertilisers and agricultural chemicals, and increases in irrigation use. In the past two decades, New Zealand pastoral farming has doubled milk production from dairying and, despite a one-third decline in ewes, lamb meat production has increased by 10% (Woodford 2006; Bolan et al. 2009). More intensive pastoral farming (increase in stocking rates or more livestock numbers per hectare) and more productivity per animal (such as increased milk production per cow, or higher lambing percentages and carcass weights) resulted in 38% increase in production between 1990 and 2003. Dairy cow numbers have almost doubled (from 2.92 million in 1981 to 5.22 million by 2006). Between 1990 and 2005, there has been a sixfold increase in N fertiliser use from 0.05 to 0.31 million tons N (MfE 2007a). It is estimated that New Zealand animals annually void almost five times more N (1.5 million tons) than the N fertiliser input (0.31 million tons N) (Saggar et al. 2005). Intensification of pastoral land use has also led to a noticeable increase in the use of irrigation in drier regions to achieve high-producing pastures (MfE 2007b). The combined increase in fertiliser use and irrigation has increased environmental pressures on waterways and groundwater. A shift to more intensive farming in some New Zealand regions has adversely affected soil health (Betteridge et al. 1999, 2002; MacKay 2009; Sparling and Schipper 2002), increased GHG emissions (MfE 2007b), decreased indigenous biodiversity (Leathwick et al. 2003) and reduced freshwater quality in lowland waters and waterways (Quinn et al. 1997). This has raised concerns about ecological sustainability of New Zealand pastoral farming and continuation of its intensification in the future (MacLeod and Moller 2006).

Phosphorus (P) is the major nutrient limiting the growth of clover-based pastures in New Zealand, and superphosphate has been the major P fertiliser in use (Saggar et al. 1993). Pasture improvement can influence P fluxes in waterways and streams by increasing P transfer from the soil, applied fertiliser and animal excretal deposition. In most intensive livestock production areas around the world P inputs are in excess of requirements (Sharpley et al. 1998; OECD 2008). As P requirements for intensively grazed pastures are relatively high and New Zealand soils are naturally P deficient (Caradus 1994; Sinclair et al. 1996), New Zealand dairy farms use large amounts of P fertilisers; this has increased the potential for P loss to waterways (Monaghan et al. 2007). In Northern England, Withers et al. (2007) showed a direct link between upland pasture improvements by liming and P fertilisation and soil P accumulation, which doubled the transfer of dissolved inorganic P and particulate P but not suspended sediment to the drainage stream. In New Zealand, sulphur (S) fertilisers are also applied on intensively grazed pastures in addition to that supplied through the commonly used single superphosphate fertiliser. Total application rate can range from 60 to 100 kg S ha^{-1} $year^{-1}$ for New Zealand dairy farms, and leaching losses of 40–70 kg S ha^{-1} $year^{-1}$ as sulphate-S have been reported (Rajendram et al. 1998).

10.3 Nutrient Inputs and Dynamics in Grazed Pastures

Historically, fertiliser applications have greatly increased pasture and animal production on many grassland soils that are inherently deficient in nutrients. The annual amount of fertiliser nutrients used in a farm system and those recycled through the uneven deposition of animal excreta are the two key factors determining the nutrient surplus, their spatial and temporal heterogeneity, their potential for loss and, therefore, the nutrient-use efficiency. Additionally, in legume-based pastures atmospheric N input through biological N_2 fixation can also contribute to significant N inputs. The amount of biological N_2 fixation depends on a number of factors, including legume species, climatic and soil conditions, nutrient supply and grazing management (Menneer et al. 2004). Estimates of N_2 fixed by legumes (mainly white and subterranean clovers) in temperate pastures throughout the world range from 10 to 270 kg N ha^{-1} $year^{-1}$ (Ledgard 2001). Biological N_2 fixation generally decreases in intensive pasture systems as inorganic N supply to the legumes increases (Saggar 2004; Saggar et al. 2009a; Fig. 10.1).

In grazed pastures, the conversion efficiency of consumed N, P and S into product is low, and a substantial amount of N, P and S (from 70 to 85%) is recycled through the direct deposition of animal excreta. The low utilisation of these pasture nutrients reflects a simple feature of the pasture–animal relationship: in most situations, pasture plants require significantly higher concentrations of N, P and S to grow at optimal rates than is needed by the grazing ruminant for amino acid and protein synthesis (Haynes and Williams 1993). The proportion of N in the urine increases with increasing N content of the diet. In most intensive high-producing pasture systems, where animal intake of N is high, more than half the N is excreted as urine (Oenema et al. 1997).

In intensive dairying systems where winters are cold (e.g. northern Europe), housing for varying periods throughout the year is common in grazing systems. This means collection and application of large quantities of manure become critical for nutrient-use efficiency, as there are many opportunities and places for N compounds to escape from animal manure management systems.

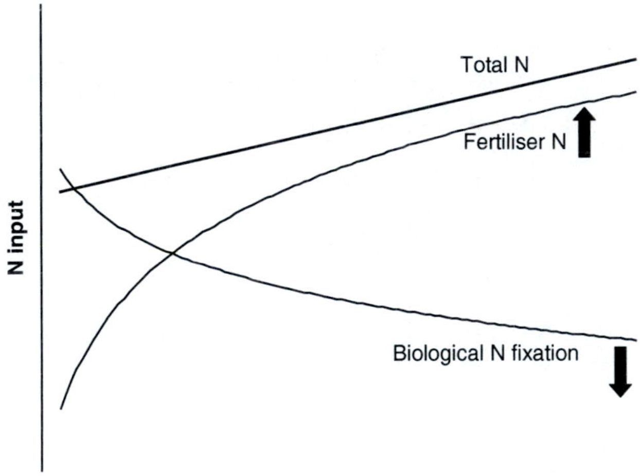

Fig. 10.1 Schematic representation of the influence of increased nitrogen (N) fertiliser application on biological nitrogen fixation (BNF) in legume-based pastures (adapted from Saggar 2004; Saggar et al. 2009a). The x-axis represents the changes in N contribution between BNF and fertiliser N

10.3.1 Nitrogen Transformation Processes in Grazed Pastures

The transformations and losses of N in managed grazed pastures have been reviewed (Bolan et al. 2004; Fig. 10.2). The N in excreta following deposition undergoes microbial mineralisation before it is released as the ammonium ion (NH_4^+) and NH_3 gas. N mineralisation is much faster from urine than from dung. N can be lost to the atmosphere by NH_3 volatilisation, or converted to nitrate (NO_3^-) through the nitrification process by nitrifying bacteria. Nitrate can then be leached and denitrified. Denitrification is the conversion of NO_3^- to gaseous N products (NO, N_2O and N_2). Denitrification rate and the relative production of NO, N_2O and N_2 depend on the availability of mineral N (NH_4^+ and NO_3^-), organic C, temperature and pH, together with processes that lower the redox potential of the soil, such as changes in soil moisture. These factors not only influence the abundance of the denitrifier community, but also affect the denitrification enzyme activity in soils (Wallenstein et al. 2006).

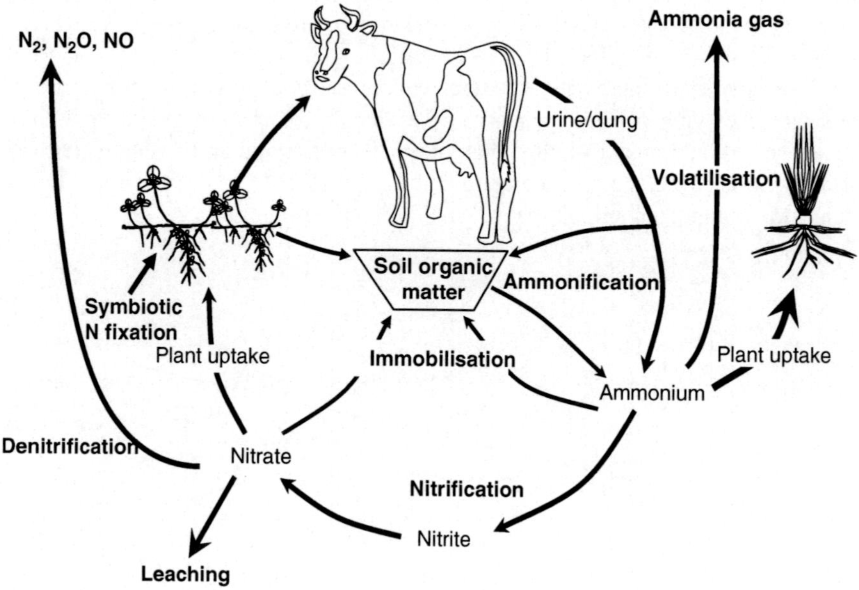

Fig. 10.2 Schematic representations of N transformations and losses in intensively managed dairy-grazed pastures (adapted from Bolan et al. 2004)

10.3.2 Nitrogen Losses

The magnitude of N input to grazed systems is generally the main factor determining the N surplus and, therefore, the potential for N losses. There are limits to how much pasture production can be increased with fertilisation. The intensively

Table 10.1 N inputs and outputs from intensive dairy farm systems in New Zealand receiving nitrogen (N) fertiliser at nil or 410 kg N ha^{-1} year^{-1}

	0 N	410 N
N inputs (kg N ha^{-1} year^{-1})		
Clover N_2 fixation	160 (80–210)	40 (15–115)
Non-symb. fixation + atm. dep.	10	10
Fertiliser N	0	410
Purchased feed	0	41
N outputs (kg N ha^{-1} year^{-1})		
Milk + meat	78 (68–83)	114 (90–135)
Transfer of excreta to lanes/sheds	53 (41–63)	77 (72–91)
Denitrification	5 (3–7)	25 (13–34)
Ammonia volatilisation	15 (15–17)	68 (47–78)
Leaching	30 (12–74)	130 (109–147)
Immobilisation of fertiliser N		70 (60–84)
N balance (kg N ha^{-1} year^{-1})	−11 (−74 to +47)	17 (−11 to +24)
Farm N surplus (kg N ha^{-1} year^{-1})	92	387
N use efficiency (product N/input N)	46%	23%

Bracketed values are range in N flows measured over 5 years (adapted from Ledgard et al. 2009)

managed pasture systems reach N saturation when the plants, microbes and soils cannot use or assimilate or retain more N, and additional N inputs are lost through both leaching and gaseous emissions. Ledgard et al. (1999) found that a threefold increase in total N inputs resulted in a fourfold increase in N surplus, a fourfold to fivefold increase in gaseous and leaching losses, and a halving of the N use efficiency (Table 10.1). The primary transformations leading to N losses are ammonia (NH_3) volatilisation, NO_3^- leaching and denitrification (N_2 and N_2O emissions).

10.3.2.1 Ammonia Volatilisation

In grazed pastures, biological degradation of animal excreta and hydrolysis of fertilisers containing urea and ammonium ions lead to the continuous formation of NH_3 in the soil, which can volatilise to the atmosphere. Jarvis et al. (1989) found that NH_3 loss from urine patches increased under high N fertilisation because more N was excreted in urine. Less NH_3 is lost from grazing systems than from animal housing systems, where the combined loss from the animal houses, manure storage and field application can be large. Jarvis and Ledgard (2002) compared NH_3 losses from two contrasting model dairy systems in the UK and New Zealand. Their study demonstrated distinct differences between the two farming systems in terms of total N input, N off-take, N surplus and NH_3 loss. These values were 1.7, 1.2, 1.8 and 2.4 times greater in the UK than in New Zealand, respectively. The greater loss of NH_3 in the UK farm is attributed mainly to the higher fertiliser N input, and the housing of animals and subsequent spreading of the manure on the farm. However, when NH_3 loss was expressed in relation to the farm N surplus, there was little difference between the two farms; NH_3 loss being approximately 20% of the surplus. Studies conducted in New Zealand and overseas and reviewed by Saggar et al. (2009b) have shown that fertilisers containing urea can lose up to 30% or more of their N through NH_3 emission if not rapidly incorporated into the soils. Compilation of the data using aspirated chambers from studies conducted in New Zealand by Sherlock et al. (2008) suggests the direct average NH_3-N emissions from urine applied to pasture soils are 15.9%. One method of reducing losses is to use a urease inhibitor (UI) that retards the hydrolysis of urea by soil urease and allows the urea to diffuse deeper into the soil. Much of the NH_3 then released would be retained by the soil (Saggar et al. 2009b).

10.3.2.2 Nitrate Leaching

A review of the research on grazed systems suggests that NO_3^- leaching increases exponentially with increased N inputs (Ledgard et al. 2009; Fig. 10.3).

Various studies have also shown that urine N makes a much greater contribution to NO_3^- leaching compared with fertiliser N (because of much larger specific rate of N deposition in urine). Urine typically contributes 70–90% of total N leaching loss (Monaghan et al. 2007). Fertiliser N is generally used efficiently by pastures, but it enhances pasture N uptake and grass-N concentrations, thereby increasing N excretion in urine and consequently the risk of N loss to the environment. Winter

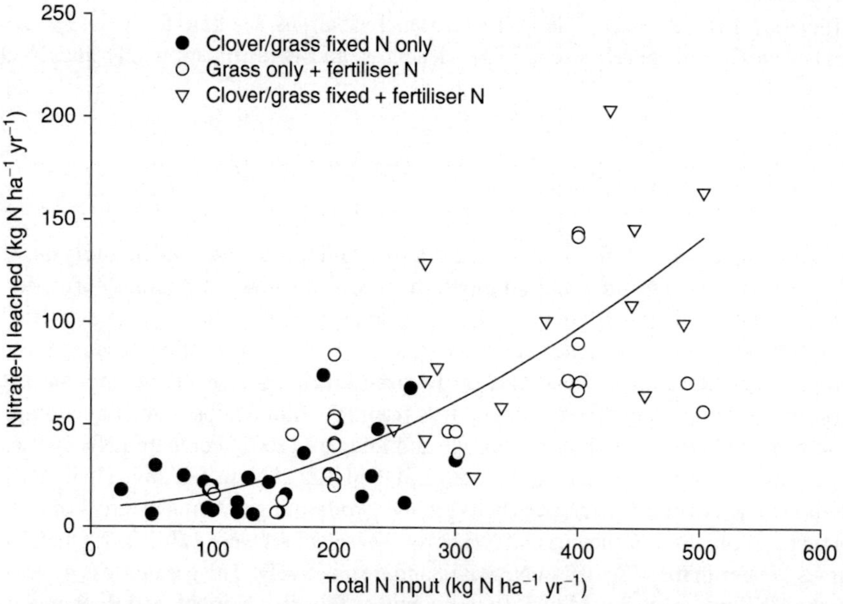

Fig. 10.3 Nitrate leaching from grazed pasture systems as affected by total N input (adapted from Ledgard et al. 2009)

leaching of N can be further exacerbated by dry summer/autumn conditions and an associated slowing down of plant growth, which results in a build-up of NO_3^- levels in soil by the end of autumn (Scholefield et al. 1993). Estimates of N leached from managed pastures vary widely, ranging from 6 to 162 kg N ha^{-1} $year^{-1}$, and this is due to the differences in N input, pasture N uptake, soil drainage, animal type and stocking rate (Stout et al. 2000).

Leaching of N forms other than NO_3^- is generally low. However, ammonium leaching can occur in some soils and may be enhanced where mitigation practices target reduced nitrification. Research also indicates that in some situations, dissolved organic N can be a significant source of leached N (Jones et al. 2004; Bolan et al. 2010).

Eriksen et al. (2004) observed higher leaching losses from grazed N-fertilised ryegrass pasture (on average 47 kg N ha^{-1} $year^{-1}$) than from grazed non-N-fertilised clover/ryegrass pasture (on average 24 kg N ha^{-1} $year^{-1}$). Over time the losses from the clover/ryegrass pasture decreased due to a reduction in N_2 fixation together with a reduction in dry matter production that in turn led to a lower grazing intensity and lower rate of recycling of animal excreta. The research summary of N leaching from grazed pastures in Fig. 10.3 shows overlap of N leaching values estimated from pastures with or without clover at similar N inputs. However, in long-term pastures, N inputs from N_2 fixation are usually less than 200 kg N ha^{-1} $year^{-1}$, thereby limiting maximum N leaching from non-N-fertilised clover/grass pastures. By contrast, N fertiliser may be used at much higher annual rates of application, with potential for high N losses.

10.3.2.3 Denitrification

The process and factors regulating denitrification are described above (see Sect. 10.3.1; Chap. 8). The annual denitrification rates in agricultural and forest soils range between 0 and 239 kg N ha^{-1} (Barton et al. 1999), with the highest rates typically occurring in irrigated, N-fertilised soils and the lowest rates occurring in native ecosystems. In New Zealand pastures, denitrification is considered the primary source of N_2O emissions as nitrification, and other aerobic transformations of urine-derived N contribute little to overall emissions (Luo et al. 1999a, b; 2008c; Saggar et al. 2004a, b, 2007b, 2009a).

10.4 Intensification Impacts on Soil Attributes

Intensively managed grazing systems can result in reductions in biodiversity, increased soil erosion and overland flow, reduced soil weight-bearing capacity, reduced soil quality and increased soil compaction. These effects are usually prominent in temperate climates under excess moisture conditions. A common concern regarding land-use intensification is the potential deterioration of soil quality or soil health. Karlen et al. (1997) defined "soil quality" as the ability of soil to function, where the soil resource is recognised as a dynamic living system comprising a balance of biological, chemical and physical processes; the two terms "soil quality" and "soil health" are considered synonymously in this book (see Chap. 1 and Preface of this book). Sparling and Schipper (2002), surveying soil quality data for 500 New Zealand soils based on land use, demonstrated a degradation in the physical properties of these soils under highly intensive land uses such as dairy farming, arable cropping and horticulture. There are many soil properties that regulate compaction and soil physical health including pore space (e.g. porosity and macroporosity), water movement (saturated and unsaturated hydraulic conductivity), resistance (penetration resistance), soil structure (aggregate size and stability) and bulk density. Soil macroporosity (or air-filled porosity) is a sensitive indicator of soil compaction (Ball et al. 2007) and soil quality. Animal treading can result in the degradation of soil physical quality through the hoof action of grazing animals (Betteridge et al. 1999, 2002; Pande et al. 2000; Ward and Greenwood 2002). These physical attributes provide the environment in which soil biological and chemical processes interact. In a recent review, MacKay (2009) identified soil erosion in hill land, compaction in low land and loss of soil organic matter in some pasture soils as additional emerging soil degradation issues of intensively managed pastoral soils.

In a review of the impacts of grazing animals on soil quality, vegetation and surface water quality in intensively managed grasslands, Bilotta et al. (2007) report that intensively managed grazing can actually lead to the degradation of both soil and vegetation by causing changes in vegetation cover and biodiversity in the pastoral sward, structural deformation of soil, changes in hydrological behaviour and deterioration of water quality within these environments. These authors quote from DEFRA (2005) and UK Environment Agency (2002) reports that ~29% of the total land area in England and Wales is intensively managed, and the damage to homes, commercial property and agricultural land from poor soil structure caused by intensification costs the UK approximately £115 million per year.

Although intensification and increased environmental damage are often associated with increased external inputs, this can also result simply from increased grazing pressures (Cuttle 2008). In terms of soil chemical and biological processes, changes in organic matter concentration, the supply and losses of nutrients, and changes in biological activity reflect the impact of intensification. Soil flora and fauna play an important role in the transformation of organic matter and in regulation of C and cycling of nutrients such as N, P and S. As farming intensifies, nutrients and chemicals are increasingly used to enhance productivity and control weeds, pests and animal diseases. For example, herbicides are used to control weeds, anti-parasitic agents are used to control gastrointestinal parasites and zinc (Zn) supplementation is used to control facial eczema in grazing animals. In addition, pathogenic organisms transferred to soil through animal excreta may transmit infection to other livestock and to humans. This section considers the impacts of intensification of pastures on key soil attributes such as soil C stocks, nutrients and physical condition.

10.4.1 Influence on Soil Carbon

Soil organic matter is important for the sustained function of agro-ecosystems as it influences chemical, biological and physical soil properties and plays a vital role in nutrient cycling (see Chap. 5). There is experimental evidence that increased utilisation of pasture biomass and increased irrigation frequency can reduce soil C content (Hoglund 1985; Lambert et al. 2000; Metherell et al. 2002). This however, is contrary to a common view that intensive pastoral agriculture can build up soil C or at best has a neutral effect. Long-term experiments in New Zealand do provide insights into the steady-state status of soil C in pastoral soils, but the soils at these sites are not without limitations. Saggar et al. (2001) found no difference in soil C levels and P status between fertilised and unfertilised soils. Process-based studies by Saggar and Hedley (2001) and Saggar et al. (1997, 2000) showed that addition of fertiliser not only increased pasture production and translocated more C to roots than non-fertilised pastures, but also enhanced the decomposition of soil organic C. Thus, increased N inputs to intensively managed pasture soils, already well supplied with N, are more likely to decrease C storage (Cuttle 2008), due to more rapid decomposition of N-enriched residues. A recent New Zealand study on soil C storage reveals that soil C has decreased on some dairy pastures but has increased on hill country pastures (Schipper et al. 2007). Investigating the relationship between the above-ground net productivity of permanent swards and soil C concentration, Bélanger et al. (1999) also found no increase in soil C concentrations from increased above-ground net productivity through N, P and K fertilisation. The loss in soil C observed with intensification mainly occurs from labile pools (Ghani et al. 2003), with implications for reduced retention of N (and other nutrients) in the soil, leading to lowered nutrient availability for plant uptake and greater losses to the environment. Soil C in sheep-grazed pastures has recently been shown to increase from the effect of increased atmospheric CO_2 concentration (Dr KR Tate pers. comm.). These results all seem to indicate the dynamic nature of soil organic matter and that several different factors, such as soil moisture, temperature, pasture growth and C input, can interact to cause changes in soil C storage. Overall, the net effect of intensification of pastoral farming on soil C can be neutral or positive or negative depending on the level of saturation of C in the soil.

10.4.2 Influence on Soil Physical Properties

As discussed, soils under pasture can accumulate soil organic matter, favouring the development of good soil structure and other properties to sustain pasture growth. However, intensification of pastoral farming can cause stress on the physical condition of pasture soils. Bilotta et al. (2007) reviewed changes in soil physical properties caused by grazing animals and showed that these can have serious implications for soil quality. They concluded that there are three main forms of soil structural change associated with grazing animals, namely compaction, pugging and poaching. In grazed pastures, animal pugging and treading damage during grazing reduce soil infiltration rates and pasture growth (Drewry et al. 2008) – a reduction that can be serious under wet soil conditions. Animal treading of pasture can also decrease soil porosity and bulk density and consequently cause an increase in mechanical impedance to root penetration and a reduction in aeration and/or an increase in water-logging of soil. This will also have a negative effect on legume growth, productivity and N_2 fixation in pasture (Menneer et al. 2004). In addition, the decrease in soil infiltration capability and hydraulic conductivity due to treading damage makes the soil more prone to ponding, and thus increases the risk of run-off losses of other nutrients, particularly P, and gaseous loss of N through denitrification from intensively grazed pastures (Monaghan et al. 2005; Bhandral et al. 2007). Changes in soil physical properties can also affect nutrient transformation processes in soil, as the changes can alter the moisture regime and influence soil respiration rates and plant nutrient uptake (Di et al. 2001). Pugging and compaction are generally more serious in areas where animals congregate, such as around paddock gateways, water troughs and in camping areas.

As discussed in this section, intensification of pastoral farming can negatively modify soil properties such as structure, permeability and soil organic matter content. These modifications change the buffering and filtering capacity of pastoral soils, for example, by increasing preferential flow and transport leading to faster and greater nutrient and contaminant leakage, favouring the potential for degradation (Ledgard and Luo 2008).

10.5 Intensification Impacts on GHG Emissions

Increased pasture production for higher per hectare animal productivity is the major goal of pastoral farmers in New Zealand, Australia, parts of South and North America, Europe, China and India. However, intensification in pastoral productivity also leads to increased emissions of the potent agricultural GHGs, CH_4 and N_2O. The global emissions of CH_4 and N_2O from grasslands-derived feeds are estimated at about 44 Tg (1 Tg $= 10^{12}$ g $=$ 1 million metric tones) CH_4 year^{-1} and 2.5 Tg N year^{-1}, comprising 18% and 20% of global CH_4 and N_2O emissions, respectively (Clark et al. 2005; Saggar et al. 2009a). These two non-CO_2 GHGs comprise about half of New Zealand's total emissions. Globally, N_2O production has increased by 17% from 1990 to 2005, and it has been assumed that N_2O emissions from agricultural practices will further increase by 35–60% by 2030 (IPCC 2007). Projections by Bouwman et al. (2005) estimate that in the next three decades, intensification involving improved management and use of fertilisers will be required to produce 30% more grass/animal feed to meet the global demand for meat and milk production. The impacts of future livestock intensification and fertiliser use on GHG emissions need to be assessed against the potential increases in grassland productivity and animal production.

Annual CH_4 emissions from enteric fermentation and animal manure are about 106 Tg (Steinfeld et al. 2006) globally. As livestock numbers grow, and livestock rearing becomes increasingly industrial, the production of manure is projected to increase by about 60% by 2030 resulting in similar proportional increases in enteric and manure CH_4 emissions (http://www.fao.org/docrep/004/y3557e/y3557e11.htm). Therefore, livestock CH_4 emissions are directly proportional to livestock intensification, except in situations where output per livestock unit is also increased causing reduced CH_4 emissions per unit of feed intake. Also increased intensification of grazed pastures has been shown to have a little impact on the soil CH_4 sink capacity in the Netherlands (van den Pol-Van Dasselaar et al. 1999) and New Zealand (Saggar et al. 2004c; Walcroft et al. 2008). In contrast, N_2O emissions have increased from the effects of intensification of livestock numbers, but these changes are complex and poorly understood. Increasing sheep stock numbers elevate soil N_2O emissions (e.g. Ma et al. 2006; Saggar et al. 2007a). The process-based model NZ–DNDC simulated the effects of increasing sheep stocking rates (5, 10, 15, 20 and 25 sheep ha^{-1}) and showed that soil N_2O emissions increased linearly with the stocking rates in a well-drained New Zealand pasture site (Saggar et al. 2007a) (Fig. 10.4).

This linear increase in N_2O emissions with increasing sheep numbers suggests that intensification of sheep farming may have little impact on emissions per stock unit. Saggar et al. (2007b) showed that more of the input N was used and less was

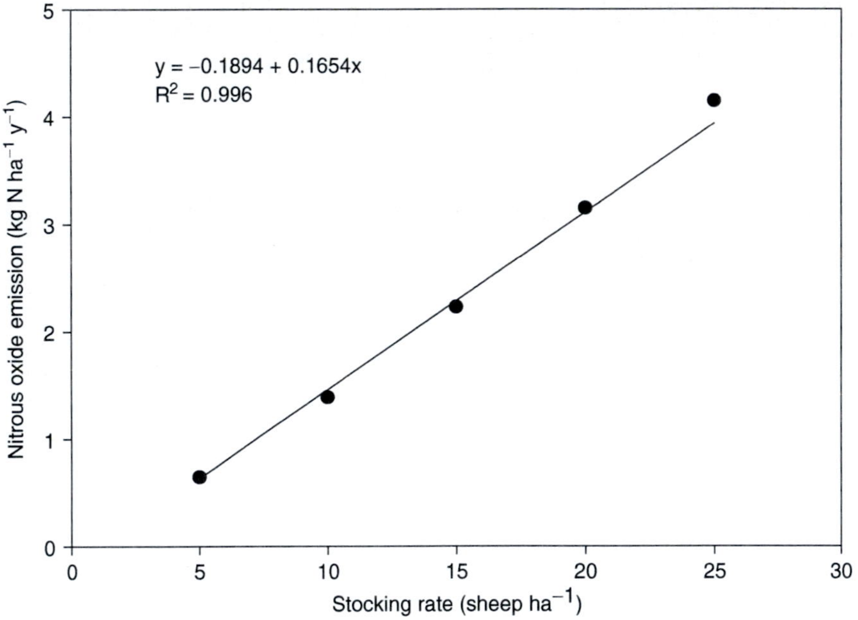

Fig. 10.4 Relationship between sheep stocking rate and nitrous oxide emissions simulated by a process-based model NZ–DNDC in a well-drained pasture site in New Zealand (data from Saggar et al. 2007a)

lost as N_2O in sheep-grazed pastures compared with dairy-grazed pastures. In steppe grassland sites in Inner Mongolia, China, a significant positive correlation was found between the stocking rate and the contribution of the growing-season emissions to the annual N_2O budget (Wolf et al. 2010). There are two main reasons for the elevated soil N_2O emissions by increasing livestock numbers. First, grazing animals excrete N in urine and dung, and N accumulates in dung and urine patches. Also synthetic N fertiliser (i.e. urea fertiliser) is often applied to enhance pasture growth for intensively managed grasslands. The excretal and synthetic N can be a source of N_2O through nitrification, denitrification and nitrifier denitrification. Second, treading and trampling by the animals cause soil compaction, making the soil more anaerobic and stimulating denitrification activity, thus facilitating N_2O production (Davidson and Firestone 1988).

It is generally assumed that there is a linear relationship between N input and direct N_2O emissions in managed agro-ecosystems (Bouwman 1996; Dobbie et al. 1999). However, there is a growing body of evidence indicating a nonlinear, exponential response of direct N_2O emissions to N input (Kim and Hernandez-Ramirez 2010). The data in this review indicated that direct N_2O emissions can increase abruptly when N input exceeds 300 kg N ha^{-1}, and there is an exponential relationship between N input and direct N_2O emissions (Fig. 10.5) and emission

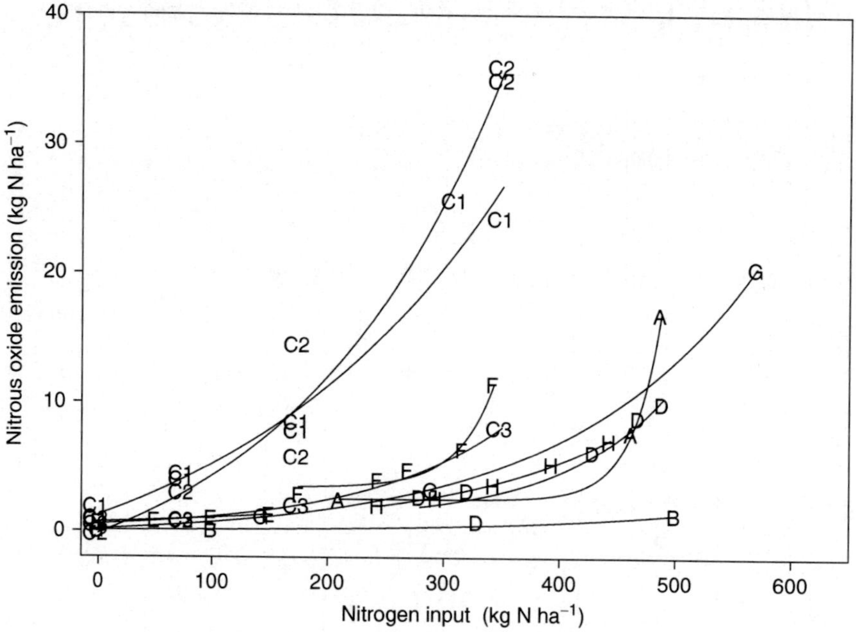

Fig. 10.5 Observed exponential relationship between N input and nitrous oxide emissions in studies conducted in grazed pasture systems. Data sources: *A* (Dobbie et al. 1999), *B* (Letica et al. 2009), *C1* (Cardenas et al. 2010; Aberystwyth site), *C2* (Cardenas et al. 2010; Aberystwyth site) and *C3* (Cardenas et al. 2010; North Yorkshire site), *D* (Hyde et al. 2006), *E* (Zhang and Han 2008), *F* (Kim et al. 2010), *G* (Singh et al. 2008) and *H* (Saggar et al. 2007a)

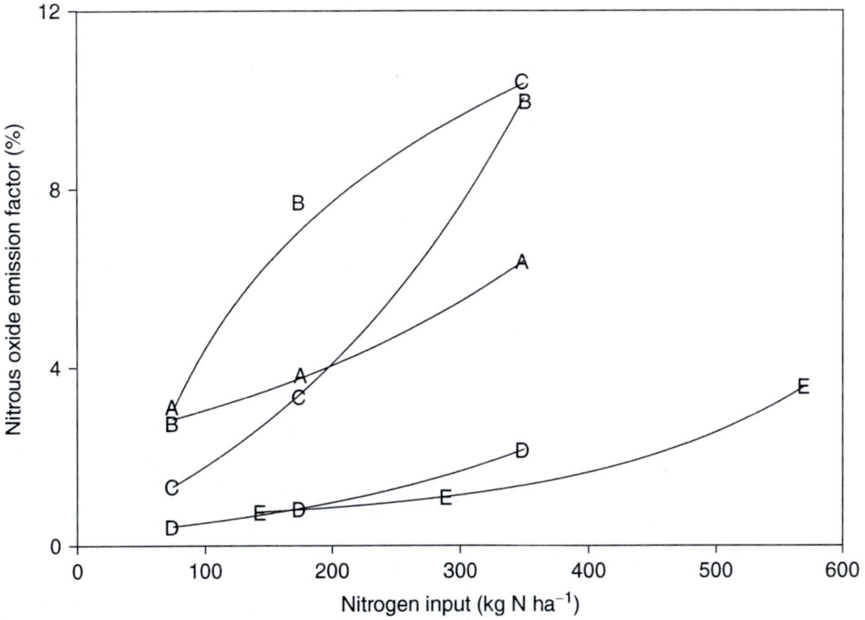

Fig. 10.6 Observed relationship between N input and nitrous oxide emission factor in studies conducted in grazed pasture systems. Data sources: *A* (Cardenas et al. 2010; Devon site), *B* (Cardenas et al. 2010; Aberystwyth site), *C3* (Cardenas et al. 2010; North Yorkshire site) and *D* (Singh et al. 2008)

factors (Fig. 10.6), attributed to excessive soil N, lower N uptake and priming effect (Kim and Hernandez-Ramirez 2010).

As the stocking density is increased, the frequency and closeness of grazing also increase. This leads to soil compaction and reduction in pore space, both of which increase N_2O emissions in grazed pastures, as found in laboratory (Uchida et al. 2008; van Groenigen et al. 2005a) and field studies (Bhandral et al. 2007; van Groenigen et al. 2005b). Repacking dairy pasture soil with four different soil aggregate sizes and four levels of soil compaction showed that the highest N_2O emissions were obtained from the smallest and most compacted aggregates (Uchida et al. 2008). Measured N_2O emissions from two pasture on well-drained and poorly drained soil grazed by dairy cows over a year following grazing events were about 2% of excretal and fertiliser N inputs (Saggar et al. 2004a), twice those determined from field-plot experiments with animal exclusion (de Klein et al. 2003). These results and those of Douglas and Crawford (1993) suggest that animal treading could accelerate N_2O emissions. Collectively, these studies (Bhandral et al. 2007; Uchida et al. 2008; van Groenigen et al. 2005a, b) show 1.3–14-fold increases in N_2O emissions with 1.1–1.4-fold increase in bulk density caused by soil compaction (Table 10.2).

Overall, intensification of pasture using high N input and stocking rate is likely to result in soil compaction, thereby causing exceptionally high gaseous and leaching losses of N. This suggests optimal N management considering stocking

Table 10.2 The effect of soil compaction on bulk density and nitrous oxide (N_2O) emission in grazed pasture systems

Site	Study type	Study periods	Soil type	N input (kg N ha^{-1})	Control (A) Bulk density (Mg m^{-3})	Control (A) N_2O emission (kg N ha^{-1})	Compacted (B) Bulk density (Mg m^{-3})	Compacted (B) N_2O emission (kg N ha^{-1})	Difference (B/A) Bulk density	Difference (B/A) N_2O emission	References
Palmerston North, New Zealand	Field	10 September to 4 December 2002	Fine sandy loam	Urine, 600	1.18	2.9	1.31	9.2	1.1	3.1	Bhandral et al. (2007)
				Nitrate, 600		4.4		61.5		14.1	
				Ammonium, 600		2.6		9.2		3.5	
				Urea, 600		2.1		9.1		4.3	
				Water, 0		1.12		2.61		2.3	
Lincoln, New Zealand	Laboratory	37 days	Silt loam	Urine, 340	0.97	13.3	1.29	32.6	1.3	2.5	Uchida et al. (2008)
					0.78	4.1	1.08	30.3	1.4	7.4	
					0.82	1.0	1.13	8.5	1.4	8.3	
					0.81	1.0	1.12	1.4	1.4	1.3	
Wageningen, The Netherlands	Laboratory	103 days	Sandy	Urine, 119[a]	1.07	1.0[a]	1.22	8.2[a]	1.1	8.2	van Groenigen et al. (2005a)
				Urine, 237[a]	1.07	1.7[a]	1.22	13.0[a]	1.1	7.6	
				Urine, 474[a]	1.07	5.4[a]	1.22	27.2[a]	1.1	5.0	
				Urine, 949[a]	1.07	9.5[a]	1.22	16.1[a]	1.1	1.7	
Wageningen, The Netherlands	Field	August 2000 to November 2001	Sandy	Urine, 186	1.54	1.7	1.6	6.2	1.04	3.6	van Groenigen et al. (2005b)
				Urine, 373		6.2		9.4		1.5	

[a]Unit: mg N kg^{-1}

rate and expected pasture yields is the key to mitigating N losses in grazed pasture systems.

10.6 Approaches to Reduce Intensification Impacts

Meat and milk products are considered global public goods because of their role in the world food supply. Concerns about food safety and security, energy security, biosecurity and traceability are gaining significance as consumers recognise the relationship between diet and health. In the last decade, there has been a rapid rise in the combined consumption of meat and milk globally, and there is an expectation that the total demand for livestock products might almost double by 2050 (Herrero et al. 2009). This increasing demand for animal products has major economic and environmental implications for countries whose economies are based primarily on livestock farming. In a carbon-constrained post-peak oil era, these countries will need to have globally competitive and sustainable livestock farming systems in place, and have strategies developed and ready for implementation to manage the risks and opportunities from global climate change. These strategies will also need to meet stringent goals for sustainability, environmental security, economy and society. They will also have to be able to adapt in response to changing circumstances. Therefore, sustainable livestock production goals will need to balance livestock production, livelihoods and environmental protection (Herrero et al. 2009).

A range of practices and technologies has been examined in New Zealand to mitigate adverse environmental effects due to intensification (de Klein and Eckard 2008; Di and Cameron 2006; Luo et al. 2008a, b, d, 2010; Saggar et al. 2009a). These practices and technologies include soil management (Uchida et al. 2008; Velthof et al. 2009; Zaman et al. 2008), the use of winter stand-off/feed pads or housing systems during high-risk periods of N and other nutrient loss (Ledgard et al. 2006; Luo et al. 2006), integration of low protein or condensed tannin forages (Luo et al. 2008a; Nielsen et al. 2003), improved management of N fertilisers (Luo et al. 2007) and the use of nitrification inhibitors (NIs) (Asing et al. 2008; Di and Cameron 2006; Singh et al. 2008; Zaman et al. 2009).

10.6.1 Soil Management

A number of studies have shown that reducing N_2O and NO_3^- losses and soil CH_4 production could be achieved by altering soil conditions including application of lime and zeolite (Zaman et al. 2007, 2008) and biochar (Spokas et al. 2009; Yanai et al. 2007), improving drainage (de Klein et al. 2003), and avoiding soil compaction (Livesley et al. 2008; Uchida et al. 2008; van Groenigen et al. 2005a, b). Among these, biochar application has attracted more interest recently, as it has been hypothesised that this can achieve C sequestration and may reduce net GHG emissions (Lehmann and Joseph 2009; Sohi et al. 2010). However, research so far is very limited, with conflicting results (Chap. 15). Biochar incorporation in soil has reduced GHG emissions in laboratory experiments conducted in Japan (Yanai et al. 2007), the USA (Spokas et al. 2009), and Australia (Singh et al. 2010). However, a recent New Zealand laboratory study (Clough et al. 2010) showed that biochar application did not affect N_2O emissions but enhanced NH_3 emissions. Further studies are clearly needed to evaluate the effect of biochar addition on GHG emissions where factors such as biochar type and soil properties are studied (Chap. 15). Renewal of grazed pastures can cause high N_2O emissions (e.g. Davies et al. 2001; Mori and Hojito 2007) with higher emissions after renewal without ploughing than with ploughing (Velthof et al. 2009). There are conflicting results on the appropriate season for mitigating N_2O emissions caused by renovation. While it is expected that pasture renovation in spring instead of autumn might offer opportunities to lower N_2O emission (Vellinga et al. 2004), Velthof et al. (2009) found higher N_2O emissions after renovation in spring than in autumn.

10.6.2 Winter Stand-Off/Feed Pads or Housing Systems

Soil compaction can be minimised through farm management practices, including reduced stocking rates and length of grazing rotation, avoiding grazing in wet soil conditions, and improving soil drainage (Greenwood and McKenzie 2001; Singleton and Addison 1999). About 7% of New Zealand N_2O emission can be reduced by following management practice avoiding soil compaction (de Klein and Ledgard 2005).

Stand-off/feed pads or housing systems have been used to reduce soil physical damage due to grazing on wet soils. They can also reduce N_2O emissions and NO_3^- leaching (Ledgard et al. 2006; Luo et al. 2006) because of lower excreta input to the soil and less soil compaction during the wet winter and early spring seasons. In a limited number of studies in New Zealand, N_2O emissions and N leaching were reduced by up to 60% when animals were held on stand-off/feed pads or in animal houses for 3–4 months during late-autumn–winter periods compared with year-round grazing (Chadwick et al. 2002; de Klein et al. 2006; Ledgard et al. 2006; Luo et al. 2008b). Use of a stand-off pad decreased total N_2O emissions per hectare of a dairy farmlet by 9%, compared with the control farm (Luo et al. 2010).

10.6.3 Integration of Low Protein or Condensed Tannin Forages

A lower proportion of N was excreted in urine and faeces when animals grazing perennial ryegrass pasture were fed supplements containing a low protein concentration and highly fermentable organic matter (Mulligan et al. 2004; Nielsen et al. 2003). Results from a study by Luo et al. (2008a) suggest that integration of low protein forage can be an effective management practice to mitigate adverse environmental effects such as N_2O emission intensity with higher stocking rates in dairy farm systems,

10.6.4 Management of N Fertilisers

Since NO_3^-, leaching and N_2O emissions following fertiliser application can be elevated in wet soils (Luo et al. 2007), strategic application of N fertilisers and farm dairy effluent to pastures under low soil moisture status can potentially reduce N losses (Luo et al. 2008b). Limiting the amount of N fertiliser applied during late-autumn/winter or early spring, when pasture growth is slow and soil is wet, can decrease N losses from grazed pastures.

10.6.5 Use of Nitrogen Transformation Inhibitors

NIs such as dicyandiamide (DCD), nitrapyrin and 3,4 dimethylpyrazole phosphate (DMPP) slow the activity of nitrifying bacteria responsible for the oxidation of NH_4^+ to NO_2^- and can thereby reduce NO_3^- leaching and N_2O emissions (Abbsi and Adams 2000; Cameron et al. 2005; Di et al. 2007). Ammonia emissions can be reduced using UIs such as [*N*-(*n*-butyl) thiophosphoric acid triamide; nBTPT] sold under the trade name Agrotain®, which reduce the rate of urea hydrolysis to NH_4^+ (Saggar et al. 2009b). Both NO_3^- leaching and N_2O emissions from urine patches can be potentially reduced by up to 70% with land application of NI to pastures (Asing et al. 2008; Di and Cameron 2006; Zaman et al. 2009). In addition, N is held in the NH_4^+ form longer, encouraging NH_4^+ uptake by pasture plants and preventing N_2O emissions from either nitrification or denitrification. However, this may also increase NH_3 emissions and potential NH_4^+-N leaching from urea fertiliser and cattle urine (Singh 2007). Results of New Zealand studies reviewed by Saggar et al. (2009b) suggest UI Agrotain reduced NH_3 emissions on average by 43% from urea and by 48% from urine. Some more recent studies have found that the combined use of NI (DCD) and (UI) (nBTPT) can be very effective in reducing NH_3 and N_2O emissions and NO_3^- leaching (Singh et al. 2008; Zaman and Blennerhassett 2010; Zaman et al. 2009).

Overall, the key mitigation options for reducing gaseous and leaching losses of N from intensive pastoral farming are: (1) N transformation inhibitors (NI and UI), (2) strategic farm effluent irrigation and (3) restricted winter grazing.

10.6.6 Enhancement of Soil Uptake of CH_4

There are as yet no cost-effective technologies and strategies available to livestock farmers to reduce enteric CH_4 emissions. As soils contain both CH_4-producing (methanogens) and CH_4-oxidising (methanotrophs: soil bacteria that use CH_4 as a sole C source) organisms, they have the ability to produce and consume CH_4 simultaneously (see Chap. 8). Globally, these methanotrophs remove about 10–44 Tg of CH_4 from the atmosphere. However, the methanotrophs are more important than this figure might indicate, as they also consume a great deal of CH_4 before it is released to the atmosphere.

Saggar et al. (2008) indicated that the key questions that need to be addressed for enhancing soil CH_4 uptake rates in the field are:

- What are the key microbiological populations and processes responsible for CH_4 oxidation in soils and can they be optimised?
- How do soil/plant/animal interactions and climate affect net CH_4 oxidation rates and the microbial populations regulating them, and what are the opportunities for enhancing oxidation rates?

A novel approach for capturing CH_4 produced by animals and animal effluent in confined locations (e.g. waste ponds and barns) being investigated is use of methanotrophs (Pratt et al. 2010). Biofiltration, using very active methanotroph populations in porous media to convert CH_4 to CO_2, appears to be a potentially effective strategy for treating CH_4 emissions from waste ponds on dairy farms. Melse and van der Werf (2005) observed an 85% removal efficiency of CH_4 emissions from piggery effluent using biofiltration. As the methanotrophs can rapidly consume atmospheric CH_4, they offer the potential to capture the enteric CH_4 from housed animals, effluent ponds and also emissions from landfills using biofilters to convert this potent gas to CO_2.

10.6.7 Carbon Sequestration

Soil C could be sequestered in grazed pasture systems using a range of management practices (Chan et al. 2010; Conant and Paustian 2002; Herrero et al. 2009; Reid et al. 2004). Conant and Paustian (2002) found that universal rehabilitation of overgrazed grasslands can sequester approximately 45 Tg C year^{-1} by cessation of overgrazing and implementation of moderate grazing intensity. It was suggested that soil C sequestration can be achieved by traditional pastoral practices and knowledge, such as managing grazing intensity and duration, improving pasture quality, reducing the frequency and/or intensity of grassland fires, and by providing pastoral farmers with food security benefits at the same time (Reid et al. 2004) and by management practices aimed at increasing N retention at the landscape level (Pineiro et al. 2010).

It is certainly true that C cycles rapidly through pastoral systems, and that farming ruminant animals do not add any "new" C to the atmosphere. However, in the process of rumination, some of the C in the atmosphere is transformed from a gas with a lower GWP (CO_2) to a gas with a higher GWP (CH_4). Some livestock farmers believe that increasing pasture production will lead to more C stored in the soil. Parson et al. (2009) report that increasing stocking rate in general should decrease the flow of C to soil, and so reduce the potential to increase soil C. However, the factors (such as higher soil fertility) that increase plant growth increase overall C flows to soil and may increase soil C. In New Zealand, the potential for significant, permanent increases in soil organic C in intensive pasture systems is limited (Whitehead et al. 2009).

10.6.8 Farm Economics

Reduced agricultural intensity can decrease N emissions and other inverse environmental effects, but would have a major impact on economic returns and farm viability. Productivity and environmental gains occur through above-mentioned practices and technologies by avoiding losses of N and other nutrients to the environment and increasing the quantity and quality of the forage produced. A recent modelling study of the production, environmental and financial impacts of intensification of New Zealand sheep and beef farming systems found increases in total N leaching and GHG emissions from intensification through both feeding maize silage and applying N fertilisers (White et al. 2010). These model estimates also show that neither method of intensification increased profitability without a small annual N application of 50 kg N ha^{-1}, especially to 75% of hill country farms. Other New Zealand studies discussed in this chapter clearly show the advantages of instigating several of these individual dairy farm management practices such as the use of winter stand-off pads, maize silage and improved N fertiliser management, in reducing N_2O emissions and N leaching (Luo et al. 2010). Results of a 3-year CO_2, N_2O and CH_4 emission measurements in an upland semi-natural grassland site grazed by cattle (Allard et al. 2007) showed that reducing fertiliser input and grazing pressure strongly reduced N_2O and CH_4 emissions per unit of land area but gradually reduced the C storage potential of the grassland. These results clearly demonstrate the need for taking into account all the three major GHGs (CO_2, N_2O and CH_4) when developing strategies to mitigate the GHG effect.

10.7 Conclusions and Future Work

Intensification of pastoral agriculture has occurred since the establishment of managed pastoral systems and will continue to occur in future to meet growing demands for food worldwide. The key drivers of intensification are the need to maintain or increase profit and return on investment, containing cost of inputs relative to the value of outputs, the availability of new knowledge and technologies, high land values and increasing international competition, and possible tariff barriers for unsustainable practices.

Intensively managed livestock farming changes the buffering and filtering capacity of our structured pastoral soils and lead to faster and greater nutrient and contaminant leakage, favouring the potential for degradation, increased nutrient losses to waterways and the atmosphere, and increased environmental pollution. Intensification of pastoral productivity also leads to increased emissions of the principal agricultural GHGs, CH_4 and N_2O. Some of the contaminants have the potential to disrupt wildlife welfare and human health. With the current pressures exerted on grassland resources, it is not possible to continue to increase productivity without causing further soil, vegetation and environmental degradation.

Concerns about food safety and security, energy security, biosecurity and traceability are gaining significance as consumers recognise the relationship between their diet and health. Agricultural policies based on sound science and robust risk assessments are needed, and research efforts must be directed towards achieving a balance between environmentally sustainable management of pastoral resources and efficient food production. Further studies are needed to quantify organic matter dynamics and consequent nutrient fluxes in soils, and associated soil-atmosphere exchange of GHGs under different livestock-based land uses. These studies need to be coupled with predicting water and solute storage and movement through soils and land systems. The information is needed to assess and/or mitigate the nutrient losses for robust models that combine farm systems expertise and provide information for developing solutions at multiple scales. Efforts to integrate remote-sensing and geographic information system capabilities need to be expanded from on-farm management systems to provide information for spatial modelling and forecasting at multiple scales. Simultaneously, strong linkages with landowners, community groups and regional authorities are needed to unify and use aspirations through sustainable practices.

Researchers and policy makers need to consider the whole food chain, and to account for multiple environmental emissions and resource efficiency, for example, as energy use under intensive livestock farming, through the use of tools such as life cycle assessment. This type of assessment can help identify potential (and unintended) issues such as pollution swapping or likely additional energy costs associated with a particular management system, processing and mitigation option. These approaches are essential not only to protect food-producing countries, but also to avoid the imposition of tariff barriers in distant markets.

Acknowledgements Funding for this work was provided by the New Zealand foundation for Research Science and Technology. Our thanks to Dr. Kevin Tate for providing very helpful suggestions and thoroughly reviewing an earlier version of this book chapter, and to Anne Austin for editing.

References

Abbsi MK, Adams WA (2000) Gaseous N emission during simultaneous nitrification–denitrification associated with mineral N fertilization to a grassland soil under field conditions. Soil Biol Biochem 32:1251–1259

Allard V, Soussana J-F, Falcimagne R, Berbigier P, Bonnefond JM, Ceschia E, D'hour P, Hénault C, Laville P, Martin C, Pinarès-Patino C (2007) The role of grazing management for the net biome productivity and greenhouse gas budget (CO_2, N_2O and CH_4) of semi-natural grassland. Agric Ecosyst Environ 121:47–58

Annon (2005) Advice on hand for farmers on nitrate leaching from soil. Country-Wide 27(1): 34

Asing J, Saggar S, Singh J, Bolan NS (2008) Assessment of nitrogen losses from urea and an organic manure with and without nitrification inhibitor, dicyandiamide, applied to lettuce under glasshouse conditions. Aust J Soil Res 46:535–541

Ball BC, Watson CA, Baddeley JA (2007) Soil physical fertility, soil structure and rooting conditions after ploughing organically managed grass/clover swards. Soil Use Manage 23: 20–27

Barton L, McLay CDA, Schipper LA, Smith CT (1999) Denitrification rates in a wastewater-irrigated forest soil in New Zealand. J Environ Qual 28:2008–2014

Bélanger G, Richards JE, Angers DA (1999) Long-term fertilization effects on soil carbon under permanent swards. Can J Soil Sci 79:99–102

Betteridge K, Mackay AD, Shepherd TG, Barker DJ, Budding PJ, Devantier BP, Costall DA (1999) Effect of cattle and sheep treading on surface configuration of a sedimentary hill soil. Aust J Soil Res 37:743–760

Betteridge K, Mackay AD, Pande TN, Costall DA, Budding PJ, Valentine I, Singleton PL, Findlayson J, Drewry JJ, Boyes M, Judge A (2002) Cattle treading on wet soils: implications for pasture growth and soil physical condition. In: Currie LD, Loganathan P (eds) Dairy farm soil management. Massey University, Palmerston North, pp 79–88

Bhandral R, Saggar S, Bolan NS, Hedley MJ (2007) Transformation of nitrous oxide emission from grassland soils as affected by compaction. Soil Till Res 94:482–492

Bilotta GS, Brazier RE, Haygarth PM (2007) The impacts of grazing animals on the quality of soils, vegetation, and surface waters in intensively managed grasslands. Adv Agron 94: 237–280

Bolan NS, Saggar S, Luo JF, Bhandral R, Singh J (2004) Gaseous emissions of nitrogen from grazed pastures: processes, measurements and modeling, environmental implications, and mitigation. Adv Agron 84:37–120

Bolan NS, Laurenson S, Luo J, Sukias J (2009) Integrated treatment of farm effluents in New Zealand's dairy operations. Bioresource Tech 100:5490–5497

Bolan NS, Adriano DC, Kunhikrishnan A, James T, McDowell R, Senesi N (2010) Dissolved organic matter: biogeochemistry, dynamics and environmental significance in soils. Adv Agron 110:1–75

Bouwman AF (1996) Direct emission of nitrous oxide from agricultural soils. Nutr Cycl Agroecosyst 46:53–70

Bouwman AF, Van Der Hoek KW, Eickhout B, Soenario I (2005) Exploring changes in world ruminant production systems. Agric Syst 84:121–153

Cameron KC, Di HJ, Moir J, Roberts A, Pellow R, Christie R (2005) Treating grazed pasture soil with a nitrification inhibitor "ECO–N" to decrease nitrate leaching. In: Currie LD, Hanly JA (eds) Developments in fertilizer application technologies and nutrient management fertilizer and lime research centre occasional report no 18. Massey University, Palmerston North, pp 93–103

Caradus JR (1994) Selection for improved adaptation of white clover to low phosphorus and acid soils. Euphytica 77:243–250

Cardenas LM, Thorman R, Ashlee N, Butler M, Chadwick D, Chambers B, Cuttle S, Donovan N, Kingston H, Lane S, Dhanoa MS, Scholefield D (2010) Quantifying annual N_2O emission fluxes from grazed grassland under a range of inorganic fertiliser nitrogen inputs. Agric Ecosyst Environ 136:218–226

Carroll ZL, Bird SB, Emmett BA, Reynolds B, Sinclair FL (2004) Can tree shelterbelts on agricultural land reduce flood risk? Soil Use Manage 20:357–359

Chadwick DR, Ledgard SF, Brown L (2002) Nitrogen flows and losses in dairy farms in New Zealand and the UK: effects of grazing management. In: Currie LD, Loganathan P (eds) Dairy farm soil management occasional report fertiliser and lime research centre. Massey University, Palmerston North, pp 319–332

Chan KY, Oates A, Li GD, Conyers MK, Prangnell RJ, Poile G, Liu DL, Barchia IM (2010) Soil carbon stocks under different pastures and pasture management in the higher rainfall areas of south-eastern Australia. Aust J Soil Res 48:7–15

Clark H, Pinares PC, De Klein CAM (2005) Methane and nitrous oxide emissions from grazed grasslands. Grassland: a global resource. In: McGilloway DA (ed) Grasslands a global resource. Academic, Wageningen, pp 279–294

Clough TJ, Berrtram JE, Ray JL, Condron LM, O'Callaghan M, Sherlock RR, Wells NS (2010) Unweathered wood biochar impact on nitrous oxide emissions from a bovine-urine-amended pasture soil. Soil Sci Soc Am J 74:852–860

Conant RT, Paustian K (2002) Potential soil carbon sequestration in overgrazed grassland ecosystems. Global Biogeochem Cy 16:1143

Cuttle SP (2008) Impacts of pastoral grazing on soil quality. In: McDowell RW (ed) Environmental impacts of pasture-based farming. CAB International, Wallingford, pp 33–74

Davidson EA, Firestone MK (1988) Measurement of nitrous oxide dissolved in soil solution. Soil Sci Soc Am J 52:1201–1203

Davies MG, Smith KA, Vinten AJA (2001) The mineralisation and fate of nitrogen following ploughing of grass and grass-clover swards. Biol Fertil Soils 33:423–434

de Klein CAM, Eckard RJ (2008) Targeted technologies for nitrous oxide abatement from animal agriculture. Aust J Exp Agric 48:14–20

de Klein CAM, Ledgard SF (2005) Nitrous oxide emissions from New Zealand agriculture – key sources and mitigation strategies. Nutr Cycl Agroecosyst 72:77–85

de Klein CAM, Barton L, Sherlock RR, Li Z, Littlejohn RP (2003) Estimating a nitrous oxide emission factor for animal urine from some New Zealand pastoral soils. Aust J Soil Res 41: 381–399

de Klein CAM, Smith LC, Monaghan RM (2006) Restricted autumn grazing to reduce nitrous oxide emissions from dairy pastures in Southland, New Zealand. Agric Ecosyst Environ 112:192–199

DEFRA (2005) Agriculture in the United Kingdom. HMSO, London

Di HJ, Cameron KC (2006) Nitrous oxide emissions from two dairy pasture soils as affected by different rates of a fine particle suspension nitrification inhibitor, dicyandiamide. Biol Fertil Soils 42:472–480

Di HJ, Cameron KC, Milne J, Drewry JJ, Smith NP, Hendry T, Moore S, Reijnen B (2001) A mechanical hoof for stimulating animal treading under controlled conditions. NZ J Agric Res 44:111–116

Di HJ, Cameron KC, Sherlock RR (2007) Comparison of the effectiveness of a nitrification inhibitor, dicyandiamide, in reducing nitrous oxide emissions in four different soils under different climatic and management conditions. Soil Use Manage 23:1–9

Dobbie KE, McTaggart IP, Smith KA (1999) Nitrous oxide emissions from intensive agricultural systems: variations between crops and seasons, key driving variables, and mean emission factors. J Geophys Res 104:26891–26899

Douglas JT, Crawford CE (1993) The response of a ryegrass sward to wheel traffic and applied nitrogen. Grass Forage Sci 48:91–100

Dr KR Tate pers. comm.

Drewry JJ (2006) Natural recovery of soil physical properties from treading damage of pastoral soils in New Zealand and Australia – a review. Agric Ecosyst Environ 114:159–169

Drewry JJ, Cameron KC, Buchan GD (2008) Pasture yield and soil physical property responses to soil compaction from treading and grazing – a review. Aust J Soil Res 46:237–256

Eriksen J, Vinther FP, Søegaard K (2004) Nitrate leaching and N_2-fixation in grasslands of different composition, age and management. J Agric Sci 142:141–151

FAO (Food and Agriculture Organisation) (2008) Global warming: climate change and farm animal welfare, executive summary. http://www.fao.org/fileadmin/user_upload/animalwelfare/GlobalWarningExecutiveSummary1.pdf. Accessed 17 September 2009

Ghani A, Dexter M, Perrott KW (2003) Hot-water extractable carbon in soils: a sensitive measurement for determining impacts of fertilisation, grazing and cultivation. Soil Biol Biochem 35:1231–1243

Greenwood KL, McKenzie BM (2001) Grazing effects on soil physical properties and the consequences for pastures: a review. Aust J Exp Agr 41:1231–1250

Haynes RJ, Williams PH (1993) Nutrient cycling and soil fertility in the grazed pasture ecosystem. Adv Agron 49:119–199

Herrero M, Thornton PK, Gerber P, Reid RS (2009) Livestock, livelihoods and the environment: understanding the trade-offs. Curr Opin Environ Sustain 1:111–120

Hoglund JH (1985) Grazing intensity and soil nitrogen accumulation. NZ Grassland Assoc 46: 65–69

Hyde BP, Hawkins MJ, Fanning AF, Noonan D, Ryan M, O'Toole P, Carton OT (2006) Nitrous oxide emissions from a fertilized and grazed grassland in the South East of Ireland. Nutr Cycl Agroecosyst 75:187–200

IPCC (Intergovernmental Panel on Climate Change) (2007) Summary for policymakers. In: Parry ML, Canziani OF, Palutikof JP, van der Linden PJ, Hanson CE (eds) Climate change 2007: impacts, adaptation and vulnerability contribution of working group II to the fourth assessment report of the intergovernmental panel on climate change. Cambridge University Press, Cambridge, pp 7–22

Jarvis SC, Ledgard S (2002) Ammonia emissions from intensive dairying: a comparison of contrasting systems in the United Kingdom and New Zealand. Agric Ecosyst Environ 92: 83–92

Jarvis SC, Hatch DJ, Roberts DH (1989) The effects of grassland management on nitrogen losses from grazed swards through ammonia volatilization; the relationship to excretal N returns from cattle. J Agric Sci 112:205–216

Jones DL, Shannon D, Murphy DV, Farrar J (2004) Role of dissolved organic nitrogen (DON) in soil N cycling in grassland soils. Soil Biol Biochem 36:749–756

Karlen DL, Mausbach MJ, Doran JW, Cline RG, Harris RF, Schuman GE (1997) Soil quality: a concept, definition, and framework for evaluation: (a guest editorial). Soil Sci Soc Am J 61: 4–10

Kemp DR, Michalk DL (2005) Grasslands for production and the environment. In: McGilloway DA (ed) Grasslands a global resource. Academic, Wageningen, pp 193–208

Kim D-G, Hernandez-Ramirez G (2010) Dependency of nitrous oxide emission factors on nitrogen input rates: a meta-analysis. In: 19th world congress of soil science, soil solutions for a changing world, Brisbane, 1–6 Aug 2010

Kim D-G, Mishurov M, Kiely G (2010) Effect of increased N use and drought on N_2O emission in a fertilized grassland. Nutr Cycl Agroecosyst 88(3):397–410

Kurtz I, O'Reilly CD, Tunney H (2006) Impact of cattle on soil physical properties and nutrient concentrations in overland flow from pasture in Ireland. Agric Ecosyst Environ 113:378–390

Lambert MG, Clark DA, Mackay AD, Costall DA (2000) Effects of fertiliser application on nutrient status and organic matter content of hill soils. NZ J Agric Res 43:127–138

Leathwick J, Wilson G, Rutledge D, Wardle P, Morgan F, Johnston K, McLeod M, Kirkpatrick R (2003) Land environments of New Zealand – Nga Taiao o Aotearoa. David Bateman, Auckland

Ledgard SF (2001) Nitrogen cycling in low input legume-based agriculture, with emphasis on legume/grass pastures. Plant Soil 228:43–59

Ledgard SF, Luo J (2008) Nitrogen cycling in intensively grazed pastures and practices to reduce whole-farm nitrogen losses. In: Organizing Committee of 2008 IGC/IRC Conference (ed) Multifunctional grasslands in a changing world, vol 1. Guangdong People's Publishing House, Guangzhou, pp 292–297

Ledgard SF, Sprosen MS, Penno JW (1999) Nitrogen inputs and losses from clover/grass pastures grazed by dairy cows: temporal variation and effects of nitrogen fertilization. Plant Soil 229: 177–187

Ledgard SF, Sprosen MS, Judge A, Lindsey S, Jensen R, Clark DA, Luo J (2006) Nitrogen leaching as affected by dairy intensification and mitigation practices in the resource efficient dairying (RED) trial. In: Currie LD, Hanly JA (eds) Implementing sustainable nutrient management strategies in agriculture. Occasional Report No. 19: Fertilizer and Lime Research Centre, Massey University, Palmerston North, New Zealand, pp 263–268

Ledgard SF, Schils R, Eriksen J, Luo J (2009) Environmental impacts of grazed clover/grass pastures. Irish J Agric Food Res 48:209–226

Lehmann R, Joseph S (2009) Biochar for environmental management: science and technology. Earthscan Publications, London, p 450

Letica SA, de Klein CAM, Hoogendoorn CJ, Tillman RW, Littlejohn RP, Rutherford AJ (2009) Short-term measurement of N_2O emissions from sheep-grazed pasture receiving increasing rates of fertiliser nitrogen in Otago, New Zealand. Anim Prod Sci 50:17–24

Livesley SJ, Kiese R, Graham J, Weston CJ, Butterbach-Bahl K, Arndt SK (2008) Trace gas flux and the influence of short-term soil water and temperature dynamics in Australian sheep grazed pastures of differing productivity. Plant Soil 309:89–103

Luo J, Tillman RW, Ball PR (1999a) Grazing effects on denitrification in a soil under pasture during two contrasting seasons. Soil Biol Biochem 31:903–912

Luo J, Tillman RW, Ball PR (1999b) Factors regulating denitrification in a soil under pasture. Soil Biol Biochem 31:913–927

Luo J, Dinnison A, Ross C, Ledgard S, Longhurst RD (2006) Control of pollutants using stand-off pads containing different natural materials. NZ Grassland Assoc 68:315–320

Luo J, Ledgard SF, Lindsey SB (2007) Nitrous oxide emissions from application of urea on New Zealand pasture. NZ J Agric Res 50:1–11

Luo J, Ledgard SF, De Klein CAM, Lindsey SB, Kear M (2008a) Effects of dairy farming intensification on nitrous oxide emissions. Plant Soil 309:227–237

Luo J, Ledgard SF, Lindsey SB (2008b) A test of a winter farm management option for mitigating nitrous oxide emissions from a dairy farm. Soil Use Manage 24:121–130

Luo J, Lindsey SB, Ledgard SF (2008c) Nitrous oxide emissions from animal urine application on a New Zealand pasture. Biol Fertil Soils 44:463–470

Luo J, Saggar S, Bhandral R, Bolan NS, Ledgard SF, Lindsey S, Sun W (2008d) Effects of irrigating dairy-grazed grassland with farm dairy effluent on nitrous oxide emissions. Plant Soil 309:119–130

Luo J, de Klein CAM, Ledgard SF, Saggar S (2010) Management options to reduce nitrous oxide emissions from intensively grazed pastures: a review. Agric Ecosyst Environ 136:282–291

Ma X, Wang S, Wang Y, Jiang G, Nyren P (2006) Short-term effects of sheep excrement on carbon dioxide, nitrous oxide and methane fluxes in typical grassland of inner Mongolia. NZ J Agric Res 49:285–297

Mackay AD (2009) Water quality and land use. New Zealand Veterinarian Association, Society of sheep and beef cattle veterinarians 39th annual conference, Rotorua, 20–22 May 2009, pp 147–154

MacLeod CJ, Moller H (2006) Intensification and diversification of New Zealand agriculture since 1960: an evaluation of current indicators of land use change. Agric Ecosyst Environ 115: 201–218

McIsaac GT, David MB, Gertner GT, Goolsby DA (2001) Eutrophication: nitrate flux in the Mississippi River. Nature 414:166–167

McKenzie FR, Jacobs JL, Ward GN (2006) Irrigated dairy pasture yield and water use efficiency responses to summer applied nitrogen. NZ Grassland Assoc 68:155–160

Melse RW, van der Werf AW (2005) Biofiltration for mitigation of methane emission from animal husbandry. Environ Sci Tech 39:5460–5468

Menneer JC, Ledgard S, McLay C, Silvester W (2004) The impact of grazing animals on N_2 fixation in legume-based pastures and management options for Improvement. Adv Agron 83: 181–241

Metherell AK, Stewart DPC, Carey PL, Moss RA (2002) Long-term irrigation improves soil quality, but decreases soil carbon sequestration. In: Stephens P, Callaghan J, Austin A (eds) Soil quality and sustainable land management conference. Landcare Research New Zealand Ltd, Palmerston North, pp 55–61

MfE (2007a) New Zealand's greenhouse inventory 1990–2005. An overview. Ministry for the Environment, Wellington. http://www.mfe.govt.nz/publications/climate/greenhouse-gas-inventory-overview-jul07/index.html. Accessed 6 July 2009

MfE (2007b) Environment New Zealand. Ministry for the Environment, Wellington

Monaghan RM, Paton RJ, Smith LC, Drewry JJ, Littlejohn RP (2005) The impacts of nitrogen fertilisation and increased stocking rate on pasture yield, soil physical condition and nutrient losses in drainage from a cattle-grazed pasture. NZ J Agric Res 48:227–240

Monaghan RM, Hedley MJ, Di HJ, McDowell RW, Cameron KC, Ledgard SF (2007) Nutrient management in New Zealand pastures – recent developments and future issues. NZ J Agric Res 50:181–202

Mori A, Hojito M (2007) Grassland renovation increases N_2O emission from a volcanic grassland soil in Nasu, Japan. Soil Sci Plant Nutr 53:812–818

Mulligan FJ, Dillon P, Callan JJ, Rath M, O'Mara FP (2004) Supplementary concentrate type affects nitrogen excretion of grazing dairy cows. J Dairy Sci 87:3451–3460

Nielsen NM, Kristensen T, Nørgaard P, Hansen H (2003) The effect of low protein supplementation to dairy cows grazing clover grass during half of the day. Livest Prod Sci 81:293–306

OECD (Organization for Economic Co-operation and Development) (2008) Environmental performance of agriculture since 1990: main report. OECD, Paris. http://statsoecdorg/viewhtmlaspx?QueryName=516&QueryType=View. Accessed 23 June 2009

Oenema O, Velthof GL, Yamulki S, Jarvis SC (1997) Nitrous oxide emissions from grazed grasslands. Soil Use Manage 13:288–295

Oliver I, Garden D, Greenslade PJ, Haller B, Rodgers D, Seeman O, Johnston B (2005) Effects of fertiliser and grazing on the arthropod communities of a native grassland in south-eastern Australia. Agric Ecosyst Environ 109:323–334

Pande TN, Valentine I, Betteridge K, Mackay AD, Horne D (2000) Pasture damage and regrowth from cattle treading. P NZ Grassland Assoc 62:155–160

Parson AJ, Rowarth JS, Newton PCD (2009) Managing pasture for animals and soil carbon. Proc NZ Grassland Assoc 71:77–84

PCE (Parliamentary Commissioner for the Environment) (2004) Growing for good: intensive farming, sustainability and New Zealand's environment. Parliamentary Commissioner for the Environment, Wellington

Pineiro G, Paruelo JM, Oesterheld M, Jobbagy EG (2010) Pathways of grazing effects on soil organic carbon and nitrogen. Rangeland Ecol Manage 63:109–119

Pratt C, Walcroft AS, Tate KR, Ross DJ, Hills MH, Roy R (2010) Managing methanotrophs in an on-farm biofilter to reduce methane emissions. In: Currie LD, Christensen CL (eds) Farming's future: minimising footprints and maximising margins occasional report no 23 fertiliser & lime research centre. Massey University, Palmerston North, pp 183–191

Quinn JM, Cooper AB, Davies-Colley RJ, Rutherford JC, Williamson RB (1997) Land use effects on habitat, water quality, periphyton, and benthic invertebrates in Waikato, New Zealand, hill-country streams. NZ J Mar Freshwater Res 31:579–597

Rajendram GS, Ledgard SF, Monaghan R, Penno JW, Sprosen MS, Ouyang L (1998) Effect of rate of nitrogen fertiliser on cation and anion leaching under intensively grazed dairy pasture. Massey University Occational Report No 11, pp 67–73

Reid RS, Thornton PK, McCrabb GJ, Kruska RL, Atieno F, Jones PG (2004) Is it possible to mitigate greenhouse gas emissions in pastoral ecosystems of the tropics? Environ Dev Sustain 6:91–109

Rotz CA, Buckmaster DR, Comerford JW (2005) A beef herd model for simulating feed intake, animal performance, and manure excretion in farm systems. J Anim Sci 83:231–242

Saggar S (2004) Changes in nitrogen dynamics of legume-based pastures with increased nitrogen fertilizer use: impacts on New Zealand's nitrous oxide emissions inventory. NZ Soil News 52: 110–117

Saggar S, Hedley CB (2001) Estimating seasonal and annual carbon inputs, and root decomposition rates in a temperate pasture following field ^{14}C pulse-labelling. Plant Soil 236:91–103

Saggar S, Hedley MJ, White RE, Gregg PEH, Perrott KW, Cornforth IS (1993) Assessment of the relative agronomic effectiveness of phosphate rocks under glasshouse conditions. Fertil Res 34:141–151

Saggar S, Hedley C, Mackay AD (1997) Partitioning and translocation of photosynthetically fixed ^{14}C in grazed hill pastures. Biol Fertil Soils 25:152–158

Saggar S, Hedley CB, Salt G, Giddens KM (2000) Influence of soil P status and of added N on C mineralisation from ^{14}C-labelled glucose. Biol Fertil Soils 32:209–216

Saggar S, Tate K, Hedley C, Perrott K, Logananthan P (2001) Are soil carbon levels in our established pastures at or near steady state? NZ Soil News 49(4):73–78

Saggar S, Andrew RM, Tate KR, Rodda NJ, Hedley CB, Townsend JA (2004a) Modelling nitrous oxide emissions from New Zealand dairy grazed pastures. Nutr Cycl Agroecosyst 68:243–255

Saggar S, Bolan NS, Bhandral R, Hedley CB, Luo J (2004b) A review of emissions of methane, ammonia, and nitrous oxide from animal excreta deposition and farm effluent application in grazed pastures. NZ J Agric Res 47:513–544

Saggar S, Clark H, Hedley C, Tate K, Carran A, Rys G (2004c) Methane emission estimates from animal dung and waste management systems in New Zealand. In: Proceedings of the trace gas workshop, Wellington, New Zealand, 18–19 March 2004, pp 98–101

Saggar S, Bolan N, Singh J, Blard A (2005) Economic and environmental impacts of increased nitrogen use in grazed pastures and the role of inhibitors in mitigating nitrogen losses. NZ Sci Rev 62:62

Saggar S, Giltrap DL, Li C, Tate KR (2007a) Modelling nitrous oxide emissions from grazed grasslands in New Zealand. Agric Ecosyst Environ 119:205–216

Saggar S, Hedley CB, Giltrap DL, Lambie SM (2007b) Measured and modelled estimates of nitrous oxide emission and methane consumption from a sheep-grazed pasture. Agric Ecosyst Environ 122:357–365

Saggar S, Tate K, Giltrap D, Singh J (2008) Soil-atmosphere exchange of nitrous oxide and methane in New Zealand terrestrial ecosystems and their mitigation options: a review. Plant Soil 309:25–42

Saggar S, Luo J, Giltrap DL, Maddena M (2009a) Nitrous oxide emission from temperate grasslands: Process, measurements, modelling and mitigation. In: Sheldon AI, Barnhart EP (eds) Nitrous oxide emissions research progress. Nova Science Publisher, New York, pp 1–66

Saggar S, Singh J, Catto W, Stafford A, Blennerhassett J, Zaman M (2009b) Desktop study of emission factors for urease inhibitors for nitrogen fertiliser. Landcare Research Contract Report LC0809/141, New Zealand Ministry of Agriculture and Forestry, PO BOx 2526 Wellington, 83p

Schipper LA, Baisden WT, Parfitt RL, Ross C, Claydon JJ, Arnold G (2007) Large losses of soil C and N from soil profiles under pasture in New Zealand during the past 20 years. Global Change Biol 13:1138–1144

Scholefield D, Tyson KC, Garwood EA, Armstrong AC, Hawkins J, Stone AC (1993) Nitrate leaching from grazed grassland lysimeters: effects of fertilizer input, field drainage, age of sward and patterns of weather. J Soil Sci 44:601–613

Sharpley A, Meisinger JJ, Breeuwsma A, Sims JT, Danie TC, Schepers JS (1998) Impact of animal manure management on ground and surface water quality. In: Hatfield JL, Stewart BA (eds) Animal waste utilization: effective use of manure as a soil resource. Ann Arbor Press, Chelsea, pp 173–242

Sherlock R, Jewell P, Clough T (2008) Review of New Zealand specific $Frac_{GASM}$ and $Frac_{GASF}$ emission factors. Report for Ministry of Agriculture and Forestry, October 2008, 52 p

Shindell DT, Faluveg G, Koch DM, Schmidt DA, Unger N, Bauer SE (2009) Improved attribution of climate forcing emissions. Science 326:716–718

Sinclair AG, Smith LC, Morrison JD, Dodds KG (1996) Effects and interactions of phosphorus and sulphur on a mown white clover/ryegrass sward 1. Herbage dry matter production and balanced nutrition. NZ J Agric Res 39:421–433

Singh J (2007) The role of inhibitors in mitigating nitrogen losses from cattle urine and nitrogen fertiliser inputs in pastures. Unpublished PhD Thesis, Massey University

Singh J, Saggar S, Giltrap DL, Bolan NS (2008) Decomposition of dicyandiamide (DCD) in three contrasting soils and its effect on nitrous oxide emission, soil respiratory activity, and microbial biomass – an incubation study. Aust J Soil Res 46:517–525

Singh BP, Hatton BJ, Singh B, Cowie A, Kathuria A (2010) Influence of biochars on nitrous oxide emission and nitrogen leaching from two contrasting soils. J Environ Qual 39:1224–1235

Singleton PL, Addison B (1999) Effects of cattle treading on physical properties of three soils used for dairy farming in the Waikato, North Island, New Zealand. Aust J Soil Res 37:891–902

Sohi SP, Krull E, Lopez-Capel E, Bol R (2010) A review of biochar and its use and function in soil. Adv Agron 105:47–82

Sparling GP, Schipper LA (2002) Soil quality at a national scale in New Zealand. J Environ Qual 31: 1848–1857

Spokas KA, Koskinen WC, Baker JM, Reicosky DC (2009) Impacts of woodchip biochar additions on greenhouse gas production and sorption/degradation of two herbicides in a Minnesota soil. Chemosphere 77:574–581

Steinfeld H, Gerber P, Wassenaar T, Castel V, Rosales M, de Haan C (2006) Livestock's long shadow: environmental issues and options. Renew Resour J 24:15–17

Stout WL, Fales SL, Muller LD, Schnabel RR, Weaver SR (2000) Water quality implications of nitrate leaching from intensively grazed pasture swards in the northeast US. Agric Ecosyst Environ 77:203–210

Sun X, Luo N, Longhurst B, Luo J (2008) Fertiliser nitrogen and factors affecting pasture responses. Open Agric J 2:35–42

Uchida Y, Clough TJ, Kelliher FM, Sherlock RR (2008) Effects of aggregate size, soil compaction, and bovine urine on N_2O emissions from a pasture soil. Soil Biol Biochem 40:924–931

UK Environment Agency (2002) Agriculture and natural resources: benefits, costs and potential solutions. Environmental Agency, Bristol

van den Pol-Van Dasselaar DA, van Beusichem ML, Oenema O (1999) Effects of nitrogen input and grazing on methane fluxes of extensively and intensively managed grasslands in the Netherlands. Biol Fertil Soils 29:24–30

van Groenigen JW, Kuikman PJ, De Groot WJM, Velthof GL (2005a) Nitrous oxide emission from urine-treated soil as influenced by urine composition and soil physical conditions. Soil Biol Biochem 37:463–473

van Groenigen JW, Velthof GL, Van Der Bolt FJE, Vos A, Kuikman PJ (2005b) Seasonal variation in N_2O emissions from urine patches: effects of urine concentration, soil compaction and dung. Plant Soil 273:15–27

Vant B, Huser B (2000) Effects of intensifying catchment land use on the water quality of Lake Taupo. Proc NZ Soc Anim Prod 6:261–264

Vellinga TV, van den Pol-van DA, Kuikman PJ (2004) The impact of grassland ploughing on CO_2 and N_2O emissions in the Netherlands. Nutr Cycl Agroecosyst 70:33–45

Velthof GL, Oudendag D, Witzke HP, Asman WAH, Klimont Z, Oenema O (2009) Integrated assessment of nitrogen losses from agriculture in EU-27 using MITERRA-EUROPE. J Environ Qual 38:402–417

Walcroft A, Price S, Tate K, Sherlock R, Whitehead D (2008) Soil methanotrophs – a novel methane mitigation technology? Landcare Research Contract Report LCO708/154. Ministry of Agriculture and Forestry, 21 p

Wallenstein MD, Myrold DD, Firestone M, Voytek M (2006) Environmental controls on denitrifying communities and denitrification rates: insights from molecular methods. Ecol Appl 16:2143–2152

Ward G, Greenwood K (2002) Research and experiences in treading and wet soil management in Victoria. In: Currie LD, Loganathan P (eds) Dairy farm soil management. Massey University, Palmerston North, pp 47–59

White TA, Snow VO, King WM (2010) Intensification of New Zealand beef farming systems. Agric Syst 103:21–35

Whitehead DC (1995) Grassland nitrogen. CAB International, Wallingford

Whitehead D, Walcroft A, Saggar S, Parker W (2009) Dairy farmers are not green. Whakamarama 26 March 2009. http://staffroom.landcareresearch.co.nz/about/newsletters/staffnews/staffnews.asp?StaffNews_Issue_ID=38. Accessed 10 April 2010

Withers PJA, Hodgkinson RH, Adamson H, Green G (2007) The impact of pasture improvement on phosphorus concentrations in soils and streams in an upland catchment in Northern England. Agric Ecosyst Environ 122:220–232

Wolf B, Zheng X, Bruggemann N, Chen W, Dannenmann M, Han X, Sutton MA, Wu H, Yao Z, Butterbach-Bahl K (2010) Grazing-induced reduction of natural nitrous oxide release from continental steppe. Nature 464:881–884

Woodford K (2006) The intensification of pastoral agriculture: some trends and implications. Prime Indus Manage 9:3–4

Yanai Y, Toyota K, Okazaki M (2007) Effects of charcoal addition on N_2O emissions from soil resulting from rewetting air-dried soil in short-term laboratory experiments: original article. Soil Sci Plant Nutr 53:181–188

Zaman M, Blennerhassett JD (2010) Effects of the different rates of urease and nitrification inhibitors on gaseous emissions of ammonia and nitrous oxide, nitrate leaching and pasture production from urine patches in an intensive grazed pasture system. Agric Ecosyst Environ 136:236–246

Zaman M, Nguyen ML, Matheson F, Blennerhassett JD, Quin BF (2007) Can soil amendments (zeolite or lime) shift the balance between nitrous oxide and dinitrogen emissions from pasture and wetland soils receiving urine or urea-N? Aust J Soil Res 45:543–553

Zaman M, Nguyen ML, Saggar S (2008) N_2O and N_2 emissions from pasture and wetland soils with and without amendments of nitrate, lime and zeolite under laboratory condition. Aust J Soil Res 46:526–534

Zaman M, Saggar S, Blennerhassett JD, Singh J (2009) Effect of urease and nitrification inhibitors on N transformation, gaseous emissions of ammonia and nitrous oxide, pasture yield and N uptake in grazed pasture system. Soil Biol Biochem 41:1270–1280

Zhang J, Han X (2008) N_2O emission from the semi-arid ecosystem under mineral fertilizer (urea and superphosphate) and increased precipitation in northern China. Atmos Environ 42: 291–302

Chapter 11
Climate Change Impacts on Soil Processes in Rangelands

David J. Eldridge, Richard S.B. Greene, and Christopher Dean

11.1 Introduction

Rangelands are terrestrial ecosystems dominated by unimproved vegetation communities that occupy about half of the land area of the globe, or about 67 million km^2 (WRI 1996). The term "rangelands" generally applies to areas with arid, semi-arid or dry sub-humid climates that are unsuitable for broad-scale farming (Harrington et al. 1984). Climatic variables, particularly rainfall and temperature, are the major drivers of ecosystem productivity and stability (and therefore of soil health) in rangelands because they directly influence soil moisture and nutrient pools. Areas receiving $<$250 mm annual rainfall are termed arid, while those with 250–500 mm are semi-arid. Rainfall is highly variable from year to year with a coefficient of variation of annual rainfall typically $>$30%.

Rangeland soils store more than 10% of terrestrial biomass carbon (C), and up to 30% of global soil organic carbon (SOC) (Schlesinger 1997; Scurlock and Hall 1998). While C sequestration rates are low in rangelands relative to those in environments that regularly support arable farming, their soils have an enormous potential to sequester C simply because of the vast area they occupy. Declines in

D.J. Eldridge (✉)
Department of Environment, Climate Change and Water, Evolution and Ecology Research Centre, School of Biological, Earth and Environmental Sciences, University of NSW, Sydney, NSW 2052, Australia
and
School of Biological, Earth and Environmental Sciences, University of NSW, Sydney, NSW 2052, Australia
e-mail: D.Eldridge@unsw.edu.au

R.S.B. Greene
Fenner School of Environment and Society, Australian National University, Canberra, ACT 0200, Australia

C. Dean
School of Biological, Earth and Environmental Sciences, University of NSW, Sydney, NSW 2052, Australia

B.P. Singh et al. (eds.), *Soil Health and Climate Change*, Soil Biology 29,
DOI 10.1007/978-3-642-20256-8_11,

plant production in rangelands resulting from lower precipitation directly affect soil organic matter (SOM) levels in the soil, given the dependence of SOC stocks on plant biomass (Dermer and Schuman 2007). In the Patagonian steppe, for example, simulations experiments indicate that effects of extreme drought on primary productivity can extend well into the following years, suggesting that droughts are likely to have an enduring effect on production (Yahdjian and Sala 2006).

Soil health in rangelands is intimately linked to vegetation cover and biomass through the production and storage of SOM. Managing healthy soils, therefore, involves managing vegetation cover, its biomass and composition. Thus, primary avenues for maximizing C retention and sequestration in rangelands are (1) manipulating grazing intensity through management of stocking rates, at various spatial scales, (2) revegetating areas of reduced native vegetation and (3) ameliorating ongoing erosion (Howden et al. 1991; McKeon et al. 1992; Walker and Steffen 1993; Conant and Paustian 2002; Henry et al. 2002). Many rangeland soils are unfortunately, heavily degraded (Ojima et al. 1993; FAO 2004), and there is even potential to lose C from otherwise, well-managed soils through continued grazing and frequent burning (see Sect. 11.5). Overgrazing or unsustainable farming of fertile patches of rangeland is likely to increase under increasing pressure from climate change.

The issues facing soil health in rangelands in the face of changing climate are extremely complex, being dependent on site history, the condition of the extant vegetation, the land management practices imposed and the dominating influence from a range of changed climatic attributes. There is also considerable uncertainty associated with the assessment of both the extant SOC stock, current greenhouse gas emissions and the potential sink in rangelands. Quantitative analyses based on actual measurements of SOC stocks and losses for many of the world's rangelands are rare (Conant and Paustian 2002).

In this chapter, we describe the main influences that a changing climate is likely to have on rangelands ecosystems worldwide, with an emphasis on soils and soil processes. While a substantial amount of our knowledge is drawn from information for Australian landscapes, we use examples from rangelands worldwide to illustrate both the soil-level perturbations of changing climate and the more general ecosystem-wide impacts of a changing climate on both soils and vegetation.

11.2 Characteristics of Rangeland Ecosystems

Rangelands are managed for a variety of uses including pastoralism, mining, tourism, conservation, native cultures, military zones and, occasionally, cropping. Globally, they provide forage production for about three-quarters of the world's domestic livestock (Rangelands Australia 2008). Although traditionally used for pastoralism, there is a growing recognition of their importance for other uses such as conservation, hunting and the provision of ecosystem goods and services, recreation and aesthetics (Grice and Hodgkinson 2002).

A distinctive feature of rangelands worldwide is that productive, resource-rich soil occurs in patches. These "fertile patches" have moderate to high levels of moisture and nutrients, and support a larger proportion of plant diversity and productivity than the intervening resource-poor matrix (Stafford Smith and Morton 1990; Bestelmeyer et al. 2006). Feedback processes reinforce the intensity of these fertile patches, which also determine the distribution and abundance of soil biota. Small changes in soil moisture and fertility result in relatively large changes in soil biota (Whitford 2002). Thus, any changes in the amount and distribution rainfall or temperatures will have substantial effects on soils and their capacity to function.

Globally, many of the world's rangelands are severely overgrazed, invaded by exotic pests or adversely affected by inappropriate management such as frequent burning or cultivation along drainage lines. While low levels of grazing may increase the incorporation and decomposition of surface-resident SOM into the surface soil layers, overgrazing leads to potential losses in belowground SOC (Dermer and Schuman 2007). In wooded rangelands, higher grazing intensity (often accompanied by frequent, low-intensity fires) has been associated with decreased abundance of shrubs, less coarse woody debris and fewer trees with hollows (Eyre et al. 2010). Overgrazing, and vehicular usage is commonplace near drainage lines (areas of alluvial soils with higher net primary productivity), and near artificial watering points. Distance from water is a primary determinant of reduced biomass, and erosion is a secondary factor (Sparrow et al. 1997). With such surface disturbances, the soil nutrients and moisture become decoupled (Sparrow et al. 2003), with likely reductions in SOC. The degraded nature of rangelands and their reliance on a relatively small proportion of the landscape for production and diversity make them vulnerable to large-scale shifts in climate, particularly changes in the amount and relative distribution of rainfall.

11.3 Climate Change Forecasts for the World's Rangelands

There is growing evidence that the earth has begun to experience the effects of a changing climate. The area of land surface experiencing protracted periods of below-average rainfall has increased from 10 to 15% in the early 1970s to greater than 30% by early 2000 (Dai et al. 2004). Climate change is already affecting South American rangelands, with a mean warming of 1–4°C forecast for the next 70 years, particularly over the tropics (Yahdjian and Sala 2008). Over the last century, the north-western USA has warmed 0.5–1.5°C, and the temperature is projected to rise an additional 2–5°C by the end of the century. This is likely to increase the frequency and variability of droughts and floods (Chambers and Pellant 2008). Climate change predictions for rangelands in the south-western USA include an increase in average temperatures by 3–4°C by 2030, and an increase of 8–11°C by 2090 (Archer and Predick 2008). Climate models forecast a 1–2°C temperature rise in arid Central Asia by 2050, particularly in winter (Lioubimtseva and Cole 2006). For the Middle East, models forecast an overall temperature increase of 1.4°C by

2050 or 4°C by 2100. The largest change in precipitation is forecast for the Eastern Mediterranean, Turkey, Syria and the Caucasus, with a decline in precipitation due to decreased storm track activity (Evans 2009). Finally, climate change is expected to increase temperatures in India by 3–6°C, with reductions in rainfall of 5–25% by 2100, particularly during winter (Prabhakar and Shaw 2008).

These global trends are expected to increase the vulnerability of arid and semi-arid rangelands to drought and fire, and represent major challenges for managing vegetation and soils. More frequent and higher-intensity rainfall events in particular, exacerbated by greater drying of surface soils, are likely to induce higher rates of runoff, sediment removal and erosion, leading to feedback effects on nutrient and SOC loss. Higher diurnal surface temperatures globally will likely increase the frequency of hot days and warm nights, decrease the frequency of frosts, increase fire risk and result in a general pattern of drying (e.g. Zaitchik et al. 2007), particularly in the mid-latitudes.

Climate projections for Australia indicate an increase in drought frequency and severity (particularly in the productive grasslands and open woodlands of the southeast); rainfall intensity and the number of dry days will also increase continentally (Stokes et al. 2008). Rainfall seasonality is also forecast to change, with significant reductions in winter rainfall in the south. Forecast change in annual rainfall for Australia varies across the continent. For example: (1) a decrease of 2–5% by 2030 over much of Western Australia, western South Australia and the south-western Northern Territory, even for a low emission scenario (IPCC 2000), (2) changes from −2 to +2% by 2030 for different regions of eastern Australia (IPCC 2000), and (3) a decrease of up to 40% for the southwest of the continent (CSIRO and BoM 2007). Rainfall is not the only climatic influence on SOC, with temperature change having a significant impact (Cowling and Shin 2006). The temperature for continental Australia is forecast to increase by 1–5°C over the next century (Williams et al. 2009).

11.3.1 Influences on Soil Through Altered Plant Processes

The short-term effects of climate change on rangelands will be to reduce plant growth rates (through reduced soil moisture), and therefore cover and biomass, altering litter production (including root litter and exudates) and therefore soil microbial communities (Table 11.1). As soil moisture is the principal driver of primary production in rangelands (Noy-Meir 1973), declines in vegetation cover will severely reduce the soil's capacity to resist erosion, further diminishing soil productivity. In higher productivity grasslands, a scenario of increased frequency of wildfire is likely (Table 11.1).

Altered climate will almost certainly be accompanied by changes in plant community structure through increases in C_3 shrubs at the expense of C_4 grasses, leading to woody thickening (see Sect. 11.3.3.1). In some areas, replacement of shrublands by annual grasslands has led to increased fire frequencies, changing these communities from a C sink to a C source (Chambers and Pellant 2008). The influences of these changes on ecosystems could be devastating, with changes affecting regional albedo levels, resulting in feedback effects on increased evapotranspiration, loss of soil moisture and ultimately rainfall decline (Chambers and Pellant 2008). There are also likely to be increased invasions of exotic species. For example, in Australia a potential threat to rangeland SOC is the introduced species buffel grass (*Cenchrus ciliaris*), which can retain high senescent biomass, increasing wildfire intensity and extent, and depleting native shrub biomass through multiple pathways (Butler and Fairfax 2003), and consequently also reducing SOC in the long term.

Table 11.1 Summary of the effects of changing climate on terrestrial processes in rangelands and the short- and long-term effects on soils and soil processes

		Indirect effects	
Climate components	Direct effects	Short term	Long term
Lower rainfall	Increased drought severity, frequency and duration	Reduced plant cover	Lower inputs of soil organic matter
Altered rainfall distribution	Reduced soil moisture	Reduced plant biomass	Altered mineralization rates
Altered rainfall intensity	Increased rainfall erosivity	Altered litter fall	Reduced soil aggregation
Increased atmospheric CO_2	Altered photosynthetic capacity	Altered microbial communities	Altered C:N ratios
Increased atmospheric temperatures	Higher soil surface temperatures	Increased soil erosion	Increased soil bulk density
		Increased frequency of wildfire	Altered spatial distribution of soil nutrients
		Woody thickening	Reduced decomposition
			Reduced infiltration
			Increased coefficient of runoff
			Altered soil microbial communities
			Reduced termite populations
			Altered plant species composition
			Changes in ratio of C_3:C_4 plants
			Reduced habitat value
			Woody thickening attenuated by wildfire
			Exotic plant (weed) invasion
			Altered cover of biological crusts

Changes in plant community structure and composition will have indirect flow-on effects to soil microbial communities, which respond to substrate chemistry (Waldrop and Firestone 2006), and this could compromise the ability of soils to retain C stocks. Conditions of high soil moisture, higher temperatures and grazing-induced disturbance could lead to a flush of microbial activity, depleting labile forms of soil C through microbial respiration (Killham 1994). This would lead to substantial reductions in soil aggregation and water-holding capacity, as well as cascading effects of increased runoff and sediment loss, further exacerbating the diminished capacity of the soil to sequester C and retain nutrients (Fig. 11.1). Soil health, climate and rangeland management are therefore intimately linked.

Changes in the C:N ratio of plant material would alter decomposition rates and the spatial distribution of soil nutrients. Perennial grasses have considerable capacity to store C as labile forms such as mucigels and polysaccharides on belowground tissue, which also supports a diverse soil microbial community (Whitford 2002). However, more labile C inherently has a faster turnover rate (i.e. lower longevity) than the C associated with the roots of woody plants (trees and shrubs), which can have different microbial associations with those of grasses. With its higher content of aliphatic suberin, waxes and lignin, the decomposition product of woody roots

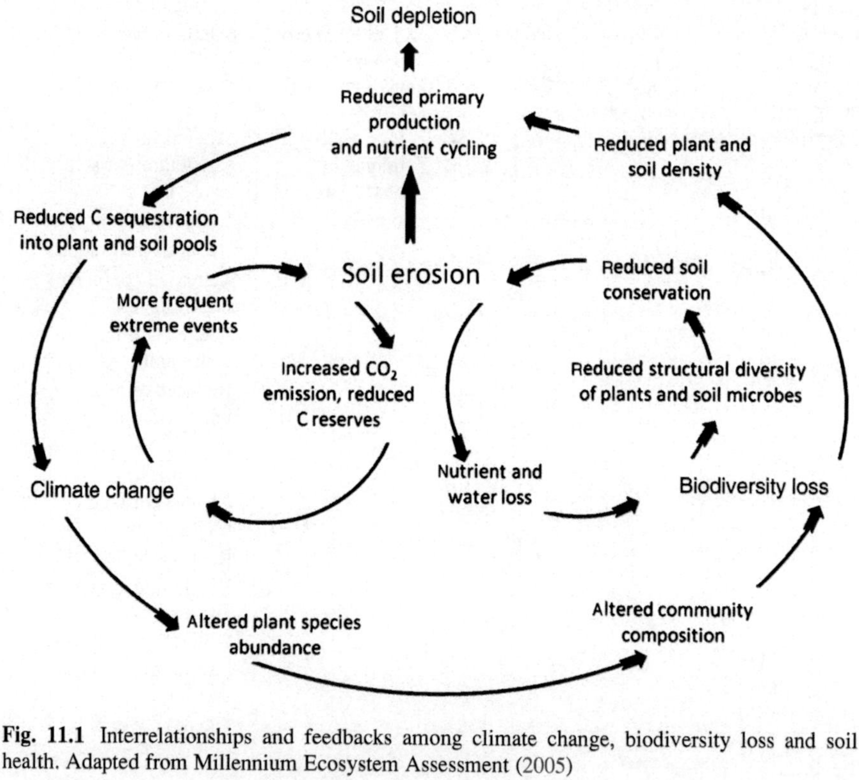

Fig. 11.1 Interrelationships and feedbacks among climate change, biodiversity loss and soil health. Adapted from Millennium Ecosystem Assessment (2005)

has a longer turnover time, thereby contributing slowly to SOC sequestration but forming a carbon pool of higher stability and hence a higher SOC stock in the long term.

11.3.2 Soil Feedbacks on Climate Change

Forecast changes in rainfall amount, distribution and intensity, increased temperatures and atmospheric CO_2 concentration will have substantial influence on: (1) retention and sequestration of C, (2) nutrient cycling, (3) resistance to erosion and (4) maintenance of functional hydrological processes (*sensu* Tongway 1995). Changes in rainfall will alter rangeland soil processes directly due to increased drought severity and soil surface temperatures, and exposure of the surface to radiation.

Accompanying changes in the rates and magnitudes of C fluxes between the atmosphere and vegetation will be changes in fluxes between vegetation and soil, with consequent flow-on effects to future climates. Positive climate change feedback may add 18% to atmospheric CO_2 levels by 2100 (Bonan 2008). Emissions from soil, reduced biomass and ecological impacts are part of this positive feedback cycle in climate change. The effects on soil processes are dependent on a large number of interacting factors including moisture, temperature, fire, human population demands and the degree of surface disturbance.

Rangeland management practices are also direct and immediate drivers in the positive feedback cycle, with deforestation (and prevention of regrowth) for cattle grazing a major source of greenhouse gas emissions, locally reduced rainfall and locally increased temperatures (and possibly higher drought severity) in Australia, Brazil and Columbia (McAlpine et al. 2009a, b). Fire is a major feedback mechanism wherever the evapotranspiration balance changes towards warmer and drier landscapes, while sufficient biomass remains to carry fire. This phenomenon applies to rangelands globally and to some higher productivity landscapes such as temperate forests.

Increasing temperature could alter the balance between influx and emission of C from the soil to produce a net emission, i.e. a net release of CO_2 back into the atmosphere. The activity of soil microbes is sensitive to changes in temperature and moisture. In areas such as Africa's Kalahari Desert, increases in temperatures could lead to a greater release of CO_2 to the atmosphere as C mineralization increases (Thomas et al. 2008). Rising temperatures will lead to significant increases in microbial respiration, resulting in an increased rate of SOM decomposition (Kirschbaum 1995). The effect could be to further exacerbate changing climate by accelerating the loss of SOC (Cox et al. 2000). Thus any increase in C sequestration resulting from the CO_2 fertilization of photosynthesis could be offset by increased microbial respiration (see Chap. 7).

11.3.3 Climate-Induced Changes in Shrublands and Grasslands

We consider now some of the interrelationships and feedbacks between climate, soils and rangeland management by examining two rangeland communities where climate-induced change is inevitable. Although Australian rangelands are unique in that they are predominantly shrubland and woodland with lesser amounts of scrub, heath and grassland, they share a common history of overgrazing effects with the Americas and southern Africa (Pickup 1998). We consider here community-level changes in desert shrublands in the western USA and the semi-arid grasslands of Australia, highlighting the importance of feedback mechanisms between altered climate, and altered vegetation and soil processes. These two case studies illustrate the close interconnections between climate, altered biodiversity and reduced ecosystem function, and therefore the effects of altered climate on rangeland soils (Fig. 11.1).

11.3.3.1 Changing Climate Exacerbates Woody Thickening

The increase in numbers of woody plants, and encroachment of woody plants into grasslands (called "woody thickening"), is a global phenomenon (Archer et al. 1995). Dramatic increases in the density and extent of shrubs such as mesquite (*Prosopis glandulosa*) and creosote bush (*Larrea tridentata*) have occurred in the western USA (Buffington and Herbel 1965). Interactions among grazing, drought, rainfall events and reduced wildfire are thought to be significant drivers of woody thickening (Archer 1994; Grover and Musick 1990). Both climatic change and the more fundamental increase in atmospheric CO_2 concentration are thought to be responsible for woody thickening (Archer et al. 1995; Brown and Thorpe 2008).

Predictions of reduced rainfall and more frequent drought for south-western USA are likely to result in reduced cover of herbaceous desert vegetation (Brown and Thorpe 2008) and woody thickening of grasslands (Archer 1994). An increase in the frequency and severity of wildfires, a further consequence of climate change, would likely favour the proliferation of shrubs by removing competition for resources by grasses as well as reducing biological soil crusts. However, fire frequency or intensity above a critical threshold will have the reverse effect of reducing woody shrub density (Hodgkinson et al. 1984).

Woody thickening changes the scale at which soil nutrients are distributed, from a fine scale, which corresponds to the distribution of former grass tussocks, to a scale consistent with the average spacing of shrub hummocks (Schlesinger and Pilmanis 1998). Water, sediment, dust and airborne nutrients tend to accumulate under shrub canopies due to wind and water processes. Shrub-free (bare) interspaces experience higher surface temperatures and evapotranspiration, reduced organic matter incorporation, increased erosion and, on fine-textured soils where water accumulates, denitrification and ammonia volatilization (Schlesinger et al. 1990). Increased infiltration below shrub canopies can result in higher concentrations of tissue nitrogen

(N) (Bhark and Small 2003) through enhanced micro-arthropod activity and greater rates of decomposition. Well-developed tap roots allow semi-arid shrubs to access water from greater depths (Archer et al. 2002), thereby establishing microbial communities and SOC sequestration at depths greater than that of the former grasses.

Land-to-atmosphere feedbacks are also likely to occur at regional scales. Conversely, reduction in woody cover has been shown to reduce local rainfall through changes in heat flux, surface roughness, evapotranspiration and decreased cumulus cloud formation (McAlpine et al. 2009a), suggesting that a denser woody cover would increase local rainfall. These are conflicting scenarios, but the most torrid and drying climatic changes will precipitate a net loss of biomass. Due to the land degradation often associated with woody thickening, there is some belief that converting thickened areas back to grasslands would increase SOC. However, woody shrubs are effective at belowground C sequestration (Hibbard et al. 2003; Bai et al. 2009) and represent an opportunity to increase C stocks, even if animal production is adversely affected. Shrub removal, however, would likely induce net C emissions.

11.3.3.2 Ecological Complexity Under Climate Change: The Semi-Arid Grasslands of South-Eastern Australia

Climate projections for the winter-dominant, semi-arid grasslands of south-eastern Australia indicate substantially less winter rainfall (80% decline) by 2070 (DECC 2008). Increased rainfall variability and frequency of high-intensity storms are likely to have substantial impacts on vegetation and soil. The effects most likely lead to replacement of relatively drought-intolerant bladder saltbush (*Atriplex vesicaria*) with an arid-adapted, drought-tolerant black bluebush (*Maireana pyramidata*) community. The cover and diversity of native grasses, herbs and forbs are also predicted to decline in response to greater soil moisture stress, and be replaced by arid-adapted Mediterranean weeds or grazing-tolerant forbs such as copperburrs (*Sclerolaena* spp.).

Reduced rates of C sequestration into aboveground and belowground pools will lead to reductions in abundance and diversity of aboveground biota, with feedback to the structure and diversity of belowground communities. Reductions in winter rainfall in these semi-arid grasslands will also increase the frequency of erosion events and the loss of soil nutrients. If grazing pressures are maintained, the risk of wind erosion of sandy soils will increase, particularly if plant cover drops below approximately 60% (Leys and Heinjus 1992). Reductions in winter rainfall, combined with greater levels of erosion, will also lead to a reduced cover of biological soil crusts (see Sect. 11.4) and further surface soil destabilization.

Soil and vegetation effects are likely to exacerbate changes in soil faunal populations, particularly termites, although we are unaware of empirical data for this area of grasslands. Research elsewhere in semi-arid grasslands indicates that termites are important components of healthy rangeland soils due to the range of ecosystem functions that they moderate. These functions are as broad as enhancing water flow into soils (Elkins et al. 1986), litter decomposition (Holt and Coventry 1988; Brown and Whitford 2003), C mineralization, nutrient recycling and subsequent plant production, particularly in low fertility soils (Parker et al. 1982; Coventry et al. 1988). Diminished grass cover will reduce abundance of termites, the main invertebrate decomposers in semi-arid and arid grasslands (Whitford 2002).

Termites are also preferred food items for a range of vertebrates and invertebrates. Replacement of grasses by exotic plants has potentially devastating bottom-up effects on semi-arid ecosystems by reducing termites, thereby reducing mineralization of N and C at landscape scales (Whitford 2002). This would appear to indicate a net reduction in C emissions with a reduced termite population. However, termites and other macro-arthropods also enhance water flow through soils by creating soil micropores (Eldridge 1994). Thus, reduced termite populations will reduce soil porosity and water storage (Whitford 2002) with consequent reduced water availability for plant growth.

11.4 The Potential of Rangeland Soils to Retain and Sequester Carbon

11.4.1 Plant Cover and the Maintenance of Healthy Soils

The most appropriate strategy for managing rangeland soils is to manage surface cover, either vascular plants (grasses, herbs, shrubs, trees) or non-vascular plants which make up the biological soil crust. Plant cover buffers the effects of wind and water on surface soils and therefore reduces the potential for erosion (Greene et al. 1994; McTainsh and Leys 1993). Below, we outline the importance of cover of both vascular plants and biological soil crusts for maintaining healthy rangeland soils, give some examples of how cover is likely to be affected by changing climate and describe the potential soil and ecosystem consequences of such changes.

Vascular plant cover has a major role in protecting soils against erosion, maintaining C stocks, and therefore improving soil health. Strategies to manage grazing in rangelands aim to manage plant cover and therefore maintain a range of critical physical, chemical and biological soil functions such as water-holding capacity, soil aggregation, surface stability and nutrient cycling. Plant cover also reduces raindrop impact and restricts the development of physically induced surface seals that impede water infiltration and prevent seedling emergence (Valentin and Bresson 1992). A major objective of rangeland management therefore is to maintain sufficient surface cover, which depends on soil type, rainfall amount and erosivity, soil moisture, slope and soil type (Greene et al. 1994). These critical cover thresholds for erosion prevention are thought to be in the range of about 40% (Greene et al 1994; Eldridge and Koen 2003).

11.4.2 Cyanobacterial Soil Crusts: Carbon Flux and Nitrogen Pools

Cyanobacteria are common components of soil crusts, along with lichens and mosses (Eldridge 2001a). Together these crusts stabilize the soil against water and wind erosion, regulate water flow into soils, provide a source of SOC and play vital roles in the maintenance and regulation of ecosystem functions. Cyanobacteria can survive soil temperatures of 50°C for prolonged periods and up to 100°C for 48 h (Rogers 1989). Cyanobacterial crusts can also sequester large volumes of CO_2, and in studies in the Mojave Desert, average net ecosystem exchange taken over a 2-year period ranged from 1 to 4 μmol-CO_2 m^{-2} s^{-2} net productivity. This is equivalent to about 1 t C ha^{-1} $year^{-1}$ (Wohlfahrt et al. 2008). Cyanobacterial crusts from soil and rock at sites in western Queensland sequester 0.5–1.8 μmol-CO_2 m^{-2} s^{-2} (Wendy Williams, personal communication, 2009). These crusts therefore have the capacity to sequester up to 1 t C ha^{-1} $year^{-1}$ (~1.7 million t C $year^{-1}$ for Australian rangelands).

Increases in dust storm frequency resulting from changing rainfall patterns (McTainsh and Lynch 1996) may alter the ability of cyanobacterial soil crusts to produce N. Cyanobacteria and cyanolichens fix substantial quantities of N in rangeland soils (Smith et al. 1990). However, increased frequency of sand storms may reduce these quantities. In western Queensland, landscape-scale deposition of coarse sand in the semi-arid woodlands is associated with an increase in soil N pools (Williams and Eldridge, unpublished data). Sand deposition leads to autolysis of N-enriched cyanobacterial cell material and therefore greater soil N pools. While stored N gradually accumulates in surface soils, long-term N production from these soils is compromised, reducing surface soil stability. Any inappropriate land management that leads to an increase in sand deposition (e.g. overgrazing) is likely to lead to long-term reductions in soil N pools.

11.4.3 Climate Change Impacts on Lichen-Dominant Crusts

Lichens cannot tolerate high summer temperatures combined with high humidity or rainfall because it reduces the photosynthetic ability of the algae component. In Australia, this intolerance to summer rainfall limits the distribution of lichen crusts to mainly winter-dominant areas in the south. Thus, while lichen crusts can tolerate surface temperatures in excess of 70°C, a temperature of 30°C for 30 min when fully hydrated is fatal (Rogers 1989). Under current climate change forecasts for Australia, lichen crusts are likely to be lost over significant areas of semi-arid southern Australia as summer rainfall is likely to increase in some areas that are currently winter dominant. The effect of lichen reduction may be catastrophic given their pivotal role in soil stability (Eldridge 2001b), and may lead to reduced landscape stability, with varied effects on water flow and wind erosion.

Compositional shifts in soil crusts, however, may mitigate against ecosystem collapse because cyanobacteria are likely to dominate the crusts as summer rainfall increases. It is possible, therefore, that cyanobacteria will play a moderating role in soil stabilization in the face of long-term climate change (Rogers 1989).

11.5 Grazing and Burning Exacerbate the Effects of Climate Change on Rangeland Soils

11.5.1 Grazing Effects on Rangeland Soils

Any negative effects of a changing climate are likely to be exacerbated by overgrazing, as stocking rate is a major driver of vegetation and soil change in rangelands (McKeon et al. 2009). Overgrazing leads to a range of soil-related problems including compaction (Thurow et al. 1988), soil fertility and nutrient decline (Snyman 1999), and loss of structural integrity (Thurow et al. 1988). Given projections of lower rainfall and higher temperatures in rangelands globally, continued grazing will likely alter the structure or plant communities, with potentially irreversible effects on ecosystem resilience.

Although grazing is a useful tool for managing vegetation and therefore influencing decomposition and soil nutrient levels, its effects on soils are site-specific (Beukes and Cowling 2003). Some grazing practices such as low-risk, opportunistic grazing may have little effect on soils. However, there are likely to be substantial soil effects under continuous grazing. Tactical grazing strategies, whereby larger numbers of animals graze smaller paddocks over shorter time periods, are thought to provide ecological benefits to the soil in some rangelands (e.g. South African grassland) where rainfall is more reliable. However, in many seasonally variable rangelands such as the semi-arid woodlands of eastern Australia or the grasslands and woodlands of the USA, trampling associated with rotational grazing has substantial adverse effects on soils such as destruction of the biological soil crust and reduction in infiltration rate (e.g. Weltz et al. 1989).

11.5.2 Savannah Burning in Subtropical Australian Rangelands

Australia's northern savannas can be a net source or sink of C depending on how they are managed. Under a regime of low-intensity biennial burning, savannahs are net C sequestrators (0.5–2.0 t C ha^{-1} $year^{-1}$), with major contributions including emissions from fires (1.6 t C ha^{-1} $year^{-1}$), sequestration due to tree growth (1.2 t C ha^{-1} $year^{-1}$) and sequestration from woody thickening (0.2 t C ha^{-1} $year^{-1}$) (Beringer et al. 2007). The present state of annual savannah burning, however, being a mixture of intense and moderate burns, contributes substantially to the national greenhouse gas budget (Williams et al. 2005). Similarly, high-intensity, late-season burns constitute an ongoing C efflux (Dyer and Stafford Smith 2003), which is often exacerbated by the consumption of large amounts of standing dead timber (Fensham 2005). This is accompanied by reductions in SOC levels and increased atmospheric aerosol concentrations. Modelling suggests that increased grass growth, a consequence of atmospheric CO_2 enrichment, coupled with increased wildfire, presents significant threats to net C sequestration in the semi-arid woodlands (Howden et al. 2001).

Careful management of livestock and grassy-fuel loads is necessary to maintain long-term sequestration rates above the present 1 t-C ha^{-1} year^{-1} (Williams et al. 2004) in the tropical savannas and to prevent this sink becoming a C source. Increased grass density may favour higher levels of grazing, ultimately promoting woody thickening, and converting open savannah into woodland (but possibly with initial erosion before SOC recovery through thickening). In Australia's semi-arid mulga woodlands, modelling indicates that century-long net sequestration and maintenance of aboveground C is only possible under a scenario of no fire and no grazing (Howden et al. 2001).

11.6 Rehabilitation of Rangeland Soils

The vastness, degraded state and relatively low human population of the world's rangelands have prompted numerous studies of the effects of management changes on C sequestration (e.g. McKeon et al. 1992; Walker and Steffen 1993; Garnaut 2008). Howden et al. (1991) estimated that rehabilitating the rangelands globally to their pre-degraded state would sequester about 7 t-C ha^{-1} as SOC, i.e. an extra 0.5 wt% organic C. Rangeland rehabilitation has long been considered uneconomical due to the low returns and long timelines involved (e.g. Perry 1974). It can, however, be profitable under soil C trading (e.g. Cerri et al. 2003). Also, woody thickening offers C sequestration even without investment in management intervention. Currently, projects assessing C sequestration in rangelands, while simultaneously improving land condition, are underway worldwide.

There are currently an estimated 4.5 Gt of SOC in the 380 Mha of Australia's commercially grazed rangelands to 0.3 m depth (with possibly another 40% to 2 m depth). Ongoing emissions from the soil due to management for grazing are ~5.3 Mt-C year^{-1} (Dean et al. 2009). Alternatively, an increase in SOC by only 1% over those rangelands would equate to 45 Mt-C. However, the period for that sequestration is in the order of two centuries (Dean et al. 2009), owing to the slow rate of SOC sequestration in semi-arid rangelands (Hibbard et al. 2003; Singh et al. 2007). When sequestration in biomass and coarse woody debris are included (from both woodland regrowth and the slower process of land rehabilitation), the sequestration is much faster and estimates for Australia range from 0.273 to 3.2 t-C ha^{-1} year^{-1} over periods of 20–140 years (Moore et al. 2001; Garnaut 2008;

Dean et al. 2009). Rehabilitation of Australian mulga-lands has been forecast to sequester up to 68 Mt-C year^{-1} over several decades (Garnaut 2008). At the national scale, however, the current trend of a net emission from rangeland soils must be slowed and reversed before sequestration is manifested.

11.7 Rangeland Soils Under a Changing Climate: Concluding Remarks

Climate change will intensify problems currently facing rangeland managers worldwide (FAO 2004; McKeon et al. 2009). Declining pasture productivity, reduced forage quality, increased livestock heat stress, more frequent weed and pest invasions, more frequent droughts, less frequent, but more intense rainfall and greater soil erosion are all likely outcomes of the projected changes in climate (Stokes et al. 2008; McKeon et al. 2009). Although the magnitude of these effects and their feedbacks are not well understood (Henry et al. 2007), they will have major implications for rangeland managers. How rangeland managers adjust to these different impacts will, in turn, have major feedback effects on the extent of climate change.

Climate change is expected to increase the vulnerability of arid and semi-arid rangelands to further degradation. With climate change effects already evident in the rangelands, and with an increasing human population, both mitigation and adaptation (to climate change) are imperative. Several avenues for mitigation in the rangelands have already been broached (Henry et al. 2002; WAG 2002; FAO 2004; Foran 2007; Dean et al. 2009; Dwyer et al. 2009; Fensham and Guymer 2009; McAlpine et al. 2009b). They include (1) protection and enhancement of SOC through, for example, management of grazing intensity, (2) reduced clearance of native vegetation, (3) regrowth of areas previously deforested for grazing, (4) planting of deeper-rooting vegetation, (5) increasing scientific knowledge of physical and ecological processes and revision of institutional frameworks for knowledge integration, (6) control of invasive plants and animals, and (7) implementation of policy initiatives aimed at C sequestration. Although requiring gross changes to current practices, these avenues can be simply distilled as constituting careful management of soil and ecological processes, biodiversity and vegetation cover. Adapting to climate change principally involves a concerted application of resources and management effort to these very same issues. Accordingly, adaptation will require action at a governmental level, establishing national and international funding to the avenues listed above, along with integrated management towards these activities (e.g. FAO 2004).

Acknowledgements We thank Ron Hacker, Brian Murphy, Erin Roger, Alex James and Jim Noble for constructive comments on earlier drafts.

References

Archer S (1994) Woody plant encroachment into southwestern grasslands and savannas: rates, patterns and proximate causes. In: Vavra M, Laycock WA, Pieper RD (eds) Ecological implications of livestock herbivory in the west. Society for Range Management, Denver, pp 13–68

Archer SR, Predick KI (2008) Climate change and ecosystems of the Southwestern United States. Rangelands 30:23–28

Archer S, Schimel DS, Holland EA (1995) Mechanisms of shrubland expansion: landuse, climate or CO2? Clim Change 29:91–99

Archer NAL, Quinton JN, Hess TM (2002) Below-ground relationships of soil texture, roots and hydraulic conductivity in two-phase mosaic vegetation in South-east Spain. J Arid Environ 52:535–553

Bai Y, Colberg T, Romo JT, McConkey B, Pennock D, Farrell R (2009) Does expansion of western snowberry enhance ecosystem carbon sequestration and storage in Canadian Prairies? Agri Ecos Environ 134:269–276

Beringer J, Hutley LB, Tapper NG, Cernusak LA (2007) Savanna fires and their impact on net ecosystem productivity in North Australia. Glob Change Biol 13:990–1004

Bestelmeyer BT, Ward JP, Herrick JE, Tugel AJ (2006) Fragmentation effects on soil aggregate stability. Rangeland Ecol Manag 59:406–415

Beukes PC, Cowling RM (2003) Non-selective grazing impacts on soil-properties of the Nama Karoo. J Range Manage 56:547–552

Bhark EW, Small EE (2003) Association between plant canopies and the spatial patterns of infiltration in shrubland and grassland of the Chihuahuan Desert, New Mexico. Ecosystems 6:185–196

Bonan GB (2008) Forests and climate change: forcings, feedbacks, and the climate benefits of forests. Science 320:1444–1449

Brown JR, Thorpe J (2008) Rangelands and climate change: a synthesis and challenges. Rangelands 30:52–53

Brown MF, Whitford W (2003) The effects of termite and straw mulch on soil nitrogen in a creosotebush (*Larrea tridentata*) dominated Chihuahuan Desert Ecosystem. J Arid Environ 53:15–20

Buffington LC, Herbel CH (1965) Vegetational changes on a semidesert grasslands range from 1858 to 1963. Ecol Monogr 35:140–164

Butler DW, Fairfax RJ (2003) Buffel Grass and fire in a gidgee and brigalow woodland: a case study from central Queensland. Ecol Manage Restor 4:120–125

Cerri CEP, Coleman K, Jenkinson DS, Bernoux M, Victoria R, Cerri CC (2003) Modeling soil carbon from forest and pasture ecosystems of Amazon, Brazil. Soil Sci Soc Am J 67: 1879–1887

Chambers JC, Pellant M (2008) Climate change impacts on Northwestern and intermountain United States Rangelands. Rangelands 30:29–33

Conant R, Paustian K (2002) Potential soil carbon sequestration in overgrazed grassland ecosystems. Global Biogeochem Cycles 16. doi:10.1029/2001GB001661

Coventry RJ, Holt JA, Sinclair DF (1988) Nutrient cycling by mound-building termites in low fertility soils of semi-arid tropical Australia. Aust J Soil Res 26:375–390

Cowling SA, Shin Y (2006) Simulated ecosystem threshold responses to co-varying temperature, precipitation and atmospheric CO_2 within a region of the Amazon. Glob Ecol Biogeog 15:553–566

Cox PM, Betts RA, Jones CD, Spall SA, Totterdell IJ (2000) Acceleration of global warming due to carbon-cycle feedbacks in a coupled climate model. Nature 408:184–187

CSIRO and Bureau of Meteorology (2007) Climate change in Australia-observed changes and predictions. CSIRO. http://www.cawcr.gov.au/file/Observedchangesprojection2007.pdf. Accessed 3 June 2010

Dai A, Trenberth KE, Qian T (2004) A global data set of Palmer Drought Severity Index for 1870–2002: relationship with soil moisture and effects of surface warming. J Hyromet 5:1117–1130

Dean C, Eldridge DJ, Harper RJ (2009) Prognosis for carbon sequestration in rangelands upon destocking. A case study of the vast Australian rangelands with reference to biogeochemistry, ecology, fire, biodiversity and climate change. Unpublished technical report. Rural Industries Research and Development Corporation, Canberra

DECC (2008) A biophysical assessment of climate change impacts in NSW. Unpublished report, Department of Environment and Climate Change, Sydney

Dermer JD, Schuman GE (2007) Carbon sequestration and rangelands: a synthesis of land management and precipitation effects. J Soil Water Conserv 62:77–85

Dwyer JM, Fensham RJ, Butler DW, Buckley YM (2009) Carbon for conservation: assessing the potential for win-win investment in an extensive Australian regrowth ecosystem. Agric Ecosyst Environ 134:1–7

Dyer R, Stafford Smith MS (2003) Ecological and economic assessment of prescribed burning impacts in semi-arid pastoral lands of northern Australia. Int J Wildland Fire 12:403–413

Eldridge DJ (1994) Nests of ants and termites influence infiltration in a semi-arid woodland. Pedobiologia 38:481–492

Eldridge DJ (2001a) Biological soil crusts of Australia. In: Belnap J, Lange O (eds) Biological soil crusts: structure, management and function. Ecological Studies 150. Springer, Berlin, pp 119–132

Eldridge DJ (2001b) Biological soil crusts and water relations in of Australian deserts. In: Belnap J, Lange O (eds) Biological soil crusts: structure, management and function. Ecological Studies 150. Springer, Berlin, pp 315–326

Eldridge DJ, Koen TB (2003) Detecting environmental change in eastern Australia: rangeland health in the semi-arid woodlands. Sci Total Environ 310:211–219

Elkins NZ, Sabol GV, Ward TJ, Whitford WG (1986) The influence of subterranean termites on the hydrological characteristics of a Chihuahuan desert ecosystem. Oecologia 68:1432–1939

Evans JP (2009) 21st Century climate change in the Middle East. Clim Change 92:417–432

Eyre TJ, Butler DW, Kelly AL, Wang J (2010) Effects of forest management on structural features important for biodiversity in mixed-age hardwood forests in Australia's subtropics. For Ecol Manage 259:534–546

FAO (2004) Carbon sequestration in dryland soils. World soil resources reports 102. Food and Agriculture Organization of the United Nations, Rome

Fensham RJ (2005) Monitoring standing dead wood for carbon accounting in tropical savanna. Aust J Bot 53:631–638

Fensham RJ, Guymer GP (2009) Carbon accumulation through ecosystem recovery. Eviron Sci Policy 12:367–372

Foran BD (2007) Sifting the future from the past: a personal assessment of trends impacting the Australian rangelands. Rangeland J 29:3–11

Garnaut R (2008) The garnaut climate change report to the commonwealth, state and territory governments of Australia. http://www.garnautreview.org.au. Accessed 3 June 2010

Greene RSB, Kinnell PIA, Wood JT (1994) Role of plant cover and stock trampling on runoff and soil erosion from semi-arid wooded rangelands. Aust J Soil Res 32:953–973

Grice AC, Hodgkinson KC (2002) Challenges for rangeland people. In: Grice AC, Hodgkinson KC (eds) Global rangelands: progress and problems. CAB International, London, pp 1–9

Grover HD, Musick HB (1990) Shrubland encroachment in southern New Mexico, USA: an analysis of desertification processes in the American southwest. Clim Change 17:305–330

Harrington GN, Wilson AD, Young MD (eds) (1984) Management of Australia's rangelands. CSIRO Australia, Melbourne

Henry BK, Danaher T, McKeon GM, Burrows WH (2002) A review of the potential role of greenhouse gas abatement in native vegetation management in Queensland's rangelands. Rangeland J 24:112–132

Henry BK, McKeon GM, Syktus J, Carter JO, Day K, Rayner D (2007) Climate variability, climate change and land degradation. In: Sivakumar MVK, Ndiang'ui N (eds) Climate and land degradation. Springer, Berlin, pp 205–221

Hibbard KA, Schimel DS, Archer S, Ojima DS, Parton W (2003) Grassland to woodland transitions: integrating changes in landscape structure and biogeochemistry. Ecol Appl 13:911–926

Hodgkinson KC, Harrington GN, Griffin GF, Noble JC, Young MD (1984) Management of vegetation with fire . In: Harrington GN, Wilson AD, Young MD (eds) Management of Australia's rangelands. CSIRO, Melbourne, pp 141–156

Holt JA, Coventry RJ (1988) The effects of tree clearing and pasture establishment on a population of mound-building termites (Isoptera) in North Queensland. Aust J Ecol 13:321–325

Howden SM, McKeon GM, Scanlan JC, Carter JO, White DH, Galbally IE (1991) Managing pastures in northern Australia to minimise greenhouse gas emissions: adaptation of an existing simulation model. Proceedings of the Australian Simulation Society, Gold Coast

Howden SM, Moore JL, McKeon GM, Carter JO (2001) Global change and the mulga woodlands of southwest Queensland: greenhouse gas emissions, impacts and adaptation. Environ Int 27:161–166

IPCC (2000) Intergovernmental Panel on Climate Change (IPCC) special report on emission scenarios. http://www.ipcc.ch/pdf/special-reports/spm/sres-en.pdf. Accessed 3 June 2010

Killham K (1994) Soil ecology. Cambridge University Press, Cambridge

Kirschbaum MUF (1995) The temperature dependence of soil organic matter decomposition and the effect of global warming on soil organic carbon storage. Soil Biol Biochem 27:753–760

Leys JF, Heinjus DR (1992) Cover levels to control soil and nutrient loss from wind erosion in the south Australian Murray Mallee. In: Haskins PG, Murphy BM (eds) People managing their land, 7th ISCO Conference Proceedings, Sydney, Australia. International Soil Conservation Organisation, Sydney, pp 626–637

Lioubimtseva E, Cole R (2006) Uncertainties of climate change in arid environments of central Asia. Rev Fish Sci 14:29–49

McAlpine CA, Syktusw J, Ryan JG, Deo RC, McKeon GM, McGowan HA, Phinn SR (2009a) A continent under stress: interactions, feedbacks and risks associated with impact of modified land cover on Australia's climate. Glob Chang Biol 15:2206–2223

McAlpine CA, Etter A, Fearnside PM, Seabrook L, Laurance WF (2009b) Increasing world consumption of beef as a driver of regional and global change: a call for policy action based on evidence from Queensland (Australia), Colombia and Brazil. Glob Environ Chang 19:21–33

McKeon GM, Howden SM, Stafford Smith DM (1992) The management of extensive agriculture: greenhouse gas emissions and climate change. In: Assessing technologies and management systems for agriculture and forestry in relation to global climate change. Intergovernmental Panel on Climate Change, response strategies working group. Australian Government Publishing Service, Canberra, pp 42–47

McKeon GM, Stone GS, Syktus JI, Carter JO, Flood NR, Ahrens DG, Bruget DN, Chilcott CR, Cobon DH, Cowley RA, Crimp SJ, Fraser GW, Howden SM, Johnston PW, Ryan JG, Stokes CJ, Day KA (2009) Climate change impacts on northern Australian rangeland livestock carrying capacity: a review of the issues. Rangeland J 31:1–29

McTainsh GH, Leys JF (1993) Soil erosion by wind in land degradation processes in Australia. In: McTainsh G, Broughton WC (eds) Land degradation processes in Australia. Longman, Melbourne, pp 188–233

McTainsh GH, Lynch AW (1996) Quantitative estimates of the effect of climate change on dust storm activity in Australia during the last glacial maximum. Geomorphology 17:263–271

Millennium Ecosystem Assessment (2005) Ecosystems and human well-being: desertification synthesis. World Resources Institute, Washington

Moore JL, Howden SM, McKeon GM, Carter JO, Scanlan JC (2001) The dynamics of grazed woodlands in southwest Queensland, Australia and their effect on greenhouse gas emissions. Environ Int 27:147–153

Noy-Meir I (1973) Desert ecosystems: environment and producers. Annu Rev Ecol Syst 4:25–51
Ojima DS, Dirks BM, Glenn EP, Owensby CE, Scurlock JO (1993) Assessment of C budget for grasslands and drylands of the world. Water Air Soil Pollut 70:95–109
Parker LW, Fowler HG, Ettershank G, Whitford WG (1982) The effects of subterranean termite removal on desert soil nitrogen and flora. J Arid Environ 5:52–59
Perry RA (1974) The future of arid land vegetation under grazing. In: Wilson AD (ed) Studies of the Australian arid zone. II Animal production. Commonwealth Scientific and Industrial Research Organization, Melbourne, pp 144–150
Pickup G (1998) Desertification and climate change – the Australian perspective. Climate Res 11:51–63
Prabhakar SVRK, Shaw R (2008) Climate change adaptation implications for drought risk mitigation: a perspective for India. Clim Change 88:113–130
Rangelands Australia (2008) What are the rangelands? http://www.rangelands-australia.com.au/frameSet1_OurRangelands.html. Accessed 9 May 2009
Rogers RW (1989) Blue-green algae in southern Australian rangeland soils. Aust Range J 11:67–73
Schlesinger WH (1997) Biogeochemistry. Geotimes 42:44
Schlesinger WH, Pilmanis AM (1998) Plant-soil interactions in deserts. Biogeochemistry 42:169–187
Schlesinger WH, Reynolds JF, Cunningham GL, Huenneke LF, Jarrell WM, Virginia RA, Whitford WG (1990) Biological feedbacks in global desertification. Science 247:1043–1048
Scurlock JMO, Hall DO (1998) The global carbon sink: a grassland perspective. Glob Change Biol 4:229–233
Singh SK, Singh AK, Sharma BK, Tarafdar JC (2007) Carbon stock and organic carbon dynamics in soils of Rajasthan, India. J Arid Environ 68:408–421
Smith GD, Lynch RM, Jacobson G, Barnes CJ (1990) Cyanobacterial nitrogen fixation in arid soils of Central Australia. FEMS Microbiol Lett 74:79–89
Snyman HA (1999) Short term effect of soil-water, defoliation and rangeland condition on productivity of a semi-arid rangeland in South Africa. J Arid Environ 43:47–62
Sparrow AD, Friedel MH, Stafford Smith MD (1997) A landscape-scale model of shrub and herbage dynamics in central Australia, validated by satellite data. Ecol Model 97:197–216
Sparrow AD, Friedel MH, Tongway DJ (2003) Degradation and recovery processes in arid grazing lands of central Australia Part3: implications at landscape scale. J Arid Environ 55:349–360
Stafford Smith DM, Morton SR (1990) A framework for the ecology of arid Australia. J Arid Environ 18:255–278
Stokes CJ, Ash A, Howden SM (2008) Climate change impacts on Australia's rangelands. Rangelands 3:40–45
Thomas AD, Hoon SR, Linton PE (2008) Carbon dioxide fluxes from cyanobacteria crusted soils in the Kalahari. Appl Soil Ecol 39:254–263
Thurow TL, Blackburn WH, Taylor CA (1988) Some vegetation responses to selected livestock grazing strategies. J Range Manage 41:108–114
Tongway D (1995) Monitoring soil productive potential. Environ Monit Assess 37:309–318
Valentin C, Bresson LM (1992) Morphology, genesis and classification of surface crusts in loamy and sandy soils. Geoderma 55:225–245
WAG (The Washington Advisory Group) (2002) Sequestering carbon emissions in the terrestrial biosphere. The energy information administration, office of integrated analysis and forecasting. United States Department of Energy, Washington
Waldrop MP, Firestone MK (2006) Response of microbial community composition and function to soil climate change. Microb Ecol 52:716–724
Walker BH, Steffen WL (1993) Rangelands and global change. Rangel J 15:95–103
Weltz M, Wood MK, Parker EE (1989) Flash grazing and trampling-effects on infiltration rates and sediment yield on a selected New-Mexico range site. J Arid Environ 16:95–100
Whitford WG (2002) Ecology of desert systems. Academic, London

Williams RJ, Hutley LB, Cook GD, Russell-Smith J, Edwards A, Chen X (2004) Assessing the carbon sequestration potential of mesic savannas in the Northern Territory, Australia: approaches, uncertainties and potential impacts of fire . Funct Plant Biol 31:415–422

Williams RJ, Zerihun A, Montagu KD, Hoffman M, Hutley LB, Chen X (2005) Allometry for estimating aboveground tree biomass in tropical and subtropical eucalypt woodlands: towards general predictive equations. Aust J Bot 53:607–619

Williams RJ, Bradstock RA, Cary GJ, Enright NJ, Gill AM, Liedloff AC, Lucas C, Whelan RJ, Andersen AA, Bowman DJMS, Clarke P, Cook GJ, Hennessy K, York A (2009) The impact of climate change on fire regimes and biodiversity in Australia – a preliminary assessment. A CSIRO unpublished report to the Australian Government. Department of Climate Change, Canberra

Wohlfahrt G, Fenstermaker LF, Arnone JA (2008) Large annual net ecosystem CO_2 uptake of a Mojave Desert ecosystem. Glob Change Biol 14:1475–1487

WRI (1996) World Resources Institute World Resources 1986: an assessment of the resource base that supports the global economy. Basic Books, New York

Yahdjian L, Sala OE (2006) Vegetation structure constrains primary production response to water availability in the Patagonian steppe. Ecology 87:952–962

Yahdjian L, Sala OE (2008) Climate change impacts on South American rangelands. Rangelands 30:34–39

Zaitchik BF, Evans JP, Geerken RA, Smith RB (2007) Climate and vegetation in the Middle East: interannual variability and drought feedbacks. J Climate 20:3924–3941

Chapter 12
Possible Impacts of Climate Change on Forest Soil Health

R. John Raison and Partap K. Khanna

12.1 Introduction

Any discussion of the impacts of climate change on forest soil health must start with a brief discussion of what is meant by climate change, and how we define soil health.

We adopt the following definition of climate change used by the Intergovernmental Panel on Climate Change (IPCC) (2007), and also include increasing atmospheric concentrations of CO_2 because of their likely major impact on the carbon cycle in forest ecosystems. Although increased N deposition is also often a consequence of the same human activities that result in greenhouse gas (GHG) emissions and climate change, and can have significant impacts on soils (e.g. Brumme and Khanna 2009; Matzner 2004; Reay et al. 2008), it is not considered in this chapter.

Climate change in IPCC usage refers to a change in the state of the climate that can be identified (e.g. using statistical tests) by changes in the mean and/or the variability of climate properties, and that persists for an extended period, typically decades or longer. It refers to any change in climate over time, whether due to natural variability or as a result of human activity. This usage differs from that in the United Nations Framework Convention on Climate Change (UNFCCC), where climate change refers to a change in the climate that is attributed directly or indirectly to human activity that alters the composition of the global atmosphere and that is in addition to natural climate variability observed over comparable time periods.

Soil health has been reviewed in detail elsewhere (e.g. Arias et al. 2005; Doran and Zeiss 2000), including appropriate indicators and assessment methods. Doran

R.J. Raison (✉)
Chief Research Scientist, CSIRO Ecosystem Sciences, PO Box 284 Canberra, ACT 2601, Australia
e-mail: John.Raison@csiro.au

P.K. Khanna
Formerly with CSIRO, now retired

B.P. Singh et al. (eds.), *Soil Health and Climate Change*, Soil Biology 29,
DOI 10.1007/978-3-642-20256-8_12,

and Zeiss (2000) described soil health as the capacity of the soil to function as a vital living system, within ecosystem and land use boundaries to sustain plant and animal productivity, maintain or enhance water and air quality, and promote plant and animal health. Here, we adopt a somewhat similar and broad interpretation of soil health – to include all soil physical, chemical and biological properties and processes that affect forest composition and productivity and associated carbon, nutrient and hydrological cycles. We note that soil health is, to some extent, a human construct that relates to goals or value judgements – very different states of soil condition can be considered "healthy" in relation to contrasting natural ecosystems or management objectives (e.g. for commercial forest production).

It is worth briefly touching on the broad range of disturbance regimes that pertain to forest soils, because this affects both how soil health may be affected by climate change, as well as the ability to manage or mitigate any adverse impacts. Forest soils, especially those under native forests, differ from soils of other land use systems in terms of their infrequent (but sometimes major) disturbance, and high organic matter content which may increase the capacity to buffer the effects of climate change. Some of the effect of climate change on soils in such forests may be slow but cumulative, and would require special and sensitive parameters to detect any change in terms of soil health. On the other hand, industrial plantations are often subject to relatively frequent and severe soil disturbance from intensive harvesting, use of slash burning, heavy fertiliser application, and other site preparation practices. These disturbed soil systems may be more vulnerable to climate change, but they may also offer more opportunities to manage such impacts as part of the overall regime of intensive forest management.

A changing climate can induce a myriad of changes in forests at the global scale, as a consequence of both direct and indirect impacts (Fig. 12.1). Many of these changes will be location specific – determined by the interactions between climate, terrain, vegetation, soil type and the degree of management intervention.

The direct effects of climate change include:

- Increased temperature and atmospheric concentration of CO_2;
- Changes in precipitation and the frequency and severity of extreme climatic events such as heatwaves, droughts, storms (rain, ice and wind), frosts, etc.

The indirect effects of climate change can be very significant (e.g. Dale et al. 2001; Lucier et al. 2009) and mediated, in particular, by:

- Changes in the frequency and impacts of outbreaks of pests and pathogens;
- Changes in fire regimes;
- Changes in vegetation growth and species composition (including from invasive species) which affect litter inputs and many associated soil processes.

There can also be management responses to the above changes, and these can also impact forest soil health. For example, where there is increased risk of wildfire under climate change, the frequency of low-intensity fire may be increased to reduce fuel loads and the impacts of future wildfires, and this may affect vegetation

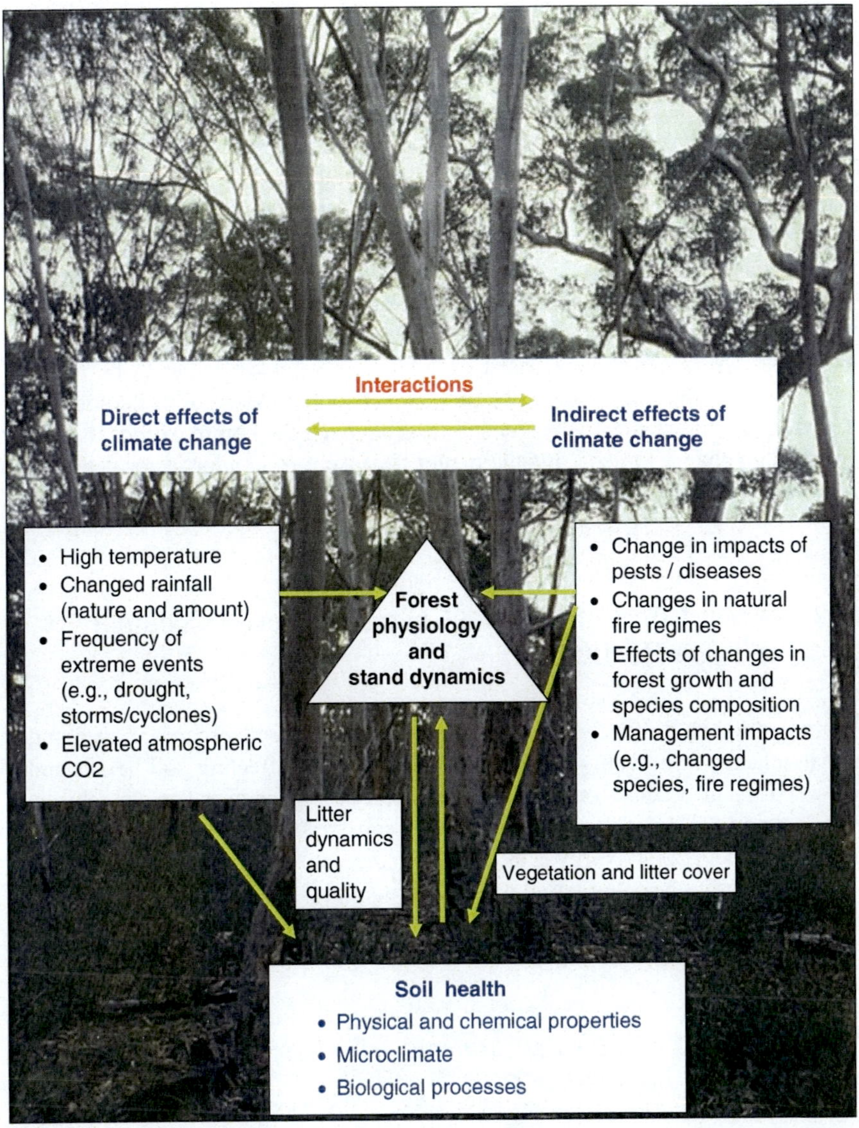

Fig. 12.1 Schematic diagram showing how both direct and indirect effects of climate change can markedly affect many forest properties and processes. These changes occur over a range of time scales and can have significant effects on the health of forest soils. The process involved, relevant time scales and the nature of the impacts on forest soil health are discussed in detail in the text

dynamics, carbon and nutrient cycling, and other forest processes that impinge on critical soil processes.

There can also be interactions between the direct and indirect effects of climate change. For example, storms with high rainfall intensity may lead to enhanced rates of soil erosion in combination with increased frequency of either wildfire or management burns. Sometimes the indirect effects of climate change can increase the susceptibility to other disturbances, e.g. in North America increased outbreaks of bark beetles under climate change can lead to tree mortality, increased fuel loads and greater fire risks (Moore and Allard 2008). These very complex, and location-specific, interactions make it very difficult to predict the future effects of climate change on forest processes related to soil health.

In this review, we provide a framework for analysing the potential effects of climate change on forest soil health (Fig. 12.1), and propose, discuss and illustrate some broad principles, recognising that location-specific circumstances can be of overriding importance. We emphasise that at this time we can only draw inference about possible future change based on review of what we judge to be relevant existing information, because we do not yet have direct measures of the long-term effects of climate change on forest soil health. Given the complexity of the topic, we focus on changes in soil organic matter because it is a major factor controlling the "fertility" of forest soils (see e.g. Knops et al. 2007; Krull et al. 2004).

12.2 Direct Effect of Climate Change on Forest Productivity and Soil Health

Any alteration to either the quantity or the quality of soil organic matter under climate change is probably the most important factor affecting soil health under forests. This is because organic matter exerts strong controls on the physical, chemical and biological properties affecting soil "fertility" (see, e.g. Knops et al. 2007; Krull et al. 2004; Norby et al. 2001). Changes in soil organic matter can be of both a qualitative or quantitative nature and may provide a positive or a negative feedback to atmospheric CO_2 concentrations, and thus to climate forcing.

12.2.1 Effects of Elevated Atmospheric Concentrations of CO_2 and Other Gases

While there are various avenues of direct effects of rising atmospheric CO_2 on forest productivity, the evidence of changes in soil organic matter and soil health is limited and variable. Elevated CO_2 concentrations in the atmosphere (Van Groenigen et al. 2006) may increase C inputs to soils through changes in plant production and vegetation composition (see reviews by Fischlin et al. 2007; Hyvönen et al. 2007), but in most elevated CO_2 experiments the role of nutrient and water limitations is not well considered, thus making findings hard to apply to field environments where these limitations usually exist to varying degrees. These nutrient and water feedbacks are critical in determining the magnitude and longevity of response to elevated CO_2 (see review by Raison et al. 2007). For example, elevated atmospheric CO_2 may decrease soil N availability as reported by Norby et al. (2010) from a FACE experiment in a deciduous Liquidambar (sweetgum) forest stand in Tennessee, USA. Isotopic evidence and N budget analysis indicated that N availability in this forest ecosystem has been declining over time, especially in the elevated CO_2 treatment, causing a decrease in tree response to elevated CO_2. Given that most of the world's forests already suffer some N limitation, a key issue is whether higher rates of N fixation such as those sometimes observed in short-term pot studies (e.g. Tobita et al. 2005) will occur over the longer term in the field under elevated CO_2.

Jastrow et al. (2005) used a meta-analysis to show that if results from 35 independent elevated CO_2 studies were combined, a net increase in soil carbon in depths ranging from 5 to 30 cm over periods of 2–9 years ranged from 2.8 to 8.4% (mean of 5.6%). Thus, an overall positive effect of elevated CO_2 on forest soil carbon storage of 44 ± 9 g C m^2 $year^{-1}$ was suggested. This increase was assigned to greater production of roots and an increase of protected C by incorporation into micro-aggregates. Six et al. (2002) observed that coarse particulate organic matter initiated aggregate formation in soils.

In contrast, some studies suggest that any additional carbon input may accelerate decomposition of stable soil carbon (Fontaine et al. 2007). Other studies (e.g. Hagedorn et al. 2003; Lichter et al. 2005) suggest that any stimulated increase in soil organic C would be allocated to rapidly cycling, labile pools with little long-term stabilisation. Furthermore, the effects of increasing temperatures which are expected to accelerate decomposition rates need to be factored in (see the following section). Any small increase of C in forest soils is difficult to measure in the short term due to (a) a high background stock of C already present in soils, and (b) high spatial and annual variability in the amount of soil organic matter under field conditions (Hagedorn et al. 2003). Bauhus et al. (2002) noted that it was impractical to assess changes in soil C stocks of 15% of the mean under Australian native forest because of very high spatial variation in soil C stocks. Any increase in soil C content in forest soils though difficult to measure may still be significant in terms of C sequestration, and in affecting emissions of other GHGs such as nitrous oxide (N_2O) and methane (CH_4). The magnitude of this depends upon many environmental and site factors discussed later in this chapter.

In conclusion, we are in a poor position to draw conclusions about what will happen to forest soil C stocks and consequently soil health under elevated CO_2 – see also the following section on interactions with temperature.

12.2.2 Effects of Temperature Changes

Commonly, an overall increase in global temperature of between 1.8 and 4°C is being predicted by the end of this century when medium severity scenarios are used for climate modelling projections (IPCC 2007). Heat waves and drought will increase in frequency, intensity and duration in many locations around the globe (Tebaldi et al. 2006), together with projected higher probabilities of extreme storm events (Leckebusch et al. 2006).

There are many interacting facets of the direct effect of temperature on forest soil health. For example, an increase in temperature is expected to increase the decomposition of organic matter under adequate rainfall conditions, and thus a positive feedback to climate change would occur. However, this effect has to be considered with respect to changes in plant-derived carbon inputs – a negative feedback to climate change. The net effect on forest soil C stocks will therefore depend on the balance between organic matter inputs and decomposition (Kirschbaum 1995). Whether a temperature increase will affect global soil carbon stocks is still a matter of strong discussion, and there is currently no consensus on this issue (Davidson and Janssens 2006; Kirschbaum 2006), due to the issues outlined below:

1. Increased temperatures may stimulate forest soil CO_2 respiration, as reported by many experimental studies involving different techniques (Rustad et al. 2001; see Chap. 7). However, a number of other belowground processes relating to CO_2 fluxes are also affected by temperature changes, which are strongly mediated by feedbacks and interactions with variables such as nutrient and water availability (for a discussion on methods and models, see Pendall et al. 2004). One of the most difficult issues to resolve in experimental studies is to be able to differentiate soil carbon loss from other CO_2 sources such as root respiration or decay of additional plant litter (Chap. 7). Most of the new methods demonstrate that all of these processes are affected by temperature. In their analysis of partitioning of belowground C flux in response to temperature changes, Litton and Giardina (2008) noted that mean annual temperature was the most important variable explaining the global patterns in belowground C processes.
2. Another issue is that other environmental constraints (such as soil protective mechanisms, soil moisture) interact with temperature to affect carbon loss (Emmett et al. 2004; see Chap. 7). These obscure the intrinsic temperature sensitivity of substrate decomposition, resulting in lower observed "apparent" temperature sensitivity, and these constraints may also be sensitive to climate change (Davidson and Janssens 2006). The magnitude of temperature response in soil warming experiments declines ("acclimates") over time, most likely due to limitation of readily available substrate supply for microbial activity (Kirschbaum 1995, 2006). However, microbial composition may also evolve to less temperature-sensitive communities, which if more tolerant to extreme conditions, as observed by Zhang et al. (2005) for prairie grass, will reduce the rate of soil carbon loss under future warming.
3. Elevated temperatures may increase the dissolved organic carbon (DOC) fraction as reported in surface waters in some studies in North America and Northern and Central Europe (Freeman et al. 2001; Worrall et al. 2003). Many other factors may be involved in such an increase such as elevated CO_2 concentrations (Freeman et al. 2004), recovery from acidification and changes in organic matter solubility (Monteith et al. 2007).

4. An increase in temperature may affect C inputs belowground. If rates of soil C mineralization increase under warming, so too will rates of N mineralization, and this may increase net photosynthesis by N-limited forests. Some of this C may be transferred belowground, thus partly compensating for respiratory loss of soil C. However, the effects of environmental change on belowground C flux and partitioning, especially the longer living roots of trees, are poorly understood (Giardina et al. 2005).
5. Elevated temperatures in combination with extreme drought in temperate areas may cause a large net loss of ecosystem C as suggested by the eddy-covariance studies across Europe during the drought year of 2003 (Ciais et al. 2005; Reichstein et al. 2007) indicating the possibility of a significant shift in soils from being carbon sinks to becoming carbon sources.

Temperature can also have other direct effects on forest soil health, e.g. by melting of permafrost that result in increased water logging, or by altering the activity of soil pathogens. A dramatic example, which is discussed in detail below, is the root pathogen *Phytophthora cinnamomi* (PC) which can cause major damage to forest ecosystems, including tree death and loss of biodiversity (Podger et al. 1990).

12.2.3 Effects of Change in the Amount and Nature of Rainfall

Any effects of climate change on the amount, intensity and seasonal pattern of rainfall will be site and region specific. Any associated extreme events such as storms, floods and droughts may have long-lasting effects on processes relating to health of forest soils. Although there are clearly some "direct" effects of changing rainfall regimes on soil hydrological conditions that can affect water table depth, duration of water logging, expression of dryland salinity and so on, almost always there is a strong interaction with vegetation (an indirect effect) that determines the degree of change and thus the implications for soil health. We stress that it is usually the interaction with other factors (particularly forest type, temperature, nutrient availability and indirect "disturbances") that determines rates of water use by vegetation, and thus the degree to which changing rainfall affects soil health. Any change in the amount and quality of soil organic matter resulting from altered NPP or soil respiration is of major importance.

An example of how the interaction between drought and insect attack can affect forest dynamics and net forest C balance, which could affect soil health over the longer term, is briefly described as a case study for an Australian native eucalypt forest in Sect. 12.3.2.

The following are some potential impacts of changing rainfall regime on forest soil health:

1. Changes in forest soil respiration and soil C stocks due to drought. This is still a controversial area because of the complexity caused by several interactive variables (see Chaps. 5 and 7). High temperatures often occur together with

low precipitation, and it is difficult to determine their interactive effects. In a temperate forest, Borken et al. (2006) reported a significant decrease in soil respiration during drought with a potential for an increase in soil carbon storage during that period. However, in Amazonian forest where throughfall was experimentally reduced, no change in soil respiration but a large decrease in plant production was observed indicating a likely net loss of carbon from the ecosystem due to reduced carbon fixation (Brando et al. 2008). Ågren et al. (1996) suggested that soil respiration is less sensitive to soil moisture limitation than is plant production, and that this would cause more carbon losses than the amount fixed during a period of drought. This decrease in carbon fixation would lead to reduced soil carbon in the long term. Bellamy et al. (2005) reported a 15% loss of soil carbon stock over the prior 20 years in the UK. Drought was suggested by Schulze and Freibauer (2005) to be the most likely contributing factor.

2. Changes in rates of soil erosion. Drought may increase wind erosion and thus losses of surface soil rich in C and nutrients, especially in Mediterranean countries where relatively high risks of desertification occur (see Chap. 11). Similarly in the mountainous areas of central Europe, expected changes in the frequency and intensity of rain events may increase soil erosion (Sauerborn et al. 1999). Flood events may erode soils but some of the eroded material is redistributed across the landscape (e.g. Quinton et al. 2006).
3. An increase in soil hydrophobicity (water repellency) may occur due to changes in microbial communities, and in some cases together with changes in forest fire frequency, due to precipitation changes. The effect of hydrophobicity on the distribution and retention of precipitation may either increase or reduce soil carbon loss depending on site conditions (Doerr et al. 2009).
4. Repeated drought in successive years and/or increased frequency of severe drought caused by reductions in rainfall below an historic threshold may be critical in causing a change in plant community composition and structure (e.g. Leuzinger et al. 2005), which could fundamentally change the production potential and soil functioning in an ecosystem.
5. Any change in the decomposition of stabilised soil organic matter is of special concern. This could result from an increase in microbial activity due to changes in rainfall and temperature, cycles of wetting and drying, or soil freezing. Physical cracking of the soil during freezing can lead to subsequent increase in release of several GHGs (Teepe et al. 2004).

12.2.4 Changing Humidity

There may be several important impacts, including direct effects on photosynthesis and water use, and indirect effects mediated through changing impacts of pests and pathogens, or increased impacts of fire where the fire climate becomes more severe. The consequences of some of these for forest soil health are discussed later in this chapter.

12.3 Indirect Effects of Climate Change, Including Management Responses

The indirect effects are mostly related to change in forest growing conditions or to disturbances that affect the functioning of forest ecosystem processes, and their consequences for the health of forest soils. The indirect effects of climate change occur via the following.

12.3.1 Changes in Forest Growth and Input of Organic Matter and Nutrients to Soils

An increase in atmospheric CO_2, and/or an improvement to growth-limiting factors which may include changing temperature (e.g. increasing temperatures in the boreal forest zone) and rainfall regimes, improved nutrient availability, or change to other site factors may increase NPP on some forest sites. In some situations, climate change will be associated with increased atmospheric inputs of nutrients, especially N and P, and this may increase growth because these nutrients limit NPP over large parts of the globe. As discussed earlier, nutrient limitation may increase under elevated CO_2 concentrations (e.g. Norby et al. 2010), so that additional sources of nutrients will become especially important in determining the longevity of response to elevated CO_2.

There will of course also be situations where climate change may have adverse effects on forest growth, notably where there will be reductions in rainfall in already water-limited regions, or where increasing frequency of major disturbance such as fire lead to forest damage (see below).

12.3.2 Change in Pest and Pathogen Impacts on Forest and Forest Soil Health

One of the potentially greatest effects of climate change on forest ecosystems is via change in tree and stand health. Clearly, any major changes in the health of the aboveground components of the forest will also have consequences for soil health. Soil changes can result from changes in vegetation cover and species composition that affect microclimate, erosion risk, the amount and quality of litter inputs, rates of N fixation or the vulnerability of the stand to disturbances such as wildfire. In some cases, it is a change in the severity of impact of soil-borne pathogens such as fungal root rots that can be the primary cause of major decline in forest health that then drives further change in soil health – an example is provided at the end of this section. The possible effects of climate change on forest health have been extensively reviewed (e.g. Chakraborty et al. 1998; Dale et al. 2001; Moore and Allard 2008; Old and Stone 2005; Volney and Fleming 2000), and there is only the scope for a brief treatment here.

Old and Stone (2005) provide a logical framework for analysis of the potential effects of climate change on forest health. They emphasise that the incidence and severity of pest and pathogen attacks on forests depend on three main groups of factors, namely: (1) forest type, species and genetic make-up, community composition (host factors), (2) climatic and edaphic characteristics and stand history, including fire (environmental factors), and (3) insects, pathogens and vertebrate pests (pest factors).

These groups of factors interact strongly with each other to determine the overall health of the stand as represented by the Forest Health Triangle (Old and Floyd 2000). Old and Stone (2005) state as follows: "when all three groups of factors favour pests and pathogens and disfavour the host tree or forest stand, then major outbreaks can occur and the ability of trees to recover from outbreaks or epidemics may be impaired. A consideration of these interactions is essential for pest and disease management, and is necessary for assessing possible outcomes of climate change on forest health. This is especially the case for interactions of pests and diseases with drought and wildfires, the occurrences of which are predicted to increase for many forested regions under changing climate".

Old and Stone (2005) also provide useful summaries of the way in which climate change affects pathogen/tree interactions (Fig. 12.2) and pest/tree interactions (Fig. 12.3).

Figures 12.2 and 12.3 illustrate direct effects of climate change, "top down effects", on pest and pathogen lifecycles and populations and on tree physiology. Biotic interactions, shown here as "bottom up" effects, are much less predictable, but these may determine the incidence, duration and severity of outbreaks. These diagrams are closely similar, except for the inclusion, in Fig. 12.3, of interactions of pest populations and their parasitoids and predators.

The importance of climatic, edaphic and biotic factors and their interactions in determining the occurrence and severity of outbreaks and epidemics of pests and diseases, as illustrated in Figs. 12.2 and 12.3, suggests that climate change could result in either increased or reduced damage to forests by particular groups of pests and pathogens.

Climate change could have many consequences for the impacts of a wide range of pests and pathogens on overall forest health. Old and Stone (2005) provide a useful "generic" summary (Table 12.1).

Below is a brief summary of how climate change might impact directly on soil conditions that change the pathogenicity of the root rot fungus *P. cinnamomi* that damages forest health across the globe (e.g. Bergot et al. 2004; Brasier and Scott 1994; Shearer and Smith 2000; Simpson and Podger 2000). This is a clear example of a direct change in soil health that severely affects tree health.

The presence of *P. cinnamomi* in native vegetation, even where susceptible tree species dominate (e.g. in coastal eucalypt forests of East Gippsland in Victoria, Australia), often is accompanied by very low levels of disease, indicating that specific environmental conditions are needed to trigger disease. *Phytophthora* spp. require periods of soil wetness for sporulation and spread of zoospores, and infection of roots of susceptible hosts (Shearer and Smith 2000). This root disease

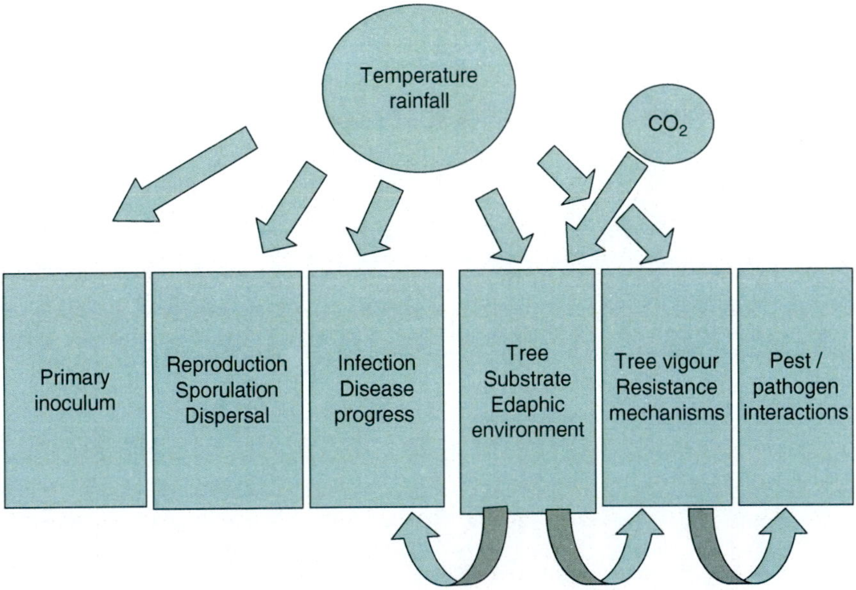

Fig. 12.2 Effects of climate and CO_2 concentrations on pathogen/tree interactions. Temperature and rainfall have major effects on primary inoculum levels, reproduction and dispersal of pathogen spores and infection processes. They also influence tree growth, phenology and the capacity of the tree to withstand infection. Enhanced CO_2 concentration will mainly affect the host tree, its vigour and susceptibility to combined attacks by pests and pathogens

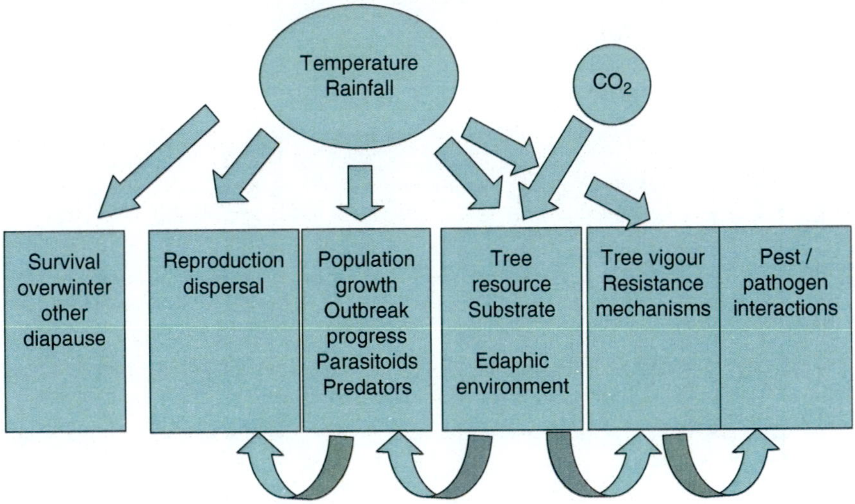

Fig. 12.3 Effects of climate change on pest/tree interactions. Temperature will have direct effects on survival, geographical range and rate of reproduction of insect pests, and the presence and impacts of their parasitoid and predators. These factors plus tree responses to temperature, rainfall and CO_2 concentrations will affect insect pest populations and the severity and duration of outbreaks. Trees suffering environmental stresses may be more susceptible to damage by invasion by both pests and pathogens

Table 12.1 Possible mechanisms by which climate change can alter the impacts of pests and pathogens on forest stands and indirectly affect forest soils

Environmental change	Effect on pest or pathogen	Effect on host	Possible outcome
Increased ambient temperature	Shorten life cycles and increase reproductive rates. Alter geographical distributions. May affect parasitoid and predator relationships	Changes in tree growth, phenology and seasonality in the short term. Changes in tree species and plant community composition and geographic distribution in the longer term	Greater survival of pests and pathogens overwinter. Earlier and more severe pest outbreaks and epidemics of disease if other conditions favourable Seasonal period of feeding extended and overwintering period reduced. This allows for additional generations per year. Outcome complicated by impacts on parasitoid and predator relationships
Increased atmospheric CO_2	Little direct effect demonstrated on pests or pathogens	Increased tree growth and potential increases in water–use efficiency. Changes in leaf and phloem chemistry with regard to nutrient content, C:N ratios, fibre content and concentrations of secondary substances	For insects, different groups, e.g. leaf chewing and phloem feeding insects, will respond differently to changes in leaf chemistry, especially increases in C:N ratios. This may result in increased defoliation by some pests and reductions in the activity of others. Generalisations across insect groups for impacts on forest health are not possible at this stage
Reduced annual rainfall	Unlikely to favour those pathogens which rely on splash dispersal of spores and significant periods of leaf wetness for infection	Reduced growth rates of trees unless compensated for by increased water use efficiency. Changed phenology, e.g. possible prolonged juvenility of foliage. Under drought conditions, trees become stressed which can induce susceptibility to stem-boring insect and canker pathogens. Reduced capacity for trees to recover from defoliation and crown dieback	Reduced incidence of defoliation by fungal pathogens. Prolonged juvenility of foliage can increase impacts of pest and disease. Trees stressed by drought suffer increased damage from suites of insect pests, e.g. psyllids and borers. Attack of stressed trees by fungi, present as latent infections in healthy trees, and by some root and butt infecting fungi, can result in crown dieback and increased tree mortality

Increased annual rainfall	Can favour defoliation by foliar pathogens, depending on season. Defoliating insects also will benefit from flushes of new foliage. Extreme rainfall events can favour root disease	Generally favours growth of forest and plantation trees, allowing them to sustain even high levels of defoliation as crowns are rapidly replaced. Extreme rainfall events can cause water logging and stress	Outcomes will depend largely on seasonal distribution of rainfall patterns in relation to tree phenology, life cycles of pests and epidemiology of individual pathogens. In general, most insect herbivores prefer young foliage as sites for oviposition and for early feeding of instars

is favoured by rainfall events causing high soil moisture levels followed by extended dry periods. There is also a critical temperature sensitivity, with infection favoured by soil temperatures above 15°C.

A serious epidemic of phytophthora-related dieback in the coastal forests of the Central and East Gippsland regions of Victoria in the 1970s (Marks et al. 1972) was partially attributed by Fagg (1987) to an unusually wet summer, which favoured root infection, followed by severe drought that resulted in death of water-stressed trees. *Phytophthora cinnamomi* causes jarrah dieback in the southwest of WA, where in 1982 it was estimated that 250,000 ha of native forest was damaged by the disease (Davison and Shearer 1989). The pathogen causes mortality in many native plant communities, including woodlands, heathland and coastal vegetation (Shearer and Smith 2000; Simpson and Podger 2000). Although impacts on susceptible forest understorey and heathland species are severe and acute, tree dieback usually proceeds slowly. Climate change could clearly alter conditions affecting the severity of impact of *P. cinnamomi*, but there are complex interactions involved and it is thus hard to predict impacts at particular locations. In Europe, Brasier and Scott (1994) predicted increased disease activity within the existing range plus some northward extension.

12.3.2.1 A Brief Case Study Describing the Interactive Effects of Drought and Insect Damage on Stand Dynamics and Changes in the C Cycle in a Native Eucalypt Forest

The combined effects of drought and insect attack can have a marked effect on the net annual C balance of Australian sub-alpine *Eucalyptus delegatensis* forests (Keith et al. 2009; Kirschbaum et al. 2007), turning the forest from a usual significant sink for C, to a weak source in extreme years. This insight comes from intensive studies at Tumbarumba in southern NSW over the period from 1997 to 2001 (years of about average annual rainfall without significant insect attack) and 2001–2005 (a period which included severe drought and significant attack from psyllid insects during 2002–2003). Details are provided in the above papers, and a summary is provided below.

From time to time, these forests are subject to severe attack by phasmatids (stick insects; *Didymuria violescens* Leach) that can result in estimated growth reductions of up to 50%, and tree death of up to 50% in some years (State Forest of NSW, unpublished). Trees can also be subject to attack by psyllids (*Cardiaspina* spp.) that suck leaf sap and cause necrosis, and which also form protective covers over their bodies that prevent sunlight from reaching the leaf. Under severe circumstances, psyllid damage can induce leaf fall (Farrow 1996). There was a severe attack of psyllids at the study site during 2002–2003, which led to significant additional loss of foliage during the winter of 2003.

During the period of 2001–2005, drought and insect attack resulted in a major increase in tree mortality, especially in the younger, denser and faster growing stands (Fig. 12.4). There was also a major reduction (range 45–80%, average 54%)

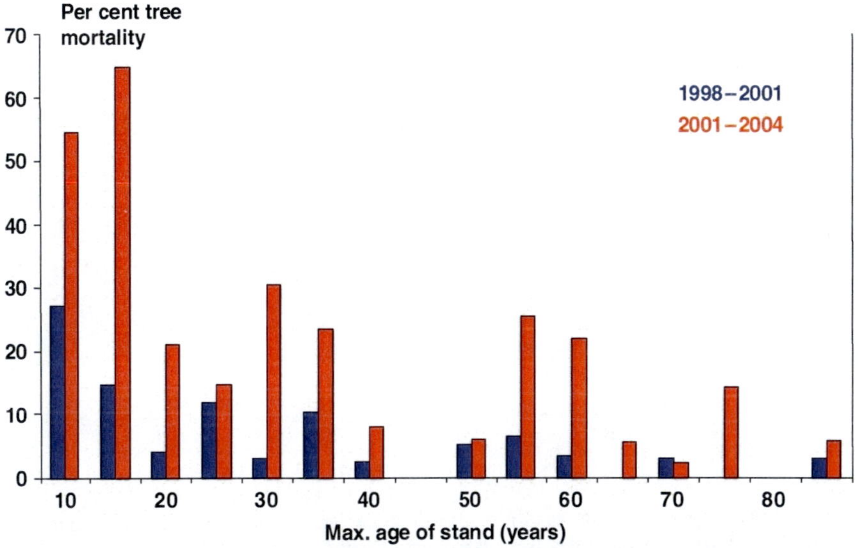

Fig. 12.4 Comparison of rates of tree mortality (% of trees in the stand that died) in native forests dominated by *Eucalyptus delegatensis* during a period of average rainfall without insect attack (1998–2001) and the following period (2001–2004) in which there was co-incident drought and severe psyllid insect attack. Stands are plotted in order of increasing stand age, with the youngest (aged 10 years) on the *left* of the figure and the oldest (>80 years) on the *right*. Climate change is likely to increase the frequency of droughts and associated insect attack, and thus disturbance to soil processes

in stand basal area increment during this period with reductions again being greatest in the faster growing stands (Fig. 12.5). Simulation modelling (Kirschbaum et al. 2007) suggested that the reduced carbon gain was primarily due to insect damage, whereas soil water stress had only very minor effects on estimated net ecosystem exchange (NEE). The greatest effect on carbon fluxes during the period of drought and insect disturbance in 2002–2003 was on reducing photosynthesis rather than or changing (in front of respiration). Water stress was generally relatively unimportant in the *E. delegatensis* forest because average annual rainfall considerably exceeds average annual tree water use, and the deep soils can hold large amounts of prior rainfall. As such, the site is quite unusual among Australian ecosystems (e.g. Roxburgh et al. 2004), and especially where soils have low water-holding capacity there can be a rapid shedding of foliage in response to drought (e.g. Pook 1985).

These changes in forest health (defoliation, reduced growth rate, tree mortality) affect both the pattern (insects attack altered the pattern of litter fall from the usual peak in summer, to significant amounts in mid-winter) and magnitude of litter inputs to the soil, and thus may alter a myriad of processes relating to rates of C and nutrient cycling. Soil temperatures will be increased with the opening of the canopy, and soil moisture is likely to also be higher because of lowering of stand leaf area – these changes are conducive to increased rates of decomposition of soil organic matter. Furthermore, there are greater risks of fire in the forests following

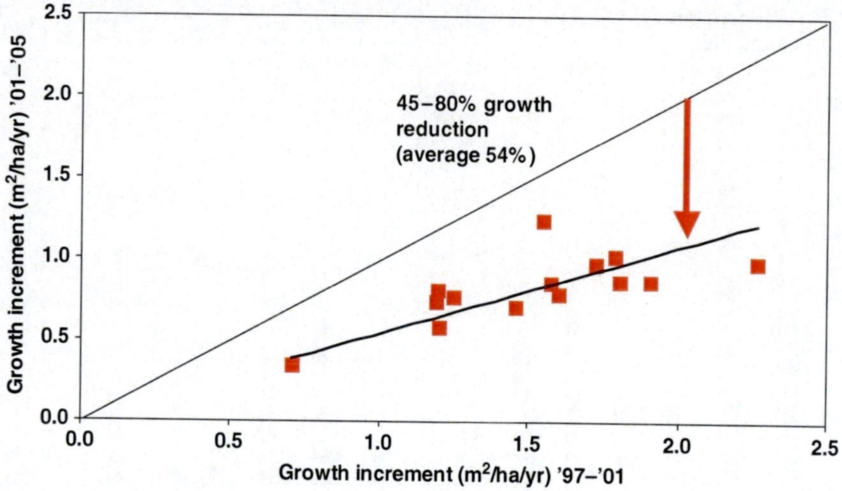

Fig. 12.5 Comparison of rate of forest growth in native forests dominated by *Eucalyptus delegatensis* during a period of near-average rainfall without major insect attack (1997–2001) and the following 4 years (2001–2005) during which there was a severe drought and attack by psyllid insects. Growth reductions ranged from 45 to 80% and were greatest in the faster growing (often younger) stands. Such marked reductions in forest growth will affect the rates of carbon and nutrient cycling through the soil

tree mortality that creates an increase in fuel loads. While it is not clearly known why the insect outbreak was so severe over the study period, it may have been caused by drought stress which as discussed earlier in this chapter often predisposes forest stands to secondary pest attack. A range of stands in the local area experienced significant tree mortality over that period (Fig. 12.4). If droughts become more frequent or severe under future climate change, there is clearly the opportunity for interactions with insects and fire to adversely affect forest health, longer term forest dynamics and ecosystem carbon storage.

12.3.3 Changes in Fire Regimes

Climate change may change the frequency, timing and intensity of fires mainly due to changes in precipitation leading to drought, lower humidity or increased frequency of lightning strikes or strong winds. Fire, depending on its frequency and severity, can have major effects on forest ecosystems, including forest soil health.

Fires in forests can be described in terms of three key variables: intensity, frequency and season of burn. Various combinations of these variables define the fire regime to which the forest is exposed. The fire regime, in combination with the characteristics of the site and species making up the forest, will determine the level of GHG emissions, and the effects of fire on nutrient cycles and other properties and processes of the forest ecosystem. Effects of fires in forests should be considered a continuum in terms of severity of impacts that is often, but not always, related to the amount of fuel consumed and fire intensity. Fuel consumption can range from a few tons of organic matter per hectare in fire in low-intensity controlled burning to hundreds of tons of organic matter in slash and wild fires (Raison et al. 1993). Any long-term increase in the frequency or intensity of fire almost always results in a reduction in forest C stocks, including soil C (Raison and Kirschbaum 2008).

Forest fires affect nutrient cycling as a consequence of nutrient losses in smoke, nutrient deposition in ash, changes in soil mineralization rates, erosion of ash and soil particles, and an increase in nutrient leaching (for a detailed review, see Raison et al. 2009). Fires, especially those occurring repeatedly at shorter duration may affect the ground and understorey vegetation, N fixation rates and plant diversity. These changes may affect fertility and health of forest soils, depending on the particular forest type and degree of perturbation of the fire regime from that under which the vegetation has developed.

One can distinguish many different types of fire regimes created by climate change and management responses to it. The following four broad categories can be described that have very different impacts on forest vegetation and on soil fertility.

12.3.3.1 Controlled Low-Intensity Burns

Fuel load in forests are frequently managed through burning which is commonly carried out under prescribed site and weather conditions. Such fires may become a greater component of fuel management in forested landscapes where climate change may lead to higher probabilities of wild fires. Controlled burns in forests remove most of the shrub and ground vegetation, a part of litter layer and part of lower tree canopy depending upon the burning conditions (Table 12.2). Commonly 8–15 t of biomass is consumed in the burn and most of that is converted to CO_2, but some of it is charred and added to soil (Raison et al. 2009). The dynamics and species composition and structure of the understorey can be affected by frequent burns. The overstorey trees are usually not defoliated, and any effects on the dynamics of the tree layer generally only occur if fires are frequent and occur over a long period of time.

Burning will cause loss of nutrients in smoke, through volatilization and particle transfer. Managed burns have been estimated to remove 60% or more of the N in the available fuel (Raison et al. 2009; Terry et al. 2004), as well as significant amounts of phosphorus (Kinako and Gimingham 1980; Raison et al. 1993). These nutrient losses, for example 50–150 kg N/ha (Table 12.2), if repeated at short intervals, may decrease forest productivity in the long term (Raison et al. 2009), but conversely in some regions the maintenance of low-nutrient conditions through burning may be effective in offsetting ecosystem eutrophication caused by elevated N deposition (Pilkington et al. 2007). The implications of controlled burning for carbon stocks can be complex. For example, Holden et al. (2007) suggest that regular fire may

Table 12.2 Characteristics of forest fires relevant to potential direct losses of nutrients and impacts on soils (modified from Raison et al. 2009). Fire regimes will differ in many regions under climate change with major consequences for soil fertility

	Controlled low-intensity burns	Wildfire	Forest regeneration	Land clearing, slash and burn
Fuel components	Mostly shrubs, ground vegetation and fine litter	Tree foliage, shrubs, woody and fine litter	Tree crowns, woody residues, shrubs and litter	Mostly trunks, crown and understorey slash
Fuel consumption ($t\ ha^{-1}$)	8–15	20–60	50–300+	100–400+
Fire intensity[a] ($kW\ m^{-1}$)	<500 (low)	501–7,000 (moderate to high)	7,001–70,000 (very high)	70,000–1,00,000+ (extreme)
Maximum temperature (°C) at 2 cm depth	<60	100	200	500
Time (h) for which soil is heated above 80% of the maximum temperature	0.1–0.3	0.1–1	0.5–2	>24
Fire frequency (years)	5–15	20–100+	15–30+	30–100+
Ash residues produced	Low on patches of area	Low to medium on most of area	Medium to high on most of area	Very high on part of the area
N losses ($kg\ ha^{-1}$)	50–150	>200	350–800	500–2,000+
Effects on vegetation and soils	Part removal of understorey, trees little affected. Cumulative effects on soils can be important if fire frequency too high	Defoliation and possible mortality of understorey and trees. Degradation of soils may also result especially if severe erosion follows fire	Much vegetation removed, recovery on decadal timescale. Potential adverse impacts on soils can be mitigated	Most vegetation removed, recovery depends on management goals. Major changes in soils under new production systems

[a]Fire intensity is described by an index as defined by Byram (1959)

induce change from heathland to grassland species, increase decomposition rates and hence decrease C accumulation. Holden et al. (2007) concluded that some degree of management (burning or cutting) is required to maintain current heathland vegetation on drier (organo-mineral) soils, but that the overall impact of burning on soil C stocks is uncertain.

Post-burn levels of soil nutrients and pH may be slightly raised by ash deposition and soil heating. Soil temperatures may also increase until vegetation cover is re-established (e.g. Raison et al. 1986). These factors may collectively lead to an increase in decomposition rates (Kim and Tanaka 2003; Stevenson et al. 1996). Direct impacts of a single fire on the soil are small. Cumulative effects of repeated fire on the nutrient cycle are likely to be more significant than those of a single fire. For purposes of reducing fuel loads, managed burning generally takes place when soils are wet.

12.3.3.2 Moderate to High-Intensity Fire

These types of fires commonly occur under uncontrolled conditions (wildfires) and are expected to increase where there is increased drought and severe fire weather resulting from climate change. Wildfires occur at unpredictable frequency and have a more severe impact on ecosystem C stocks because they can consume much of the aboveground biomass and forest floor, as well as part of the upper organic-rich surface soil. These fires may or may not kill trees. In fire-adapted forests, trees with thicker and protective bark recover by sprouting from either buds on branches or the trunk. The dynamics of the understorey vegetation are often markedly changed. The soil may be bared and subjected to significant heating and input of ash. Exposure of bare soil following severe burns also increases susceptibility to physical erosion of both soil and nutrient-rich ash. Soil microclimate remains changed for 1–2 years after the fire, while the understorey vegetation and tree canopy recover. Atmospheric losses of nutrients are high (e.g. >200 kg of N/ha), and the total losses of nutrients due to erosion may permanently affect the productivity of the site (Raison et al. 2009).

12.3.3.3 High-Intensity Management Burns Used to Enhance Forest Regeneration

Such fires are used to kill most of the vegetation on a site to initiate a myriad of changes that lead to stand replacement. These fires consume most of the tree components, including crowns, stems, woody residues, shrubs and herbaceous ground cover and the litter layer. Climate change may, or may not, affect the frequency of such fires. Surface soils are subjected to high temperatures causing higher losses of easily volatile C and N. Losses of other nutrients depend on the fate of ash; any wind or water erosion would result in major losses. The soil environment is thus significantly changed in the short term and for many years as the forest recovers. N fixing understorey vegetation may in some cases be markedly stimulated as part of the processes of forest recovery.

12.3.3.4 High-Intensity Slash Fires Used to Clear Forest for Agriculture

These fires are often used to remove felled forest biomass as part of land use change, especially in the tropics. Cleared vegetation is sometimes pushed into piles before burning – in such situations, the soil will be subjected to considerable heating and ash input, especially under large piles of woody slash (leading to the creation of "ashbeds"). Fire impacts on the soil will be highly spatially variable because of variations in heating and ash input. Vegetation dynamics and soil microclimate are markedly changed over the time scale of decades.

It is important to briefly mention peat fires which are very different from fires consuming aboveground fuels, and which can have major adverse effects on the fertility of peat soils. Peat fires are different because of oxygen limitations and the smouldering nature of combustion. The nature of peat fires can vary greatly across temperate, boreal and tropical biomes. Peat fires can burn for months or years and consume 30 cm or more of the peat layer depending on moisture conditions. In the northern hemisphere, peat fires are normally only significant in drought years. In the tropics, peat fires can be very extensive following drainage and exposure of the peat after deforestation, especially in El Nino years (e.g. van der Werf et al. 2008). Often extensive killing of tree root systems and forest trees can occur, with major subsequent effects on the soil environment. The longer term effects of fires on peat soils have been relatively little studied, especially in the tropics.

Raison et al. (2009), while cautioning against over-generalisation, provided a synthesis of the effects of fire on forest nutrient cycles and soil fertility. They suggested the following situations where fire may pose a significant risk to the maintenance of forest nutrient cycles:

- Where forests are already nutrient-limited and where natural inputs of nutrients, especially of N and P, are small;
- Where a high proportion of site nutrient pools are held in combustible vegetation or dead organic matter;
- Where fuel mass (and fuel nutrient content) re-accumulate rapidly after burning, and there is frequent fire;
- Where fire intensity and the mass of fuel combusted are high;
- Where fire induces increases in erosion of ash and/or surface soil (typically when there are erodible soils, steep slopes and high rain intensities);
- Where fire reduces rates of N fixation (e.g. by changing the composition of understorey vegetation);
- Where the frequency of fire increases compared to historical rates, due to climate change, population pressure or to changed fire management policies.

They stressed that while *there are some well established principles, local studies are required to refine understanding of the relationship between fire, nutrient cycling, and soil fertility, and to guide fire management.*

12.4 Linkages Between Forest Management Practices and Climate Change That May Affect Forest Soil Health

A number of forest management practices such as regeneration by planting with different species, intensive harvesting, slash management, fire management and fertilisation may interact with climate change and affect forest soil health. There are numerous permutations and is impossible to provide a comprehensive review here. The following few examples are provided for some key potential interactions:

- Increased use of management burning to reduce fire risks under projected increased severity of fire weather caused by climate change. As discussed above, repeated fire has the potential to adversely affect soil health via a number of mechanisms. Fire regimes need to be developed that are compatible with local sustainable management objectives (Raison et al. 2009).
- New species that are better adapted to climate change may be introduced to plantation forests. These new species, and associated management systems, may affect the health of forest soils. Although the role of the vegetation on soil formation has been researched for a long time, there is a lack of generalised understanding on the effect of tree species across various soil and site types (Augusto et al. 2002; Binkley and Menyailo 2005). With respect to the effects of tree species on soil organic matter and the C cycle, effects on forest floor C stocks may occur rapidly, but long-term C sequestration in soils may be restricted due to slow formation of stabilised C in the mineral soil. Production and turnover of belowground biomass (Jobbágy and Jackson 2000; Vesterdal et al. 2002) are the major processes affecting C stocks in mineral soils, and this area is poorly understood.
- Major disturbances such as wildfire or severe storms may lead to intensive salvage harvest of timber resources. Both the disturbance itself and any subsequent intensive harvesting operation have the potential to damage (via soil physical change, erosion, depletion of organic matter and nutrients) soils. Great care is needed in the specification and implementation of site-specific management practices that mitigate risks to soil resources. These are often specified in management plans or codes of sustainable forest management practice.
- Forest management practices may be modified to reduce the impact of anticipated increased disturbances on forests and soils. For example, appropriate selection of tree species, thinning regime and felling operations for a site may be required to minimise the risks from wind throw. Wind throw has a major effect on forest soils by destroying soil structure, increasing erosion and fire risk, and probably the loss of protected soil C through increased decomposition. By increasing stability of stands through well-designed thinning regimes (including

no-thinning in high risk stands) and carefully planned harvesting (decreased length of exposed edges), and planting deep rooted trees are some of the possible measures to protect such sites (Gardiner and Quine 2000).

- Intensive production (e.g. whole-tree harvesting) and short-rotation harvesting systems may be extensively deployed to produce feedstocks for bioenergy as part of GHG mitigation strategies as a political response to climate change. Such systems can be very demanding on soil resources and can lower soil fertility (Richardson et al. 2002). To be sustainable, careful management including supplementary nutrient inputs may be required, especially when foliage and roots are removed together with stump extraction from nutrient-poor sites. There will be a need for compensatory fertiliser application on nutrient-poor sites (Sverdrup and Rosen 1998; Richardson et al. 2002; Raulund-Rasmussen et al. 2008). Even on more fertile soil types, it will be beneficial to retain foliage on the site (Samuelsson 2002). Experimental evidence of the effects of slash and of stump extraction on soil carbon dynamics and site productivity has been reported. Many short-term studies show no negative effect of harvest residue removal on growth (Roberts et al. 2005), whereas in the long term negative growth impacts have been reported. For example, in Northern Sweden growth declines in Scots pine stands were only observed 12–24 years after whole-tree harvesting on nutrient-poor sites (Egnell and Valinger 2003). Zabowski et al. (2008) found that total N and C in the surface mineral soil and forest floor decreased in five stands in the Pacific Northwest of America 22–29 years after stump removal. Lower foliar N, though non-significant, was also observed.
- Climate change may alter the net emissions of non-CO_2 GHGs (N_2O and CH_4) from forest soils. Nitrous oxide is an important GHG, which destroys ozone in the stratosphere by catalytic reactions (Crutzen 1970). The global warming potential (GWP) of N_2O is about 300 times higher than that of CO_2. Methane has a GWP of 25, and forest soils can either be a source or a sink for methane depending on soil environmental conditions (Allen et al. 2009). Climate change can affect the rates of non-CO_2 GHG exchange by forest soils. Forest soils tend to be good sinks for CH_4 because the trees help keep the water table well below the surface and allow methanotrophs to grow. If soils were to become more waterlogged, the balance shifts from methanotrophs to anaerobic CH_4-producing bacteria (methanogens) and the soil becomes a CH_4 source (see Chap. 8). In addition to water content of the soil, soil temperature and the concentration of nitrogen can be crucial factors in determining whether a particular soil will act as a sink for CH_4. The increased deposition of nitrogen from the atmosphere due to human activity can reduce or completely inhibit CH_4 oxidation in soil. Soil warming, especially in Northern latitudes, is likely to increase soil N mineralization rates and thus the likelihood of greater N_2O emissions from soils (Brumme and Borken 2009). Management practices that can minimise such effects of climate change include maintenance of forest cover, minimising soil disturbance and the judicious use of N fertilisers. Methods for estimating the effects of forest management and other disturbances such as fire on the emissions of GHG from forests are described by the IPCC (2006).

12.5 Concluding Remarks

Very complex interactions determine the nature and magnitude of climate change impacts on forest soils. Soil organic matter is a key variable affecting the health and fertility of forest soils that is very sensitive to climate change and forest management. Change in soils can result from both the direct and indirect effects of climate change, and these can also be either moderated or magnified by management responses to the risks or opportunities created by climate change.

Further complexity occurs because there can also be interactions between the direct and indirect effects of climate change, and impacts can be cumulative over time. Major disturbance such as storms, wildfires or attack from pests or pathogens can have significant effects on forest and soil health within a very short period of time. Direct effects of climate change, or those caused by successional processes, may occur more slowly. Many of these changes will be location specific and are determined by the interactions between climate, terrain, vegetation, soil type and the degree of management intervention.

The continuing increase in atmospheric concentrations of CO_2 may be interacting with the other components of climate change and having a major impact on the physiology of the world's forests. The magnitude and longevity of forest growth stimulation and direction of any changes in forest soil health by elevated CO_2 are poorly understood, but are expected to be very location specific, and determined by complex constraints imposed by the availability of water and nutrients, and by temperature.

Fire is a significant disturbance factor in many of the world's forests, and both the frequency and severity of fire are expected to increase with climate change. Potential impacts of increased fire (including increases in management burning) on forest soils and other components of forest ecosystems are high. There is, however, good opportunity to mitigate adverse impacts of fire using scientifically based, regionally appropriate fire management practices (e.g. Raison et al. 2009).

One of the potentially most significant effects of climate change on forests may be via increased attack by pests and pathogens. The consequences for forest health generally and for forest soils can be catastrophic. The incidence and severity of such attack depend on the interaction of host (tree), environmental and pest factors – all of which can be affected by climate change. Stress factors such as drought, which is predicted to worsen over wide areas in the future, can predispose forests to attack by pests and pathogens. Greater investment in monitoring of the health of forests and in strategic interventions to protect forests in high-risk regions may be needed in the future.

Because many of the effects of climate change on soil health involve complex interactive processes (Fig. 12.1), development and use of appropriate models are needed to help understand them. An appropriate analogy is the modelling efforts underway to quantify the role that forest ecosystems play in global C cycling under a changing climate. Terrestrial ecosystem models are used for this purpose (e.g. Cox et al. 2000; Schimel et al. 2000; Thornton et al. 2002; Ise and Moorcroft 2006).

While there has been much progress already made, these models are still limited by high uncertainties about underlying climate–process relationships (e.g. Giardina and Ryan 2000; Grace and Rayment 2000), incomplete understanding of key biological processes such as feedbacks relating to responses to elevated CO_2 and of controls on C allocation in forests (e.g. Friedlingstein et al. 1999; Landsberg 2003; Litton et al. 2007).

Likewise, there are critical gaps in our knowledge of how climate change will affect forests and forest soil health. Modelling efforts, while still very incomplete, are of great value in identifying key knowledge gaps and in guiding further research, and in helping define possible indicators for monitoring change in forest soil health under future climate change.

Acknowledgement We thank Dr. Ken Old, formerly of CSIRO, for providing us with access to unpublished work (Old and Stone 2005), and for permission to reproduce parts of it in this chapter.

References

Ågren GL, Kirschbaum MUF, Johnson DW, Boasatt E (1996) Ecosystem physiology – soil organic matter. In: Breymeyer AI, Hall DO, Melillo JM, Ågren GL (eds) Global change: effects on coniferous forests and grasslands. Wiley, New York, pp 207–228

Allen DE, Mendham D, Singh BP, Cowie AL, Wang W, Dalal RC, Raison RJ (2009) Nitrous oxide and methane emissions from soil are reduced following afforestation of pasture lands in three contrasting climatic zones. Aust J Soil Res 47:443–458

Arias ME, González-Pérez JA, González-Vila FJ, Ball AS (2005) Soil health – a new challenge for microbiologists and chemists. Intern Microbiol 8:13–21

Augusto L, Ranger J, Binkley D, Rothe A (2002) Impact of tree species on soil solutions in acidic conditions. Ann For Sci 59:233–253

Bauhus J, Khanna PK, Hopmans P, Weston C (2002) Is soil carbon a useful indicator of sustainable forest management? A case study from native eucalypt forests of south-eastern Australia. For Ecol Manage 171:59–74

Bellamy PH, Loveland PJ, Bradley RI, Lark RM, Kirk GJD (2005) Carbon losses from all soils across England and Wales 1978–2003. Nature 437:245–248

Bergot M, Cloppet E, Pérarnaud V, Déqué M, Marçais B, Ceprez-Loustau M-L (2004) Simulation of potential range expansion by *Phytophthora cinnamomi* under climate change. Glob Change Biol 10:1539–1552

Binkley D, Menyailo O (2005) Gaining insights on the effects of trees on soils. In: Binkley D, Menyailo O (eds) Tree species effects on soils: implications for global change. Springer, New York, pp 1–16

Borken W, Savage K, Davidson EA, Trumbore S (2006) Effects of experimental drought on soil respiration and radiocarbon efflux from a temperate forest soil. Glob Change Biol 12:177–193

Brando PM, Nepstad DC, Davidson EA, Trumbore SE, Ray D, Camargo P (2008) Drought effects on litterfall, wood production and belowground carbon cycling in an Amazon forest: results of a throughfall reduction experiment. Philos Trans R Soc Lond B Biol Sci 363(1498):1839–1848

Brasier CM, Scott JK (1994) European oak declines and global warming: a theoretical assessment with special reference to the activity of *Phytophthora cinnamomi*. Bull OEPP 24:221–232

Brumme R, Borken W (2009) N_2O emission from temperate beech forest soils. In: Brumme R, Khanna PK (eds) Functioning and management of European beech ecosystems. Ecological studies 208. Springer, Berlin, pp 353–367

Brumme R, Khanna PK (eds) (2009) Functioning and management of European beech ecosystems. Ecological studies 208. Springer, Berlin

Byram GM (1959) Combustion of forest fuels. In: Davis KP (ed) Forest fire: control and use. McGraw-Hill, New York, pp 61–89

Chakraborty S, Murray GM, Mangarey PA, Yonow T, O'Brien RG, Croft BJ, Sivasithamparam K, Old KM, Dudzinski MJ, Sutherst RW, PenroseLJ AC, Emmett RW (1998) Potential impact of climate change on plant diseases of economic significance to Australia. Australas Plant Pathol 27:15–35

Ciais P et al (2005) Europe-wide reduction in primary productivity caused by the heat and drought in 2003. Nature 437:529–533

Cox PM, Betts RA, Jones CD, Spall SA, Totterdell IJ (2000) Acceleration of global warming due to carbon-cycle feedbacks in a coupled climate model. Nature 408:184–187

Crutzen PJ (1970) The influence of nitrogen oxides on the atmospheric ozone content. Quart J R Meteorol Soc 96:320–325

Dale VH, Joyce LA, Mcnulty S, Neilson RP, Ayres MP, Flannigan MD, Hanson PJ, Irland LC, Lugo AE, Peterson CJ, Simberloff D, Swanson FJ, Stocks BJ, Wotton BM (2001) Climate change and forest disturbances. Bioscience 51:723–734

Davidson EA, Janssens IA (2006) Temperature sensitivity of soil carbon and feedbacks to climate change. Nature 7081:165–173

Davison EM, Shearer BL (1989) Phytophthora spp. in indigenous forests in Australia. N Z J For Sci 19:277–289

Doerr SH, Shakesby RA, MacDonald LH (2009) Soil water repellancy: a key factor in post-fire erosion (Chapter 7). In: Cerda A, Robichaud PR (eds) Fire effects on soils and restoration strategies. Science publishers, Enfield, p 589

Doran JW, Zeiss MR (2000) Soil health and sustainability: managing the biotic component of soil quality. Appl Soil Ecol 15:3–11

Egnell G, Valinger E (2003) Survival, growth, and growth allocation of planted Scots pine trees after different levels of biomass removal in clear-felling. For Ecol Manage 177:65–74

Emmett BA, Beier B, Estiarte M, Tietema A, Kristensen HL, Williams D, Penuelas J, Schmidt I, Sowerby A (2004) The response of soil processes to climate change: results from manipulation studies of shrublands across an environmental gradient. Ecosystems 7:625–637

Fagg PC (1987) Establishment and early growth of indigenous eucalypts sown on forest sites infested with *Phytophthora cinnamomi* in East Gippsland. Aust For Ecol Manage 20:53–78

Farrow RA (1996) Insect pests of eucalypts on farms and in plantations. Identification Leaflet No.4. Sawflies. CSIRO, Australia

Fischlin A, Midgley GF, Price JT, Leemans R, Gopal B, Turley C, Rounsevell MDA, Dube OP, Tarazona J, Velichko AA (2007) Ecosystems, their properties, goods, and services. In: Parry ML, Canziani OF, Palutikof JP, van der Linden PJ, Hanson CE (eds) Climate change 2007: impacts, adaptation and vulnerability. Contribution of Working Group II to the Fourth Assessment Report of the Intergovernmental Panel on Climate Change. Cambridge University Press, Cambridge, pp 211–272

Fontaine S, Barot S, Barre P, Bdioui N, Mary B, Rumpel C (2007) Stability of organic carbon in deep soil layers controlled by fresh carbon supply. Nature 450:277–280

Freeman C, Evans CD, Monteith DT, Reynolds B, Fenner N (2001) Export of organic carbon from peat soils. Nature 412:785

Freeman C, Ostle NJ, Fenner N, Kang H (2004) A regulatory role for phenol oxidase during decomposition in peatlands. Soil Biol Biochem 36:1663–1667

Friedlingstein P, Joel G, Field C, Fung I (1999) Toward an allocation scheme for global terrestrial carbon models. Glob Change Biol 5:755–770

Gardiner BA, Quine CP (2000) Management of forests to reduce the risk of abiotic damage – a review with particular reference to the effects of strong winds. For Ecol Manage 135:261–277

Giardina CP, Ryan MG (2000) Evidence that decomposition of organic carbon in mineral soils do not vary with temperature. Nature 6780:858–861

Giardina CP, Coleman MD, Hancock JE, King JS, Lilleskov EA, Loya WM, Pregitzer KS, Ryan MG, Trettin CC (2005) The response of belowground carbon allocation in forests to global change. In: Binkley D, Menyailo O (eds) Tree species effects on soils: implications for global change. NATO Science Series. Kluwer Academic Publishers, Dordrecht, pp 119–154

Grace J, Rayment M (2000) Respiration in the balance. Nature 404:819–820

Hagedorn F, Sprinnler D, Bundt M, Blaser P, Siegwolf R (2003) The input and fate of new C in two forest soils under elevated CO2. Glob Change Biol 9:862–872

Holden J, Shotbolt L, Bonn A, Burt TP, Chapman PJ, Dougill AJ, Fraser EDG, Hubacek K, Irvine B, Kirkby MJ, Reed MS, Prell C, Stagl S, Stringer LC, Turner A, Worrall F (2007) Environmental change in moorland landscapes. Earth Sci Rev 82:75–100

Hyvönen R, Agren GI, Linder S, Persson T, Cotrufo MF, Ekblad A, Freeman M, Grelle A, Janssens IA, Jarvis PG, Kellomäki S, Lindroth A, Loustau D, Lundmark T, Norby RJ, Oren R, Pilegaard K, Ryan MG, Sigurdsson BD, Strömgren M, van Oijen M, Wallin G (2007) The likely impact of elevated [CO2], nitrogen deposition, increased temperature and management on carbon sequestration in temperate and boreal forest ecosystems: a literature review. New Phytol 173:463–480

IPCC (2006) Chapter 2 'Generic methodologies applicable to multiple land-use categories' Volume 4. In: 2006 IPCC Guidelines for National Greenhouse Gas Inventories. IPCC Secretariat, Geneva

IPCC (2007) Climate change 2007: the physical science basis; contribution of Working Group I to the Fourth Assessment. Report of the Intergovernmental Panel on Climate Change. UNEP, New York, 996 p

Ise T, Moorcroft PR (2006) The global-scale temperature and moisture dependencies of soil organic carbon decomposition: an analysis using a mechanistic decomposition model. Biogeochemistry 80:217–231

Jastrow JD, Miller RM, Matamala R, Norby RJ, Boutton TW, Rice CW, Owensby CE (2005) Elevated carbon dioxide increases soils carbon. Glob Change Biol 11:2057–2064

Jobbágy EG, Jackson RB (2000) The vertical distribution of soil organic carbon and its relation to climate and vegetation. Ecol Appl 10:423–436

Keith H, Leuning R, Jacobsen KL, Cleugh HA, van Gorsel E, Raison RJ, Medlyn BE, Winters A, Keitel C (2009) Multiple measurements constrain estimates of net carbon exchange by a Eucalyptus forest. Agric For Meteorol 149:535–558

Kim Y, Tanaka N (2003) Effect of forest fire on the fluxes of CO2, CH4 and N2O in boreal forest soils, interior Alaska. J Geophys Res 108(D1):8154. doi:10.1029/2001JD000663

Kinako PDS, Gimingham CH (1980) Heather burning and soil erosion on upland heaths in Scotland. J Environ Manage 10:277–284

Kirschbaum MUF (1995) The temperature dependence of soil organic matter decomposiotion and the effect of global warming on soil organic C storage. Soil Biol Biochem 27:753–760

Kirschbaum MUF (2006) The temperature dependence of organic-matter decomposition – still a topic of debate. Soil Biol Biochem 38:2510–2518

Kirschbaum MUF, Keith H, Leuning R, Cleugh HA, Jacobsen KL, van Gorsel E, Raison RJ (2007) Modelling net ecosystem carbon and water exchange of a temperate *Eucalyptus delegatensis* forest using multiple constraints. Agric For Meteorol 145:48–68

Knops JMH, Naeem S, Reich PB (2007) The impact of elevated CO2, increased nitrogen availability and biodiversity on plant tissue quality and decomposition. Glob Change Biol 13:1960–1971

Krull ES, Skjemstad JO, Baldock JA (2004). Functions of soil organic matter and the effect on soil properties. GRDC report, Project CSO 00029. http://www.grdc.com.au/growers/res_summ/cso00029/contents.html. Sourced Nov., 2010

Landsberg J (2003) Modelling forest ecosystems: state of the art, challenges, and future directions. Can J For Res 33:385–397

Leckebusch GC, Koffi B, Ulbrich U, Pinto G, Spangehl T, Zacharias S (2006) Analysis of frequency and intensity of European winter storm events from a multi- model perspective, at synoptic and regional scales. Climate Res 31:59–74

Leuzinger S, Zotz G, Asshoff R, Körner C (2005) Responses of deciduous forest trees to severe drought in Central Europe. Tree Physiol 25:641–650

Lichter J, Barron SH, Bevacqua CE, Finzi AC, Iriving KF, Stemmler EA, Schlesinger WH (2005) Soil carbon sequestration and turnover in a pine forest after six years of atmospheric CO2 enrichment. Ecology 86:1835–1847

Litton CM, Giardina CP (2008) Below-ground carbon flux and partitioning: global patterns and response to temperature. Funct Ecol 22:941–954

Litton CM, Raich JW, Ryan MG (2007) Review: carbon allocation in forest ecosystems. Glob Change Biol 13:2089–2109

Lucier A, Ayres M, Karnosky D, Thompson I, Loehle C, Percy K, Sohngen B (2009) Forest responses and vulnerabilities to recent climate change. In: Seppala R, Buck A, Katila P (eds) Adaptation of forests and people to climate change – a global assessment report. IUFRO World Series, Vol 22. IUFRO, Helsinki, 224 pp

Marks GC, Kassaby FY, Reynolds ST (1972) Dieback in the mixed hardwood forests of Eastern Victoria; a preliminary report. Aust J Bot 20:141–151

Matzner E (2004) Biogeochemistry of forested catchments in a changing environment, a German case study, Ecological Studies 172. Springer, Berlin

Monteith DT, Stoddard JL, Evans CD, de Wit HA, Forsius M, Hogasen T, Wilander A, Skjelkvale BL, Jeffries DS, Vuorenmaa J, Keller B, Kopacek J, Vesely J (2007) Dissolved organic carbon trends resulting from changes in atmospheric deposition chemistry. Nature 450:537–540

Moore BA, Allard GB (2008) Forest health and biosecurity working papers FBS/34E. Forest Resources Development Service, Forest Management Division. FAO, Rome

Norby RJ, Cotrufo MF, Ineson P, O'Neill EG, Canadell JG (2001) Elevated CO2, litter chemistry, and decomposition: a synthesis. Oecologia 127:153–165

Norby RJ, Warren JM, Iversen CM, Medlyn BE, McMurtrie RE (2010) CO_2 enhancement of forest productivity constrained by limited nitrogen availability. Proceedings of the National Academy of Sciences USA 107:19,368–19,373

Old KM, Floyd RB (2000) Threats to plantations from pests and diseases. In: Nambiar S, Cromer R, Brown AG (eds) Restoring tree cover in the Murray Darling Basin. CSIRO Forestry and Forest Products, Canberra, 60 pp (ISBN 0643063196)

Old KM, Stone C (2005) Vulnerability of Australian forest carbon sinks to pests and pathogens in a changing climate: a discussion paper prepared for the Australian Greenhouse Office. Australian Greenhouse Office, Canberra

Pendall E, Bridgham S, Hanson PJ, Hungate B, Kicklighter DW, Johnson DW, Law BE, Luo Y, Megonigal JP, Olsrud M, Ryan MG, Wan S (2004) Below-ground process responses to elevated CO2 and temperature: a discussion of observations, measurement methods, and models. New Phytol 162:311–322

Pilkington MG, Caporn SJM, Carroll JA, Cresswell N, Phoenix GK, Lee JA, Emmett BA, Sparks T (2007) Impacts of burning and increased nitrogen deposition on nitrogen pools and leaching in an upland moor. J Ecol 95:1195–1207

Podger FD, Mummary DC, Palzer CR, Brown MJ (1990) Bioclimatic analysis of the distribution and damage to native plants in Tasmania by *Phytophthora cinnamomi*. Aust J Ecol 15: 281–289

Pook EW (1985) Canopy dynamics of *Eucalyptus maculata* 3. Effects of drought. Aust J Bot 33: 65–80

Quinton JN, Catt JA, Wood GA, Steer J (2006) Soil carbon losses by water erosion: experimentation and modeling at field and national scales in the UK. Agric Ecosyst Environ 112:87–102

Raison RJ, Kirschbaum MUF (2008) Fire and the carbon balance of terrestrial ecosystems. Geophys Res Abstr 10:EGU2008-A-02946

Raison RJ, Woods PV, Jakobsen BF, Bary GAV (1986) Soil temperatures during and following low-intensity prescribed burning in a *Eucalyptus pauciflora* forest. Aust J Soil Res 24:33–47

Raison RJ, O'Connell AM, Khanna PK, Keith H (1993) Effects of repeated fires on nitrogen and phosphorus budgets and cycling processes in forest ecosystems. In: Trabaud L, Prodon R (eds)

Fire in Mediterranean ecosystems. Ecosystem Research Report 5:347–363. Commission of the European Communities, Brussels

Raison RJ, Eamus D, Gifford RM, McGrath J (2007) The feasibility of forest free air CO_2 enrichment (FACE) experimentation in Australia. Australian Greenhouse Office, Canberra (during 2010 renamed the Department of Climate Change and Energy Efficiency), 110pp. (ISBN: 987-1-921297-68-7)

Raison RJ, Khanna PK, Romanya J, Serrasolses I (2009) Effect of fire on forest nutrient cycles. In: Cerda A, Robichaud P (eds) Fire effects on soils and restoration strategies. Science Publisher, Enfield, pp 225–256

Raulund-Rasmussen K, Stupak I, Clarke N, Callesen I, Helmisaari HS, Karltun E, Varnagiryte-Kabasinskiene I (2008) Effects of very intensive forest biomass harvesting on short and long term site productivity. In: Röser D, Asikainen A, Raulund-Rasmussen K, Stupak I (eds) Sustainable use of forest biomass for energy. A synthesis with focus on the Baltic and Nordic region. Springer, Dordrecht, pp 29–78

Reay DS, Dentener F, Smith P, Grace J, Feely R (2008) Global nitrogen deposition and carbon sinks. Nat Geosci 1:430–437. doi:10.1038/ngeo230

Reichstein M et al (2007) Determinants of terrestrial ecosystem carbon balance inferred from European eddy covariance flux sites. Geophys Res Lett 34:L01402. doi:10.1029/2006GL027880

Richardson J, Björheden R, Hakkila P, Lowe AT, Smith CT (eds) (2002) Bioenergy from sustainable forestry: guiding principles and practices. Kluwer Academic, Dordrecht

Roberts SD, Harrington CA, Terry TA (2005) Harvest residue and competing vegetation affect soil moisture, soil temperature, N availability, and Douglas-fir seedling growth. For Ecol Manage 205: 333–350

Roxburgh SH, Barrett DJ, Berry SL, Carter JO, Davies ID, Gifford RM, Kirschbaum MUF, McBeth BP, Noble IR, Parton WG, Raupach MR, Roderick ML (2004) A critical overview of net primary productivity model estimates for the Australian continent. Funct Plant Biol 31: 1043–1059

Rustad L, Campbell CJ, Marion MG, Norby NR, Mitchell MM, Hartley HA, Cornelissen CJ, Gurevitch GJ, Gcte N, Gcte N (2001) A meta-analysis of the response of soil respiration, net nitrogen mineralization, and aboveground plant growth to experimental ecosystem warming. Oecologia 126:543–562

Sauerborn P, Klein A, Botschek J, Skowronek A (1999) Future rainfall erosivity derived from large-scale climate models – methods and scenarios for a humid region. Geoderma 93:269–276

Schimel D, Melillo J, Tian H, McGuire AD, Kicklighter D, Kittel T, Rosenbloom N, Running S, Thornton P, Ojima D, Parton W, Kelly R, Sykes M, Neilson R, Rizzo B (2000) Contribution of increasing CO2 and climate to carbon storage by ecosystems in the United States. Science 287(5460):2004–2006

Schulze D, Freibauer A (2005) Carbon unlocked from soils. Nature 437:205–206

Shearer BL, Smith IW (2000) Diseases of eucalypts caused by soilborne species of *Phytophthora* and *Pythium*. In: Keane PJ, Kile GA, Podger FD, Brown BN (eds) Diseases and pathogens of eucalypts. CSIRO, Collingwood, pp 259–291

Simpson JA, Podger FD (2000) Management of eucalyptus diseases-options and constraints. In: Keane PJ, Kile GA, Podger FD, Brown BN (eds) Diseases and pathogens of eucalypts. CSIRO, Collingwood, pp 411–425

Six J, Callewaert P, Lenders S, De Gryze S, Morris SJ, Gregorich EG, Paul EA, Paustian K (2002) Measuring and understanding carbon storage in afforested soils by physical fractionation. Soil Sci Soc Am J 66:1981–1987

Stevenson AC, Rhodes AN, Kirkpatrick AH, MacDonald AJ (1996) The determination of fire histories and an assessment of their effects on moorland soils and vegetation. Research, Survey and Monitoring Report No.16. Scottish Natural Heritage, Edinburgh

Sverdrup H, Rosen K (1998) Long-term base cation mass balances for Swedish forests and the concept of sustainability. For Ecol Manage 110:221–236

Swedish National Board of Forestry (2002) Recommendations for the Extraction of Forest Fuel and Compensation Fertilising. Samuelsson H (Ed.) Skogsstyrelsen National Board of Forestry, Jönköping, Sweden, 25 pp

Tebaldi C, Hayhoe K, Arblaster JM, Meehl GA (2006) Going to the extremes :an intercomparsion of model-simulated historical and future changes in extreme events. Climate Change 79: 185–211

Teepe R, Brumme R, Beese F, Ludwig B (2004) Nitrous oxide emission and methane consumption following compaction of forest soils. Soil Sci Soc Am J 68:605–611

Terry AC, Ashmore MR, Power SA, Allchin EA, Heil GW (2004) Modelling the impacts of atmospheric nitrogen deposition on Calluna-dominated ecosystems in the UK. J Appl Ecol 41: 897–909

Thornton PE, Law BE, Gholz HL, Clark KL, Falge E, Ellsworth DS, Goldstein AH, Monson RK, Hollinger D, Falk M, Chen J, Sparks JP (2002) Modeling and measuring the effects of disturbance history and climate on carbon and water budgets in evergreen needleleaf forests. Agric For Meteorol 113:185–222

Tobita H, Kitao M, Koike T, Maruyama Y (2005) Effects of elevated CO2 and nitrogen availability on nodulation of *Alnus hirsuta* Turcz. Phyton 45:125–131

Van Der Werf GR, Dempewolf J, Trigg SN, Randerson JT, Kasibhatla PS, Giglio L, Murdiyarso D, Peters W, Morton DC, Collatz GJ, Dolman AJ, DeFries RS (2008) Climate regulation of fire emissions and deforestation in equatorial Asia. Proc Natl Acad Sci USA 105:20350–20355

Van Groenigen KJ, Six J, Hungate BA, de Graaff MA, van Breemen N, Van Kessel C (2006) Element interaction limit carbon storage. Proc Nat Acad Sci USA 103:6571–6574

Vesterdal L, Ritter E, Gundersen P (2002) Change in soil organic carbon following afforestation of former arable land. For Ecol Manage 169:137–147

Volney WJA, Fleming RA (2000) Climate change and impacts of boreal forest insects. Agric Ecosyst Environ 82:283–294

Worrall F, Burt T, Shedden R (2003) Long term records of riverine dissolved organic matter. Biogeochemistry 64:165–178

Zabowski D, Chambreau D, Rotramel N, Thies WG (2008) Long-term effects of stump removal to control root rot on forest soil bulk density, soil carbon and nitrogen content. For Ecol Manage 255:720–727

Zhang W, Parker K, Luo Y, Wan S, Wallace LL (2005) Soil microbial responses to experimental warming and clipping in a tall grass prairie. Glob Change Biol 11:266–277

Chapter 13
Rehabilitated Mine-Site Management, Soil Health and Climate Change

Natasha C. Banning, Briony M. Lalor, Andrew H. Grigg, Ian R. Phillips, Ian J. Colquhoun, Davey L. Jones, and Daniel V. Murphy

13.1 Introduction

Natural and anthropogenic disturbances to terrestrial ecosystems have the potential to cause changes in plant species composition and abundance, to disrupt the spatial organisation of landscapes and therefore have the potential to affect ecosystem sustainability. Ecological restoration has been defined as "the process of assisting the recovery of an ecosystem that has been degraded, damaged or destroyed" (Society for Ecological Restoration 2004). Several different terms including rehabilitation, reclamation, reallocation, mitigation and recovery are used to describe management activities associated with the overall goal of fully restoring ecosystems (Hobbs and Norton 1996). In this chapter, we have used the term rehabilitation, which is commonly used within the mining industry, to describe all activities that aim, as a minimum, to return or develop some form of vegetative cover on the disturbed land that is protective, productive, aesthetically pleasing or valuable in a conservation sense (Grant 2006).

Common aims of mine-site rehabilitation are to build or re-establish a soil profile, to re-establish aboveground and belowground biodiversity, and ultimately to establish a self-sustaining ecosystem (i.e. one that requires no further

N.C. Banning • D.V. Murphy (✉)
Soil Biology Group, School of Earth and Environment, The University of Western Australia, 35 Stirling Highway, Crawley, WA 6009, Australia
e-mail: daniel.murphy@uwa.edu.au

B.M. Lalor
Department of Natural Resource Sciences, McGill University, 21, 111 Lakeshore Road, Ste-Anne-de-Bellevue, Canada H9X 3V9

A.H. Grigg • I.R. Phillips • I.J. Colquhoun
Alcoa of Australia Ltd, Huntly Mine, PO Box 172 Pinjarra, WA 6208, Australia

D.L. Jones
School of the Environment, Natural Resources and Geography, Bangor University, Gwynedd LL57 2UW, UK

B.P. Singh et al. (eds.), *Soil Health and Climate Change*, Soil Biology 29,
DOI 10.1007/978-3-642-20256-8_13,

management input). Soil biological function is central to achieving these aims. Soil fauna and micro-organisms carry out many ecosystem processes encompassing the breakdown of organic matter, nutrient cycling and formation of soil structure through stabilisation of soil aggregates and enhanced water retention. Soil decomposer organisms are essential for the functioning of terrestrial ecosystems as they carry out most biochemical processes that directly regulate nutrient availability for primary production as well as other soil heterotrophic activity (Killham 1994). Direct interactions between plant roots and soil micro-organisms are also of fundamental importance to plant growth and survival and may be beneficial to the plant, such as myccorrhizal or rhizobial symbioses or detrimental, such as the case for some pathogenic fungi. Microbial colonisation of denuded soil is an essential prerequisite to ecosystem rehabilitation, and measures of microbial community size, activity, composition and diversity are being increasingly recognised as providing key indicators of rehabilitation success (Harris 2003). In this chapter, we focus on highlighting the research related to soil micro-organisms and primarily those whose activity involves chemical decomposition of plant material and nutrient cycling/recycling, as these are fundamental ecosystem processes. In this context, we review how rehabilitation success is measured; either by comparison to a reference site or by assessment of the return of ecosystem functions, and propose that an understanding of functioning in sustainable ecosystems will be increasingly important in a rapidly changing climate.

The type of rehabilitation practices and management that can be undertaken in large industrial-scale projects and how these influence the emerging ecosystem attributes is illustrated using two case studies from Western Australia: (1) the rehabilitation of jarrah (*Eucalyptus marginata* Donn ex Sm.) forest following bauxite mining for alumina, and (2) the rehabilitation of bauxite-processing residue storage areas generated by refining of the bauxite ore to alumina. Both of these activities have been carried out on a large scale (Fig. 13.1) by Alcoa of Australia Ltd. (referred to throughout as Alcoa). The jarrah forest rehabilitation case study provides an example of a well-documented rehabilitation project with a long history that has evolved using the surrounding native forest as a reference for defining a suite of ecosystem targets. These targets have then been used to assess rehabilitation success, which has also been termed "completion" (the point at which the rehabilitated land can be returned to management control by the state). Bauxite ore is located close, usually less than 1 m, to the soil surface, and its removal requires a surface mining approach involving complete removal of all vegetation (Koch 2007). The rehabilitation practices (including landscaping, topsoil return, surface contouring, seeding, planting and fertilisation) carried out at the early establishment stage are strong determinants of the final ecosystem attributes, and these are discussed with particular attention to their effect on soil properties and microbial function. Management of older rehabilitation (including stand thinning and prescription burning) can also change the successional development trajectory of rehabilitation sites and is likely to become an increasingly important tool in the context of regional climate change predictions.

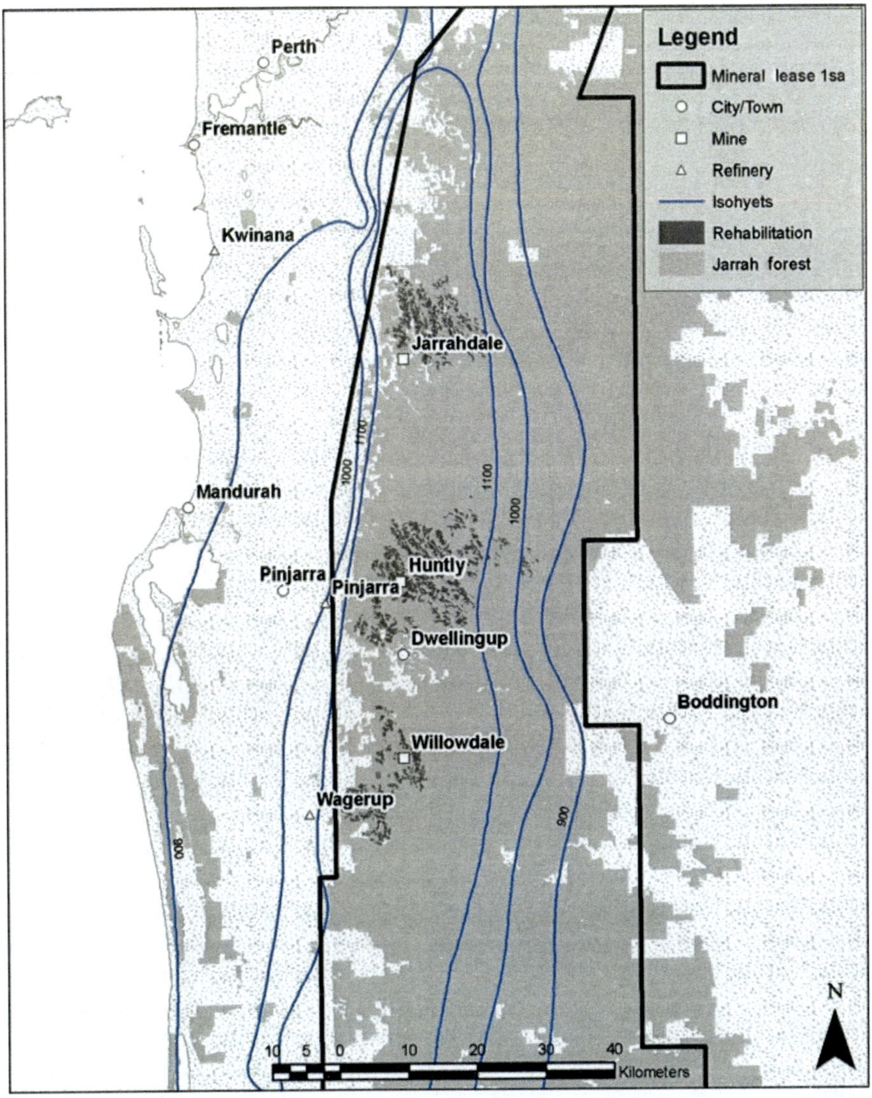

Fig. 13.1 Map illustrating the region of southwest Western Australian covered by jarrah (*Eucalyptus marginata*) forest and areas of past or current mining and forest rehabilitation carried out within the mineral lease by Alcoa of Australia. Isohyets are shown in mm per annum

The second case study of bauxite residue storage area rehabilitation provides an example where rehabilitation requirements are not as well defined. The residue sand used for rehabilitation is essentially a nascent substrate that is highly alkaline, saline, sterile and without organic matter, and there is no pre-disturbance or historical ecosystem to use as a reference ecosystem. In this case, an analogue natural ecosystem, with some similarities in soil properties, has been selected to provide a target for the rehabilitation. The final section of the chapter discusses how the broad design principles of rehabilitation projects may need to be adapted to suit a rapidly changing climate.

13.2 Evaluation of Rehabilitation Performance

Any strategy intended to evaluate the success of rehabilitation of disturbed ecosystems requires criteria for judging rehabilitation progress. There are two main approaches to evaluate the success of rehabilitation. The first is by comparison to a chosen reference or analogue site, and the second is to assess the return of ecological functions and sustainability, often referred to as "ecosystem services" (Cairns 2000).

13.2.1 Comparison to Reference Sites

Choosing appropriate reference sites is essential to correctly assess the progress and success of rehabilitation efforts. Rehabilitation practices commonly respect the history of the site and aim to return ecosystem vegetation structure, biodiversity and ecological processes, such as soil nutrient cycling, to pre-disturbance conditions. Thus, for mine-site rehabilitation, the most appropriate reference site is in general provided by the surrounding or adjacent non-mined ecosystem. However, whether the non-mined ecosystems chosen as reference sites represent the full variation that was previously present in the mined ecosystems must always be considered. Where this type of reference ecosystem does not exist or the soil disturbance is so drastic that the rehabilitation site is essentially a completely nascent substrate (e.g. some coal mines, mine tailings or mineral processing wastes), an analogue ecosystem may be chosen that possesses attributes of the desired final ecosystem. An analogue ecosystem typically has similar landscape characteristics, soil physical and chemical properties, and many of the vegetation species required in the mature rehabilitation sites (Ludwig et al. 2004; White and Walker 1997).

Effective use of pre-disturbance reference sites as benchmarks for rehabilitation success requires a fundamental understanding of how mature ecosystems function. The development of rehabilitation ecosystems has been likened to natural ecological succession, albeit on a more rapid time scale. The classical view of succession was of an orderly process of community development that is reasonably directional (i.e. towards a stable "climax" state) and hence may be predicted (Odum 1969). However, more dynamic models of succession have been developed recognising that ecosystems naturally fluctuate between "quasi-stable" states (Odum et al. 1995; Walker and del Moral 2003). Oscillations between quasi-stable states of an ecosystem can be caused by stress (such as drought or seasonal temperature extremes) or more transient disturbance events (such as fire, flood or disease) (Grant 2006; Hobbs and Harris 2001). Hence, there is unlikely to be a single stable state that represents a target end-point for a rehabilitated ecosystem. Rather, the study of reference ecosystems can provide a range of threshold values of key ecosystem processes within this natural oscillation, outside of which the ecosystem would no longer be able to sustain itself.

Understanding how a reference or analogue ecosystem is likely to respond to predicted climate change scenarios is also necessary to determine whether it is appropriate for rehabilitation efforts to be focussed on returning a historical ecosystem composition. Harris et al. (2006), in considering woodland conservation in the UK, postulate "Is it appropriate to consider a temperate woodland restoration end-point in an area likely to be flooded by rising sea level"? Of course, such questions may not in reality be as clear-cut due to uncertainty in global climate modelling and the greater uncertainty in the direction, extent and timing of climate changes on a regional basis (Charles et al. 1999). Nevertheless, it is increasingly being recognised that rehabilitation planning and practice must take into consideration future projected environmental changes rather than solely focussing on the re-establishment of the ecosystem composition present before mining or other anthropogenic disturbance. Thus, a fundamental understanding of natural ecosystem functions and processes will become increasingly important. A more suitable way to assess rehabilitation performance may be to focus on the extent of recovery of these functions and processes.

13.2.2 Assessment of Ecosystem Functions and Sustainability

The "functions" of an ecosystem include key ecological processes (e.g. biogeochemical cycling and energy transfer) that are essential for the maintenance and long-term sustainability of ecosystem health (Bell 2001). de Groot et al. (2002) described and categorised ecosystem functions (e.g. climate regulation, water supply, habitat functions) which can provide "goods and services" (e.g. maintenance of a favourable climate, provision of drinking water, maintenance of biological and genetic diversity) which are useful to human society. For example, the key ecosystem function of nutrient regulation requires the interaction of soil biological, chemical, physical and hydrological processes to maintain the equilibrium of nutrient cycles. For an ecosystem to be self-sustaining, microbial cycling and recycling of nutrients, particularly nitrogen (N), phosphorous (P) and sulphur (S), from organic matter entering the soil system must be able to meet plant nutrient demands (Bardgett 2005). In general, sustainability in developing ecosystems requires that nutrient cycling becomes more conservative (i.e. increased capacity to retain nutrients within the system; Odum 1969). The re-establishment of soil pools (e.g. organic matter, specific nutrients) and soil processes (e.g. rate of organic matter decomposition by soil heterotrophs, nitrate production by soil nitrifiers and subsequent nitrate loss by leaching or nitrous oxide production) that regulate these functions can be measured directly and used to assess whether the rehabilitation is heading toward, or likely to achieve, a sustainable state.

13.2.2.1 Ecosystem Resistance and Resilience

Stable ecosystems are able to withstand normal ecosystem disturbances to a greater extent than less developed ecosystems. The two key aspects of stability are resistance and resilience. Resistance is defined as the inherent capacity of an ecosystem or ecosystem process to withstand an immediate stress, and resilience is the extent of recovery of ecosystem properties after a disturbance (Wardle et al. 2000; Degens 1998; Westman 1986). Studies of ecosystems recovering post-disturbance have used the concept of ecosystem resistance and resilience as a framework to compare the sensitivity of components of ecosystems to disturbance (e.g. fire) and to determine how and when an ecosystem is likely to decouple from a pattern of steady-state oscillations. An understanding of the resilience properties of an ecosystem being restored may aid in predicting the performance of rehabilitation efforts. This concept has been applied to many different ecosystems and also at different scales within these ecosystems, from the landscape level (Ludwig et al. 2004), to plant communities (Li et al. 2006) and soil ecosystems (Fitter et al. 2005; Orwin et al. 2006).

13.2.2.2 Belowground Biodiversity

The relationships between belowground biodiversity, ecosystem functioning and soil management have emerged as central issues in this ecological and environmental debate (see Chap. 8). Ecosystem processes in soil such as decomposition and nutrient cycling are to a large extent carried out by soil micro-organisms. However, soil micro-organisms are very diverse and only a small percentage of them have been taxonomically characterised (Torsvik et al. 2002). There have been a number of studies representing different land uses and vegetation communities that have attempted to link microbial community structure with soil functions such as C and N cycling (Balser and Firestone 2005; Cookson et al. 2007; Waldrop and Firestone 2006). These studies suggest that there may be relationships between components of the microbial community and soil processes, particularly processes tightly linked to specific microbial phylogeny (e.g. nitrification) (Cookson et al. 2005; Waldrop and Firestone 2006). However, because broad soil processes such as C mineralisation, N ammonification and immobilisation are likely to be carried out by a large proportion of the microbial community, determining direct relationships may be more difficult due to functional redundancy within the microbial population.

This raises the question: how important is the return of a comparable microbial community structure or diversity to soil function? It has been suggested that rehabilitation of degraded lands requires a return of the diversity of ecosystem functions to that of pre-disturbance levels, and that this may be of greater importance than species diversity per se (Neher 1999). It may be that the assessment of microbial functional redundancy provides a better indicator of the stability and function of soils. For example, in their study of a soil reclamation gradient in Brazil, Yin et al. (2000) found that bacterial functional redundancy, determined by assessing the diversity of microbial species actively growing in response to the addition of individual C substrates, increased along a disturbance gradient from denuded mine spoil to preserved forest soils. They suggest that an increase in the diversity of bacteria capable of degrading a particular substrate (i.e. functional redundancy) may represent the development of a more stable ecosystem. However, not enough is currently known about the majority of soil micro-organisms to be able to categorise them into functional guilds. Furthermore, little is known of how microbial diversity within functional groups influences ecosystem functions and stability (Degens et al. 2001).

13.3 Mine-Site Rehabilitation Practices and Their Effect on Soil Health

13.3.1 Case Study 1: Bauxite Mining and Rehabilitation in the Jarrah Forest

Australia is the world's largest producer of alumina which is recovered from the processing of bauxite ore. Alcoa's mining operations in Western Australia are the largest single source of alumina in the world, able to supply 15% of the international market. Alcoa has conducted bauxite mining and rehabilitation in Western Australia's jarrah (*E. marginata*) forest since 1963 with the first rehabilitation completed in 1966. The jarrah forest covers an area of approximately 3.3 million hectares (Dell et al. 1989), of which 712,900 hectares fall within a mineral lease held by Alcoa (Fig. 13.1). The forest consists of an overstorey dominated by jarrah and marri (*Corymbia calophylla*) trees, a mid-storey of bull banksia (*Banksia grandis*), sheoak (*Allocasuarina fraseriana*) and grasstree (*Xanthorrhoea preisii*) and a dense, highly diverse understorey. The forest has been divided into 19 vegetation communities based on understorey composition which is related to differences in landscape position and underlying soil properties (Havel 1975). The region is recognised internationally as a biodiversity hotspot (Myers et al. 2000), supporting over 1,200 species of vascular plants as well as 29 mammal species, 45 reptile species, 17 frog species, 4 fish species and 150 bird species (Williams and Mitchell 2001). The forest provides numerous other community services, including timber production and recreation, and most of the northern jarrah forest is an important water catchment area.

Bauxite mining involves the complete clearing of vegetation, removing the topsoil (0–15 cm), blasting of the cemented duri-crust layer and removal of the

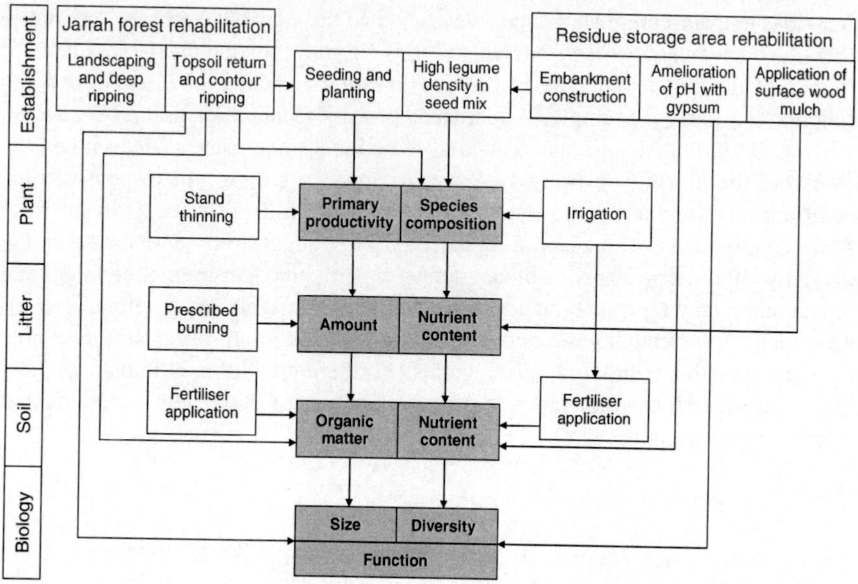

Fig. 13.2 Schematic overview of the influence of management practices on ecosystem attributes in post-mining jarrah forest rehabilitation (case study 1) and bauxite residue storage area rehabilitation (case study 2)

bauxite ore which averages 3–4 m in depth and is usually located less than 1 m below the soil surface (Koch 2007). The bauxite ore is then crushed and transported to the refineries located nearby to be processed into alumina. Thus, bauxite mining represents a drastic disturbance which impacts on both aboveground and belowground components of the ecosystem. Alcoa presently operates two mines in the northern jarrah forest at Huntly and Willowdale (Fig. 13.1). A third mine at Jarrahdale ceased production in 1998 and has now been deemed, by the Western Australian State Government, as rehabilitated. To date these operations have cleared, mined and rehabilitated 15,000 hectares of forest. The goal of current rehabilitation practices after mining is to return a self-sustaining jarrah forest ecosystem, planned to enhance or maintain water, timber, recreation and conservation values. An overview of rehabilitation practices and their effect on soil biological attributes is given in Fig. 13.2 and described in more detail in the sections below.

13.3.1.1 Landscaping and Topsoil Return

Once mining of an area (pit) has been completed, the pit walls are shaped with a bulldozer to integrate the pit with, and prevent run-off into, the surrounding undulating landscape. Deep ripping to a depth of 1.8 m breaks up the underlying compacted pit floor to enable water infiltration and tree root development. Forest topsoil (0–15 cm) collected in mined areas is directly returned to rehabilitation areas to provide an initial plant seed bank and to return soil organic matter and micro-organisms (Koch 2007; Ward 2000). Topsoil movement is managed to minimise spread of the plant "dieback" disease caused by the soil-borne pathogen *Phytophthora cinnamomi*. The presence of this pathogen is mapped before the commencement of mining activities, which allows contamination of non-diseased soils to be prevented (Colquhoun and Kerp 2007).

Where possible, soil is collected and spread directly without stockpiling as soil storage causes a decline in both plant seed and microbial cell viability. For instance, Sawada (1996) found that soil microbial biomass was around 85% lower in 2–3-year-old rehabilitation sites that had used previously stockpiled soil compared to "direct- return" soil. Analysis of topsoils (0–5 cm; <4 mm fraction) recently replaced onto mined sites within the Huntly region found that the soil had, on average, 60% less soil organic matter and 88% less soil microbial biomass than surrounding native forest soils (Banning et al. 2008). Soil organic matter and microbial biomass re-establishes over time in these rehabilitation sites, although recovery of organic matter (as measured by total organic C) is slower than the recovery of microbial biomass and in 18–26 years old rehabilitation remains below that of the non-mined forest mean (Fig. 13.3; Banning et al. 2008; Jasper 2007). However, as rehabilitation practices have changed over time (further details described below), the future trajectory of the younger rehabilitation, which represents current practices, may not be accurately predicted by the older sites of chronosequence studies.

After topsoil replacement, mine pits are ripped across the slope to assist in controlling erosion. The distinct undulations (approximately 0.5 m wide and 0.25 m deep) that result from ripping cause subsequent plant litter to accumulate in the furrows. This structure still remains in the oldest of the rehabilitated forest soils causing spatial heterogeneity in soil organic matter (Ward 2000), nutrients (Todd et al. 2000) and microbial processes such as N cycling rates (Banning et al. 2008). As such, the size of the microbial biomass and associated C and nutrient cycling rates are larger in the furrow compared to mound soils (Table 13.1). Assessment of microbial community level physiological profiles (CLPPs), which is based on C substrate utilisation patterns, found that there was a difference between mound and furrow soil CLPPs in 3-year-old rehabilitation but not in 16-year-old rehabilitation (Lalor et al. 2007). Combined, these findings suggest that the "litter trap" effect of the furrows supports a larger microbial biomass but has less effect on the microbial community substrate utilisation pattern, than to substrate response magnitude. This suggests that contour ripping may have less effect on the microbial community structure than its size. This conclusion is supported by Grayston et al. (2004) who suggested that microbial biomass size is altered by the quantity of resources available, while microbial community structure is better related to the quality of available organic matter.

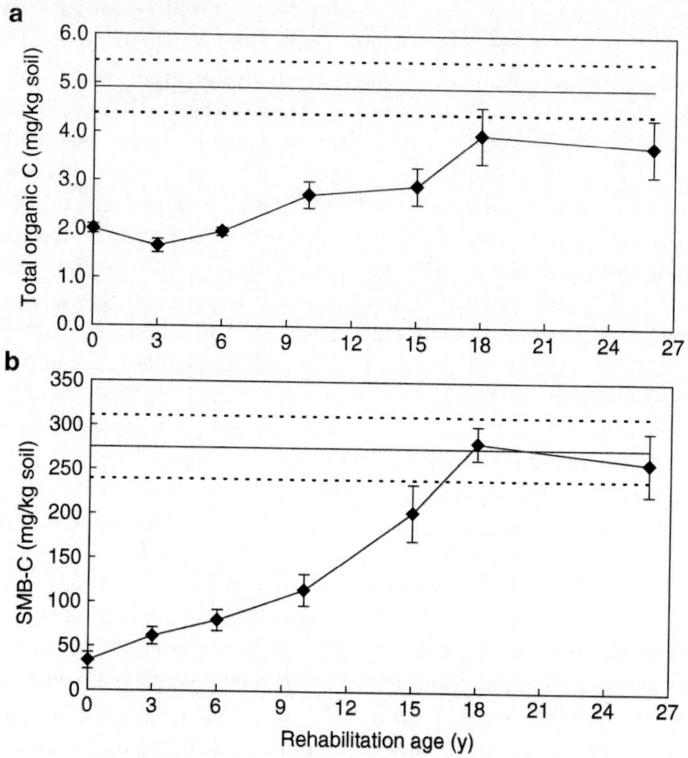

Fig. 13.3 Changes in total organic C (**a**) and microbial biomass C (SMB-C) (**b**) in a chronosequence of post-mining jarrah forest rehabilitation topsoil (0–5 cm). Mean values ($n = 5$) $\pm$ 1SE are shown. *Horizontal solid line* represents the mean value from representative non-mined forest sites ($n = 15$); *dashed lines show* $\pm$1SE. Data derived from Banning et al. (2008)

Table 13.1 Soil attributes for carbon (C) and nitrogen (N) of mound and furrow soils (0–5 cm) from 15-year-old rehabilitated and adjacent non-mined forest reference soils

Soil attribute	Rehabilitation		Non-mined
	Mound	Furrow	Forest
Total soil C (%)	2.1 a	4.3 b	6.5 b
Total soil N (%)	0.06 a	0.1 b	0.1 b
Soil C:N ratio	35 a	43 b	65 c
Microbial biomass C (mg C kg^{-1})	132 a	315 b	398 b
C respiration (mg CO_2–C kg^{-1} day^{-1})	7.3 a	10.4 b	9.5 b
N mineralisation (mg N kg^{-1} day^{-1})	2.5 a	5.1 b	5.0 b

Letters indicate significant differences at $P < 0.05$

13.3.1.2 Seeding and Planting

Rehabilitation practices and goals for Alcoa's jarrah forest rehabilitation have changed considerably over time (Gardner and Bell 2007). Early rehabilitation techniques involved planting exotic pine trees with no understorey seeding. In the early 1980s, rehabilitation sites were planted with a mix of native indigenous and eastern Australian dieback-resistant eucalypt species with an understorey comprised mainly of native legume species. Since 1988, a native overstorey has been established in rehabilitation sites by seeding or planting jarrah and marri (*Corymbia calophylla*). Alcoa's current rehabilitation objectives include returning the richness of native plant species in the rehabilitated forest, and this has largely been achieved. Rehabilitation from 2000 onwards returned around 80–100% of the number of plant species in reference jarrah forest sites (Fig. 13.4). Since 1994, even the worst 10% of monitored sites have achieved the 50% species richness target set out in the completion criteria defined by government (Koch 2007).

The rehabilitation strategy has been to establish the target plant species richness from the start of the rehabilitation process as research has shown that vegetation succession follows the initial floristic composition model; therefore, unassisted recruitment of native plant species over time is extremely slow (>20 years) or nonexistent (Norman et al. 2006). This has been achieved through a combination of improved topsoil handling, seed collection from the native plant populations in the surrounding forest, seed heat, hormone or smoke treatments pre-broadcast to improve germination rates, planting of nursery-grown seedlings and plants grown from tissue culture. Although plant richness has been successfully returned, the vegetation structure that develops in rehabilitation forest is different to the surrounding forest due to differences such as an increased density of leguminous species and decreased density of "resprouter" species (Norman et al. 2006). Improving vegetation structure similarity has been highlighted as a future challenge for jarrah forest rehabilitation (Koch and Hobbs 2007). The re-establishment of plant community richness, but with a different vegetation structure, is reflected in

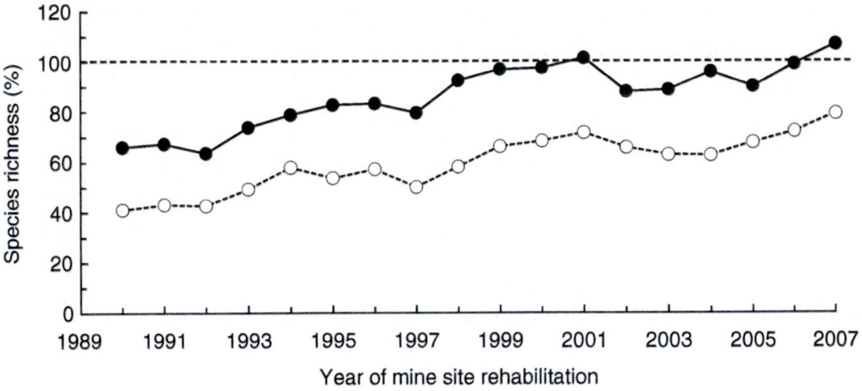

Fig. 13.4 Plant species richness in 15-month-old rehabilitated areas expressed as a percentage of corresponding non-mined forest reference plots. *Solid circles* represent the mean species richness and *open circles* the lowest 10% of monitoring plots (80 m^2). Data have been updated from that reported for 1990–2004 by Koch (2007)

the belowground community dynamics. The microbial (bacterial and fungal) diversity in the topsoil has been found to recover, after an initial decrease, in 6–14-year-old rehabilitated forest soils, but the microbial community structure was found to remain different to that in the non-mined forest soils (Banning, unpublished data; Glen et al. 2008).

13.3.1.3 Fertiliser Application and Use of Legumes

In order to reach the productivity levels of the pre-mining forest, it is necessary to input nutrients into the system following losses of plant residues and soil organic matter through the mining process. N and P are the nutrients most limiting to the growth of trees in the jarrah forest (Abbott and Loneragan 1986) and in rehabilitated areas (Ward and Pickersgill 1985). During the first year of rehabilitation, di-ammonium phosphate fertiliser containing potassium and micronutrients (currently 280 kg ha^{-1}) is broadcast applied by helicopter; no further fertiliser is applied and longer term nutrient cycling is dependent on the establishment of self-sustaining belowground biological functions.

Establishing legumes has been shown to be the most effective way to accumulate N on rehabilitation areas (Koch 1987; Ward 2000; Ward and Pickersgill 1985). Legume residues decompose more rapidly and release more N than wide C:N ratio plant material and as such can support more net primary production (Chapin et al. 1994). The soil C:N ratio of older rehabilitation (14–26 years) is significantly lower than that in forest reference soils (Fig. 13.5a), and this can be attributed to the high-density leguminous understorey in rehabilitation (Banning et al. 2008) and possible enhancement of N_2 fixation rates by P fertiliser application (Hingston et al. 1982). In addition to fixing atmospheric N_2 through symbiotic rhizobia, fast growing legumes rapidly establish vegetative cover which decreases erosion, increases soil organic matter, modifies the micro-climate at the soil surface, provides a resource for fauna and favourably affect soil physical properties (Chapin et al. 1994).

A dominance of microbial immobilisation over potential N loss pathways (e.g. leaching of nitrate produced by nitrification, nitrous oxide emissions produced by denitrifiers or via nitrifier denitrification) is required to ensure that N is retained in the system and not lost. While there is considerable variation, inorganic N concentrations in jarrah forest soils are generally low (<10 mg N kg^{-1}). Banning et al. (2008) found that the potential for nitrification (assessed as the percentage of inorganic N present as nitrate) decreased with rehabilitation age, trending towards that of the more fertile ("T" type Havel classification; Havel 1975) native forest soils (Fig. 13.5b). Nitrous oxide emissions have not been directly measured in these jarrah forest soils, although evidence from Western Australian agricultural ecosystems (Barton et al. 2008) would suggest that emissions would be low due to water and nitrate limitations.

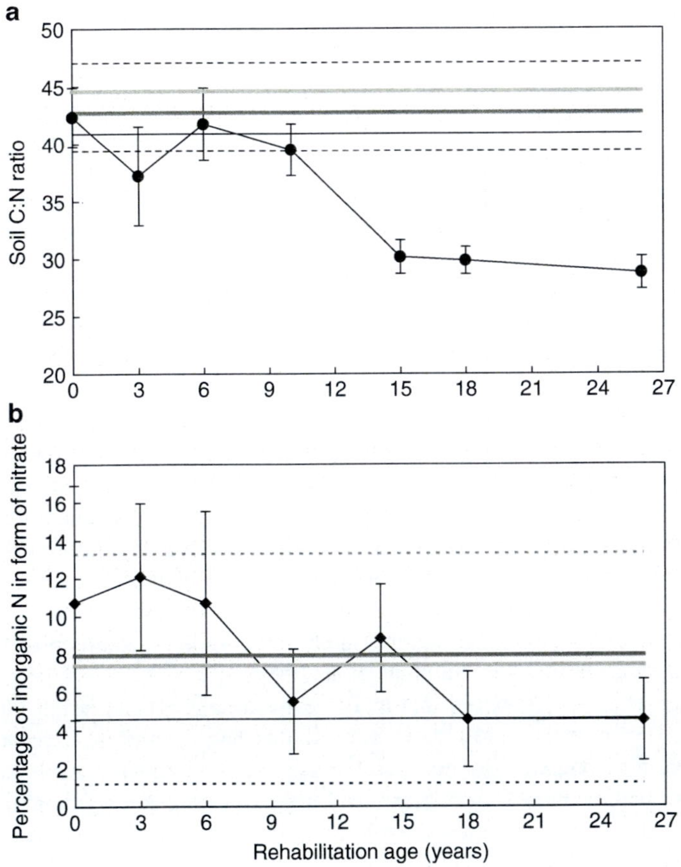

Fig. 13.5 Changes in soil C:N ratio (**a**) and proportion of inorganic N present as nitrate (**b**) in the surface 0–5 cm of a chronosequence of post-mining jarrah forest rehabilitation topsoil. Mean values ($n = 5$) ± 1SE are shown. *Horizontal lines* represent means from the surrounding native forest from areas of three different Havel site-vegetation classifications: SP (*thick dark grey*), S (*thick light grey*) and TS (*thin black*). The two *dashed lines* represent +1SE from the upper and −1SE from the lower forest means. Data derived from Banning et al. (2008)

13.3.1.4 Prescribed Burning

Prescribed burning is conducted in the jarrah forest by the Department of Environment and Conservation during spring or autumn on a burn rotation between 5 and 12 years (Grant et al. 2007). Alcoa's completion criteria state that rehabilitated forest must have the potential to be integrated into these broad-scale prescription burning practices, and therefore must be resilient and able to exhibit self-recovery following fire. The jarrah forest that Alcoa is attempting to restore is very resilient to disturbance by fire, at least in terms of aboveground flora (Bell and Heddle 1989). However, fire is a major factor in the development of plant communities as some species of plants are intolerant of fire, particularly when they are young (Shea et al. 1979). Fire can also have important rejuvenating effects on soil health such as increased nutrient and cation availability and pH which can enhance tree seedling regeneration and plant litter and humus decomposition (Wardle et al. 1998). Prescribed burning of rehabilitation forests of 12–15 years of age is desirable due to the development of high fuel loads and a mid-storey of senescent legume species (Grant et al. 2007; Morley et al. 2004). Research has demonstrated that rehabilitation forest burnt at this age is highly resilient, in terms of vegetation structure and soil seed stores, and the recommended practice is to carry out low- to moderate-intensity spring burns in 13–15-year-old rehabilitation (Grant et al. 2007).

The impact of fire on belowground processes in forest ecosystems depends largely on the intensity of the fire and can be patchy across the landscape due to the heterogeneity in litter distribution (Certini 2005). High intensity fires (usually wildfire) can result in the death of soil biota, alteration of organic matter and net loss of nutrients, particularly of N via volatilisation and ash transport (Smith et al. 2008). In contrast, low intensity fires may increase nutrient availability via ash deposition and soil heating (Bauhus et al. 1993). However, low intensity fires still combust plant biomass and surface organic matter, which are likely to produce short-term (<1 year) changes in the quality and quantity of organic matter available for microbial utilisation and thereby alter the microbial community structure (Gonzalez-Perez et al. 2004; Hamman et al. 2007).

Cookson et al. (2008) found that in 12-year-old rehabilitated jarrah forest soil, total C and N contents were greater in areas that had been prescription burnt 12 months earlier compared to non-burnt areas. Soil microbial CLPPs were not affected, or had recovered, from prescribed fire suggesting belowground microbial function is also resilient to prescribed burning in rehabilitated jarrah forest. However, Banning and Murphy (2008) found that microbial heterotrophic activity in 12-year-old rehabilitated forest soil was less resistant to subsequent heat-induced disturbance when it had been prescribed burnt compared to when it had not. This was found to be related to differences in microbial community structure between the prescribed burnt and non-burnt soils. Thus, the frequency of fire events is likely to affect the resilience of the soil microbial community to heat-induced disturbance.

13.3.1.5 Stand Thinning

Establishment of jarrah and marri trees from seed in some rehabilitated sites has been extremely high in the past (3,000–8,000 stems ha^{-1}), well above the current target of 1,300 stems ha^{-1}, and is much higher than that of the surrounding native forest (average 600 stems ha^{-1}) (Grant et al. 2007). While achieving rapid vegetation cover is beneficial in the early stages of rehabilitation development, this practice can lead to increased competition for water and nutrients, understorey suppression and decreased tree stem height and diameter (Koch and Ward 2005). Since 2004, sites with stem densities above 2,500 stems ha^{-1} have been thinned at

9 months of age by herbicide spraying; however, it has been estimated that around 4,000 ha of rehabilitation forest has more than 2,500 stems ha^{-1} and requires manual thinning (Grant 2006).

Manual thinning can result in nutrient transfer from the aboveground canopy biomass to the soil surface, if the cut down trees are left on site. Green litter from thinned trees contains higher concentrations of nutrients that are more readily decomposed than litter returned to the forest floor after senescence (O'Connell 1997). Conversely, the woody residue left on the ground after harvesting is likely to be more recalcitrant than litter dominated by foliage, as it has a greater C:N ratio and contains less nutrients (O'Connell 1997). Therefore, the quantity and quality of organic matter and nutrients available to soil fauna and micro-organisms within thinned and non-thinned rehabilitated areas may vary markedly. This was the case in the study by Cookson et al. (2008) where stand thinning from 3,000–8,000 stems ha^{-1} to 600–800 stems ha^{-1} in rehabilitated jarrah forest resulted in a greater litter mass 16 months later and a shift in microbial community substrate utilisation patterns. An increased fuel load following thinning can also result in a higher fire risk and is an important consideration for prescribed burning of thinned rehabilitation. Grigg et al. (2010) recommended that for 10–13-year-old jarrah rehabilitation, prescribed burning in autumn should not be carried out following thinning. The study found that while the fire intensity during prescribed burning of thinned forest with an increased fuel load remained low to moderate, the reinstatement of a dense, well-aerated and elevated fuel layer poses a future fire risk. Thinning has also been found to affect soil N cycling up to at least 2 years post-treatment, by increasing the potential for nitrification, particularly when combined with prescription burning (Banning et al. 2007; Fig. 13.6). Further research is required to determine how much of the nitrate produced in surface soils in response to thinning, or burning and thinning combined, is leached beyond the plant root capture zone.

13.3.1.6 Regional Climate Change Implications

Under increased atmospheric CO_2 scenarios, rainfall is expected to decline in the southwest Western Australian region. Modelled mean streamflow from the tributaries entering the Swan River (located approximately 80 km north of the mined jarrah forest area) decreased by 12 and 24% with 1.5× CO_2 and 2× CO_2 scenarios (Evans and Schreider 2002). Mpelasoka et al. (2008) projected increases in soil-moisture-based drought frequency in southwest Western Australia by 2030 relative to the 1975–2004 baseline period. This has implications for the capture of drinking water as the jarrah forest is used as a catchment for water for metropolitan Perth, the capital of Western Australia. Alcoa's mining operations are mainly in the higher rainfall zone to the west of the 1,100 mm/annum rainfall isohyet (Fig. 13.1). Paired catchment studies and modelling have demonstrated increases in streamflow volumes following vegetation removal during the mining phase followed by progressive decreases with rehabilitation age, below that of non-mined forest (Croton and Reed 2007). Rehabilitated forest areas have on average higher stand densities than the native

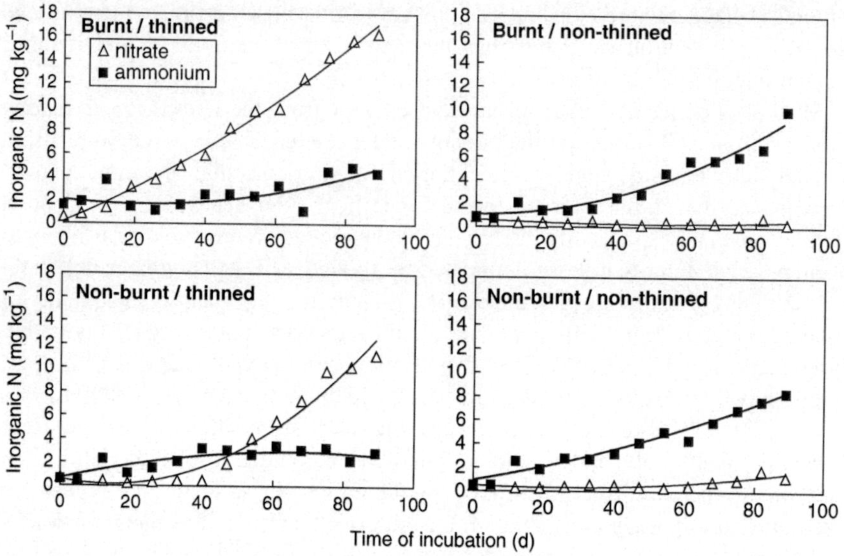

Fig. 13.6 Ammonium and nitrate accumulation during incubation of 12-year-old rehabilitation forest soil from replicated plots ($n = 3$) that had undergone stand thinning 2 years before sampling followed by prescription burning 6 months later (burnt/thinned), thinning only (non-burnt/thinned), burning only (burnt/non-thinned) or remained untreated (non-burnt/non-thinned)

jarrah forest (as described in above section) and there is little self-thinning. Thus, planting seedlings at lower density and thinning of already existing tree stands are being implemented or trialled. However, increases in stream flow volume in response to thinning of rehabilitated forest may be short-lived; a 4-year time period before return to pre-thinning flows was reported by Croton and Reed (2007). Thus, continued monitoring of the effect of thinning practices, not just on hydrological impacts but also on soil C and N cycling, will be required.

A further implication of a decline in annual rainfall in combination with increased temperatures is a potential increase in wildfire frequency or intensity (Hennessy et al. 2005). Although prescription burning is carried out in the jarrah forest (described in Sect. 13.3.1.4), this practice may not be able to mitigate an increased wildfire risk due to changes in climate. Thus, it is important to understand the resistance and resilience of rehabilitated forests to wildfire. In January 2006, a wildfire passed through the Willowdale region, burning areas of jarrah forest rehabilitation between 5 and 20 years old (all with a predominately jarrah and marri overstorey) as well as non-mined forest. Five months after the fire, soil (0–5 cm) was sampled from five replicate sites of burnt 7–8-year-old (1998–1999), 12–13-year-old (1993–1994) and 18–20-year-old (1986–1988) rehabilitation and non-mined forest. Non-burnt sites of the same rehabilitation age or equivalent forest type were selected in adjacent areas, north and south of the fire boundary. Non-mined non-burnt sites had not been burnt for at least 15 years before sampling. The fire was found to have caused significant declines in microbial

Table 13.2 Soil properties significantly ($P < 0.05$) altered by wild fire that occurred 5 months earlier in rehabilitated mine-sites (7–20 years old) and adjacent jarrah forest reference sites at Willowdale, Western Australia

	Microbial biomass C (mg kg^{-1})	Total C substrate utilisation (mg kg^{-1} ha^{-1})	Inorganic N (mg kg^{-1})	Available P (mg kg^{-1})	Soil pH ($CaCl_2$)
Mine-site rehabilitation[a]					
7–8 years old: non-burnt	253[b]	50	1	17[b]	4.7
7–8 years old: burnt	243[b]	29	18	16[b]	5.2
12–13 years old: non-burnt	518	86	2	13	4.6
12–13 years old: burnt	311	47	27	19	5.6
18–20 years old: non-burnt	615	93	6	7	4.6
18–20 years old: burnt	297	46	40	18	5.9
Forest reference					
Non-burnt	462	65	8	2	5.1
Burnt	278	48	30	17	6.1

[a]Properties of soil (0–5 cm) sampled from within furrows of contour rip lines shown. Soil sampled from mounds of contour rip lines showed a similar, but smaller, effect of burning (data not shown)
[b]No significant ($P > 0.05$) burning effect for 7–8-year-old rehabilitation

biomass and microbial activity, measured by response to C substrate addition, but significant increases in inorganic N, available P and soil pH (Table 13.2). The effect of the fire on soil properties was of similar direction and magnitude in the rehabilitation to the non-mined forest, with the exception of 7–8-year-old rehabilitation in which there was no difference between the microbial biomass or available P in burnt and non-burnt areas (Table 13.2). That the fire had little impact on soil microbial biomass in the youngest rehabilitation sites is most likely a function of lower fire intensity (due to lower aboveground biomass and plant litter on the soil surface). Although longer time recovery of these forest areas need monitoring, this study suggested that the wildfire impacts on rehabilitation forest soils correspond with those expected from a non-mined forest soil.

13.3.2 Case Study 2: Rehabilitation of Bauxite-Processing Residue Storage Areas

The extraction of alumina generates large quantities of bauxite-processing residue (i.e. non-soluble ore components), approximately two tonnes per three tonnes of bauxite. The three refineries operated by Alcoa in Western Australia (Kwinana, Pinjarra and Wagerup; shown in Fig. 13.1) each process up to 30,000 tonnes of bauxite per day. This quantity of highly alkaline waste poses a major management problem for the industry. Alcoa refineries fractionate their residue into coarse (bauxite residue sand) and fine (bauxite residue mud) fractions before deposition in residue storage areas. The sand (with a clay plus silt content of approximately 7%) is used to construct embankments that can be up to 80 m in height, and the residue mud (suspended in water with a solids content of about 55%) is then pumped into the storage areas. Establishing vegetation on operating residue storage area embankments is an essential component of progressive closure of residue storage areas, and critical for controlling dust emissions, managing alkaline-water discharge and minimising the visual and economic impact of the residue storage area on the surrounding community. It is planned that the residue sand will be used as the primary growing media for establishing a vegetation cover over the entire residue storage area upon closure of the facility. Over the expected life of the Western Australian refineries, more than 2,000 ha of residue storage area will require rehabilitation. To date, approximately 100 ha of Alcoa's residue storage area have been rehabilitated, with variable results and uncertain long-term maintenance requirements. Internationally, there is only very limited knowledge concerning the long-term growth and sustainability of residue rehabilitation (Courtney et al. 2009; Courtney and Timpson 2004; Wehr et al. 2006). As such, a fundamental understanding of the processes involved with plant–soil-–microbe–water–nutrient relationships in bauxite residue sand is required to provide effective, long-term rehabilitation of both operating and (future) closed residue storage areas.

13.3.2.1 Bauxite Residue Sand Amendment

Rehabilitation of bauxite residue sand represents a significant challenge as this material is highly alkaline (pH 10–12; primarily due to extremely high soluble carbonate and hydroxide concentrations), saline, extremely sodic and has poor water retention capacity ($K_{sat} \approx 20$ m day^{-1}). Residue sand contains almost nil or very low concentrations of organic C, total N, P and trace elements (Cu, Zn, Mn). The bauxite residue sand has essentially been sterilised during the refining process which involves digestion by concentrated hot sodium hydroxide. These conditions will constrain plant growth, limit biological activity and therefore nutrient cycling, and encourage leaching of nutrients during rainfall events.

An overview of rehabilitation practices is included in Fig. 13.2. Following embankment construction, current rehabilitation practice involves lowering the surface pH by the incorporation of phospho-gypsum (a phosphate-rich by-product of superphosphate fertiliser manufacturing) at 225 t ha^{-1} to approximately 1.5 m depth. Di-ammonium phosphate fertiliser (2,751 kg ha^{-1}; which includes a range of other nutrients such as K, Mg and trace elements) is also incorporated before seeding to initiate nutrient cycling. However, due to the alkalinity of residue sand, significant N loss through ammonia volatilisation is likely to occur (losses have been estimated to be up to 83% of the applied N in residue sand amended to a pH of 9 over 7 days; Chen et al. 2010). A 30 mm depth of wood mulch is applied to the sand surface for the purpose of dust suppression but is also likely to provide a significant source of colonising micro-organisms.

As the refining process produces a sterilised residue, a potential limiting factor in successful rehabilitation is the development of a diverse and functional microbial community in the relatively hostile geochemical conditions of residue sand. Interestingly, this does not seem to be the case as both bacterial and fungal diversity have been found to increase rapidly (within 6 months) in the bauxite residue sand (Fig. 13.7a; Banning et al. 2010). The microbial biomass in residue sand under rehabilitation up to 3 years of age has been found to remain very low (<50 mg

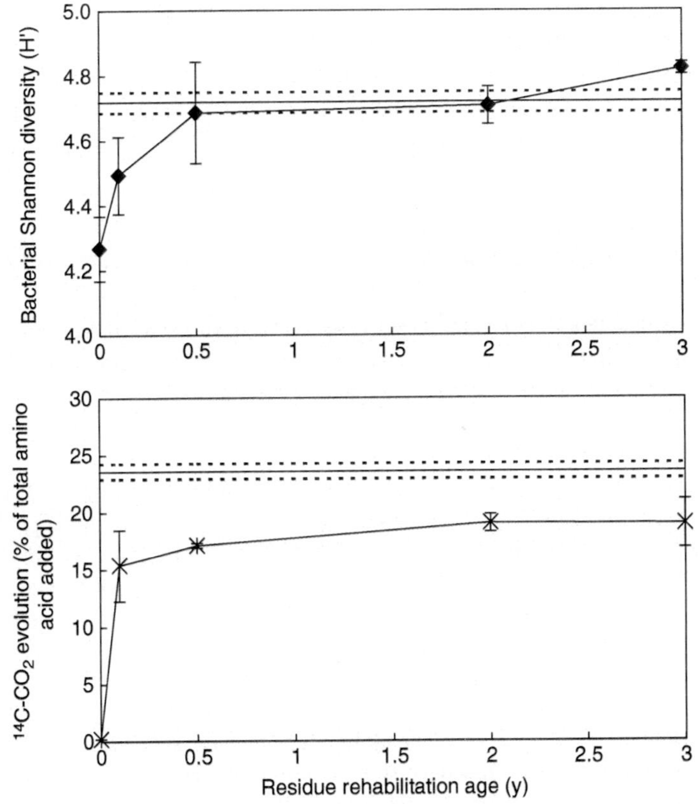

Fig. 13.7 Shannon diversity (H′) of bacterial communities derived from DNA-based amplified ribosomal intragenic spacer analysis (**a**) and amino acid mineralisation capacity measured by the amount of $^{14}CO_2$ evolved over 6 h following addition of a ^{14}C-labelled amino acid mixture (**b**) in bauxite residue sand from a pre-amendment state (0 year old) to 3 years post-amendment and rehabilitation under native vegetation. Mean values ($n = 3$) ± 1SE are shown. *Horizontal solid lines* represent mean values from a coastal sand analogue site with *dashed lines* showing ±1SE. For further methodological description, see Banning et al. (2010)

biomass C kg^{-1}), most likely due to considerable organic matter limitation. However, the microbial population present in residue sand rehabilitation, greater than 6 months of age, is able to rapidly mineralise added labile substrate (amino acids), indicating that there is no microbial constraint to this function occurring (Fig. 13.7b).

Use of organic manures and composts has also been trialled as an alternative fertiliser strategy to provide nutrients such as N, as well as providing the associated benefits of organic matter such as development of soil structure, improving water retention and providing micro-organisms (Jones et al. 2010). However, the success of compost additions in the field has been highly variable and trials to date have not indicated a benefit to plant growth (Eastham et al. 2006). One explanation may be that a C substrate-limited microbial population may benefit from organic additions to the detriment of early plant establishment. Although plants and micro-organisms are often mutually dependent upon one another, they also compete for the same nutrient pool, and this competition can negatively affect plant growth (Clode et al. 2009; Wardle 2002). Understanding the balance between competitive and mutually beneficial plant–microbe interactions in bauxite residue sand is therefore important for the development of management strategies that aim for sustainable vegetation growth.

13.3.2.2 Selection of Ecosystem Targets for Bauxite Residue Rehabilitation

Bauxite residue storage areas in Western Australia have been established on areas with naturally occurring acidic soils, and as such the surrounding native vegetation does not provide a suitable resource for plant species, and is unlikely to represent an achievable target ecosystem for rehabilitation goals. The residue storage areas are also surrounded by a variety of land uses including urban and agricultural. Thus, unlike the mining and rehabilitation practices in the jarrah forest, re-establishment of the ecosystem present before its removal is not a realistic goal. In the past, some residue sand embankments have been vegetated with grasses, although this has not been considered a suitable long-term "end-point" ecosystem for all residue storage areas. However, the strategy of initial grass establishment before further rehabilitation is under investigation (Fey et al. 2010).

An alternative strategy adopted by Alcoa was to attempt to identify an ecosystem analogue with similarities in general soil physical and chemical characteristics and climate (Outback Ecology 2006). The Quindalup Dune coastal system of southwest Western Australia was identified as providing a suitable analogue as it is comprised of calcareous sands (approximately 94% sand, 4% clay and 2% silt) resulting in a free-draining, relatively infertile, naturally alkaline (with a pH in water between 7 and 9) soil (McArthur 2004). It was established that around 40 plant species native to the Quindalup Dune sands were capable of growing in residue sand (Bell et al. 1993; Jasper et al. 2000). The seed mix currently used in residue sand embankment rehabilitation contains 55 plant species native to these coastal ecosystems. To what extent the residue sand rehabilitation is likely to become more similar to the natural analogue requires further investigation of plant-water dynamics, plant root distribution (Gwenzi et al. 2009), rates of nutrient cycling, and the presence and activity of specific functional microbial groups (e.g. mycorrhizal fungi, nitrifiers and N_2 fixers).

13.4 Mine-Site Rehabilitation Management in a Changing Climate

The case studies described in this chapter highlight the considerable impact of rehabilitation practices on shaping ecosystems' plant composition and diversity, soil structure, the spatial distribution of nutrients and belowground microbial diversity and function. Consequently, there is considerable scope within rehabilitation projects to adopt strategies that would allow early adaptation to changes in climate. An overview of the potential impacts of likely climate change scenarios on ecosystem attributes is given in Fig. 13.8. Climate change effects on the soil biota (see Chap. 8) and the processes they carry out may be direct (e.g. changes in soil temperature and water content) or indirect, through changes to the composition, productivity and structure of the vegetation (see Chap. 12). It has been suggested

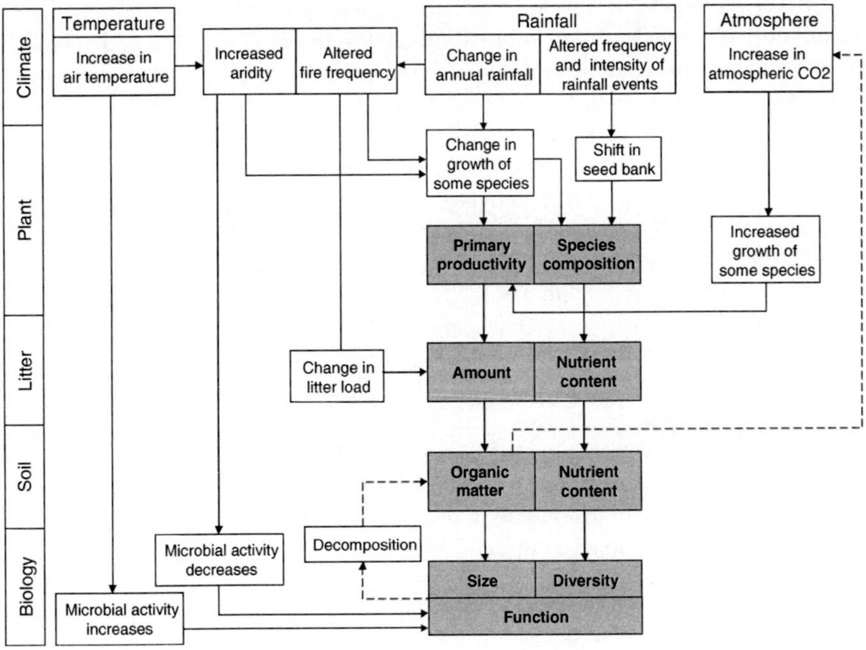

Fig. 13.8 Schematic overview of the potential impacts of likely climate change scenarios on ecosystem attributes

that the resulting feedback effects between soil biota, soil nutrient dynamics and the aboveground community will be important in determining the overall response of terrestrial ecosystems to climate change (Bardgett 2005). A key concern is that increases in air temperature may directly affect soil microbial activity, enhancing the turnover of soil organic C, leading to the release of CO_2 into the atmosphere (see Chap. 7) and potentially increased loss of dissolved organic C in run-off waters (Jenkinson et al. 1991). However, experimental studies have shown limited direct effects of small rises (1–3°C) in temperature on soil microbial activity (Bardgett 2005). In arid or semi-arid environments, it is likely that the combination of increases in temperature and decreases in rainfall may result in more frequent or more intense fire events, as highlighted in Fig. 13.8, which can have a large effect on many ecosystem properties (discussed earlier in this chapter).

The effects of elevated atmospheric concentrations of CO_2 on an ecosystem are most likely to occur via changes in the productivity and composition of plant communities, which in turn influence soil microbial growth, activity and community structure and therefore decomposition and nutrient supply rates (see Chaps. 7 and 8). An increase in plant productivity, and therefore plant-derived soil C, may lead to an increase in microbial mineralisation rates and the availability of nutrients such as N. However, growth of the microbial populations may also lead to greater N immobilisation in an N-limited environment and progressively greater N limitation. The experimental evidence on the effects of elevated atmospheric CO_2 concentrations is mixed (Bardgett 2005), and the complexity of the interactions and the aboveground and belowground feedbacks is such that general predictions are still difficult to make. More information on how these climate change scenarios are likely to affect specific vegetation assemblages and the ability of an ecosystem to acclimate or adapt will be needed to implement planning for climate change into rehabilitation practices (Harris et al. 2005).

Nonetheless, it may be possible for rehabilitation programmes to adopt the broad principle of maximising genetic diversity among and within plant species, a concept referred to as "future-proofing" (Harris et al. 2006). This would be a fundamental shift for many rehabilitation strategies which commonly use only species endemic to a region in recognition of the fact that local species are in general better adapted to the local environment and also provide better habitat to allow the return of native fauna (Harris et al. 2006; Koch and Samsa 2007). Furthermore, maintenance of the genetic integrity of the site has been considered essential to recovering the conservation value of an ecosystem (Koch and Hobbs 2007). However, the ability of rehabilitation practitioners to introduce a high diversity of plant species at the early establishment stage rather than relying on natural recruitment and succession, which commonly relies on only a few "coloniser" plants, could improve the ability of the ecosystem to adapt to future climate changes (Rice and Emery 2003). A balance would need to be achieved between the introduction of local populations that are adapted to the local environment (with consideration given to the genetic variation within the source populations) and the introduction of new species and populations that aim to increase the adaptive capacity of the ecosystem.

Another broad principle for rehabilitation programmes that is particularly significant in the context of climate change is the adoption of practices to maximise soil organic C. Many disturbed or degraded soils have a lower organic C than the potential organic C content of a soil as determined by factors such as the soil type and mineralogy, depth and bulk density as well as net primary productivity and aboveground vs. belowground allocation (Ingram and Fernandes 2001). For mine-sites, this can be achieved through the types of soil handling and rehabilitation practices described in this chapter, as well as potentially though the addition of organic amendments such as composts, biosolids or other waste organic materials (Farrell et al. 2011). Maximising the genetic diversity of plants during the rehabilitation of ecosystems may also affect C capture, as it has been suggested, although not explicitly tested, that there may be a positive relationship between plant diversity and plant productivity (and consequently soil C supply) under elevated CO_2 conditions (Bardgett 2005).

Future research will need to explore whether it may be advantageous to extend the principle of maximising genetic diversity in rehabilitated ecosystems to the belowground community in addition to the aboveground plant community. While it has been established that shifts in microbial community structure or diversity can occur in response to ecosystem disturbances and rehabilitation practices or land-use change, to what extent these shifts affect microbial function and ecosystem stability is not well understood (Girvan et al. 2005; Wardle et al. 2000). Furthermore, with the exception of cases where microbial inoculants are added to a soil as part of a rehabilitation strategy (e.g. Krishna et al. 2005), there are no mechanisms for the direct manipulation of the whole soil microbial community. However, knowledge of the dominant edaphic factors driving microbial community structure and maintaining diversity, such as pH (Fierer and Jackson 2006; Hackl et al. 2004) and organic matter quantity/quality, which in turn is related to plant community composition and diversity (Grayston et al. 2004; Zhou et al. 2002), mineralogy (Gleeson et al. 2006), and their interactive effects may enable rehabilitation strategies to be used which indirectly manipulate microbial community structure and diversity.

13.5 Conclusions

Traditionally, ecosystem rehabilitation aspires to return the ecosystem that was present before mining or other anthropogenic disturbance. Where there is no pre-disturbance ecosystem, an analogue may be chosen which possesses attributes of the desired rehabilitated ecosystem. Criteria or benchmarks for measuring the success of rehabilitation efforts are developed based on attributes of the reference sites in the target ecosystem, such as plant species composition and richness or soil nutrient pools. However, if rehabilitation projects take into account predicted future climate change scenarios, this approach may not be sufficient to ensure long-term ecosystem sustainability. Consideration needs to be given to maximising aboveground, and potentially belowground, genetic diversity in rehabilitated ecosystems, beyond the diversity present pre-disturbance, to improve the ecosystem's adaptive capacity to future climate changes. Furthermore, measuring the success of rehabilitation requires not only comparison to attributes of a reference or analogue ecosystem, but also measurement and understanding of the ecosystem processes and functions that engender ecosystem sustainability. As soil micro-organisms are responsible for the bulk of nutrient cycling, measurement of processes such as soil organic C and N cycling and characterisation of the populations involved in carrying out these functions are fundamental to determining the sustainability of rehabilitated ecosystems. There are many rehabilitation practices, as outlined in this chapter, which shape the final attributes of a rehabilitated ecosystem, both aboveground and belowground. Therefore, there is scope to adopt rehabilitation strategies that maximise the sustainability of rehabilitated ecosystems and their adaptive capacity to changes in climate.

Acknowledgements We would like to acknowledge financial support from the Australian Research Council Linkage Grants scheme with industry funding from Alcoa of Australia Ltd. (project numbers LP0454127 and LP0776593). We thank Yoshi Sawada and Anna Byrne for valued technical support and reviewers of this chapter for their comments.

References

Abbott I, Loneragan WA (1986) Ecology of Jarrah (*Eucalyptus marginata*) in the northern Jarrah forest of Western Australia. Department of Conservation and Land Management Bulletin, Perth

Balser TC, Firestone MK (2005) Linking microbial community composition and soil processes in a California annual grassland and mixed-conifer forest. Biogeochemistry 73:395–415

Banning NC, Phillips IR, Jones DL, Murphy DV (2011) Development of microbial diversity and functional potential in bauxite residue sand under rehabilitation. Restor Ecol 19:78–87

Banning NC, Murphy DV (2008) Effect of heat-induced disturbance on microbial biomass and activity in forest soil and the relationship between disturbance effects and microbial community structure. Appl Soil Ecol 40:109–119

Banning NC, Willers J, Grant CD, Grigg A, Murphy DV (2007) Combined thinning and burning impacts soil N cycling in rehabilitation jarrah (*Eucalyptus marginata*) forest. In: Xu ZH et al (eds) The proceedings of the international symposium on forest soils and ecosystem health. Centre for Forestry and Horticultural Research, Griffith University, Brisbane, pp 13–14

Banning NC, Grant CD, Jones DL, Murphy DV (2008) Recovery of soil organic matter, organic matter turnover and nitrogen cycling in a post-mining forest rehabilitation chronosequence. Soil Biol Biochem 40:2021–2031

Bardgett RD (2005) The biology of soil: a community and ecosystem approach. Oxford University Press, New York

Barton L, Kiese R, Gatter D, Butterbach-Bahl K, Buck R, Hinz C, Murphy DV (2008) Nitrous oxide emissions from a cropped soil in a semi-arid climate. Glob Chang Biol 14:177–192

Bauhus J, Khanna P, Raison RJ (1993) The effect of fire on carbon and nitrogen mineralization and nitrification in an Australian forest soil. Aust J Soil Res 31:621–639

Bell LC (2001) Establishment of native ecosystems after mining -Australian experience across diverse biogeographic zones. Ecol Eng 17:179–186

Bell DT, Heddle EM (1989) Floristic, morphologic and vegetational diversity. In: Dell B et al (eds) The Jarrah Forest: a complex mediterranean ecosystem. Kluwer Academic, Dordrecht

Bell DT, Wilkins CF, Moezel PG, Ward SC (1993) Alkalinity tolerance of woody species used in bauxite waste rehabilitation, Western Australia. Restor Ecol 1:51–58

Cairns J Jr (2000) Setting ecological restoration goals for technical feasibility and scientific validity. Ecol Eng 15:171–180

Certini G (2005) Effects of fire on properties of forest soils: a review. Oecologia 143:1–10

Chapin FS, Walker LR, Fastie CL, Sharman LC (1994) Mechanisms of primary succession following deglaciation at Glacier Bay, Alaska. Ecol Monogr 64:149–175

Charles SP, Bates BC, Whetton PH, Hughes JP (1999) Validation of downscaling models for changed climate conditions: case study of southwestern Australia. Climate Res 12:1–14

Chen C, Phillips I, Wei L, Xu Z (2010) Behaviour and dynamics of di-ammonium phosphate in bauxite processing residue sand in Western Australia – I. NH_3 volatilisation and residual nitrogen availability. Environ Sci Pollut Res Int 17:1098–1109

Clode PL, Kilburn MR, Jones DL, Stockdale EA, Cliff JB, Herrmann AM, Murphy DV (2009) In situ mapping of nutrient uptake in the rhizosphere using nanoscale secondary ion mass spectrometry. Plant Physiol 151:1751–1757

Colquhoun IJ, Kerp NL (2007) Minimizing the spread of a soil-borne plant pathogen during a large-scale mining operation. Restor Ecol 15:S85–S93

Cookson WR, Abaye DA, Marschner P, Murphy DV, Stockdale EA, Goulding KWT (2005) The contribution of soil organic matter fractions to carbon and nitrogen mineralization and microbial community size and structure. Soil Biol Biochem 37:1726–1737

Cookson WR, Osman M, Marschner P, Abaye DA, Clark I, Murphy DV, Stockdale EA, Watson CA (2007) Controls on soil nitrogen cycling and microbial community composition across land use and incubation temperature. Soil Biol Biochem 39:744–756

Cookson WR, O'Donnell AJ, Grant CD, Grierson PF, Murphy DV (2008) Impact of ecosystem management on microbial community level physiological profiles of post-mining forest rehabilitation. Microb Ecol 55:321–332

Courtney RG, Timpson JP (2004) Nutrient status of vegetation grown in alkaline bauxite processing residue amended with gypsum and thermally dried sewage sludge – a two year field study. Plant Soil 266:187–194

Courtney R, Mullen G, Harrington T (2009) An evaluation of revegetation success on bauxite residue. Restor Ecol 17:350–358

Croton JT, Reed AJ (2007) Hydrology and bauxite mining on the Darling Plateau. Restor Ecol 15: S40–S47

de Groot RS, Wilson MA, Boumans RMJ (2002) A typology for the classification, description and valuation of ecosystem functions, goods and services. Ecol Econ 41:393–408

Degens BP (1998) Microbial functional diversity can be influenced by the addition of simple organic substrates to soil. Soil Biol Biochem 30:1981–1988

Degens BP, Schipper LA, Sparling GP, Duncan LC (2001) Is the microbial community in a soil with reduced catabolic diversity less resistant to stress or disturbance? Soil Biol Biochem 33:1143–1153

Dell B, Havel JJ, Malajczuk N (1989) The Jarrah forest: a complex mediterranean ecosystem. Kluwer Academic, Dordrecht

Eastham J, Morald T, Aylmore P (2006) Effective nutrient sources for plant growth on bauxite residue. Water Air Soil Pollut 176:5–19

Evans J, Schreider S (2002) Hydrological impacts of climate change on inflows to Perth, Australia. Clim Change 55:361–393

Farrell M, Healey JR, Godbold DL, Nason MA, Tandy S, Jones DL (2011) Modification of fertility of soil materials for restoration of acid grassland habitat. Restor Ecol (in press)

Fey M, Kaur N, Phillips I (2010) Inorganic pathway to soil-like conditions in alkaline residue from bauxite refining. 19th World Congress of Soil Science. International Union of Soil Sciences, Brisbane

Fierer N, Jackson RB (2006) The diversity and biogeography of soil bacterial communities. Proc Natl Acad Sci USA 103:626–631
Fitter AH, Gilligan CA, Hollingworth K, Kleczkowski A, Twyman RM, Pitchford JW (2005) Biodiversity and ecosystem function in soil. Funct Ecol 19:369–377
Gardner JH, Bell DT (2007) Bauxite mining restoration by Alcoa World Alumina Australia in Western Australia: social, political, historical, and environmental contexts. Restor Ecol 15: S3–S10
Girvan MS, Campbell CD, Killham K, Prosser JI, Glover LA (2005) Bacterial diversity promotes community stability and functional resilience after perturbation. Environ Microbiol 7:301–313
Gleeson D, McDermott F, Clipson N (2006) Structural diversity of bacterial communities in a heavy metal mineralized granite outcrop. Environ Microbiol 8:383–393
Glen M, Bougher NL, Colquhoun IJ, Vlahos S, Loneragan WA, O'Brien PA, Hardy GESJ (2008) Ectomycorrhizal fungal communities of rehabilitated bauxite mines and adjacent, natural jarrah forest in Western Australia. For Ecol Manage 255:214–225
Gonzalez-Perez JA, Gonzalez-Vila FJ, Almendros G, Knicker H (2004) The effect of fire on soil organic matter–a review. Environ Int 30:855–870
Grant CD (2006) State-and-transition successional model for bauxite mining rehabilitation in the jarrah forest of Western Australia. Restor Ecol 14:28–37
Grant CD, Norman MA, Smith MA (2007) Fire and silvicultural management of restored bauxite mines in Western Australia. Restor Ecol 15:S127–S136
Grayston SJ, Campbell CD, Bardgett RD, Mawdsley JL, Clegg CD, Ritz K, Griffiths BS, Rodwell JS, Edwards SJ, Davies WJ, Elston DJ, Millard P (2004) Assessing shifts in microbial community structure across a range of grasslands of differing management intensity using CLPP, PLFA and community DNA techniques. Appl Soil Ecol 25:63–84
Grigg AH, Norman MA, Grant CD (2010) Prescribed burning of thinning slash in regrowth stands of jarrah (*Eucalyptus marginata*) following bauxite mining in south-west Australia. Int J Wildland Fire 19:737–745
Gwenzi W, Veneklaas EJ, Phillips I, Bleby TM, Hinz C (2009) Spatial distribution of fine roots on a rehabilitated bauxite residue disposal area in Western Australia. In: Fourie AB, Tibbett M (eds) Mine closure 2009. Australian Centre for Geomechanics, Perth, pp 317–327
Hackl E, Zechmeister-Boltenstern S, Bodrossy L, Sessitsch A (2004) Comparison of diversities and compositions of bacterial populations inhabiting natural forest soils. Appl Environ Microbiol 70:5057–5065
Hamman ST, Burke IC, Stromberger ME (2007) Relationships between microbial community structure and soil environmental conditions in a recently burned system. Soil Biol Biochem 39:1703–1711
Harris JA (2003) Measurements of the soil microbial community for estimating the success of restoration. Eur J Soil Sci 54:801–808
Harris JA, Grogan P, Hobbs RJ (2005) Restoration ecology and the role of soil biodiversity. In: Bardgett RD et al (eds) Biological diversity and function in soils. Cambridge University Press, Cambridge, pp 319–342
Harris JA, Hobbs RJ, Higgs E, Aronson J (2006) Ecological restoration and global climate change. Restor Ecol 14:170–176
Havel J (1975) Site-vegetation mapping in the norther jarrah forest (Darling Range). 1. Definition of site-vegetation types. Bulletin 86. Forest Department Western Australia
Hennessy K, Lucas C, Nicholls N, Bathols J, Suppiah R, Ricketts J (2005) Climate change impacts on fire weather in South-eastern Australia. CSIRO Marine and Atmospheric Research, Aspendale
Hingston FJ, Malajczuk N, Grove TS (1982) Acetylene reduction (N_2-fixation) by jarrah forest legumes following fire and phosphate application. J Appl Ecol 19:631–645
Hobbs RJ, Harris JA (2001) Restoration ecology: repairing the earth's ecosystems in the new millennium. Restor Ecol 9:239–246

Hobbs RJ, Norton DA (1996) Towards a conceptual framework for restoration ecology. Restor Ecol 4:93–110
Ingram JSI, Fernandes ECM (2001) Managing carbon sequestration in soils: concepts and terminology. Agric Ecosyst Environ 87:111–117
Jasper DA (2007) Beneficial soil microorganisms of the jarrah forest and their recovery in bauxite mine restoration in Southwestern Australia. Restor Ecol 15:S74–S84
Jasper D, Lockley I, Ward S, White A (2000) Current approaches and future challenges for rehabilitation of bauxite residue. In: Bell RW, Brion A (Eds) Proceedings of Remade Lands 2000, International Conference on the Remediation and Management of Degraded Lands, 30th Nov.-2nd Dec., Fremantle, Australia, pp 71–72
Jenkinson DS, Adams DE, Wild A (1991) Model estimates of CO_2 emissions from soil in response to global warming. Nature 351:304–306
Jones BEH, Haynes RJ, Phillips IR (2010) Effect of amendment of bauxite processing sand with organic materials on its chemical, physical and microbial properties. J Environ Manage 91:2281–2288
Killham K (1994) Soil ecology. Cambridge University Press, Cambridge
Koch J (1987) Nitrogen accumulation in a rehabilitated bauxite-mined area in the Darling Range, Western Australia. Aust For Res 17:59–72
Koch JM (2007) Alcoa's mining and restoration process in South Western Australia. Restor Ecol 15:S11–S16
Koch JM, Hobbs RJ (2007) Synthesis: is Alcoa successfully restoring a jarrah forest ecosystem after bauxite mining in Western Australia? Restor Ecol 15:S137–S144
Koch JM, Samsa GP (2007) Restoring Jarrah forest trees after bauxite mining in Western Australia. Restor Ecol 15:S17–S25
Koch JM, Ward SC (2005) Thirteen year jarrah (*Eucalyptus marginata*) growth on restored bauxite mines in south-west Australia. Aust For 68:176–185
Krishna P, Sudhakara Reddy M, Patnaik SK (2005) *Aspergillus tubingensis* reduces the pH of the bauxite residue (red mud) amended soils. Water Air Soil Pollut 167:201
Lalor BM, Cookson WR, Murphy DV (2007) Comparison of two methods that assess soil community level physiological profiles in a forest ecosystem. Soil Biol Biochem 39:454–462
Li J, Duggin JA, Loneragan WA, Grant CD (2006) Grassland responses to multiple disturbances on the New England Tablelands in NSW, Australia. Plant Ecol 193:39–57
Ludwig JA, Tongway DJ, Bastin GN, James CD (2004) Monitoring ecological indicators of rangeland functional integrity and their relation to biodiversity at local to regional scales. Austral Ecol 29:108–120
McArthur WM (2004) Reference soils of south-western Australia. Department of Agriculture, Perth
Morley S, Grant C, Hobbs R, Cramer V (2004) Long-term impact of prescribed burning on the nutrient status and fuel loads of rehabilitated bauxite mines in Western Australia. For Ecol Manage 190:227–239
Mpelasoka F, Hennessy K, Jones R, Bates B (2008) Comparison of suitable drought indices for climate change impacts assessment over Australia towards resource management. Int J Climatol 28:1283–1292
Myers N, Mittermeier RA, Mittermeier CG, da Fonseca GAB, Kent J (2000) Biodiversity hotspots for conservation priorities. Nature 403:853–858
Neher D (1999) Soil community composition and ecosystem processes: comparing agricultural ecosystems with natural ecosystems. Agroforest Syst 45:159–185
Norman MA, Koch JM, Grant CD, Morald TK, Ward SC (2006) Vegetation succession after bauxite mining in Western Australia. Restor Ecol 14:278–288
O'Connell AM (1997) Decomposition of slash residues in thinned regrowth eucalypt forest in Western Australia. J Appl Ecol 34:111–122
Odum EP (1969) The strategy of ecosystem development. Science 164:262–270
Odum WE, Odum EP, Odum HT (1995) Nature's pulsing paradigm. Estuaries 18:547–555

Orwin KH, Wardle DA, Greenfield LG (2006) Ecological consequences of carbon substrate identity and diversity in a laboratory study. Ecology 97:580–593

Outback Ecology (2006) Coastal soils and vegetation communities as analogues for bauxite residue rehabilitation. Unpublished report for Alcoa of Australia, March 2006

Rice KJ, Emery NC (2003) Managing microevolution: restoration in the face of global change. Front Ecol Environ 1:469–478

Sawada Y (1996) Indices of microbial biomass and activity to assess minesite rehabilitation, 3rd International and the 21st Annual Minerals Council of Australia Environmental Workshop. Vol. 1. Minerals Council of Australia, Newcastle, Australia, pp 223–236

Shea SR, McCormick J, Portlock CC (1979) The effect of fires on regeneration of leguminous species in the northern Jarrah forest of Western Australia. Aust J Ecol 4:195–205

Smith NR, Kishchuk BE, Mohn WW (2008) Effects of wildfire and harvest disturbances on forest soil bacterial communities. Appl Environ Microbiol 74:216–224

Society for Ecological Restoration International Science and Policy Working Group (2004) The SER International Primer on Ecological Restoration Society for Ecological Restoration International, Tuscon

Todd MCL, Adams MA, Grierson PF (2000) Mineralisation of nitrogen in a chronosequence of rehabilitated bauxite mines. Aust J Soil Res 38:435–451

Torsvik V, Ovreas L, Thingstad TF (2002) Prokaryotic diversity – magnitude, dynamics, and controlling factors. Science 296:1064–1066

Waldrop MP, Firestone MK (2006) Response of microbial community composition and function to soil climate change. Microb Ecol V52:716–724

Walker L, del Moral R (2003) Primary succession and ecosystem rehabilitation. Cambridge University Press, Cambridge

Ward SC (2000) Soil development on rehabilitated bauxite mines in south-west Australia. Aust J Soil Res 38:453–464

Ward SC, Pickersgill GE (1985) Biomass and nutrient distribution in eucalypt plantation growing on rehabilitated bauxite mines. Aust J Ecol 10:111–124

Wardle DA (2002) Communities and ecosystems: linking the aboveground and belowground components. Princeton University Press, New Jersey

Wardle DA, Zackrisson O, Nilsson M (1998) The charcoal effect in Boreal forests: mechanisms and ecological consequences. Oecologia 115:419–426

Wardle DA, Bonner KI, Barker GM (2000) Stability of ecosystem properties in response to aboveground functional group richness and composition. Oikos 89:11–23

Wehr JB, Fulton I, Menzies NW (2006) Revegetation strategies for bauxite refinery residue: a case study of Alcan Gove in Northern Territory, Australia. Environ Manage 37:297–306

Westman WE (1986) Resilience: concepts and measures. In: Dell B, Hopkins AJM, Lamont BB (eds) Resilience in mediterranean-type ecosystems. Dr W Junk Publishers, Dordrecht, pp 5–19

White PS, Walker JL (1997) Approximating nature's variation: selecting and using reference information in restoration ecology. Restor Ecol 5:338–349

Williams K, Mitchell D (2001) Jarrah forest 1 (JF1 – Northern Jarrah forest subregion), A Biodiversity Audit of Western Australia's 53 Biogeographical Subregions in 2002. Department of Environment and Conservation, Perth, pp 369–381

Yin B, Crowley D, Sparovek G, De Melo WJ, Borneman J (2000) Bacterial functional redundancy along a soil reclamation gradient. Appl Environ Microbiol 66:4361–4365

Zhou JZ, Xia BC, Treves DS, Wu LY, Marsh TL, O'Neill RV, Palumbo AV, Tiedje JM (2002) Spatial and resource factors influencing high microbial diversity in soil. Appl Environ Microbiol 68:326–334

Part IV
Organic Farming, Biochar and Bioenergy Systems

Chapter 14
Organic Farming: Pros and Cons for Soil Health and Climate Change

Elizabeth Stockdale

14.1 Introduction

Soil health was a key founding principle for the development and practice of organic farming, as evident in one of its original precepts (here in the words of Lady Eve Balfour 1943) that "the health of soil, plant, animal and man is one and indivisible". Modern organic farming represents a merging of different streams of thought from a number of sources (Paster and Boeringa 1980; Merrill 1983; Harwood 1990; Tate 1994; Conford 2001; Heckman 2006). Briefly, the early organic movement focused strongly on issues of human nutrition and health, as well as the promotion of soil fertility through the use of composts and other organic fertilisers. Pesticides did not become a major issue in organic agriculture until the publication of "Silent Spring" (Carson 1963) generated widespread public concern. The Limits to Growth report of the Club of Rome, and the energy crisis of 1973 drew attention to the sustainability of resource use (Lockeretz 1990). During the 1980s and 1990s, other issues also increased in importance, in particular nature and biodiversity conservation, animal welfare, social justice issues relating to fair trade with developing countries and most recently the potential of organic agriculture to contribute to rural development and reductions in greenhouse gas emissions.

As described in this book, increasing concentrations of CO_2 and other greenhouse gases in the Earth's troposphere are leading to global warming and changes in precipitation patterns, which will impact soil processes and agricultural systems (IPCC 2007). The changing climate, together with the changes in global markets, is driving short to medium-term modifications in farming systems at local and regional levels. It is therefore timely to consider the implications of climate change for maintenance and enhancement of productivity and soil health within organic farming systems. In this chapter, I will also consider what role organic farming

E. Stockdale (✉)
School of Agriculture, Food and Rural Development, Newcastle University, Agriculture Building, Newcastle Upon Tyne NE1 7RU, UK
e-mail: e.a.stockdale@ncl.ac.uk

B.P. Singh et al. (eds.), *Soil Health and Climate Change*, Soil Biology 29,
DOI 10.1007/978-3-642-20256-8_14, © Springer-Verlag Berlin Heidelberg 2011

systems can play to support mitigation and adaptation actions in response to climate change, as a result of either increased expansion of organic farming systems or adoption of practices from organic farming by conventional farming systems.

14.2 Definition of Organic Farming Systems

Intensification of conventional agricultural production over the last century has been marked by increased use of mechanical and manufactured inputs and increased specialisation of production and has resulted in spectacular increases in productivity. In contrast and almost contemporaneously, the organic farming movement has developed principles and recommendations for farm management from a biological/ecological conception of nature reflecting an underpinning recognition of the importance of the relationships and interactions between organisms – plant and animals, both aboveground and within the soil. Consequently, regulation of the agro-ecosystem through biological processes is central to organic farming systems, while these processes have become somewhat marginalised in the management of conventional farming systems (Giller et al. 1997). The objectives of environmental, social and economic sustainability at local and global scales lie at the heart of the underpinning principles governing organic farming globally (IFOAM 2005; Kristiansen et al. 2006). In parallel with the global agreement of guiding principles, organic farming systems are now legally defined by reference to production standards in many countries (Vogl et al. 2005; Courville 2006). These standards allow organic products to be assured within the global food market.

As organic farming has developed, acceptable production practices have been recorded in technical guides and handbooks. Detailed descriptions of the practices of organic farming are available (e.g. Lampkin 1990; Siebeneicher 1993; Kristiansen et al. 2006), and therefore I will give only a simplified overview here. In organic cropping systems, rotations/mixtures are designed with a strong awareness of the crops' impacts on soil physical condition and their biological interactions. The use of manufactured fertilisers is prohibited (or at least significantly restricted), and hence there is a focus on the inclusion of N-fixing crops and nutrient recycling to reduce the need for external inputs. Consequently in comparison to conventional systems where nutrients are often supplied in immediately plant-available forms, organic farming systems place a greater reliance on chemical and biological processes within the soil to release nutrients in forms available for plant uptake (Watson et al. 2002a) and standards often refer specifically to the need to maintain the biological activity of the soil, alongside fertility (e.g. Defra 2006). Crop health is maintained through complex interactions and feedback among soil, crops, pests and inputs. Crop rotation and cultivation practices have been identified by farmers as key disease control strategies in organic farming systems (Park and Lohr 2005), for example, by separating crop hosts so that soil-borne pathogen levels are diminished. Pest control strategies are largely preventative, rather than reactive;

the balance of cropped and uncropped areas, crop species and variety choice and the temporal and spatial pattern of the crop rotation seek to maintain a diverse population of pests and their natural enemies and disrupt the life cycle of pest species. Weeds are controlled at manageable levels by managing the rotation (crop choice, timing of sowing, cultivations, etc.) to ensure that direct control measures (e.g. hand weeding) can succeed in preventing crop losses (Barberi 2002; Turner et al. 2007). Consequently, in comparison to conventional systems where external inputs are often the primary tactical response to pest, disease and weed control, in organic farming systems external inputs are permitted only as supplementary tools but only to particular crops and with prior independent consent.

The implementation of the principles of organic farming across the world in diverse climates leads to a great variety in the types of farming systems which produce organic food. The resulting products include all types ranging from vegetables, meat, bread and milk to organic cola and ready meals. During the 1990s, policy support for organic farming increased globally as a result of a gradual convergence of policy goals for land management with the underlying objectives of organic agriculture, including environmental protection, animal welfare, resource use sustainability, food quality and safety, financial viability and social justice (Lampkin et al. 1999). Recent years have seen very rapid growth in organic farming, particularly in Europe and the USA, but also in many other regions of the world including China, Latin America and Africa. Worldwide, organic agriculture occupies 31 million ha of certified crop and pasture lands and more than 62 million ha of certified wild harvested areas (FAO 2007). However, data showing the growth of the organically farmed area or the size of the market for organic food globally hide great variability within and between countries, as a result of the regional distribution of farm types and current farming intensity. Where farming practice is low intensity with relatively few external inputs, conversion is also often less onerous. Equally there are, in general, a larger proportion of mixed farms managed organically than specialised dairy or arable units; mixed farms are able to integrate more easily the practices required by the organic standards.

In the late twentieth century, increasing awareness of the environmental impact of conventional farming systems (e.g. the impact of pesticides on wildlife or fertilisers on water supplies, loss of soil through erosion) led to the development of modified practices and changes within all farming systems. Doran (2002) grouped these approaches into four key integrating management strategies that are likely to improve soil health and increase sustainability: maintaining soil organic matter; minimising soil erosion; balancing production and environmental outcomes; and improving utilisation of renewable resources (see Chap. 1 also). Reduced use, rather than elimination, of certain chemical inputs is a key factor distinguishing low-input sustainable agriculture (USA) and integrated farming systems (Europe) from organic farming; reduced tillage approaches are more common in alternative conventional than in organic systems. The requirements for legal definition in the marketplace and associated requirements for robust independent certification also provide a clear dividing line between organic food systems and other alternative farming systems. The issue of the relative contributions of organic, low-input and integrated systems to soil health and their role in adapting to and mitigating climate change is outside the scope of this review; where comparisons are made, it is prevalent current conventional practices which are compared with organic farming.

14.2.1 Impacts of Organic Farming on Crop Yields

Comparison of organic and conventional farming and food systems is fraught with difficulties – there is no clear definition of what conventional is, and even within legally defined organic systems there is significant variation in practices. Before 2000, most studies seemed to suggest that in temperate cropping systems, where most studies had been carried out, organic arable crops yielded around 60–80% those of conventional systems (see review by Stockdale et al. 2001). More recent studies in cropping systems trials suggest that organic systems can be as productive as comparable conventional systems (Posner et al. 2008). However, yields in organic systems are more weather dependent and in organic row crops (such as maize, soybean) only approach conventional yields in years, where all the planned mechanical weed control opportunities were not disrupted by weather conditions (Posner et al. 2008). Yields of forage crops are often similar to conventional crops with no difference in feed quality (Stockdale et al. 2001; Posner et al. 2008). Where organic management practices have significantly altered soil conditions, then yield increases are also seen when compared with conventional production. For example, where compost additions almost doubled soil organic matter content, vegetable production in intensive organic irrigated Mediterranean increased significantly (Melero et al. 2006). In developing countries, the UNDP (1992) concluded that organic farming methods seem able to provide outputs, with fewer external resources, supplying a similar income per labour day similar to those of high input conventional approaches where bought in inputs often substitute for labour. Studies commonly show large increases in yield, compared with current cropping practices, where local farmers adopt organic farming systems, reaching levels similar to those of high input systems in tropical environments (FAO 2007). Direct comparisons of yields are often difficult outside controlled trials because of the differences in the farming systems adopted under high input compared with organic management, for example cultivar selection, rotation design and nutrient management, length of time under organic management, and differences between farms in the management skill of the farmer. Furthermore, absolute yields are subject to considerable annual and site to site variability due to disease and pest outbreaks, soil type, seasonal weather patterns, etc. However, at the scale of cropping systems, organic systems often contain fewer cash crops, due to the use of land for green manures and increased integration of livestock into cropping systems; this reduces absolute yields from organic farming systems relative to those from conventional production (see detailed consideration of these issues in Kirchmann and Bergström 2008).

The role of organic systems in supporting development in the tropics and meeting global food demand has been studied, much discussed and disputed (Pretty et al. 2003; Badgley et al. 2007; Hudson Institute 2007; Kirchmann and Bergström 2008). A significant number of critical voices raise concerns that organic agriculture is not capable of meeting the world's growing food needs due to lower productivity per unit area (e.g. Borlaug 2000; Trewavas 2002). However, recent work using the International Food Policy Research Institute's International Model for Policy Analysis of Agricultural Commodities and Trade (IMPACT) and extensive farming systems data showed that even at high levels of conversion to organic agriculture (up to 50%) in Europe and North America, there would be relatively little impact on the availability of food and price changes would be limited (FAO 2007). Posner et al. (2008) also found that there was no reason that yields would decrease significantly under organic management in temperate climates; good farmer management of both conventional and organic systems is the key to sustained, good yields. Barron (2006) found that in developing countries, organic farming delivered increased yields, yield stability and improved livelihoods for small-scale farmers where there was both limited access to technology-driven adaptation and restricted opportunity for commodity production for international markets. Generally in such regions, socio-economic factors operating at regional scale constrain farming system development as much as local climate. Better use of on-farm resources rather than farming system per se is the key to improving farm system productivity. For the case of sub-Saharan Africa, conversion to organic farming practices of up to 50% of the agricultural area was estimated to increase food availability and decrease food import dependency, with negligible changes in prices and no changes in current malnutrition rates (FAO 2007).

14.2.2 Description of Case Study Systems

Watson et al. (2008) highlight the need to ground all future comparative work in improved understanding of all farming systems and their environments. Consequently, the discussion of the impacts of organic farming systems on soil health, and of climate change on organic farming, in the remainder of this chapter will be made with reference to two contrasting particular (but generalised) model organic cropping systems:

1. Cool temperate mixed cropping systems common in Northern Europe and America
2. Intensive irrigated crop and vegetable production under Mediterranean climates common in California and Southern Europe

Figures 14.1 and 14.2 identify the key components and practices of these representative systems and highlight distinctions from comparative conventional practice.

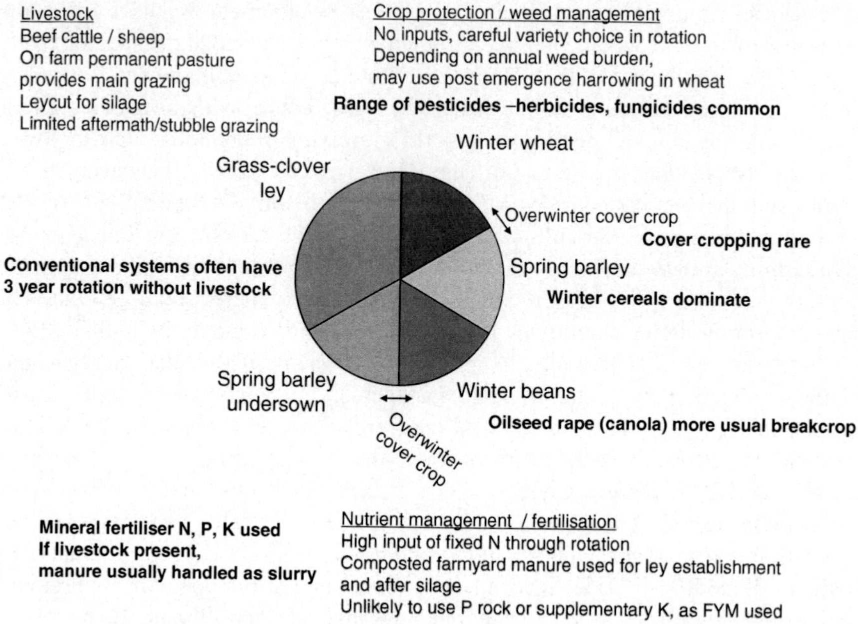

Fig. 14.1 Typical 6-year organic rotation common in cool temperate climates, forming part of a mixed farming system. Segments show approximate length of each cropping phase within the rotation. Text in *italics* identifies key distinguishing practices used in conventional systems

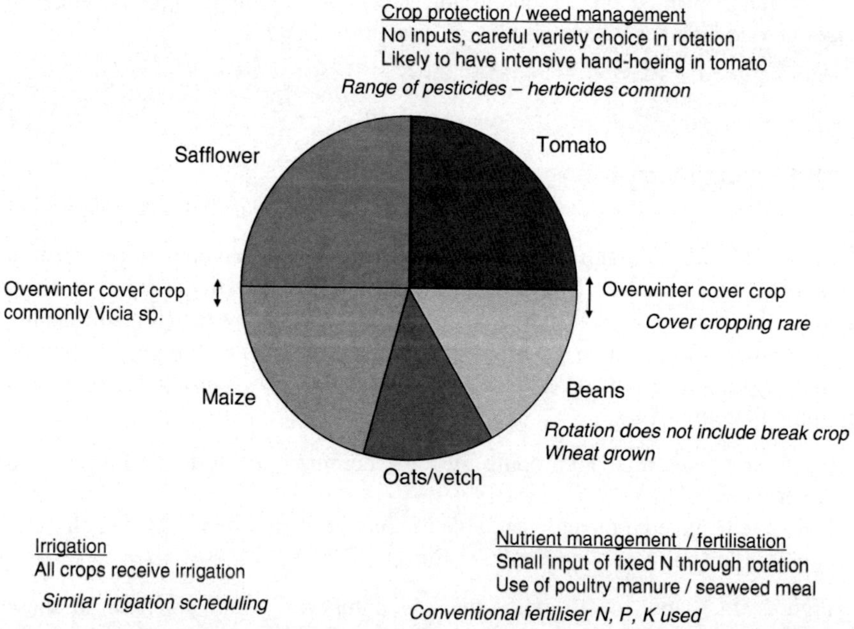

Fig. 14.2 Typical 4-year organic rotation common in Mediterranean climates, adapted from descriptions of systems in Clark et al. (1998). Segments show approximate length of each cropping phase within the rotation. Text in *italics* identifies key distinguishing practices used in conventional systems

14.3 Impacts of Organic Cropping Systems on Soil Properties and Processes

Most studies considering the impact of organic farming systems have compared organic with paired conventional systems. There are also a number of long-term cropping systems experiments that have been used to study changes in soil properties through time (Raupp et al. 2006). Most of these studies have been carried out in temperate rain-fed arable systems with many fewer studies in other key farming systems and climates. Published research into organic farming, however, is increasing rapidly (Watson et al. 2008), and it is expected that these gaps in systems studied will reduce. To be able to draw conclusions that relate to the system or other factors used for comparison such as soil texture, there is a crucial dependence on the premise that these are the only varying factors (Yeates et al. 1997), and consequently any comparisons between farming systems are fraught with difficulty as truly paired systems are very difficult to find in practice. Comparison of crop rotations is an interesting example. As soon as the crops or even varieties within a rotation are changed, the impacts of that rotation on soil properties and processes are likely to change (Stockdale et al. 2006), irrespective of whether the farming system is organic or conventional. However, it is also a fact that under given soil and climatic constraints, the most productive choice of crops and varieties in a rotation will differ depending on whether the system is managed conventionally or organically. Thus, are any differences between the biophysical aspects of the rotation due to the system or the rotation? It is also important to separate out those aspects of the system which need to be assessed at the whole systems level, i.e. those which are dominated by interactions or large-scale ecological processes (e.g. impacts on farm energy use), and those which can be reliably compared at the small plot scale (e.g. changes in soil organic matter) (Watson et al. 2008). In future the application of systems approaches to statistical comparison derived from ecology and/or biometrics may increase the robustness of comparisons at systems levels and allow the identification of key variables driving the observed differences.

In this section, I will briefly review the current evidence for the impacts of organic cropping systems on soil health and then draw this together with reference to the model systems outlined in Figs. 14.1 and 14.2. Time since conversion to organic management is a key factor affecting the extent to which differences between farming systems are observed, and, where possible, in the following sections I have only included comparisons, where the farming systems have been in place for at least one full rotational cycle.

14.3.1 Impacts on Soil Organic Matter and Its Fractions

Increases in organic matter in soils under organic management are widely reported (Reganold et al. 1993; Scow et al. 1994; Nguyen et al. 1995; Loes and Ogaard 1997;

Clark et al. 1998; Drinkwater et al. 1998; Shepherd et al. 2002). Where organic farming practices significantly increase inputs of organic matter to a cropping system compared to the conventional equivalent (e.g. Melero et al. 2006), then significant increases in soil organic matter are not unexpected. Higher levels of C in the "light fraction" of soils have been measured in organic farming systems (Wander et al. 1995; Bending et al. 2004; Leite et al. 2007), have been linked to inputs of fresh organic materials as crop residues, etc., and are thought to represent a more biologically active pool (Haynes 2005). Many organic farming systems have an increased proportion of leys and green manures in crop rotations compared with their conventional equivalents, and these include the regular incorporation of crop residues and manures into the soil. Nonetheless, increased inputs of organic matter to soil in organic systems compared with conventional systems are not a certainty (Shepherd et al. 2002). Furthermore, the size of the pool of soil organic matter is critically dependent on differences in the quality as well as the quantity of organic matter input (Janssen 1984). For example, inputs of organic carbon are closely matched in the DOK systems comparison trial, although the quality of the organic carbon is changed as a result of manure management (Mäder et al. 2002); as the degree of humification of the incorporated manure increases (as a result of increased duration and improved composting), higher levels of soil organic matter are measured (Birkhofer et al. 2008). Changes in the amount and quality of soil organic matter drive/underpin many of the other changes in soil biological and physical properties discussed in detail in the following sections (see Chap. 2 also).

14.3.2 Impacts on Soil Biota: Population Size, Activity and Diversity

A range of positive and negative effects on belowground ecology have been observed as a result of the application of contrasting management systems (see Chap. 8), but Hole et al. (2005) found more reports of positive impacts of organic farming systems on belowground species than studies where there were no difference or negative impacts. Bengtsson et al. (2005), as part of a meta-analysis considering all aspects of biodiversity in organic farming systems, showed that despite the considerable heterogeneity among studies, soil organisms were generally more abundant in organic farming systems. Effects on bacterial biomass and activity were unclear, whereas positive impacts on earthworm, collembolan, mite and fungal populations were confirmed. For groups that can be resolved to the species level, for example collembola, carabids, differential effects of systems are found for different species. In general, given a particular climate/soil combination, it is the amount and quality of organic matter inputs which interact with management disturbance such as tillage or grazing to control the size, structure and diversity of the soil ecosystem. Stockdale and Watson (2009) recently reviewed the evidence for the impact of organic farming systems on soil biota and hence only a summary is presented here.

The main factors leading to the increases in populations and activity observed in organic arable systems are the size and quality of the carbon inputs (Bossio et al. 1998; Gunapala and Scow 1998; Berkelmans et al. 2003; Fließbach et al. 2007). Breland and Eltun (1999) proposed that the higher returns of organic matter, as a result of the inclusion of ley and green manure crops in an organic arable rotation, were able to offset the impact of increased tillage. However, it is important to note that the differences between phases of the rotation in any system can be of the same magnitude as differences between organic and conventional systems (Elmholt 1996; Watson et al. 1996). Interaction between tillage and organic matter inputs within the constraints set by climate and soil texture rather than management system per se dominantly controls the size and activity of the soil microbial biomass in arable systems (Cookson et al. 2007; Esperschütz et al. 2007).

Higher arbuscular mycorrhizal colonisation has been shown in organic than conventional paired cropping systems (e.g. Sattelmacher et al. 1991; Ryan et al. 1994; Bending et al. 2004). Gosling et al. (2006) summarised evidence from 13 studies showing greater root colonisation, larger numbers of spores and greater diversity of arbuscular mycorrhizal fungi in organic than conventional systems. Increased inputs of organic matter, increased diversity in crops and lower levels of available soil P were considered to be the main differences between the organic and conventional systems that led to the differences in arbuscular mycorrhizal fungi. Earthworm populations are generally higher in organic than conventional systems (Reganold et al. 1993; Mäder et al. 2002). Peigné et al. (2009) found that earthworm population size in organic systems is controlled by the interacting effects of soil moisture regime, inputs of organic matter/manure and tillage regime. Organic farming has also been shown to increase the abundance and species richness of arable weeds (e.g. Roschewitz et al. 2005). This clearly has benefits for plant species diversity and can therefore impact on aboveground insect populations as a result of increases in niche diversity (Clough et al. 2007).

In pastoral systems, much smaller differences in soil biodiversity are seen between organic and conventional systems, and differences in management are also often smaller. Taken together the data collected in grassland systems suggest that decomposition pathways in low intensity and/or organically managed grassland are likely to be more complex/diverse than under high-intensity conventional grassland (Yeates et al. 1997; Eason et al. 1999; Mulder et al. 2003; Oehl et al. 2003). Long-term system studies have found fundamental differences in soil food web structure between conventional and less intensive cropping systems (Brussaard 1994, The Netherlands; Crossley et al. 1989, USA; Andren et al. 1990, Sweden). It has been suggested that with high management intensity, the bacterial community dominates the microbial component, while, in less intensive systems, the fungal community is the dominant microbial component. Such differences influence nutrient cycling and have implications for the efficient use of nutrient inputs and losses of nitrate by leaching (De Ruiter et al. 1994; Pankhurst et al. 1994).

14.3.3 Impacts on Soil Structure, Aggregate Stability and Erosion Risk

Measurable changes in soil physical properties can take decades to establish. Anecdotal evidence of increased ease of cultivation on conversion to organic farming systems is readily gathered from organic farmers. Several authors have measured higher aggregate stability and increased porosity under organic arable cropping than under a conventional comparison (e.g. Jordahl and Karlen 1993; Siegrist et al. 1998; Pulleman et al. 2003; Papadopoulos et al. 2009). Stability of soil structure at the millimetre to centimetre scale is largely driven by the short-term management of crop residues and fresh organic matter inputs (Pulleman et al. 2003; Papadopoulos et al. 2009). Incorporation of organic matter and stabilisation within soil microaggregates is more strongly related to the long-term history of organic matter management at any site (Papadopoulos et al. 2009).

Increases in the depth of the A horizon of the soil have also been recorded for organic systems, which may result from decreased bulk densities and/or decreased soil losses through erosion (Reganold et al. 1993; Gerhardt 1997). Lower rates of run-off and soil erosion have been measured in organic systems (Logsdon et al. 1993; Reganold 1987). However, several research studies have found no significant differences in the soil physical properties between organic and conventional farming systems (e.g. Droogers and Bouma 1996). Siegrist et al. (1998) also found no differences in soil particle detachment between organic and conventional systems. Changes in soil structure may lead to relatively small changes in aeration and water holding capacity of soils under organic management. However, Droogers and Bouma (1996) showed using simulation of water-limited yields that consequent improved plant-soil water dynamics can result in improved yields. Increased early season infiltration and water holding capacity have been suggested as the key factors supporting greater yield stability in drought years in organic cropping systems (Lotter et al. 2003).

14.3.4 Impacts on pH, Nutrient Status and Eutrophication Potential

Increased soil pH has been measured in some organic systems (Scow et al. 1994; Clark et al. 1998). However, frequent and/or long-term cropping with legumes is known to increase rates of soil acidification (Yan et al. 1996) and consequently in unlimed plots, Kirchmann et al. (2007) measured a more rapid decline in pH in an organic than in a control conventional system. Application of lime on acid soils is not restricted under the organic regulations, and maintenance of optimum pH levels for the cropping system is a key strategy for both organic and conventional farming systems.

Clark et al. (1998) showed that most changes in soil chemical properties could be predicted from a consideration of the inputs and outputs of nutrients to the system.

Nutrient budgets are a useful tool to summarise the nutrient status of complete systems (Watson and Atkinson 1999). Depending on farm management and the balance of imports and exports of nutrients, nitrogen, phosphorus and potassium budgets range from deficit to surplus in organic farming systems (Fagerberg et al. 1996; Nguyen et al. 1995; Nolte and Werner 1994; Wieser et al. 1996; Watson et al. 2002b). Balanced budgets are often more easily achieved in mixed farming systems (Berry et al. 2003).

Nitrogen budgets are generally positive for both conventional and organic systems (Goss and Goorahoo 1995; Halberg et al. 1995; Nguyen et al. 1995), while organic farming systems may be limited by nitrogen deficiency during crop growth (Berry et al. 2002); such budgets indicate that there is often an annual surplus of nitrogen, which may be lost by nitrate leaching. De Neve et al. (2006) measured lower nitrate concentrations in soils on organic farms; Smolik et al. (1995) also measured reduced nitrate loadings in an organic system. However, Kristensen et al. (1994) measured similar nitrate concentrations in the soil in autumn in organic and conventional farms using manure and suggested that there was no inherent difference in the leaching risk from conventional and organic farming systems. Eltun (1995) measured nitrate leaching losses directly from organic farming and found they were half of those measured in paired conventional systems. Such startling differences are not common in the literature, but following a comprehensive review, Kirchmann and Bergström (2001) found that average leaching of nitrate from organic farming systems over a crop rotation period was lower than in conventional agriculture. However, leaching losses are strongly related to the previous crop, cultivations and the presence of winter green cover and hence can vary significantly throughout any rotation. Much larger leaching losses of N (2–3 times the total N loss) have been measured following the cultivation of grass-clover leys than at other points in the rotation (Watson et al. 1993). Stopes et al. (2002) found that leaching losses vary by a factor of seven when comparing seasons and points in the rotation. Management practices (cultivation timing and crop establishment) are therefore critically important immediately following the incorporation of "green" legume crops to prevent losses of nitrogen, which will restrict the yield of the following crop and also cause significant environmental impact through nitrate leaching. Reduced emissions of ammonia from livestock housing in organic systems may occur due to the increased use of straw for bedding. However, handling and spreading manure can result in significant losses of ammonia (Hartung 1992). In livestock systems, the nitrogen surplus and eutrophication potential tends to increase with increasing stocking density (Berry et al. 2003; see Chap. 10). Taken overall, there is no evidence that organic farming systems have a higher risk of nitrate leaching than conventional systems; indeed organic cropping systems may show a slightly lower eutrophication potential on an area basis (Di and Cameron 2002). When leaching amounts are expressed in relation to crop yields, Korsaeth (2008) found that there was no significant difference in leaching loss between conventional and organic systems. However, modifications to the conventional cropping system were able to reduce leaching per unit product below those of the organic system (Korsaeth 2008). Eutrophication potential cannot be simply expressed in terms of the cropping system (organic vs. conventional); instead the crop rotation design, amounts and timing of N input, and the use of countermeasures such as catch crops and timing of tillage need to be known to predict leaching risk (Kirchmann and Bergström 2001).

14.3.5 Impacts on Greenhouse Gas Emissions

Carbon dioxide emissions and energy use are usually closely linked. Detailed comparisons of the energy efficiency of production systems have been carried out since the 1980s (e.g. Pimentel et al. 1983). A range of approaches and methods have been used, however, and consequently the data are difficult to compare directly; Gomiero et al. (2008) provide a critical summary.

On-farm energy requirements (linked to tractor hours and hence fuel use) are often higher on organic than conventional farms. Wood et al. (2006) found that on-site energy requirements were 22% higher for organic than conventional farms in Australia and linked this to weed control and manure spreading operations coupled with lower labour intensity on the organic farms. Improved soil structure (see Sect. 14.3.3) may reduce fuel use during cultivation in organic systems, and there are a range of opportunities for organic farming systems to further reduce direct energy consumption. Ahlgren et al. (2009) demonstrate that it is theoretically possible for an organic farm to be self-sufficient in tractor fuel (derived from biomass grown on the farm); however, the proposed solutions are of high technological complexity and not yet implementable at commercial scale or reasonable cost. The indirect use of energy embedded in inputs such as fertilisers, herbicides and pesticides is less than half that on comparator conventional farms (Wood et al. 2006). Consequently, many studies show that the overall energy use of organic farms is lower than a conventional comparator, whether assessed on an area or per tonne of product basis (see summary of analyses to date in Wood et al. 2006; Gomiero et al. 2008). However, due to variations in management for particular crops in different systems, simple comparisons between systems are not easy; for example, organic potato crops may show both lower and higher energy consumption compared to conventional systems per tonne of potatoes (Gomiero et al. 2008).

In addition to direct and indirect CO_2 emissions as a result of energy use, agriculture is a major contributor to the global emissions of methane and nitrous oxide (50 and 60% of global anthropogenic emissions, respectively; IPCC 2007) (see Chaps. 9 and 10). However, there is very little measured data available to assess the overall greenhouse warming potential of organic farming systems in comparison with conventional practice. Wood et al. (2006) showed that the greenhouse gas intensity of conventional farms was on average twice that of organic farms; however, this excluded emissions of methane from animals and emissions of N_2O from soils and fertilisers. Nitrogen surpluses in organic farming systems tend to be lower than those for conventional systems, which indicate a lower potential for gaseous nitrogen emissions from soil. However, estimated and measured N_2O

emissions show no clear differences between organic and conventional farming systems (Flessa et al. 1995; Reitmayr 1995; Syvasalo et al. 2006). Lower stocking rates in organic farms are likely to result in lower methane and N_2O emissions per hectare, even though emissions per animal are likely to be similar in organic and conventional systems.

14.3.6 Interactions at a Farming Systems Level

Although soil forming/land capability factors (climate, geology, topography, ecosystem) determine the boundary conditions for soil quality in any location as a result of their impact on inherent soil properties, soil health tends to be enhanced where management and land-use decisions take the multiple functions of soil into account and to deteriorate where decisions focus solely on one function, most often crop productivity (Doran 2002). Hence, we might expect better soil health, and consequently better overall environmental performance, in organic farming systems. Sandhu et al. (2008) calculated the total value of ecosystem services delivered in conventional and organic arable landscapes in New Zealand; while the overall value of ecosystem services delivered by organic systems was higher on average than for conventional systems, there was significant overlap on a field by field basis (Fig. 14.3), so that some conventional fields deliver more ecosystem services than organic fields.

Low-input approaches have been developed and implemented, particularly in degraded ecosystems, using farmer participatory approaches which also increase the skill and understanding of the local farming community (e.g. Scoones and Thompson 1994). In degraded areas, application of composts has been shown to increase crop yields compared with typical field practice and with matched inputs through chemical fertilisers (e.g. doubling of yields compared to field with no inputs and small increases compared to use of chemical fertilisers in Tigray, Ethiopia; Edwards 2007). Other minor impacts such as reduced crop pest problems

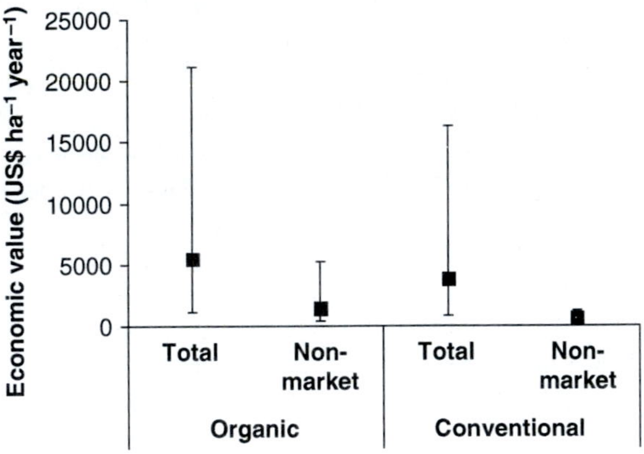

Fig. 14.3 Mean and range of economic value (total and non-market) of ecosystem services measured in organic and conventional fields ($n = 29$). Redrawn from data provided Table 1 in Sandhu et al. (2008). Non-market ecosystem services were significantly higher in organic fields

and increased soil moisture retention have also been noted where compost is used (Edwards 2007). However, there has been little evaluation in practice of the management options that are recommended as more sustainable: "For the most part sustainable agriculture projects assume that the practices they promote will improve sustainability without ever measuring the results to see if this is actually the case" (Holt-Giménez 2002). Most of these recommended low-input strategies advocated for all agricultural systems, for example use of cover crops, green manures, animal manure and reduced tillage, are already embedded in organic farming systems and lead to increases in soil organic matter levels compared to the levels typical of intensive conventional management approaches, where residues are removed/burnt, tillage is intensive and no inputs of organic materials are made (Matson et al. 1997). Reduced tillage approaches have been adopted in many arable farming systems (see Chap. 9); however, they are not widely used in organic farming systems as a result of the need to use tillage-based seedbed preparation approaches for weed management where herbicides are not available. Best practice options for soil management are not restricted to organic farming. Even within organic systems which, according to their principles, have a more integrated approach to crop production, often a narrow input-substitution approach to soil management is taken (as an example, see the farmer perceptions revealed in Kaltoft 1999). Low-input strategies require intensive use of information to design effective rotations and management strategies that ensure synchronisation of nutrient release with plant demand, given the wider quality of nutrient sources available; productive and sustainable low-input systems require high levels of farmer knowledge. The decisions of land managers, driven by their underlying values, perceptions and level of understanding, are often the critical factor determining soil health.

Integrating the information gained from paired farm comparison and field trials for the model farming systems (Figs. 14.1 and 14.2) indicates that while overall yields are reduced in organic farming systems, on average there can be significant improvements in soil health as a result of the adoption of organic farming practices (Figs. 14.4 and 14.5). These can lead to significant improvements in resilience under current climatic conditions. In Nicaragua, the introduction of sustainable land management practices (contour ploughing, contour ditches and bunds, incorporation of legumes in rotations and as inter-crops, use of compost, live fences and woodlots) was shown in the study following Hurricane Mitch to have reduced topsoil losses very significantly compared to conventional neighbours (Holt-Giménez 2002). In dry temperate climates (Australia), Wood et al. (2006) showed much lower water dependence of organic farms compared to the conventional comparator – but differences between the farming types are strongly dependent on crop/livestock type. It has been suggested that organically managed crops have higher resistance to drought conditions and higher cropping system yield stability – Gomiero et al. (2008) drew together available data and suggest that yields can be 70–90% greater than conventional systems under severe drought. Yield resilience is a very important benefit under more marginal conditions for crop growth (Fig. 14.5), especially where irrigation inputs are fixed (or decreasing); Melero et al. (2006) found

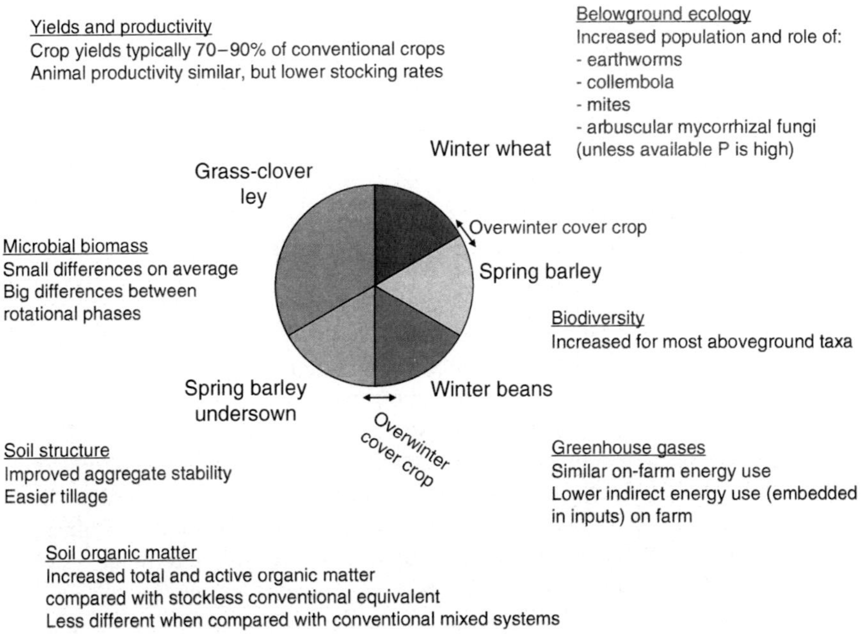

Fig. 14.4 Impacts on soil health expected from the implementation of a typical 6-year organic rotation common in cool temperate climates, forming part of a mixed farming system

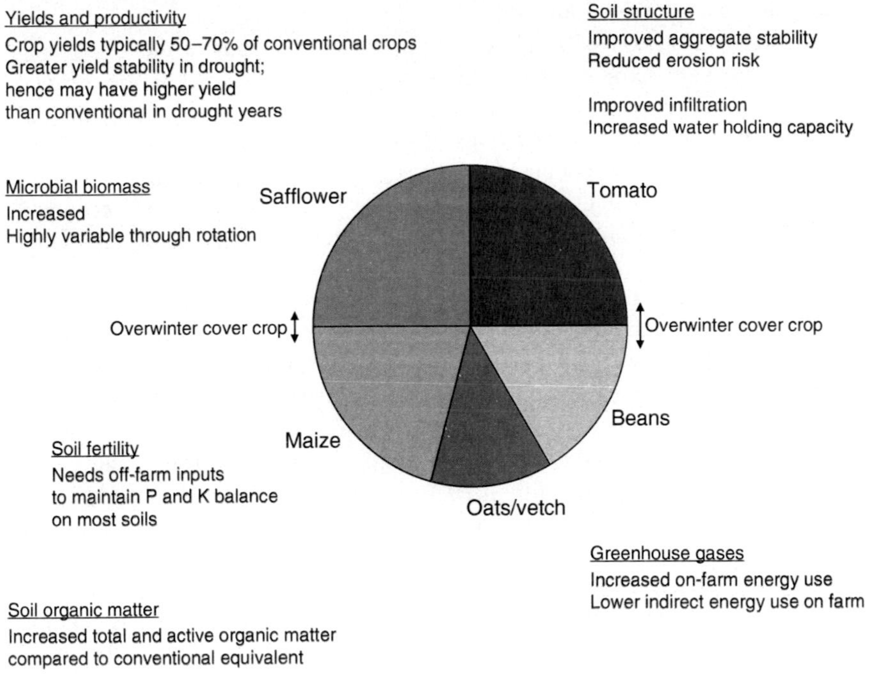

Fig. 14.5 Impacts on soil health expected from the implementation of a typical 4-year organic rotation common in Mediterranean climates

that increasing soil organic matter through compost application in organic systems increased yields of melon/watermelon crops compared to the comparative conventional system where compost was not applied.

14.4 Impacts of Climate Change on Farming Systems and Relevant Adaptation Measures: The Role of Organic Farming

It is well known that soil and environmental conditions interact to define regional patterns in agricultural systems creating agro-ecoregions (Graef et al. 2005), such as the Corn Belt (USA), Wheat belt (Western Australia), etc. In the UK, Gabriel et al. (2009) have studied the spatial aggregation of organic farming and found that currently conversion to organic farming is less common in the UK where soil types and climates are strongly suited to arable production and most common within areas where conventional farming is typified by small, mixed and dairy farming systems; increased ruralisation also favours conversion to organic production systems. It appears that current uptake of organic (and other alternative) agricultural systems is driven by a combination of factors in both the physical and socio-economic environment which interact with personal farmer motivational factors (Kaltoft 1999). Hendrickson et al. (2008) reviewed the role of environmental drivers in determining the uptake of low-input practices (as outlined above in Sect. 14.2; Doran 2002) within conventional systems in the Great Plains of the USA, using surrogate indicators (crop diversity and pasture/cropland ratio). Different trends in uptake of practices were seen in different states linked to annual rainfall. Hendrickson et al. (2008) suggested that there are other factors, such as landform, length of growing season coupled to the availability of irrigation that predispose farmers to adopt integrated low-input systems as a strategy to provide economic stability in variable conditions. While climatic regions set some constraints to the adoption of particular practices or agricultural systems, it is clear that the direct impacts of climate change will be only one driver of change in farming systems; changes in markets and agricultural policy will have an equally significant impact.

Modelling of climate change indicates that for most regions there will be changes in both means and extremes of temperature and rainfall which will affect the designation of current agro-ecoregions (e.g. Lobell et al. 2008). However, it is important to note that in any region, day length will not be affected and consequently, for some crops, suitability regions may not simply move towards higher latitudes. Porter and Semenov (2005) showed using experimental data and modelling that negative impacts on the yield and quality of wheat would occur more often than under current conditions. They also showed that increased occurrence of short periods of extreme temperatures or precipitation would have significant impacts on crop development especially C and N sink formation and activity. Changes in climate at any location will also modify the ability of the crops to respond to other factors, for example increased nutrient supply or improved plant protection (Porter and Semenov 2005). Porter and Semenov (2005) showed by modelling that the predicted changes in climate will have relatively little impact on average cereal yields in the UK; the largest impacts on yield are expected to occur where crops are currently being grown close to the edge of the climatic optimum, for example wheat may not be cultivable in the Mediterranean by 2050. In each region climate/soil interactions will hold the key in determining the suitability of current cropping patterns and farming systems under changing climates. Furthermore, direct climate change impacts on crop choice and rotation design will be largely indistinguishable for organic and conventional systems. Nevertheless, organic systems are likely to be more resilient to the anticipated impacts of climate change on temperature and rainfall, due to changes in organic matter and soil structure that currently increase resilience to drought (see Sect. 14.3.6).

Climate change factors will impact on the physiological processes, activity and phenology of pests and their natural enemies (Stacey 2003). However, accurate prediction of likely impacts depends on how all the species involved (plant, pest and predator/parasite) react to changing patterns of temperature, humidity and cloud cover (Stacey 2003). This may lead to different patterns of disease/pest risks, new pest and diseases and failure in existing methods of pest control including biological control. Stacey (2003) suggests that increased variability in weather patterns will result in increased difficulty for farmers in designing management practices, which reduce pest control impacts on yield and/or quality due to the complex interactions between biological and environmental factors. Climate change impacts on pest and disease risk will be largely indistinguishable for organic and conventional systems; however, the multi-factorial approach to develop networks of partial solutions for pest and disease management routinely used in organic systems are likely to become more common in conventional systems.

Action taken at farm scale can modify the potential impacts of climate change; farm decision making can reduce the vulnerability of farming systems to climate change impacts (Reidsma et al. 2009). However, there will be interactions between inherent land properties (land capability) and management in determining the local impact of climate change. While for many farmers, the introduction of sustainable land management practices very significantly reduced the topsoil losses that resulted from Hurricane Mitch (Holt-Giménez 2002), on slopes >30% even these practices were not sufficient to confer agro-ecological resistance in storms with high rainfall intensity. Consequently, for some areas the only safe option may be to withdraw land from cultivation in the light of the predicted increased frequency of extreme events. Modelling suggests that there may be increasing abandonment in currently marginal areas; however, variability in the scale and geographical location of predicted impacts (e.g. the extent of the wheat cultivation zone) between modelling scenarios is high (e.g. in Europe, Audsley et al. 2006; Berry et al. 2006).

In temperate Northern European climates, the predictions are that changes in climate will lead to increases in crop yields overall and that the range of crops available for cultivation will increase (Ewert et al. 2005); so there will be relatively little impact to the crop rotation in the first of our case study systems (Fig. 14.6).

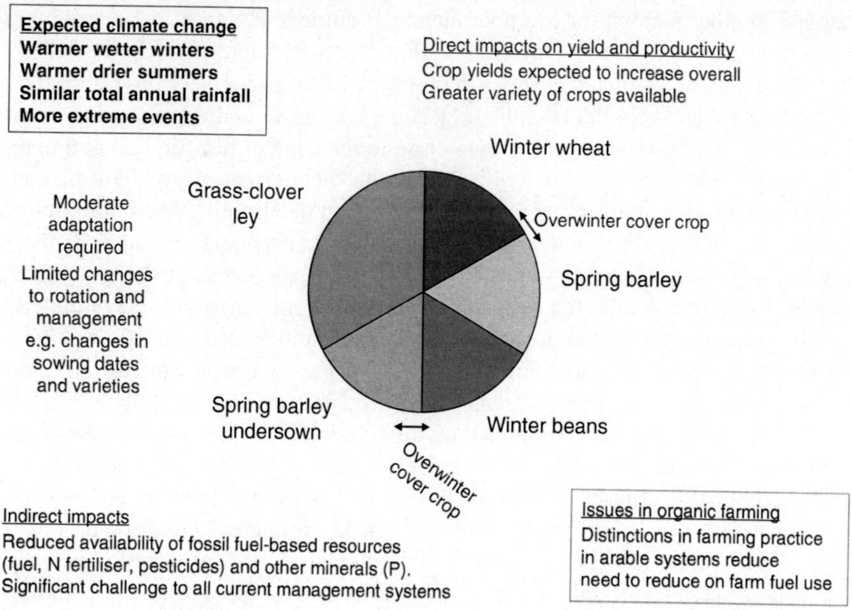

Fig. 14.6 Predicted climate change and its direct and indirect impacts for a typical 6-year organic rotation common in cool temperate climates, forming part of a mixed farming system. Issues for organic farming systems highlighted

The simplest adaptation options available to farmers, which include changes in varieties and sowing dates, may be sufficient even under the changed climates in temperate regions. However, under Mediterranean conditions, water shortages will drive agricultural change increasingly restricting yield and the land available for cropping (Metzger et al. 2006). Irrigation is currently a key strategy to reduce yield variability of high value crops in Mediterranean conditions (Reidsma et al. 2009). However, reducing rainfall is also likely to increase competition for water for irrigation, and adaptation strategies may need to include the increased use of wastewater in crop irrigation. Due to infrastructure limitations, this is likely to significantly reduce the area of irrigated cropping. With much lower amounts of water for irrigation, drought will severely restrict the range of crops available for cultivation in the Mediterranean case study system (Fig. 14.7) whether in organic or conventional systems. In marginal rainfall areas, one of the common approaches currently taken by farmers is to vary the amount of land used for crop production with land fallowed where insufficient water reserves are considered to have been accumulated (Hendrickson et al. 2008). In these areas that are marginal for cropping, increased organic matter levels in organic farming systems increase soil moisture retention and may allow more frequent cropping. Under Mediterranean conditions, it will be increasingly important to adopt new approaches and to design rotations and management strategies that are able to exploit ecological synergies in the use of water and nutrients (Tanaka et al. 2005). The focus in organic systems of

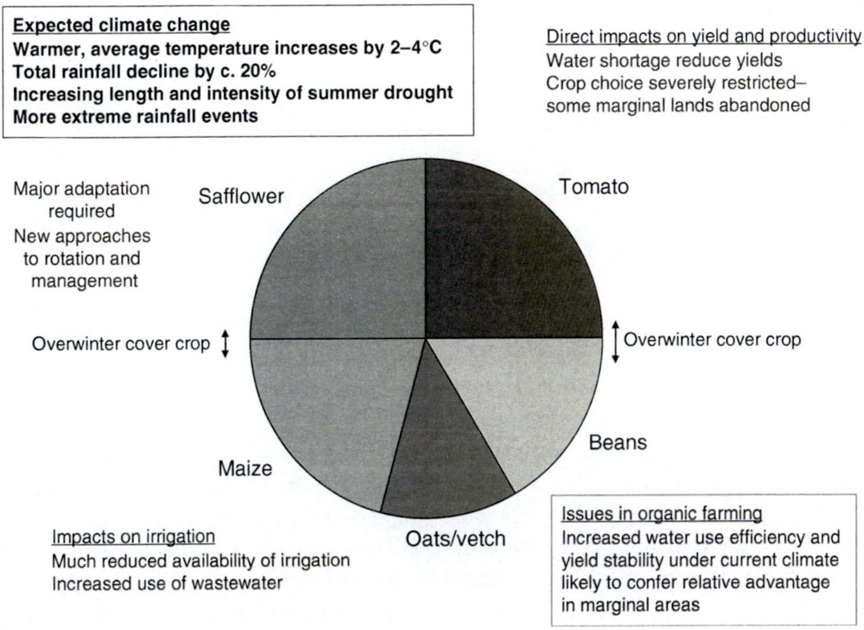

Fig. 14.7 Predicted climate change and its direct and indirect impacts for a typical 4-year organic rotation common in Mediterranean climates. Issues for organic farming systems highlighted

designing ecologically adapted rotations under current climatic constraints may give organic farmers a head start. Opinions on the role of using gene modification (GM) to improve crop stress resistance including drought tolerance range from extremely optimistic to very sceptical (Marris 2008). Hervé and Serrah (2009) also highlighted that most of these studies are still a long way from producing crop varieties that could greatly improve drought adaptation or water productivity in the field, because the traits targeted are very complex. If such an approach was successful, current organic standards would prohibit GM varieties for organic systems, and hence there may be greater divergence in the varieties cultivated in organic and conventional systems in the future than has been the case to date.

Alongside the changes to the climate itself, it seems almost certain that fossil fuel-based resources, including nitrogen fertiliser and pesticides, will become increasingly expensive and scarce (Kirschenmann 2007). Increasing scarcity of other mineral resources, for example phosphate rock, also seems likely. Consequently, a search for appropriate agronomic techniques may mean an increased focus on the approaches used in the past (Hanson et al. 2007) and now only used commonly in reduced input systems. Examples of these approaches include diverse crop rotations, crop mixtures and local recycling of waste to land. Hence, recommendations for the development of productive farm systems for the future highlight principles that are largely familiar to organic farmers, but nonetheless challenging. Examples of recommendations are outlined by Kirschenmann (2007):

Post modern farms will likely need to:

i) be energy conserving;
ii) feature both biological and genetic diversity;
iii) be largely self-regulating and self-renewing;
iv) be knowledge intensive;
v) operate on biological synergies;
vi) employ adaptive management;
vii) feature ecological restoration rather than choosing between extraction and preservation, and;
viii) achieve optimum productivity by featuring multiproduct, nutrient-dense, synergistic production on limited acreage.

Many of the approaches/management strategies for improving soil health and agricultural sustainability recommended by Doran (2002), described above and also see Chap. 1, are already well understood by both farmers and agronomists. However, the economic costs of better management practices (including negative short-term impacts on productivity) might mean low rates of adoption of these practices by farmers. Appropriate policy structures are essential therefore to enable action for medium- to long-term effectiveness to be adopted in the short term (Holt-Giménez 2002). The increased policy focus on carbon sequestration in agricultural systems as part of climate change mitigation strategies may provide a mechanism that directs support to farmers and also promotes adaptation (see Chaps. 1 and 5; Smith et al. 2000).

Can the adoption of organic farming systems help to reduce vulnerability to climate change of the farm, its ecosystem services and the community in which it is embedded? Such a question requires a consideration not only of the risk resulting from changes in the physical environment but also the adaptability of the social and economic structures, which help to determine the capacity of the system to anticipate, cope and recover (Berry et al. 2006). Tompkins and Adger (2004) argue for the urgent development of adaptive systems for working with natural resources under current climatic conditions to increase the ability to adapt to climate change and other factors, such as changes in market demand in the future. Consequently, developing farmers' skills to understand and manage their farming systems in relation to climatic variation now will enhance resilience to climate change through knowledge based rather than technologically reliant innovation. One of the biggest advantages organic farmers may have is not only that they have built resilience into their soils through increasing organic matter levels, but also that they themselves have already been converted to an adaptive way of working, which encourages considered strategic responses rather than quick-fix solutions. For example, Lohr and Park (2002) observed higher willingness and capacity to innovate when comparing organic farmers with conventional farmers in approaches to insect management. Preventing land degradation, supporting biodiversity and the supply of a range of ecological goods and services may be as important as maintaining yield in the light of climate change; a focus on short-term protection of productivity may lead to medium- to long-term yield loss and misleading conclusions in assessments of the sustainability of farming systems (Gomiero et al. 2008). Adaptation will also require a change of focus in agricultural research and extension so that the environment is no longer viewed as an adversary to be overcome through inputs and mechanisation (Hendrickson et al. 2008), rather the environment may need to be viewed as a partner in agro-ecosystem development.

14.5 Conclusions

Maintaining soil health is central to the principles and regulations which govern organic farming systems. In order to create locally adapted profitable systems, many organic farmers have converted from input/technology-driven management to adaptive management approaches. This combination of practice and philosophy means that organic farming systems are inherently well placed to cope with and adapt to climate change: maintenance of soil health and application of adaptive management are key foundational steps for successful adaptation of farming systems in the light of climate change. However, variability within farming systems means that all organic farming systems do not currently achieve these ideals, and that all conventional farming systems do not fail to achieve them. There is currently a lack of data for many farming practice/climate combinations even under current climate conditions to allow comparison of conventional and organic systems. Research and development in organic farming systems, which has been gradually increasing, is likely to underpin developments in farming practice that can be exploited by all farmers to support adaptation to climate change.

References

Ahlgren S, Baky A, Bernesson S, Nordberg Å, Norén O, Hansson P-A (2009) Tractive power in organic farming based on fuel cell technology – energy balance and environmental load. Agric Syst 102:67–76

Andren O, Lindberg T, Paustian K, Rosswall T (1990) Ecology of arable land – organism, carbon and nitrogen cycling. Ecol Bull 40:1–222

Audsley E, Pearn KR, Simota C, Cojocaru G, Koutsidou E, Rounswell MDA, Trnka M, Alexandrov V (2006) What can scenario modelling tell us about future European scale land use, and what not? Environ Sci Policy 9:148–162

Badgley C, Moghtader J, Quintero E, Zakem E, Chappell JM, Avilés-Vásquez K, Samulon A, Perfecto I (2007) Organic agriculture and the global food supply. Renew Agric Food Syst 22:86–108

Balfour E (1943) The living soil and the Haughley experiment. Faber & Faber, London

Barberi P (2002) Weed management in organic agriculture: are we addressing the right issues? Weed Res 42:177–193

Barron S (2006) Building resilience for an unpredictable future. How organic agriculture can help farmers adapt to climate change. Report of the sustainable development department. Food and Agriculture Organisation, Rome

Bending GD, Turner MK, Rayns F, Marx MC, Wood M (2004) Microbial and biochemical soil quality indicators and their potential for differentiating areas under contrasting agricultural management regimes. Soil Biol Biochem 36:1785–1792

Bengtsson J, Ahnstrom J, Weibull AC (2005) The effects of organic agriculture on biodiversity and abundance: a meta-analysis. J Appl Ecol 42:261–269

Berkelmans R, Ferris H, Tenuta M, Van Bruggen AHC (2003) Effects of long-term crop management on nematode trophic levels other than plant feeders disappear after 1 year of disruptive soil management. Appl Soil Ecol 23:223–235

Berry PM, Sylvester-Bradley R, Philipps L, Hatch DJ, Cuttle SP, Rayns FW, Gosling P (2002) Is productivity of organic farms restricted by the supply of available nitrogen? Soil Use Manage 18:248–255

Berry PM, Stockdale EA, Sylvester-Bradley R, Philipps L, Smith KA, Lord EI, Watson CA, Fortune S (2003) N, P and K budgets for crop rotations on nine organic farms in the UK. Soil Use Manage 19:112–118

Berry PM, Rounsevell MDA, Harrison PA, Audsley E (2006) Assessing the vulnerability of agricultural land use and species to climate change and the role of policy in facilitating adaptation. Environ Sci Policy 9:189–204

Birkhofer K, Bezemer TM, Bloem J, Bonkowski M, Christensen S, Dubois D, Ekelund F, Fließbach A, Gunst L, Hedlund K, Mäder P, Mikola J, Robin C, Setala H, Tatin-Froux F, Van der Putten WH, Scheu S (2008) Long-term organic farming fosters below and aboveground biota: Implications for soil quality, biological control and productivity. Soil Biol Biochem 40:2297–2308

Borlaug N (2000) Ending world hunger. The promise of biotechnology and the threat of antiscience zealotry. Plant Physiol 124:487–490

Bossio DA, Scow KM, Gunapala N, Graham KJ (1998) Determinants of soil microbial communities: effects of agricultural management, season, and soil type on phospholipid fatty acid profiles. Microbial Ecol 36:1–12

Breland TA, Eltun R (1999) Soil microbial biomass and mineralization of carbon and nitrogen in ecological, integrated and conventional forage and arable cropping systems. Biol Fertil Soils 30:193–201

Brussaard L (1994) Soil ecology of conventional and integrated farming systems. Agric Ecosyst Environ 51(i–ix):1–267

Carson R (1963) Silent spring. Hamish Hamilton, London

Clark MS, Ferris H, Klonsky K, Lanini WT, van Bruggen AHC, Zalom FG (1998) Agronomic, economic and environmental comparison of pest management in conventional and alternative tomato and corn systems in northern California. Agric Ecosyst Environ 68:51–71

Clough Y, Holzschuh A, Gabriel D, Purtauf T, Kleijn D, Kruess A, Steffan-Dewenter I, Tscharntke T (2007) Alpha and beta diversity of arthropods and plants in organically and conventionally managed wheat fields. J Appl Ecol 44:804–812

Conford P (2001) The origins of the organic movement. Floris Books, Edinburgh

Cookson WR, Osman M, Marschner P, Abaye DA, Clark I, Murphy DV, Stockdale EA, Watson CA (2007) Controls on soil nitrogen cycling and microbial community composition across land use and incubation temperature. Soil Biol Biochem 39:744–756

Courville S (2006) Organic standards and certification. In: Kristiansen P, Taji A, Reganold J (eds) Organic agriculture a global perspective. CSIRO Publishing, Collingwood, pp 201–220

Crossley DA Jr, Coleman DC, Hendrix PF (1989) The importance of the fauna in agricultural soils: research approaches and perspectives. Agric Ecosyst Environ 27:47–55

De Neve S, van Dem BA, Slutel S, Hofman G (2006) Soil nutrient status of organic farms in Flanders: an overview and a comparison with the conventional situation. Biol Agric Hortic 24:217–235

De Ruiter PC, Neutel AN, Moore JC (1994) Modelling food webs and nutrient cycling in agroecosystems. Trends Ecol Evol 9:378–383

Defra, UK Department for Environment, Food and Rural Affairs (2006) Compendium of UK organic standards. September 2006. http://www.defra.gov.uk/farm/organic/standards/pdf/compendium.pdf. Accessed 8 Feb 2009

Di HJ, Cameron KC (2002) Nitrate leaching in temperate agroecosystems: sources, factors and mitigating strategies. Nutr Cycling Agroecosyst 64:237–256

Doran JW (2002) Soil health and global sustainability: translating science into practice. Agric Ecosyst Environ 88:119–127

Drinkwater LE, Waggoner P, Sarrantonio M (1998) Legume based cropping systems have reduced carbon and nitrogen losses. Nature 396:262–265

Droogers P, Bouma J (1996) Biodynamic vs conventional farming effects on soil structure expressed by simulated potential productivity. Soil Sci Soc Am J 60:1552–1558

Eason WR, Scullion J, Scott EP (1999) Soil parameters and plant responses associated with arbuscular mycorrhizas from contrasting grassland management regimes. Agric Ecosyst Environ 73:245–255

Edwards S (2007) Role of organic agriculture in preventing and reversing land degradation. In: Sivakumar MVK, Ndiangui N (eds) Climate and land degradation. Environmental science and engineering book Series. Springer, Berlin, pp 523–536

Elmholt S (1996) Microbial activity, fungal abundance, and distribution of penicillium and fusarium as bioindicators of a temporal development of organically cultivated soils. Biol Agric Hortic 13:123–140

Eltun R (1995) Comparison of nitrogen leaching in ecological and conventional cropping systems. Biol Agric Hortic 11:103–114

Esperschütz J, Gattinger A, Mäder P, Schloter M, Fließbach A (2007) Response of soil microbial biomass and community structures to conventional and organic farming systems under identical crop rotations. FEMS Microbiol Ecol 61:26–37

Ewert F, Rounsevell MDA, Reginster I, Metzger MJ, Leemans R (2005) Future scenarios of European agricultural land use I. Estimating changes in crop productivity. Agric Ecosyst Environ 107:101–116

Fagerberg B, Salomon E, Jonsson S (1996) Comparisons between conventional and ecological farming systems at Ojebyn – nutrient flows and balances. Swed J Agric Res 26:169–180

FAO, Food and Agriculture Organisation of the United Nations (2007) Report of the international conference on organic agriculture and food security, May 2007, OFS/2007/REP. FAO, Rome

Flessa H, Dorsch P, Beese F (1995) Seasonal-variation of N_2O and CH_4 fluxes in differently managed arable soils in Southern Germany. J Geophys Res Atmos 100:23115–23124

Fließbach A, Oberholzer H-R, Gunst L, Mäder P (2007) Soil organic matter and biological soil quality indicators after 21 years of organic and conventional farming. Agric Ecosyst Environ 118:273–284

Gabriel D, Carver SJ, Durham H, Kunin WE, Palmer RC, Sait SM, Stagl S, Benton TG (2009) The spatial aggregation of organic farming in England and its underlying environmental correlates. J Appl Ecol 46:323–333

Gerhardt RA (1997) A comparative analysis of the effects of organic and conventional farming systems on soil structure. Biol Agric Hortic 14:139–157

Giller KE, Beare MH, Lavelle P, Izac AMN, Swift MJ (1997) Agricultural intensification, soil biodiversity and agroecosystem function. Appl Soil Ecol 6:3–16

Gomiero T, Paoletti MG, Pimentel D (2008) Energy and environmental issues in organic and conventional agriculture. Crit Rev Plant Sci 27:239–254

Gosling P, Hodge A, Goodlass G, Bending GD (2006) Arbuscular mycorrhizal fungi and organic farming. Agric Ecosyst Environ 113:17–35

Goss MJ, Goorahoo D (1995) Nitrate contamination of groundwater: measurement and prediction. Fertil Res 42:331–338

Graef F, Schmidt G, Schröder W, Stachow U (2005) Determining ecoregions for environmental and GMO monitoring networks. Environ Monit Assess 108:189–203

Gunapala N, Scow KM (1998) Dynamics of soil microbial biomass and activity in conventional and organic farming systems. Soil Biol Biochem 30:805–816

Halberg N, Kristensen ES, Kristensen IS (1995) Nitrogen turnover on organic and conventional mixed farms. J Agric Environ Ethic 8:30–51

Hanson JD, Liebig MA, Merrill SD, Tanaka DL, Krupinsky JM, Stott DE (2007) Dynamic cropping systems: increasing adaptability amid an uncertain future. Agron J 99:939–943

Hartung J (1992) Emission and control of gases and odorous substances from animal housing and manure stores. Zentralbl Hyg Umweltmed 192:389–418

Harwood RR (1990) A history of sustainable agriculture. In: Edwards CA, Lal R, Madden P, Miller RH, House G (eds) Sustainable agricultural systems. Soil and Water Conservation Society, Ankeny, pp 3–19

Haynes RJ (2005) Labile organic matter fractions as central components of the quality of agricultural soils: an overview. Adv Agron 85:222–268

Heckman J (2006) A history of organic farming: transitions from Sir Albert Howard's war in the soil to the USDA National Organic Programme. Renew Agric Food Syst 21:143–150

Hendrickson JR, Liebig MA, Sassenrath GF (2008) Environment and integrated agricultural systems. Renew Agric Food Syst 23:304–313

Hervé P, Serrah R (2009) Gene technology and drought: a simple solution for a complex trait? Afr J Biotechnol 8:1740–1749

Hole DG, Perkins AJ, Wilson JD, Alexander IH, Grice F, Evans AD (2005) Does organic farming benefit biodiversity? Biol Conserv 122:113–130

Holt-Giménez E (2002) Measuring farmers' agroecological resistance after Hurricane Mitch in Nicaragua: a case study in participatory, sustainable land management impact monitoring. Agric Ecosyst Environ 93:87–105

Hudson Institute (2007) Organic Abundance report fatally flawed. http://www.cgfi.org/cgficommentary/organic-abundance-report-fatally-flawed. Accessed 12 Dec 2008

International Federation of Organic Agriculture Movements, IFOAM (2005) The principles of organic agriculture. Bonn, Germany. http://www.ifoam.org/about_ifoam/principles/index.html. Accessed 28 Dec 2007

IPCC, Intergovernmental Panel on Climate Change (2007) Climate change 2007: synthesis report. In: Pachauri RK, Reisinger A (eds) Contribution of working groups I, II and III to the fourth assessment report of the intergovernmental panel on climate change. IPCC, Geneva. http://www.ipcc.ch/publications_and_data/publications_ipcc_fourth_assessment_report_synthesis_report.htm. Accessed 14 Aug 2009

Janssen BH (1984) A simple method for calculating decomposition and accumulation of "young" soil organic matter. Plant Soil 76:297–304

Jordahl JL, Karlen DL (1993) Comparison of alternative farming systems. III. Soil aggregate stability. Am J Alternat Agric 8:27–33

Kaltoft P (1999) Values about nature in organic farming practice and knowledge. Sociol Ruralis 39:39–53

Kirchmann H, Bergström L (2001) Do organic farming practices reduce nitrate leaching? Commun Soil Sci Plan 32:997–1028

Kirchmann H, Bergström L (eds) (2008) Organic crop production – ambitions and limitations. Springer, Dordrecht

Kirchmann H, Bergström L, Kätterer T, Mattsson L, Gesslein S (2007) Comparison of long-term organic and conventional crop-livestock systems on previously nutrient depleted soil in Sweden. Agron J 99:960–972

Kirschenmann FL (2007) Potential for a new generation of biodiversity in agroecosystems of the future. Agron J 99:373–376

Korsaeth A (2008) Relations between nitrogen leaching and food productivity in organic and conventional cropping systems in a long-term field study. Agric Ecosyst Environ 127:177–188

Kristensen SO, Mathiasen J, Lassen J, Madsen HB, Reenberg A (1994) Comparison of the leachable inorganic nitrogen content in organic and conventional farming systems. Acta Agric Scand B Soil Plant 44:19–27

Kristiansen P, Taji A, Reganold J (2006) Organic agriculture. A global perspective. CSIRO Publishing, Collingwood

Lampkin N (1990) Organic farming. Farming Press, Ipswich

Lampkin N, Foster C, Padel S, Midmore P (1999) The policy and regulatory environment for organic farming in Europe. Organic farming in Europe: economics and policy, vol 7. University of Hohenheim, Hohenheim

Leite LFC, Mendonca ES, Machado LOA (2007) Influence of organic and mineral fertilisation on organic matter fractions of a Brazilian acrisol under maize/common bean intercrop. Aust J Soil Res 45:25–32

Lobell DB, Burke MB, Tebaldi C, Mastrandrea MD, Falcon WP, Naylor RL (2008) Prioritizing climate change adaptation needs for food security in 2030. Science 319:607–610

Lockeretz W (1990) Major issues confronting sustainable agriculture. In: Francis CA, Flora CB, King LD (eds) Sustainable agriculture in temperate zones. Wiley, New York, pp 423–438

Loes AK, Ogaard AF (1997) Long-term changes in extractable soil phosphorus (P) in organic dairy farming systems. Plant Soil 237:321–332

Logsdon SD, Radke JK, Karlen DL (1993) Comparison of alternative farming systems. I. Infiltration techniques. Am J Alternat Agric 8:15–20

Lohr L, Park TA (2002) Choice of insect management portfolios by organic farmers: lessons and comparative analysis. Ecol Econ 43:87–99

Lotter DW, Seidel R, Leibhart W (2003) The performance of organic and conventional cropping systems in an extreme climate year. Am J Alternat Agric 18:146–154

Mäder P, Fließbach A, Dubois D, Gunst L, Fried P, Niggli U (2002) Soil fertility and biodiversity in organic farming. Science 296:1694–1697

Marris E (2008) Water: more crop per drop. Nature 452:273–277

Matson PA, Parton WJ, Power AG, Swift MJ (1997) Agricultural intensification and ecosystem properties. Science 277:504–509

Melero S, Porras JCR, Herencia JF, Madejon E (2006) Chemical and biochemical properties in a silty loam soil under conventional and organic management. Soil Till Res 90:162–170

Merrill MC (1983) Eco-agriculture: a review of its history and philosophy. Biol Agric Hortic 1:181–210

Metzger MJ, Rounsevell MDA, Acosta-Michlik L, Leemans R, Schrotere D (2006) The vulnerability of ecosystem services to land use change. Agric Ecosyst Environ 114:69–85

Mulder C, De Zwart D, Van Wijnen HJ, Schouten AJ, Breure AM (2003) Observational and simulated evidence of ecological shifts within the soil nematode community of agroecosystems under conventional and organic farming. Funct Ecol 17:516–525

Nguyen ML, Haynes RJ, Goh KM (1995) Nutrient budgets and status in 3 pairs of conventional and alternative mixed cropping farms in Canterbury, New Zealand. Agric Ecosyst Environ 52:149–162

Nolte C, Werner W (1994) Investigations on the nutrient cycle and its components of a biodynamically-managed farm. Biol Agric Hortic 10:235–254

Oehl F, Sieverding E, Ineichen K, Mäder P, Boller T, Wiemken A (2003) Impact of land use intensity on the species diversity of arbuscular mycorrhizal fungi in agroecosystems of Central Europe. Appl Environ Microbiol 69:2816–2824

Pankhurst CE, Doube BM, Gupta VVSR, Grace PR (eds) (1994) Soil biota – management in sustainable farming systems. CSIRO Publishing, Melbourne

Papadopoulos A, Bird NRA, Whitmore AP, Mooney SJ (2009) Investigating the effects of organic and conventional management on soil aggregate stability using X-ray computed tomography. Eur J Soil Sci 60:360–368

Park TA, Lohr L (2005) Organic pest management decisions: a systems approach to technology adoption. Agric Econ 33:467–478

Paster G, Boeringa R (eds) (1980) Special issue – alternative methods of agriculture – description, evaluation and recommendations for research. A translation of parts of the original Dutch report Alternatieve-Landbouw-Methoden prepared in 1976 by the committee for research into biological methods of agriculture. Agric Environ 5:1–199

Peigné J, Cannavaciuolo M, Gautronneau Y, Giteau JL, Cluzeau D (2009) Earthworm populations under different tillage systems in organic farming. Soil Till Res 104:207–214

Pimentel D, Berardi G, Fast S (1983) Energy efficiency of farming systems: organic and conventional agriculture. Agric Ecosyst Environ 9:359–372

Porter JR, Semenov MA (2005) Crop response to climatic variation. Philos Trans Roy Soc B 360:2021–2035

Posner JL, Baldock JO, Hedtcke JL (2008) Organic and conventional production systems in the Wisconsin integrated cropping systems trial I: productivity 1990–2002. Agron J 100:253–260

Pretty JN, Morison JIL, Hine RE (2003) Reducing food poverty by increasing agricultural sustainability in developing countries. Agric Ecosyst Environ 95:217–234

Pulleman M, Jongmans A, Marinissen J, Bouma J (2003) Effects of organic versus conventional arable farming on soil structure and organic matter dynamics in a marine loam in the Netherlands. Soil Use Manage 19:157–165

Raupp J, Pekrun C, Oltmans M, Köpke U (2006) Long-term field experiments in organic farming. ISOFAR, International Society of Organic Agricultural Research Scientific Series. Dr Köster, Berlin

Reganold JP (1987) Long term effects of organic and conventional farming on soil erosion. Nature 330:370–372

Reganold JP, Palmer AS, Lockhart JC, Macgregor AN (1993) Soil quality and financial performance of biodynamic and conventional farms in New Zealand. Science 260:344–349

Reidsma P, Ewer F, Lansink AO, Leemans R (2009) Vulnerability and adaptation of European farmers: a multi-level analysis of yield and income responses to climate variability. Reg Environ Change 9:25–40

Reitmayr T (1995) Entwicklungen eines rechnergestuetzten Kennzahlensystems zur oekonomischen und oekologischen Beurteilung von agrarischen Bewirtschaftungsformen. – Dargestellt an einem Beispiel. Sonderheft 147. Agrarwirtschaft, Frankfurt am Main.

Roschewitz I, Gabriel D, Tscharntke T, Carsten T (2005) The effects of landscape complexity on arable weed species diversity in organic and conventional farming. J Appl Ecol 42:873–882

Ryan MH, Chilvers GA, Dumaresq DC (1994) Colonization of wheat by VA-mycorrhizal fungi was found to be higher on a farm managed in an organic manner than on a conventional neighbor. Plant Soil 160:33–40

Sandhu HS, Wratten SD, Cullen R, Case B (2008) The future of farming: the value of ecosystem services in conventional and organic arable land. An experimental approach. Ecol Econ 64:835–848

Sattelmacher B, Reinhard S, Pomikalko A (1991) Differences in mycorrhizal colonization of rye (*Secale-Cereale* L) grown in conventional or organic (biological-dynamic) farming system. J Agron Crop Sci 167:350–355

Scoones I, Thompson J (eds) (1994) Beyond farmer first. Intermediate Technology Publications, London

Scow KM, Somasco O, Gunapala N, Lau S, Venette R, Ferris H, Miller R, Shennan C (1994) Transition from conventional to low-input agriculture changes soil fertility and biology. Calif Agric 48:20–26

Shepherd MA, Harrison R, Webb J (2002) Managing soil organic matter – implications for soil structure on organic farms. Soil Use Manage 18:284–292

Siebeneicher GE (1993) Handbuch für den biologischen Landbau. Das Standwerk für alle Richtungen und Gebiete. Naturbuch-Verlag, Augsburg

Siegrist S, Schaub D, Pfiffner L, Mäder P (1998) Does organic agriculture reduce erodibility? The results of a long term field study on loess in Switzerland. Agric Ecosyst Environ 69:253–264

Smith P, Powlson DS, Smith JU, Falloon P, Coleman K (2000) Meeting Europe's climate change commitments: quantitative estimates of the potential for carbon mitigation by agriculture. Global Change Biol 6:525–539

Smolik JD, Dobbs TL, Rickerl DH (1995) The relative sustainability of alternative, conventional and reduced-till farming systems. Am J Alternat Agric 10:25–35

Stacey DA (2003) Climate and biological control in organic crops. Int J Pest Manage 49:205–214

Stockdale EA, Watson CA, Black, HIJ Philipps L (2006) Do farm management practices alter below-ground biodiversity and ecosystem function? Implications for sustainable land

management. JNCC Report No. 354, Joint Nature Conservation Committee, Peterborough, ISSN 0963 8901
Stockdale EA, Watson CA (2009) Biological indictors of soil quality in organic farming systems. Renew Agric Food Syst 24:308–318
Stockdale EA, Lampkin NH, Hovi M, Keatinge R, Lennartsson EKM, Macdonald DW, Padel S, Tattersall FH, Wolfe MS, Watson CA (2001) Agronomic and environmental implications of organic farming systems. Adv Agron 70:261–327
Stopes C, Lord EI, Philipps L, Woodward L (2002) Nitrate leaching from organic farms and conventional farms following best practice. Soil Use Manage 18:256–263
Syvasalo E, Regina K, Turtola E, Lemola R, Esala M (2006) Fluxes of nitrous oxide and methane, and nitrogen leaching from organically and conventionally cultivated sandy soil in western Finland. Agric Ecosyst Environ 113:342–348
Tanaka DL, Anderson RL, Rao SC (2005) Crop sequencing to improve use of precipitation and synergize crop growth. Agron J 97:385–390
Tate WB (1994) The development of the organic industry and market: an international perspective. In: Lampkin NH, Padel S (eds) The economics of organic farming: an international perspective. CAB International, Wallingford, pp 11–25
Tompkins EL, Adger WN (2004) Does adaptive management of natural resources enhance resilience to climate change? Ecol Soc 9:5
Trewavas A (2002) Malthus foiled again and again. Nature 418:668–670
Turner RJ, Davies G, Moore H, Grundy AC, Mead A (2007) Organic weed management: a review of the current UK farmer perspective. Crop Prot 26:377–382
United Nations Development Programme, UNDP (1992) Benefits of diversity. An incentive towards sustainable agriculture. Environment and Natural Resources Group. UNDP, New York
Vogl CR, Kilcher L, Schmidt H (2005) Are standards and regulations of organic farming moving away from small farmers' knowledge? J Sustain Agric 26:5–26
Wander MM, Hedrick DS, Kaufman D, Traina SJ, Stinner BR, Kehrmeyer SR, White DC (1995) The functional-significance of the microbial biomass in organic and conventionally managed soils. Plant Soil 170:87–97
Watson CA, Atkinson D (1999) Using nitrogen budgets to indicate nitrogen use efficiency and losses from whole farm systems: a comparison of three methodological approaches. Nutr Cycling Agroecosyst 53:259–267
Watson CA, Fowler SA, Wilman D (1993) Soil inorganic N and nitrate leaching on organic farms. Biol Agric Hortic 10:189–195
Watson CA, Ritz K, Younie D, Franklin M (1996) Nitrogen and soil biomass dynamics in ley/arable crop rotations. Aspects Appl Biol 47:43–50
Watson CA, Atkinson D, Gosling P, Jackson LR, Rayns FW (2002a) Managing soil fertility in organic farming systems. Soil Use Manage 18:239–247
Watson CA, Bengtsson H, Ebbesvik M, Loes AK, Myrbeck A, Salomon E, Schroder J, Stockdale EA (2002b) A review of farm-scale nutrient budgets for organic farms as a tool for management of soil fertility. Soil Use Manage 18:264–273
Watson CA, Walker RL, Stockdale EA (2008) Research in organic production systems – past, present and future. J Agric Sci 146:1–19
Wieser I, Hess J, Lidenthal T (1996) Stickstoff- Phosphor- und Kaliumbilanzen okologisch wirtschaftender Grunlandbetriebe im oberosterreichischen Vorlalpengebiet. Bodenkultur 47:81–88
Wood R, Lenzen M, Dey C, Lundie S (2006) A comparative study of some environmental impacts of conventional and organic farming in Australia. Agric Syst 89:324–348
Yan F, Schubert S, Mengel K (1996) Soil pH changes during legume growth and application of plant residues. Biol Fertil Soil 23:236–242
Yeates GW, Bardgett RD, Cook R, Hobbs PJ, Bowling PJ, Potter JF (1997) Faunal and microbial diversity in three Welsh grassland soils under conventional and organic management regimes. J Appl Ecol 34:453–470

Chapter 15
Biochar in Soil for Climate Change Mitigation and Adaptation

David Waters, Lukas Van Zwieten, Bhupinder Pal Singh, Adriana Downie, Annette L. Cowie, and Johannes Lehmann

15.1 Introduction

Global environmental change, including land degradation, loss of biodiversity, changes in hydrology and changes in climate patterns resulting from enhanced anthropogenic emission of greenhouse gases, will have serious consequences for world food security, particularly affecting the more vulnerable socio-economic sectors (Ericksen et al. 2009; Lal 2010). The World Bank suggests that at least a doubling of cereal yields and a 75% increase in meat production by 2030 are required to maintain the current level of nutrition globally (Fresco 2009). This poses a quandary. To significantly increase food production when large areas of agricultural lands will be adversely affected by climate change or converted into

D. Waters
Wagga Wagga Agricultural Institute, NSW Department of Primary Industries, PMB, Wagga Wagga, NSW 2650, Australia

L. Van Zwieten (✉)
Wollongbar Primary Industries Institute, NSW Department of Primary Industries, 1243 Bruxner Highway, Wollongbar, NSW 2477, Australia
e-mail: lukas.van.zwieten@industry.nsw.gov.au

B.P. Singh
Forest Science Centre, NSW Department of Primary Industries, P.O. Box 100, Beecroft, NSW 2119, Australia

A. Downie
Pacific Pyrolysis, 56 Gindurra Rd, Somersby, NSW 2250, Australia

A.L. Cowie
National Centre for Rural Greenhouse Gas Research, University of New England, Armidale, NSW 2351, Australia

J. Lehmann
Department of Crop and Soil Sciences, Cornell University, Ithaca, NY 14853, USA

B.P. Singh et al. (eds.), *Soil Health and Climate Change*, Soil Biology 29,
DOI 10.1007/978-3-642-20256-8_15, © Springer-Verlag Berlin Heidelberg 2011

forestry for C sinks may not be possible unless new technologies and sustainable practices are rapidly adopted. The application of biochar to agricultural soils may play a crucial role in global climate change mitigation through the reduction of greenhouse gas production and the sequestering of atmospheric carbon in soils (Gaunt and Cowie 2009; Lehmann 2007; McCarl et al. 2009; McHenry 2009; Read 2009). The agronomic benefits of biochar in soils (Chan et al. 2007; Steiner et al. 2008a, b) could assist in the adaptation of agriculture to meet rising demands for food and fibre. Furthermore, improving soil health with biochar application may increase resilience of agricultural systems and enable the continuation of farming on marginal lands. Application of biochar to soil has been shown to have many advantages including enhanced soil health characteristics, reduced metal contamination risks and consequently increased plant growth (Chan et al. 2007; Namgay et al. 2010; Reichenauer et al. 2009); as well as reduced greenhouse gas emissions from soil (Singh et al. 2010a; Van Zwieten et al. 2009; Yanai et al. 2007).

The competing and often conflicting demands of land use primarily stem from growing populations requiring housing and food, coupled with community desires for greater allocation of land to ecological reserves and the increasing production of energy crops to displace greenhouse gas emitting fossil fuels (Koomen et al. 2005; Simon and Wiegmann 2009). In addition to the challenge of a changing climate, the increasing claim for this scarce land use resource will force the necessity for greater productivity from less land, meaning farmers will need to undertake activities that result in significant yield increases. Land managers from more developed countries have historically had greater access to technological innovations and training, thereby improving the productivity of agricultural systems compared with those from developing countries. With the escalating effects of climate change technological adaptation will become increasingly vital to sustainably augment production systems globally (Bryan et al. 2009; Jones and Thornton 2009).

"Black carbon" (BC), a heterogeneous mix of carbonaceous materials formed from the incomplete combustion of biomass (Hammes et al. 2008; Schmidt and Noack 2000), is found in the most stable pool of soil organic carbon (SOC) (i.e. the component that resists microbial decomposition and mineralisation to CO_2). This BC may be derived from natural events such as biomass burning in wildfires (producing charcoal, consisting of partly charred organic matter through to completely carbonised submicron particles of soot) or through human activities (referred to as biochar).

Biochar can be manufactured through the pyrolysis of biomass (Lehmann and Rondon 2006), which condenses aliphatic carbon into more stable aromatic carbon, while releasing combustible gases (H_2, CH_4, CO) that can be used to heat the kiln with surplus for bioenergy. Rudimentary biochar production systems have been used for over 2,000 years, and when applied to soils biochar has demonstrated sustained productivity increases. A well-known example of ancient soil amendment with charcoal is the Terra Preta – dark earth-soils of the Amazon. These low fertility tropical soils were amended in pre-Columbian times by indigenous Amerindians through the addition of carbonised organic matter, believed to be from their cooking hearths (Glaser et al. 2001).

Modern biochar production uses a range of technologies including fast pyrolysis, gasification and/or carbonisation (Bridgwater 2003). These processes can be applied at different scales from small cooking stoves often used in developing countries through to more advanced industrial systems which include full gas recovery for integrated bioenergy production (Brown 2009). The sustainability credentials of each of these systems including efficiency of resource utilisation, emissions control, life cycle greenhouse gas balance and environmental sustainability need to be assessed on a case-by-case basis. The production process will influence the properties of the biochar and therefore the way it behaves and interacts in a soil (Downie et al. 2009; Glaser et al. 2002; Joseph et al. 2010; Novak et al. 2009; Singh et al. 2010a). To date agronomic benefits from biochar application have been demonstrated for biochars produced from a limited range of production systems (mainly small industrial scale, demonstration-level, pyrolysis or gasification units involving co-production of biochar, bio-oil and/or syngas) and applied to limited soil/plant systems. However, further research is required to quantify the impacts of biochar produced from a range of small (including mobile units) to large industrial scale biochar production systems and conditions, and then applied to contrasting soil/plant systems.

The global potential for annual sequestration of atmospheric CO_2 through biochar application has been estimated at the billion-tonne scale (Gt/year) under present day scenarios (Laird et al. 2009). The greenhouse gas mitigation potential from the application of biochar to agricultural systems may vary widely with variation in biomass feedstock, production technologies, product utilisation methods and environmental conditions.

15.2 Biochar Properties for Soil Health and Climate Change

15.2.1 Biochar Stability

The stability of organic matter in soils is determined by its ability to resist microbial and/or chemical decomposition, through chemical transformations and physical interactions with soil minerals (Lehmann et al. 2007; Rasse et al. 2006; Skjemstad et al. 1996). BC, as either charcoal or biochar, has a predominantly condensed aromatic structure that is known to be highly resistant to microbial decomposition (Baldock and Smernik 2002). Additionally, interactions of biochar with soil minerals could further increase stability of biochar in soil (Brodowski et al. 2006), further contributing to long-term carbon sequestration (Lehmann et al. 2009), while also adding to the health and production outcomes of soil systems.

Published studies have reported soil residence time of charcoal and biochars in timescales ranging from decades to centuries to millennia (Cheng et al. 2008b; Hamer et al. 2004; Hammes et al. 2008; Kuzyakov et al. 2009; Major et al. 2010; Skjemstad et al. 1996; Titiz and Sanford 2007; Zimmerman 2010). The stability of biochar depends on the type of biomass feedstock, charring conditions (temperature, heating time), biochar particle size, and edaphic and climatic conditions under which biochar oxidises (Kuzyakov et al. 2009; Lehmann et al. 2009; Nguyen and Lehmann 2009; Nguyen et al. 2010; Singh and Cowie 2008, 2010; Zimmerman 2010). In general, the proportion of aryl-C to aliphatic-C in biochar increases with increasing charring or pyrolysis temperature (Baldock and Smernik 2002; McBeath and Smernik 2009; Nguyen et al. 2010). The lability and density of the biomass feedstock and its mineral content may also influence the decomposition rate of biochar in soil (Nguyen et al. 2010; Singh and Cowie 2008, 2010).

Spectroscopic and surface chemistry analyses have proven useful to evaluate biochar–mineral interactions and oxidation status of biochar along a decomposition continuum (Cheng et al. 2006, 2008b; Liang et al. 2008). However, these approaches do not quantify turnover time, necessary to evaluate the residence time of biochar in soil. The rate of biochar decomposition may vary according to the stability of the oxidisable component, i.e. initial rapid decomposition of surface-oriented labile components of the biochar particle (e.g. aliphatic-C) followed by slow decomposition of condensed aromatic-C, which dominates the core structure of biochar. This warrants long-term studies to accurately estimate the mean residence time of biochar in soil (Kuzyakov et al. 2009; Nguyen and Lehmann 2009). Furthermore, biochars can potentially stimulate decomposition of native soil organic matter (i.e. humic and labile components) possibly by enhancing microbial activity (Hamer et al. 2004; Wardle et al. 2008). However, application of biochar may also lead to a decline in the decomposition of other organic matter components, through the possible enhancement of soil aggregation (Liang et al. 2010). The "priming effect" of biochar on organic matter decomposition in soil needs to be accounted for to determine the magnitude of biochar decomposed. Carbon isotope methods ($\delta^{13}C$, or $^{14}C/^{13}C$ labelling) can be used to identify sources of C decomposed in biochar–soil systems (Kuzyakov et al. 2009). These methods can be relatively easy to manage in the laboratory, providing optimal conditions for biochar decomposition. However, in the field, presence of plant roots, rhizosphere processes and variable environmental conditions provide challenges to identifying C sources with a limited number of isotopes (Major et al. 2010).

15.2.2 Nutrient and Liming Values of Biochar

Some biochars are a potential source of nutrients (Table 15.1). The nutrient content of biochar is largely determined by biomass feedstocks (Gaskin et al. 2008; Singh et al. 2010b; Table 15.1). Feedstocks with higher nutrient contents such as animal manures will produce biochars with greater nutrient value, compared with plant feedstocks (Singh et al. 2010b). Pyrolysis temperature also affects nutrient value: for example, analysis of two biochars produced under different temperatures (400 and 500°C) from the same poultry litter feedstock revealed a higher N percentage (3.47%) and lower P percentage (3.01%) for the lower temperature product

Table 15.1 Nutrient content of selected biochars

Biochar source	N (%)	P (%)	K (%)	Ca (%)	CEC (cmol/kg)	C (%)	pH water	C:N	EC (dS/m)	Production temperature	References
Green wastes	0.18	0.07	0.82	<0.01	24	36	9.4[a]	200	3.2	450°C	Chan et al. (2007)
Poultry litter	2	2.5	–	–	–	38	9.9[a]	19	5.6	450°C	Chan et al. (2008)
Poultry litter	3.47	3.01	5.11	4.27	61.1	39.2	10.1	11.3		400°C	Gaskin et al. (2008)
Poultry litter	3.09	3.59	5.86	5.04	38.3	39.2	9.74	12.7		500°C	Gaskin et al. (2008)
Bark of *Acacia mangium*	1.04				37.14	39.8	7.4	38		260–360°C	Yamato et al. (2006)
Paper mill sludge and wood (1:1)	0.48		0.22	6.2	9.0	50	9.4	104		550°C	Van Zwieten et al. (2010a)
Paper mill sludge and wood (1:2)	0.31		1.0	11.0	18.0	52	8.2	168		550°C	Van Zwieten et al. (2010a)
Soybean cake	7.82					58.81		7.5		550°C	Uzun et al. (2006)
Pinus ponderosa bark	<0.01	<0.01	–	–	34.5	71.5	4.81	–	1.12	350°C	Gundale and DeLuca (2007)
Cow manure/*Pinus* spp. (3:1)	1.2	0.3	1.9	1.0	–	73.3	9.4	61	–	500°C	Kolb et al. (2009)
Pinus taeda chips	0.255	0.015	0.145	0.171	7.27	73.9	7.55	290		400°C	Gaskin et al. (2008)
Pinus taeda chips	0.223	0.014	0.145	0.185	5.03	81.7	8.3	366		500°C	Gaskin et al. (2008)
Eucalyptus deglupta wood	0.57	0.06	–	–	4.7	82.4	7.00	144	–	350°C	Rondon et al. (2007)
Eucalyptus saligna wood	0.22	0.03	0.27	0.98	–	85.1	9.4	387	–	400–500°C	Kimetu et al. (2008)
Eucalyptus saligna wood	0.26	0.02	0.24	2.13	3.48	83.6	8.82	322	0.32	550°C	Singh et al. (2010a)
Tectona grandis, *Pterocarpus macrocarpus*	0.3		3.1	4.4	10.7	87	7.5	290		Earth mound	Asai et al. (2009)

[a]pH measured in 1:5 soil/0.01 M $CaCl_2$ extract

compared with the higher temperature product (3.09% and 3.59% respectively, Table 15.1) (Gaskin et al. 2008). Furthermore, the concentration of C and N may increase with increasing pyrolysis temperature in plant-based biochars, but the C and N concentrations may decrease with increasing pyrolysis temperature for mineral-rich feedstocks, such as manure or papermill sludge, because less-volatile elements, including P, K, Ca and Mg, are concentrated as the volatiles are lost (Gaskin et al. 2008; Singh et al. 2010b). However, information on forms and bioavailability of nutrients present in biochars is scarce, and some research has shown that feedstock type and pyrolysis temperature can significantly influence bioavailable fraction of nutrients in biochars (Gaskin et al. 2008; Singh et al. 2010b).

Many biochars have a neutral to alkaline pH value (Table 15.1) and can provide some benefit in neutralising acidic soils. Van Zwieten et al. (2010a) reported liming values of 33 and 29% for two papermill waste biochars (compared to carbonate). Singh et al. (2010b) found that the $CaCO_3$ equivalence of biochars increased with increasing pyrolysis temperature.

15.2.3 *Surface Charge Properties*

Cation exchange capacity (CEC) is a measure of the ability of a substrate to retain positively charged ions through electrostatic forces. Biochar has been associated with the enhancement in CEC of some amended soils (Glaser et al. 2001; Van Zwieten et al. 2010a), thereby increasing the availability and retention of plant nutrients in soil and potentially increasing nutrient use efficiency. However, biochars from different feedstocks and produced under differing pyrolysis conditions may differ in surface charge properties. Furthermore, the method for determining CEC of biochars is far from standardised, and methods applied to soils may not be appropriate to biochars (Singh et al. 2010b). Of the two biochars from the same peanut hull biomass, the biochar produced at 500°C had a lower CEC (4.63 cmol/kg) compared with that produced at 400°C (14.2 cmol/kg) (Gaskin et al. 2008). The reduction in surface functional groups was suggested as the cause of lower CEC in the biochar produced at higher temperatures. The decline in the acidic functional groups on biochar surfaces has been reported to be greatest between 300 and 400°C (Guo and Rockstraw 2007). Liang et al. (2006) reported that the high charge density (CEC/specific surface area) of "aged" biochar resulted from oxidation of the particles and adsorption of organic matter to biochar surfaces. An increase in the charge density on biochar surfaces as biochar interacts with soil over time (e.g. Cheng et al. 2008a) could be responsible for enhanced cation retention and consequently reduced leaching from amended soils (Singh et al. 2010a). However, more research on the chemical interactions of differing biochars and soils, as well as the implications for soil nutrient retention, is needed.

15.3 Impacts of Biochar on Soil Health and Plant Growth

Biochar application can potentially influence a number of physical, chemical and biological properties of soil due to the inherent characteristics of biochar, and properties that develop over time through oxidation of biochar surfaces and interaction with plant–soil–microbial components. Some potential impacts of biochar application on soil health, soil carbon dynamics, nutrient use efficiency and plant growth are described below, and the benefits to plant–soil systems are summarised in Fig. 15.1.

15.3.1 Soil Physical Health

Increases in SOC contents often contribute to enhanced soil aggregate stability (e.g. Albiach et al. 2001; Chan et al. 2003; Neufeldt et al. 2002) which can result from interactions of carbon functional groups and clay mineral surfaces (Lehmann et al. 2008). Evidence suggests a close interaction between biochar particles and clay mineral surfaces, which may aid in the occlusion of biochar particles within newly formed soil aggregates (Brodowski et al. 2006; Liang et al. 2008). Implications that

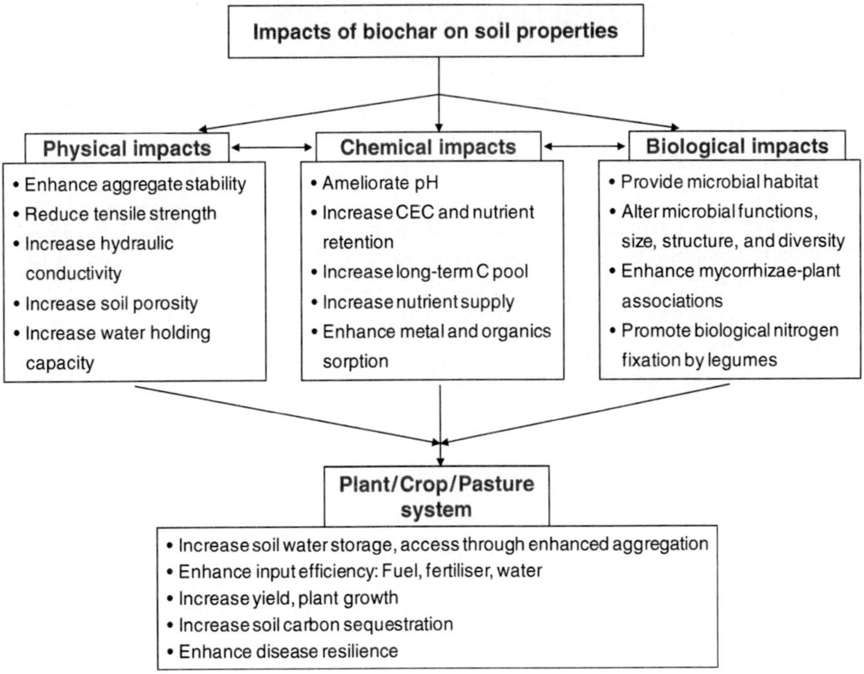

Fig. 15.1 Potential impacts of biochar application to plant–soil systems

biochar may contribute to the physical stabilisation of other soil organic matter, through aggregation (Liang et al. 2008), could also suggest an enhanced soil structure.

Biochar has also been associated with the enhancement of other soil physical properties such as soil water retention, saturated hydraulic conductivity and porosity. A study of available soil moisture in three soil types (sand, loam, clay) with 15, 30 and 45% wood biochar additions reported increases in the sandy soils, no change in the loam and a decrease in the clay soil (Tryon 1948). Similarly, Glaser et al. (2002) reported an 18% increase in field capacity for high BC Anthrosol soils compared to low BC surrounding soils, and attributed this to the increased surface area and porous structure of the char particle. In a study of soils under charcoal kilns in Ghana, saturated hydraulic conductivity and total porosity were increased and bulk density decreased compared to adjacent field soils (Oguntunde et al. 2008). Biochar was reported to enhance saturated hydraulic conductivity and water-holding capacity in upland rice production in Northern Laos (Asai et al. 2009), indicating a greater potential for efficient water use and improved soil productivity. The improvement in aggregation, water retention, saturated hydraulic conductivity and porosity from different biochar-amended soils could lead to better plant water use efficiency and consequently more resilient plant systems, and needs greater investigation.

15.3.2 Soil Chemical Health

The addition of biochar to soils can have a positive effect on soil chemical processes. Studies of the Terra Preta soils in the Amazon Basin have revealed significantly higher CEC per unit of SOC, attributed to the high level of "aged" biochar-like carbon in the Anthrosol soils (Glaser et al. 2001; Liang et al. 2006). In a pot trial, two papermill biochars (10 t/ha) increased the CEC and pH of a Ferralsol; however, there was no effect on a calcarosol (Van Zwieten et al. 2010a). The addition of a pecan biochar to a sandy Norfolk soil at rates of 1 and 2% did not change the soil's inherent CEC, although soil pH was raised more than one unit over the two incubation periods (Novak et al. 2009). These latter authors suggested minimal surface oxidation due to high pyrolysing temperatures may be the reason for the unchanged CEC.

Increased retention of plant available nutrients in soils as a result of biochar application could have significant agronomic and environmental benefits. Increased retention of inorganic nutrients such as ammonium and potassium within the soil profile may reduce fertiliser requirements. Furthermore, reducing nutrient losses from leaching may slow soil acidification (Helyar et al. 1990) and eutrophication of waterways. Although mechanisms for increasing soil nutrient retention have recently been explored (Liang et al. 2006; Major et al. 2009), greater understanding of the impacts of biochar on different soil types and in different climatic conditions is still required.

15.3.3 Soil Biological Health

Many studies have reported a positive response of soil micro-organisms to biochar amendments (O'Neill et al. 2009; Pietikäinen et al. 2000; Steiner et al. 2008a; Thies and Rillig 2009; Warnock et al. 2007; Zackrisson et al. 1996), although overall soil productivity outcomes from these interactions are mostly undocumented. Microbe/biochar interactions could include the attraction of microbes to the products of biochar adsorption such as other organic matter fractions, soil mineral components and nutrients, and extracellular enzymes (Thies and Rillig 2009).

Several studies have reported increased N mineralisation and nitrification through biological processes with charcoal amendment in forest soils (Berglund et al. 2004; DeLuca et al. 2002; MacKenzie et al. 2008). It has been suggested that the adsorption of phytotoxic phenolic compounds by charcoal in forest soils reduces the inhibition of nitrifying micro-organisms in these soils (Berglund et al. 2004; MacKenzie and DeLuca 2006; Zackrisson et al. 1996), or reduces the presence of organic compounds that could stimulate N immobilisation (DeLuca et al. 2006). In agricultural soils, N mineralisation and nitrification may be reduced by biochar addition due to either N immobilisation by N-poor and labile biochar (i.e. a high C/N ratio), or adsorption of ammonium (Lehmann et al. 2006). A study on the effect of a manure-pine biochar in four soils from Wisconsin reported enhanced microbial biomass and activity, as well as decreased extractable N with increasing biochar rates in the three agricultural soils (Kolb et al. 2009). However, Kolb et al. (2009) recorded the highest extractable N in the coniferous forest soil with the highest biochar rates. Pietikäinen et al. (2000) reported that charcoal adsorbed up to 42% of dissolved organic carbon from a litter extract, which consequently attracted and harboured micro-organisms.

Biochar may enhance the symbiotic associations of mycorrhizal fungi (MF) and terrestrial plants. Demonstrations of the positive response of plant growth and nutrient availability as a result of enhanced MF colonisation following BC additions in soils have been reported (Makoto et al. 2010). Root growth and aboveground biomass of *Larix gmelinii* (Gmelin larch) both increased with applied BC alone, and were greatest when BC was applied with MF. Phosphorus concentration in needles of the larch seedlings was also highest from the application of biochar with MF, indicating increased plant uptake, due to the utilisation of phosphate by the MF and seedling root/BC contact. A trial of maize amended with Acacia bark charcoal in Indonesia recorded increases in plant root mass and colonisation rates of MF (Yamato et al. 2006). A review of biochar–mycorrhizal interactions reported numerous positive responses, such as increases in soil nutrient availability and enhanced disease resistance, but also noted that a few studies reported a negative effect on MF with biochar addition, possibly from a reduction in plant available nutrients (Gaur and Adholeya 2000; Warnock et al. 2007).

Biochar has also been implicated in the enhancement of biological N_2 fixation (BNF) of *Phaseolus vulgaris* (Rondon et al. 2007). This study reported a BNF increase of 49% and 78% with 30 and 60 g/kg biochar additions, respectively.

However, a 90 g/kg biochar application increased BNF only by 30% above the control due to lower total biomass production and plant N uptake. Rondon et al. (2007) stated that greater boron and molybdenum availability were the main reasons for the increase in BNF. While some evidence exists for the improvement of plant–soil systems from BC/microbe interactions, this field of research is currently largely unexplored.

15.3.4 Turnover of SOC

Biochar addition to soils may influence the net carbon balance of systems. A stepwise increase in total soil carbon due to direct biochar addition is expected (Chan et al. 2007; Novak et al. 2009; Van Zwieten et al. 2010a). For example, a study of incubations of a Norfolk loamy sand amended with four rates of pecan shell biochar (0, 0.5, 1.0 and 2.0%) revealed increases in total SOC with increasing biochar rates (Novak et al. 2009). In a pot trial of *Raphanus sativus* with the addition of two poultry manure biochars (10, 25, 50 t/ha), total SOC increased compared to the controls (Chan et al. 2008). Furthermore, Liang et al. (2010) reported a greater incorporation of added plant carbon (sugarcane residue) into the intra-aggregate fraction in the *terra preta* soils as compared to the control soil (oxisol), indicating enhanced stabilisation of added carbon in the soil enriched with biochar-like organic matter. Additionally, in the studied *terra preta* soils, biochar-like carbon was found to reside primarily in organo-mineral (heavy) rather than free (light) fractions (Liang et al. 2010). However, another study reported that 72–75% of the light fraction of organic matter in an agricultural soil in Ontario was BC from the previously burnt C_3 forest, and that the turnover of the light fraction with BC was 2.5 times slower than without BC (Murage et al. 2007), suggesting a net reduction in the turnover rate of the light fraction in the presence of BC. In a cropping trial from Brazil, the loss of SOC over 20 months was reduced from biochar-amended soils (4–8% C) in comparison to soils amended with chicken manure, compost, or non-amended control plots (27, 27, and 25% C loss) (Steiner et al. 2007). In a study of historical charcoal blast furnace sites across the eastern half of the USA (Cheng et al. 2008b), organic carbon in the BC-containing soils was more stable, with a lower labile fraction and longer half-life of the recalcitrant component, compared to adjacent non-BC soils.

However, as noted previously in the stability section (Sect. 15.2.1), the overall increase of SOC due to biochar addition may sometimes be partly offset or even negated by the increased turnover of native/labile C (Hamer et al. 2004; Steinbeiss et al. 2009; Wardle et al. 2008). A 10-year study of mesh bags mixed with biochar and humus in a boreal forest site recorded a greater loss of carbon mass, compared with mesh bags of biochar or humus alone (Wardle et al. 2008). However, it was unclear as to the exact source of the carbon losses, or to their specific fate (i.e. leaching or emission). These losses occurred predominantly in the first year of mesh mixing and in the absence of a mineral component and soil profile; so there is some uncertainty as to the effect of biochar on humus in this instance (Lehmann and Sohi 2008). Another study investigating the influence of biochar on decomposition rates of litters of different quality mixed in a cambisol found no difference in the rate of decay between separate and combined mixtures of these substrates over 240 days of incubation (Abiven and Andreoli 2010). Clearly, further research is needed to generalise the effect of biochar on decomposition of relatively labile forms of organic carbon in soil and to advocate the role of biochar in offsetting global CO_2 emissions (Woolf et al. 2010).

The complexity of interactions between biochar and soil, and consequences of these for carbon sequestration, appear to revolve around the type of biochar, its degree of ageing and the extent of interaction with minerals and organic matter components in soil (Brodowski et al. 2005; Liang et al. 2008). It may well be that as biochar ages in a soil, increasing interactions with soil mineral components may help protect the labile and recalcitrant components of biochar from further biotic and abiotic oxidation. The occlusion of biochar particles within soil mineral aggregation has also been demonstrated in a study of a long-term agricultural field experiment in Germany (Brodowski et al. 2006). It was suggested that biochar could act as a binding agent in micro-aggregation. Further studies involving the identification and influence of specific biomass feedstocks and biochar production conditions to the mechanisms of biochar–soil interactions, as well as processes leading to stabilisation of biochar and other forms of organic matter in biochar-amended soil, are needed to assess the overall influence of biochar on the net soil carbon balance. In particular, the biochemical (e.g. microbial activity, aromaticity) and physicochemical (aggregation, sorption) factors affecting turnover of various forms of SOC need further investigation.

15.3.5 Nutrient Use Efficiency

There have been several reports of increases in fertiliser use efficiency with the addition of biochar to soils. A glasshouse study of the agronomic response of wheat, soybean and radish to the application of paper mill waste biochar in a ferrosol and calcarosol, revealed an increase in biomass of wheat (250% of fertilised control), as well as soybean and radish, with fertiliser plus biochar in the ferrosol (Van Zwieten et al. 2010a). The authors reported significantly increased N uptake for the wheat treatment and suggested an improvement in fertiliser use efficiency. However, the results of biochar and fertiliser amendments in the calcarosol were variable, with increased soybean growth but reduced wheat and radish growth. In an upland rice production system in Northern Laos, treatments with wood biochar reported higher grain yields and improved response to fertiliser treatments (Asai et al. 2009), although the authors noted that the positive yield response was dependent on adequate soil nitrogen. In another study, the application of a low nutrient biochar derived from timber increased the retention of N in soil and increased uptake of N into crop biomass (Steiner et al. 2008a, b).

When biochar amendments are combined with fertilisers, the effect is often synergistic, most likely due to increased plant nutrients and nutrient use efficiency from greater retention (Hossain et al. 2010). In a cropping trial in Brazil, wood charcoal and NPK fertiliser together significantly improved plant growth and doubled grain production of *Oryza sativa* and *Sorghum bicolor*, compared with NPK fertiliser alone (Steiner et al. 2007). These authors also reported higher plant available nutrients for following crops, despite the greater nutrient export from the higher plant yields of the biochar-amended plots. In a pot trial of *R. sativus*, a combination of 50 or 100 t/ha addition of green waste biochar and N fertiliser increased dry matter by approximately two times, compared with the N fertiliser treatment only, and 3.7 times compared to the biochar treatment only (Chan et al. 2007).

15.3.6 Plant Growth and Yield

The production of plant biomass through photosynthesis removes CO_2 from the atmosphere, and therefore any increase in plant biomass (carbon stock) due to biochar additions in soil systems will contribute to the mitigation of rapidly rising atmospheric CO_2 levels. Specifically, biochar either increases plant nutrient availability or enhances the soil environment (e.g. CEC, soil pH, aeration) and therefore may indirectly contribute to enhanced plant growth (e.g. Chan et al. 2008; Lehmann et al. 2003; Steiner et al. 2007; Zackrisson et al. 1996).

Some studies have reported increased plant nutrient availability and crop yield with the addition of BC alone. In a cropping trial (*Vigna unguiculata* and *O. sativa*) in Amazon Basin archaeological Anthrosol soils with high carbon levels and Ferralsols with added wood biochar, significantly increased phosphorus, calcium, manganese and zinc availability was found, with a 38–45% increase in biomass of the two crops in the Anthrosol (Lehmann et al. 2003). In a *Zea mays* trial of degraded cropping soils in Western Kenya, the authors noted that the application of biochar doubled crop yield, and furthermore suggested that the improvement could not be explained by biochar nutrient availability alone (Kimetu et al. 2008). Hence, despite the low nutrient status of some biochars, biochars generally appear to increase nutrient availability through increased ion retention in soils (Liang et al. 2006; Tryon 1948) and therefore potentially enhance plant yields. Biochar applications produced from manures may directly contribute high levels of nutrients to soils. Chan et al. (2008) reported yield increases of *R. sativus* with the application of 10, 25 and 50 t/ha of poultry manure biochar alone.

However, there have been variable results from the addition of some biochar types in particular soils. Van Zwieten et al. (2010a) reported reduced growth in wheat and radish with the addition of a paper mill sludge biochar in a calcarosol. In a pot trial of *R. sativus* in an Alfisol, a 10 t/ha green waste biochar and N fertiliser amendment resulted in a biomass decrease of 30%. This latter study also reported biomass increases at higher biochar rates. Other studies have reported a decline in soil N availability with wood biochar addition, potentially causing reduced yields (Asai et al. 2009). The inconsistency of plant response, ranging from small declines to large increases, would indicate a need for further research to verify the different plant responses to different biochars under varying soil conditions.

15.4 Role of Biochar in Climate Change Mitigation and Adaptation

The previous sections of this chapter have demonstrated the considerable potential for biochar to enhance the fertility and productivity of agricultural systems, as well as provide a stable form of carbon for sequestration in soil. As the sustainability of agriculture becomes increasingly threatened by climate change (Chap. 1), tools such as biochar will be needed to enhance resilience and productivity of these systems, so that world food supply can satisfy demand. Changes to world rainfall patterns may see declines in some of the major food producing areas of the world (Howden et al. 2007); the role of biochar in enhancing moisture retention may prove critical to maintaining production in these locations. Increases in soil health and crop productivity may have a range of resultant environmental, social and greenhouse gas balance implications. For example, higher crop productivity due to improved soil health could result in less use of land for the same yield, thereby reducing the need to produce food on more marginal land, and potentially increasing production per unit of gaseous emission. Enhanced crop productivity from biochar application may also reduce the rate of land clearing and deforestation, or encourage the rejuvenation of degraded land, again with significant positive ecological, social and economic consequences. A schematic of potentially interrelated ecosystem benefits of biochar production/application systems, including enhanced climate change mitigation and adaptation and improved performance of plant–soil systems, is presented in Fig. 15.2, and evidence for some of these benefits is described below.

15.4.1 Mitigation of N_2O Gas Emissions from Soil

Soil represents a significant source of the greenhouse gas nitrous oxide (N_2O). The microbial processes nitrification and denitrification are largely responsible for production of N_2O in soil (Chen et al. 2008; Dalal et al. 2003; Yanai et al. 2007). As the global warming potential of N_2O is 298 times greater than the equivalent mass of CO_2 in the atmosphere (Forster et al. 2007), technologies to minimise soil N_2O emissions need to be implemented to meet demands for climate change mitigation. Some recent studies have provided evidence that emissions of N_2O

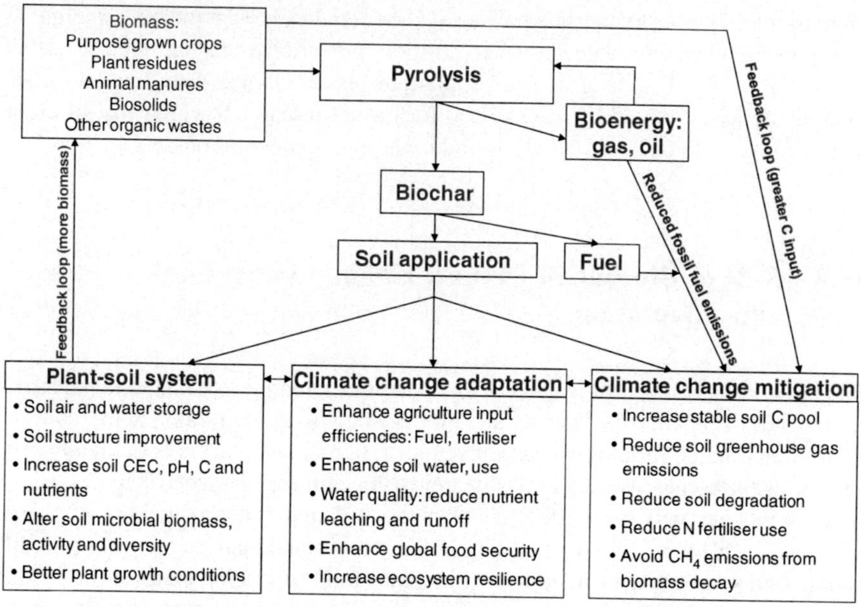

Fig. 15.2 Potential ecosystem benefits from biochar production/application systems

may be reduced by biochar application to soil (Singh et al. 2010a; Spokas and Reicosky 2009; Van Zwieten et al. 2010b; Yanai et al. 2007). The magnitude of reduction in N_2O emissions is dependent on soil type, biochar type and application rate, soil moisture content, and biochar ageing (Singh et al. 2010a; Spokas and Reicosky 2009; Van Zwieten et al. 2010b). However, in some cases, emissions of N_2O from soil can also be increased or not affected by the presence of biochar. For example, Clough et al. (2010) showed short-term increases in N_2O emissions in a pasture soil following biochar application (at 4.3% w/w) in the presence of ruminant urine; however, no significant differences were observed in cumulative N_2O emissions over the 53-day laboratory incubation between the biochar plus urine and urine-only treatments. Likewise, Spokas and Reicosky (2009) found that application of a high nitrogen compost-amended biochar (at 10% w/w) resulted in high N_2O emissions from three different soils and Spokas et al. (2009) found no significant differences in soil N_2O emissions at biochar application rates of 2–10% (w/w). The exact mechanisms for observed effects of biochar on N_2O emissions remain largely unexplored (Van Zwieten et al. 2009). Singh et al. (2010a) found that effectiveness of biochars in reducing soil N_2O emissions can increase over time, and hypothesised that this may be due to increased sorption capacity of biochars through oxidative reactions on biochar surfaces with ageing. Thus, in addition to its potential long-term soil carbon sequestration value, biochar application could provide considerable greenhouse gas mitigation benefit if reductions in N_2O emissions are found to apply broadly (Van Zwieten et al. 2009).

15.4.2 Reduced N Fertiliser Requirements

As the demand for food increases through wealth and population pressures, so too does the need for resources such as fertilisers and water. Nitrogen in particular is a resource that is poorly managed (Spiertz 2010), and more effort is needed to ensure that N supply matches N demand (see Chap. 6). As less than 50% of soil nitrogen can be used by the crop (Baligar et al. 2001), technologies that improve N use efficiency will have implications for productivity and emissions. A large portion of N is lost through mechanisms such as leaching (Olarewaju et al. 2009), or loss by denitrification and ammonia (NH_3) volatilization (Khalil et al. 2009). As the manufacture of nitrogen fertiliser releases more than 3 t CO_2e per t N (West and Marland 2002), technologies that can reduce the frequency and quantity of N application will result in lower emissions from the resulting reduction in fertiliser application. The evidence for increases in N fertiliser use efficiency with biochar amendments is reviewed in Sect. 15.3.5.

15.4.3 Biofuel Production

With mounting evidence for global warming from anthropogenic emissions of greenhouse gases, alternative forms of energy to reduce society's dependence on fossil fuels are required. The production of biofuels, from the chemical or thermal conversion of biomass (Bridgwater 2003), is currently being promoted as an alternative energy source that may help to reduce reliance on fossil fuel and avoid CO_2 emissions. The chemical and thermal pathways that produce biochar result in the co-production of combustible gas and/or oil which can be used for bioenergy production. It has been estimated that agricultural lands in the USA could provide enough manure through feedlot and intensive dairies to supply 0.7 billion US dollars of energy in terms of barrel of oil equivalents, based on a 20% thermochemical conversion factor of biomass (Ro et al. 2009). The energy output of pyrolysis has been favourably compared to that of the production of ethanol from corn. Even when pyrolysis is optimised for biochar production, energy output is 2–7 MJ, per MJ of fossil energy input (Gaunt and Lehmann 2008) compared with 1–2 MJ for corn to ethanol (Cherubini et al. 2009). The future decline in world fossil fuel reserves may enhance the relative merits of sustainable energy technologies such as pyrolysis with the added benefits of biochar application to soils.

15.4.4 Soil Structure Improvements

Well-structured soils are generally characterised by stable aggregation, high saturated hydraulic conductivity, low tensile strength and often high water-holding capacity. These qualities are all desirable as they assist in maintaining soil and plant productivity. Amendments such as biochar that may assist in the efficient capture, storage and utilisation of water in soils through structural improvement will become increasingly vital with any decline in rainfall as a consequence of a changing climate.

A poorly structured soil can present a substantial challenge for plant root development due to physical constraint associated with higher bulk densities and high soil tensile strength. Soil structure can be improved through the accumulation of soil organic matter (e.g. Perie and Ouimet 2008; Ruehlmann and Korschens 2009), with the more labile forms increasing the stability of macro-aggregates and less labile forms increasing the stability of micro-aggregates (Tisdall and Oades 1982). The evidence for better soil structure and enhanced physical properties (see Sect. 15.3.1) suggests that biochar may be a useful tool to mitigate climate change outcomes such as reduced rainfall, or extreme weather events (e.g. floods). Increased soil water use efficiency can help mitigate the impact of reduced annual rainfall on plant growth, while soil and nutrient losses from erosion during extreme weather events can be greatly reduced by increased soil aggregate stability and decreased surface runoff through enhanced infiltration.

As soil tensile strength and compaction increase, so does the requirement for greater cultivation draught capacity and frequency of tillage (O'Sullivan and Simota 1995). It could thus be anticipated that as biochar amendments can reduce soil tensile strength in a hard setting soil (Alfisol) as reported by Chan et al. (2007), and in a Norfolk loamy sand (Busscher et al. 2010), it would be reasonable to suggest that biochar could, in some soils, also reduce cultivation requirements, and hence reduce fuel usage. However, there is little direct evidence for overall enhancement of soil aggregation by biochar application, and the timeframe required; this aspect needs further research, especially as part of long-term assessment of potential agronomic and environmental benefits of biochar application in field studies.

15.4.5 Ecological Resilience

Appropriate biological functioning in soil systems can contribute to climate change adaptation through improvements in nutrient availability (Geisseler et al. 2009; Lavelle 1988), disease suppression (Larkin 2008) and aggregate stability (Lee and Foster 1991; Rillig and Mummey 2006). Many studies have reported increased microbial biomass in response to biochar amendments (O'Neill et al. 2009; Steiner et al. 2008a; Warnock et al. 2007; see Sect. 15.3.3). Biochar may enhance the symbiotic associations between MF and terrestrial plants, strengthening the plant's adaptability to climate change. Furthermore, biochar could provide long-term storage of carbon in soils while enhancing soil productivity, thereby enhancing the sustainability of agro-systems.

15.4.6 *Net Mitigation Benefits*

As indicated above, biochar may deliver mitigation benefits to terrestrial systems through several routes: stabilisation of soil organic matter, thus reducing its rate of oxidation while also decreasing soil erosion through improved aggregation; production of bioenergy that can displace fossil energy emissions (see Chap. 16); reduction in N_2O emissions from soil and fertilisers; reduction in fuel requirement for cultivation; increased carbon stock in plants and soil (Woolf et al. 2010). Furthermore, some biomass feedstocks, when used in biochar production, may deliver added benefit through avoided emissions: biomass that would have been deposited in landfill would have released methane (CH_4), while decomposition of manures can release CH_4 and N_2O gases (Gaunt and Cowie 2009). Therefore, the production and sequestration of biomass C in the form of biochar (with co-production and utilisation of bioenergy to offset fossil fuel emissions) could help slow climate change through the net removal of CO_2 from the atmosphere and avoiding emissions in the order of 1.0–1.8 Mt CO_2e/year at current levels of feedstock availability (Woolf et al. 2010). Gaunt and Cowie (2009) estimated net emissions reduction of 130–5,900 kg CO_2e/t feedstock for biomass residues (straw, manure and greenwaste), with variation arising from differences in feedstock properties, conventional use of feedstock and fossil energy source displaced. Roberts et al. (2010) also calculated an emissions reduction of 800–900 kg CO_2e/t biomass for similar biomass feedstocks (corn stover and yard waste). However, the mitigation benefit was much reduced for purpose grown biomass: Roberts' estimates ranged from a reduction of 440 kg CO_2e/t feedstock to an increase of 36 kg CO_2e/t feedstock, depending on the method used to estimate emissions from land use change (Roberts et al. 2010).

The sequestration benefit coupled with the creation of carbon neutral fuel (emissions from the burnt fuel are balanced by C sequestered in its production) could potentially reduce American emissions of CO_2 by 10% (Laird 2008). Globally, the potential mitigation benefit from biochar has been estimated at between 0.7 and 2.6 Gt C/year by 2050 (Laird et al. 2009).

15.5 Implementing Biochar Globally

The biochar supply chain includes biomass sourcing, conversion technology, product distribution and use. Due to the widely distributed nature of many biomass sources, biochar will need to be converted efficiently and economically from local biomass resources for distribution into regional agricultural soils.

The conversion of biomass to energy on a global scale will require a range of systems that use available resources while recognising regional socio-economic constraints and desired outcomes. For example, while up-scaling of "industrialised" biochar systems will be attractive to investors in developed countries, there is potential to implement clean, efficient biochar solutions at a small scale in developing nations to improve community welfare and reduce greenhouse gas emissions (Bailis 2009; Ewing and Msangi 2009). Existing thermal conversion technologies may be enhanced through the development of more sustainable sources of biomass (e.g. plantation biomass) and the implementation of modern kilns. Advantages of these more efficient systems may include rapid carbonisation, reduced gaseous emissions and higher yield of biochar from a greater potential range of feedstocks. The production outcomes from these systems will vary according to local resources and needs, ranging from small-scale production of biochar for fuel to larger industrialised production of liquid/gas energy and biochar for soil amendment.

The quantity of wood charcoal traded by the global forestry industry in 2008 was 49.35 million tonnes, of which more than half was produced in Africa and only 0.6% in Europe (FAO 2010). While the sustainability of this biomass source needs to be secured to remove the threat of net deforestation, with technological modernisation, growth in this industry could supply the energy needs of communities while meeting the expanding demand for biochar in agriculture. The relatively small amounts of charcoal produced in industrialised nations, under strict environmental regulatory control, are generally for specialised applications which can afford the more expensive process technology to ensure product meets specification.

The large-scale production of biochar for a low-value agricultural market requires the commercialisation of a new generation of clean and safe thermal conversion technology. The additional costs of the various regional environment and planning regulations can perhaps be offset through the lower cost of feedstock that comes with the efficient conversion of waste residue feedstocks to biochar. Commercial viability may also be assisted if the technology enables the utilisation of co-products such as bio-oils and gas for energy generation. Examples of commercial biochar production systems that have been able to demonstrate that large-scale reliable, economically viable and environmentally sustainable supply of product are largely non-existent at this time.

15.6 Future Directions

Biochar research is in its infancy; further investment in research is needed to understand the mechanisms of its impacts, particularly in relation to N_2O emissions, nutrient retention and interactions with soil constituents such as native organic matter and minerals in a range of soil type, vegetation systems and climatic conditions. As the impacts of biochar on soil processes may change over time, there is a need for long-term studies to assess biochar's potential to provide the projected benefits. Measures to secure sustainable feedstock supply and novel biochar processing technologies are needed to ensure that biochar production delivers net environmental benefits. Measures could include certification against an agreed standard, similar to the sustainability certification undertaken in the forestry sector and being developed for bioenergy. Government incentives for commercial demonstration are needed to enable the technologies to become an acceptably low-risk proposition in a free market economy. With the possibility of multiple environmental benefits from its use, biochar-amended systems may become a vital tool to mitigate climate change and enhance the sustainability and productive capacity of global terrestrial systems.

References

Abiven S, Andreoli R (2010) Charcoal does not change the decomposition rate of mixed litters in a mineral cambisol: a controlled conditions study. Biol Fertil Soils. doi:10.1007/s00374-010-0489-1

Albiach R, Canet R, Pomares F, Ingelmo F (2001) Organic matter components and aggregate stability after the application of different amendments to a horticultural soil. Biores Technol 76:125–129

Asai H, Samson BK, Stephan HM, Songyikhangsuthor K, Homma K, Kiyono Y, Inoue Y, Shiraiwa T, Horie T (2009) Biochar amendment techniques for upland rice production in Northern Laos 1. Soil physical properties, leaf SPAD and grain yield. Field Crops Res 111:81–84

Bailis R (2009) Modeling climate change mitigation from alternative methods of charcoal production in Kenya. Biomass Bioenergy 33:1491–1502

Baldock JA, Smernik RJ (2002) Chemical composition and bioavailability of thermally altered *Pinus resinosa* (Red pine) wood. Org Geochem 33:1093–1109

Baligar VC, Fageria NK, He ZL (2001) Nutrient use efficiency in plants. Commun Soil Sci Plant Anal 32:921–950

Berglund LM, DeLuca TH, Zackrisson O (2004) Activated carbon amendments to soil alters nitrification rates in Scots pine forests. Soil Biol Biochem 36:2067–2073

Bridgwater AV (2003) Renewable fuels and chemicals by thermal processing of biomass. Chem Eng J 91:87–102

Brodowski S, Amelung W, Haumaier L, Abetz C, Zech W (2005) Morphological and chemical properties of black carbon in physical soil fractions as revealed by scanning electron microscopy and energy-dispersive X-ray spectroscopy. Geoderma 128:116–129

Brodowski S, John B, Flessa H, Amelung W (2006) Aggregate-occluded black carbon in soil. Eur J Soil Sci 57:539–546

Brown R (2009) Biochar production technology. In: Lehmann J, Joseph S (eds) Biochar for environmental management: science and technology. Earthscan, London, pp 127–146

Bryan E, Deressa TT, Gbetibouo GA, Ringler C (2009) Adaptation to climate change in Ethiopia and South Africa: options and constraints. Environ Sci Policy 12:413–426

Busscher WJ, Novak JM, Evans DE, Watts DW, Niandou MAS, Ahmedna M (2010) Influence of Pecan biochar on physical properties of a Norfolk loamy sand. Soil Sci 175:10–14

Chan KY, Heenan DP, So HB (2003) Sequestration of carbon and changes in soil quality under conservation tillage on light-textured soils in Australia: a review. Aust J Exp Agric 43:325–334

Chan KY, Van Zwieten L, Meszaros I, Downie A, Joseph S (2007) Agronomic values of greenwaste biochar as a soil amendment. Aust J Soil Res 45:629–634

Chan KY, Van Zwieten L, Meszaros I, Downie A, Joseph S (2008) Using poultry litter biochars as soil amendments. Aust J Soil Res 46:437–444

Chen D, Li Y, Grace P, Mosier A (2008) N2O emissions from agricultural lands: a synthesis of simulation approaches. Plant Soil 309:169–189

Cheng C-H, Lehmann J, Thies JE, Burton SD, Engelhard MH (2006) Oxidation of black carbon by biotic and abiotic processes. Org Geochem 37:1477–1488

Cheng C-H, Lehmann J, Engelhard MH (2008a) Natural oxidation of black carbon in soils: changes in molecular form and surface charge along a climosequence. Geochim Cosmochim Acta 72:1598–1610

Cheng C-H, Lehmann J, Thies JE, Burton SD (2008b) Stability of black carbon in soils across a climatic gradient. J Geophys Res 113:G02027

Cherubini F, Bird ND, Cowie A, Jungmeier G, Schlamadinger B, Woess-Gallasch S (2009) Energy- and greenhouse gas-based LCA of biofuel and bioenergy systems: key issues, ranges and recommendations. Resour Conserv Recycling 53:434–447

Clough TJ, Bertram JE, Ray JL, Condron LM, O'Callaghan M, Sherlock RR, Wells NS (2010) Unweathered wood biochar impact on nitrous oxide emissions from a bovine-urine-amended pasture soil. Soil Sci Soc Am J 74:852–860

Dalal RC, Wang WJ, Robertson GP, Parton WJ (2003) Nitrous oxide emission from Australian agricultural lands and mitigation options: a review. Aust J Soil Res 41:165–195

DeLuca TH, Nilsson MC, Zackrisson O (2002) Nitrogen mineralization and phenol accumulation along a fire chronosequence in northern Sweden. Oecologia 133:206–214

DeLuca TH, MacKenzie MD, Gundale MJ, Holben WE (2006) Wildfire-produced charcoal directly influences nitrogen cycling in ponderosa pine forests. Soil Sci Soc Am J 70:448–453

Downie A, Crosky A, Munroe P (2009) Physical properties of biochar. In: Lehmann J, Joseph S (eds) Biochar for environmental management: science and technology. Earthscan, London, pp 13–32

Ericksen PJ, Ingram JSI, Liverman DM (2009) Food security and global environmental change: emerging challenges. Environ Sci Policy 12:373–377

Ewing M, Msangi S (2009) Biofuels production in developing countries: assessing tradeoffs in welfare and food security. Environ Sci Policy 12:520–528

FAO (2010) FAOSTAT statistical database. Food and Agriculture Organisation of the United Nations, Rome

Forster P, Ramaswamy V, Artaxo P, Berntsen T, Betts R, Fahey DW, Haywood J, Lean J, Lowe DC, Myhre G, Nganga J, Prinn R, Raga GMS, Van Dorland R (2007) Changes in atmospheric constituents and in radiative forcing. In: Solomon S, Qin D, Manning M et al (eds) Climate change 2007: the physical science basis. Contribution of working group I to the fourth assessment report of the intergovernmental panel on climate change. Cambridge University Press, Cambridge

Fresco LO (2009) Challenges for food system adaptation today and tomorrow. Environ Sci Policy 12:378–385

Gaskin JW, Steiner C, Harris K, Das KC, Bibens B (2008) Effect of low-temperature pyrolysis conditions on biochar for agricultural use. Trans ASABE 51:2061–2069

Gaunt JL, Cowie AL (2009) Biochar, greenhouse gas accounting and emissions trading. In: Lehmann J, Joseph S (eds) Biochar for environmental management: science and technology. Earthscan, London, pp 317–340

Gaunt JL, Lehmann J (2008) Energy balance and emissions associated with biochar sequestration and pyrolysis bioenergy production. Environ Sci Technol 42:4152–4158

Gaur A, Adholeya A (2000) Effects of the particle size of soil-less substrates upon AM fungus inoculum production. Mycorrhiza 10:43–48

Geisseler D, Horwath WR, Doane TA (2009) Significance of organic nitrogen uptake from plant residues by soil microorganisms as affected by carbon and nitrogen availability. Soil Biol Biochem 41:1281–1288

Glaser B, Haumaier L, Guggenberger G, Zech W (2001) The "Terra Preta" phenomenon: a model for sustainable agriculture in the humid tropics. Naturwissenschaften 88:37–41

Glaser B, Lehmann J, Zech W (2002) Ameliorating physical and chemical properties of highly weathered soils in the tropics with charcoal – a review. Biol Fertil Soils 35:219–230

Gundale M, DeLuca T (2007) Charcoal effects on soil solution chemistry and growth of Koeleria macrantha in the ponderosa pine/Douglas-fir ecosystem. Biol Fertil Soils 43:303–311

Guo Y, Rockstraw DA (2007) Activated carbons prepared from rice hull by one-step phosphoric acid activation. Micropor Mesopor Mater 100:12–19

Hamer U, Marschner B, Brodowski S, Amelung W (2004) Interactive priming of black carbon and glucose mineralisation. Org Geochem 35:823–830

Hammes K, Smernik RJ, Skjemstad JO, Schmidt MWI (2008) Characterisation and evaluation of reference materials for black carbon analysis using elemental composition, colour, BET surface area and C-13 NMR spectroscopy. Appl Geochem 23:2113–2122

Helyar KR, Cregan PD, Godyn DL (1990) Soil acidity in New-South-Wales – current pH values and estimates of acidification rates. Aust J Soil Res 28:523–537

Hossain MK, Strezov V, Chan KY, Nelson PF (2010) Agronomic properties of wastewater sludge biochar and bioavailability of metals in production of cherry tomato (*Lycopersicon esculentum*). Chemosphere 78:1167–1171

Howden SM, J-Fo S, Tubiello FN, Chhetri N, Dunlop M, Meinke H (2007) Adapting agriculture to climate change. Proc Natl Acad Sci U S A 104:19691–19696

Jones PG, Thornton PK (2009) Croppers to livestock keepers: livelihood transitions to 2050 in Africa due to climate change. Environ Sci Policy 12:427–437

Joseph SD, Camps-Arbestain M, Lin Y, Munroe P, Chia CH, Hook J, van Zwieten L, Kimber S, Cowie A, Singh BP, Lehmann J, Foidl N, Smernik RJ, Amonette JE (2010) An investigation into the reactions of biochar in soil. Aust J Soil Res 48:501–515

Khalil MI, Buegger F, Schraml M, Gutser R, Richards KG, Schmidhalter U (2009) Gaseous nitrogen losses from a cambisol cropped to spring wheat with urea sizes and placement depths. Soil Sci Soc Am J 73:1335–1344

Kimetu J, Lehmann J, Ngoze S, Mugendi D, Kinyangi J, Riha S, Verchot L, Recha J, Pell A (2008) Reversibility of soil productivity decline with organic matter of differing quality along a degradation gradient. Ecosystems 11:726–739

Kolb SE, Fermanich KJ, Dornbush ME (2009) Effect of charcoal quantity on microbial biomass and activity in temperate soils. Soil Sci Soc Am J 73:1173–1181

Koomen E, Kuhlman T, Groen J, Bouwman A (2005) Simulating the future of agricultural land use in the Netherlands. Tijdschr Econ Soc Ge 96:218–224

Kuzyakov Y, Subbotina I, Chen HQ, Bogomolova I, Xu XL (2009) Black carbon decomposition and incorporation into soil microbial biomass estimated by C-14 labeling. Soil Biol Biochem 41:210–219

Laird DA (2008) The charcoal vision: a win win win scenario for simultaneously producing bioenergy, permanently sequestering carbon, while improving soil and water quality. Agron J 100:178–181

Laird DA, Brown RC, Amonette JE, Lehmann J (2009) Review of the pyrolysis platform for coproducing bio-oil and biochar. Biofuels Bioprod Bioref 3:547–562

Lal R (2010) Managing soils for a warming earth in a food-insecure and energy-starved world. J Plant Nutr Soil Sci 173:4–15

Larkin RP (2008) Relative effects of biological amendments and crop rotations on soil microbial communities and soilborne diseases of potato. Soil Biol Biochem 40:1341–1351

Lavelle P (1988) Earthworm activities and the soil system. Biol Fertil Soils 6:237–251

Lee KE, Foster RC (1991) Soil fauna and soil structure. Aust J Soil Res 29:745–775

Lehmann J (2007) Bio-energy in the black. Front Ecol Environ 5:381–387

Lehmann J, Rondon M (2006) Bio-char soil management on highly weathered soils in the humid tropics. In: Uphoff N (ed) Biological approaches to sustainable soil systems. CRC Press, Boca Raton

Lehmann J, Sohi S (2008) Comment on "fire-derived charcoal causes loss of forest humus". Science 321:1295. doi:10.1126/science.1160005

Lehmann J, Pereira da Silva J, Steiner C, Nehls T, Zech W, Glaser B (2003) Nutrient availability and leaching in an archaeological Anthrosol and a Ferralsol of the Central Amazon basin: fertilizer, manure and charcoal amendments. Plant Soil 249:343–357

Lehmann J, Gaunt J, Rondon M (2006) Bio-char sequestration in terrestrial ecosystems – a review. Mitig Adapt Strategy Global Change 11:395–419

Lehmann J, Kinyangi J, Solomon D (2007) Organic matter stabilization in soil microaggregates: implications from spatial heterogeneity of organic carbon contents and carbon forms. Biogeochemistry 85:45–57

Lehmann J, Solomon D, Kinyangi J, Dathe L, Wirick S, Jacobsen C (2008) Spatial complexity of soil organic matter forms at nanometre scales. Nature Geosci 1:238–242

Lehmann J, Czimczik CI, Laird DL, Sohi S (2009) Stability of biochar in soil. In: Lehmann J, Joseph S (eds) Biochar for environmental management: science and technology. Earthscan, London, pp 183–205

Liang B, Lehmann J, Solomon D, Kinyangi J, Grossman J, O'Neill B, Skjemstad JO, Thies J, Luizao FJ, Petersen J, Neves EG (2006) Black carbon increases cation exchange capacity in soils. Soil Sci Soc Am J 70:1719–1730

Liang B, Lehmann J, Solomon D, Sohi S, Thies JE, Skjemstad JO, Luizao FJ, Engelhard MH, Neves EG, Wirick S (2008) Stability of biomass-derived black carbon in soils. Geochim Cosmochim Acta 72:6069–6078

Liang BQ, Lehmann J, Sohi SP, Thies JE, O'Neill B, Trujillo L, Gaunt J, Solomon D, Grossman J, Neves EG, Luizao FJ (2010) Black carbon affects the cycling of non-black carbon in soil. Org Geochem 41:206–213

MacKenzie M, DeLuca T (2006) Charcoal and shrubs modify soil processes in ponderosa pine forests of western Montana. Plant Soil 287:257–266

MacKenzie MD, McIntire EJB, Quideau SA, Graham RC (2008) Charcoal distribution affects carbon and nitrogen contents in forest soils of California. Soil Sci Soc Am J 72:1774–1785

Major J, Steiner C, Downie A, Lehmann J (2009) Biochar effects on nutrient leaching. In: Lehmann J, Joseph S (eds) Biochar for environmental management: science and technology. Earthscan, London, pp 271–287

Major J, Lehmann J, Rondon M, Goodale C (2010) Fate of soil-applied black carbon: downward migration, leaching and soil respiration. Global Change Biol 16:1366–1379

Makoto K, Tamai Y, Kim YS, Koike T (2010) Buried charcoal layer and ectomycorrhizae cooperatively promote the growth of *Larix gmelinii* seedlings. Plant Soil 327:143–152

McBeath AV, Smernik RJ (2009) Variation in the degree of aromatic condensation of chars. Org Geochem 40:1161–1168

McCarl BA, Peacocke C, Chrisman R, Kung C-C, Sands R (2009) Economics of biochar production, utilisation and greenhouse gas offsets. In: Lehmann J, Joseph S (eds) Biochar for environmental management: science and technology. Earthscan, London, pp 341–357

McHenry MP (2009) Agricultural bio-char production, renewable energy generation and farm carbon sequestration in Western Australia: certainty, uncertainty and risk. Agric Ecosyst Environ 129:1–7

Murage EW, Voroney P, Beyaert RP (2007) Turnover of carbon in the free light fraction with and without charcoal as determined using the ^{13}C natural abundance method. Geoderma 138:133–143

Namgay T, Singh B, Singh BP (2010) Influence of biochar on the availability of As, Cd, Cu, Pb and Zn to maize (*Zea mays* L.). Aust J Soil Res 48:638–647

Neufeldt H, Resck DVS, Ayarza MA (2002) Texture and land-use effects on soil organic matter in Cerrado Oxisols, Central Brazil. Geoderma 107:151–164

Nguyen BT, Lehmann J (2009) Black carbon decomposition under varying water regimes. Org Geochem 40:846–853

Nguyen BT, Lehmann J, Hockaday WC, Joseph S, Masiello CA (2010) Temperature sensitivity of black carbon decomposition and oxidation. Environ Sci Technol 44:3324–3331

Novak JM, Busscher WJ, Laird DL, Ahmedna M, Watts DW, Niandou MAS (2009) Impact of biochar amendment on fertility of a Southeastern coastal plain soil. Soil Sci 174:105–112

O'Neill B, Grossman J, Tsai M, Gomes J, Lehmann J, Peterson J, Neves E, Thies J (2009) Bacterial community composition in Brazilian anthrosols and adjacent soils characterized using culturing and molecular identification. Microb Ecol 58:23–35

O'Sullivan MF, Simota C (1995) Modelling the environmental impacts of soil compaction: a review. Soil Till Res 35:69–84

Oguntunde PG, Abiodun BJ, Ajayi AE, van de Giesen N (2008) Effects of charcoal production on soil physical properties in Ghana. J Plant Nutr Soil Sci 171:591–596

Olarewaju O, Adetunji M, Adeofun C, Adekunle I (2009) Nitrate and phosphorus loss from agricultural land: implications for nonpoint pollution. Nutr Cycle Agroecosyst 85:79–85

Perie C, Ouimet R (2008) Organic carbon, organic matter and bulk density relationships in boreal forest soils. Can J Soil Sci 88:315–325

Pietikäinen J, Kiikkilä O, Fritze H (2000) Charcoal as a habitat for microbes and its effect on the microbial community of the underlying humus. Oikos 89:231–242

Rasse DP, Dignac MF, Bahri H, Rumpel C, Mariotti A, Chenu C (2006) Lignin turnover in an agricultural field: from plant residues to soil-protected fractions. Eur J Soil Sci 57:530–538

Read P (2009) Policy to address the threat of dangerous climate change: a leading role for biochar. In: Lehmann J, Joseph S (eds) Biochar for environmental management: science and technology. Earthscan, London, pp 393–404

Reichenauer TG, Panamulla S, Subasinghe S, Wimmer B (2009) Soil amendments and cultivar selection can improve rice yield in salt-influenced (tsunami-affected) paddy fields in Sri Lanka. Environ Geochem Health 31:573–579

Rillig M, Mummey DL (2006) Mycorrhizas and soil structure. New Phytol 171:41–53

Ro KS, Cantrell KB, Hunt PG, Ducey TF, Vanotti MB, Szogi AA (2009) Thermochemical conversion of livestock wastes: carbonization of swine solids. Bioresour Technol 100:5466–5471

Roberts KG, Gloy BA, Joseph S, Scott NR, Lehmann J (2010) Life cycle assessment of biochar systems: estimating the energetic, economic, and climate change potential. Environ Sci Technol 44:827–833

Rondon M, Lehmann J, Ramírez J, Hurtado M (2007) Biological nitrogen fixation by common beans (*Phaseolus vulgaris* L.) increases with bio-char additions. Biol Fertil Soils 43:699–708

Ruehlmann J, Korschens M (2009) Calculating the effect of soil organic matter concentration on soil bulk density. Soil Sci Soc Am J 73:876–885

Schmidt MWI, Noack AG (2000) Black carbon in soils and sediments: analysis, distribution, implications, and current challenges. Global Biogeochem Cy 14:777–793

Simon S, Wiegmann K (2009) Modelling sustainable bioenergy potentials from agriculture for Germany and Eastern European countries. Biomass Bioenergy 33:603–609

Singh BP, Cowie AL (2008) A novel approach, using ^{13}C natural abundance, for measuring decomposition of biochars in soil. In: Currie L, Yates L (eds) Carbon and nutrient management in agriculture, fertilizer and lime research centre workshop proceedings. Massey University, Palmerston North, p 549

Singh BP, Cowie A (2010) The mean turnover time of biochar in soil varies depending on biomass source and pyrolysis temperature. In: 19th world congress of soil science, soil solutions for a changing world, Brisbane, 1–6 Aug 2010. http://www.iuss.org/19th%20WCSS/symposium/pdf/0326.pdf. Accessed 1 Dec 2010

Singh BP, Hatton BJ, Singh B, Cowie A, Kathuria A (2010a) Influence of biochars on nitrous oxide emission and nitrogen leaching from two contrasting soils. J Environ Qual 39:1224–1235

Singh B, Singh BP, Cowie AL (2010b) Characterisation and evaluation of biochars for their application as a soil amendment. Aust J Soil Res 48:516–525

Skjemstad JO, Clarke P, Taylor JA, Oades JM, McClure SG (1996) The chemistry and nature of protected carbon in soil. Aust J Soil Res 34:251–271

Spiertz JHJ (2010) Nitrogen, sustainable agriculture and food security. A review. Agron Sustain Dev 30:43–55

Spokas KA, Reicosky DC (2009) Impacts of sixteen different biochars on soil greenhouse gas production. Ann Environ Sci 3:179–193

Spokas KA, Koskinen WC, Baker JM, Reicosky DC (2009) Impacts of woodchip biochar additions on greenhouse gas production and sorption/degradation of two herbicides in a Minnesota soil. Chemosphere 77:574–581

Steinbeiss S, Gleixner G, Antonietti M (2009) Effect of biochar amendment on soil carbon balance and soil microbial activity. Soil Biol Biochem 41:1301–1310

Steiner C, Teixeira W, Lehmann J, Nehls T, de Macêdo J, Blum W, Zech W (2007) Long term effects of manure, charcoal and mineral fertilization on crop production and fertility on a highly weathered Central Amazonian upland soil. Plant Soil 291:275–290

Steiner C, Das KC, Garcia M, Förster B, Zech W (2008a) Charcoal and smoke extract stimulate the soil microbial community in a highly weathered xanthic Ferralsol. Pedobiologia 51:359–366

Steiner C, Glaser B, Teixeira WG, Lehmann J, Blum WEH, Zech W (2008b) Nitrogen retention and plant uptake on a highly weathered central Amazonian Ferralsol amended with compost and charcoal. J Plant Nutr Soil Sci 171:893–899

Thies J, Rillig M (2009) Characteristics of biochar: biological properties. In: Lehmann J, Joseph S (eds) Biochar for environmental management: science and technology. Earthscan, London, pp 85–106

Tisdall JM, Oades JM (1982) Organic matter and water-stable aggregates in soils. Eur J Soil Sci 33:141–163

Titiz B, Sanford RLJ (2007) Soil charcoal in old-growth rain forests from sea level to the continental divide. Biotropica 39:673–682

Tryon EH (1948) Effect of charcoal on certain physical, chemical and biological properties of forest soils. Ecol Monogr 18:81–115

Uzun BB, Pütün AE, Pütün E (2006) Fast pyrolysis of soybean cake: product yields and compositions. Bioresour Technol 97:569–576

Van Zwieten L, Singh BP, Joseph S, Kimber S, Cowie A, Chan KY (2009) Biochar and emissions of non-CO2 greenhouse gases from soil. In: Lehmann J, Joseph S (eds) Biochar for environmental management: science and technology. Earthscan, London, pp 227–249

Van Zwieten L, Kimber S, Morris S, Chan K, Downie A, Rust J, Joseph S, Cowie A (2010a) Effects of biochar from slow pyrolysis of papermill waste on agronomic performance and soil fertility. Plant Soil 327:235–246

Van Zwieten L, Kimber S, Morris S, Downie A, Berger E, Rust J, Scheer C (2010b) Influence of biochars on flux of N_2O and CO_2 from ferrosol. Aust J Soil Res 48:555–568

Wardle DA, Nilsson M-C, Zackrisson O (2008) Fire-derived charcoal causes loss of forest humus. Science 320:629

Warnock D, Lehmann J, Kuyper T, Rillig M (2007) Mycorrhizal responses to biochar in soil – concepts and mechanisms. Plant Soil 300:9–20

West TO, Marland G (2002) A synthesis of carbon sequestration, carbon emissions, and net carbon flux in agriculture: comparing tillage practices in the United States. Agric Ecosyst Environ 91:217–232

Woolf D, Amonette JE, Street-Perrott FA, Lehmann J, Joseph S (2010) Sustainable biochar to mitigate global climate change. Nat Commun 1(5):1–9. doi:10.1038/ncomms1053

Yamato M, Okimori Y, Wibowo IF, Anshori S, Ogawa M (2006) Effects of the application of charred bark of *Acacia mangium* on the yield of maize, cowpea and peanut, and soil chemical properties in South Sumatra, Indonesia. Soil Sci Plant Nutr 52:489–495

Yanai Y, Toyota K, Okazaki M (2007) Effects of charcoal addition on N_2O emissions from soil resulting from rewetting air-dried soil in short-term laboratory experiments. Soil Sci Plant Nutr 53:181–188

Zackrisson O, Nilsson M-C, Wardle DA (1996) Key ecological function of charcoal from wildfire in the Boreal forest. Oikos 77:10–19

Zimmerman AR (2010) Abiotic and microbial oxidation of laboratory-produced black carbon (biochar). Environ Sci Technol 44:1295–1301

Chapter 16
Bioenergy Systems, Soil Health and Climate Change

Brendan H. George and Annette L. Cowie

16.1 Introduction

Currently fossil fuels supply most of the world's energy needs, in processes that move carbon from geologic pools to the atmosphere. In 2007, use of fossil fuels released an estimated 28.8 Gt of carbon dioxide equivalent (CO_2-e) into the atmosphere (International Energy Agency 2009). The increasing atmospheric concentration of carbon dioxide (CO_2) and other greenhouse gases (GHGs) influences climate (Solomon et al. 2007), and this process may be accentuated if climate and carbon cycle feedback loops continue to develop (Friedlingstein 2008; Gregory et al. 2009).

It is generally accepted that we need to move to a "low-carbon future" with reduced reliance on fossil fuels for energy. Bioenergy can play a significant role, meeting some of our energy needs while reducing carbon emissions and even sequestering significant amounts of carbon. While there is the potential for bioenergy systems to provide very significant amounts of energy across the world (Bauen et al. 2010), intensifying production systems and changing land use and land management to produce biomass may impact on soil health and soil carbon in particular (Lal et al. 2003). The question of how bioenergy systems should be managed to achieve optimal outcomes is not yet clearly resolved (e.g. Blanco-Canqui 2010; Cowie et al. 2006; Delucchi 2010; Lal et al. 2003). And while bioenergy systems, if thoughtfully designed, can contribute to *mitigating* climate change, we also need to consider some of the issues in *adapting* to climate change to sustain production.

The impacts of bioenergy systems on soil health and climate change are relevant at various scales:

B.H. George (✉) • A.L. Cowie
National Centre for Rural Greenhouse Gas Research, University of New England and Department of Primary Industries, Armidale, NSW 2351, Australia
e-mail: brendan.george@industry.nsw.gov.au

B.P. Singh et al. (eds.), *Soil Health and Climate Change*, Soil Biology 29,
DOI 10.1007/978-3-642-20256-8_16, © Springer-Verlag Berlin Heidelberg 2011

- At the local scale, where bioenergy systems may influence physical, chemical and biological properties of soils, affecting productivity, soil health and carbon sequestration;
- At the catchment scale where bioenergy systems may affect regional hydrology and water quality, influencing carbon stocks and potentially soil health;
- At a global scale where biomass production may cause competition for land, leading to deforestation and consequential loss of carbon stocks, sometimes in locations distant to biomass production.

Conversely, intensification and expansion of bioenergy could involve revegetation of marginal lands potentially enhancing terrestrial carbon stocks. To manage potential sustainability impacts on these various scales will require the development of robust and clear assessment protocols (van Dam et al. 2010). This is an area of significant and evolving policy and will have significant impacts on bioenergy systems and their development to meet local and global needs.

16.2 Definition of Bioenergy, Soil Health and Climate Change

Bioenergy is a composite word, combining "biomass" and "energy". Bioenergy is the production of heat, power and liquid fuels (termed biofuels) from biomass. Biomass is plant matter or derived from plant matter, and can be sourced from processing and post-consumer residues, in-field residues from forestry and agriculture, and purpose-grown energy crops. In this chapter, we consider the impacts on soil health of the latter two sources of biomass. Similar to Raison and Khanna (2011), we use the description of "soil health" offered by (Doran and Zeiss 2000) as *the capacity of the soil to function as a vital living system, within ecosystem and land use boundaries to sustain plant and animal productivity, maintain or enhance water and air quality and promote plant and animal health.* Furthermore, we include all significant soil physical, chemical and biological properties and processes that affect bioenergy systems and are in turn impacted by utilisation of land for biomass energy production.

The potential for bioenergy to mitigate climate change is a major driver of the recent policy development and industry expansion. Following the Intergovernmental Panel on Climate Change (IPCC) definition, we define climate change as *a statistically significant variation in either the mean state of the climate or in its variability, persisting for an extended period (typically decades or longer). Climate change may be due to natural internal processes or external forcings, or to persistent anthropogenic changes in the composition of the atmosphere or in land use* (IPCC 2001).

In much of the literature reference is made to the potential for biomass production in marginal or degraded soils (or landscapes). We use the United Nations Environment Programme (UNEP) definition of "land degradation" as *a long-term loss of ecosystem function and services, caused by disturbances from which the system cannot recover unaided.* Thus degraded land has lower productivity as a result of human actions, sometimes interacting with natural factors (e.g. overgrazing in drought).

"Marginal land" is land on which the value of produce barely exceeds production costs. It could be marginal because it is degraded or there may be other constraints such as soil type or climate. "Marginal" is context related and is subject to current technologies, plant breeds, agronomic practices, market values, etc. Thus, an area may be considered marginal with respect to one land use but not another.

16.3 Policy Drivers Increasing Bioenergy

Bioenergy generally costs significantly more than existing fossil fuel-based energy sources (Duer and Christensen 2010), but it is promoted for various reasons including:

- Energy security;
- Environmental benefits, particularly mitigation of climate change;
- Rural and regional development;
- Human health.

The development of bioenergy is encouraged by policies including renewable energy targets, industry development grants and emissions trading (Capon et al. 2010; Vivid Economics 2010). However, there are concerns about potential impacts of the expansion and intensification of large-scale bioenergy systems including:

- The "food versus fuel" debate – it has been asserted that utilisation of food for fuel [particularly corn (*Zea mays*) for ethanol] has led to rising food prices (Pimentel et al. 2009), but others claim that this impact has been exaggerated (Baffes and Haniotis 2010) or can be avoided by integrating food, feed and biomass production (Dale et al. 2010). Nevertheless, the stronger competition for resources as the human population grows could be a significant challenge for future bioenergy development (Bartle and Abadi 2010; Fischer et al. 2009; Pimentel et al. 2008; Solomon 2010; Tilman et al. 2009).
- The actual net GHG balance of bioenergy systems – some systems may deliver minimal GHG savings, when production-chain emissions and impacts on terrestrial carbon stocks are accounted (Fargione et al. 2008; Halleux et al. 2008; Tyner et al. 2010).
- Indirect land use change (iLUC) – use of agricultural land for biomass production may lead indirectly to deforestation, with implications for soil health and climate change (Delucchi 2010; Fischer et al. 2009; Schlamadinger et al. 2007).
- Local environmental impacts – for example, impacting air quality due to increased emissions in production of energy from biomass crops and biofuel refining; reduced water supply where production of biofuels uses significant water; and soil health (Blanco-Canqui 2010; Lasch et al. 2010; Lattimore et al. 2009).

To secure the confidence of policy makers, on behalf of society, the bioenergy industry needs to demonstrate that production can meet energy needs as well as environmental and social objectives. Sustainability frameworks are being developed to give confidence that bioenergy production will deliver energy security and GHG mitigation while minimising negative impacts such as soil degradation (Fritsche et al. 2010; O'Connell et al. 2009; van Dam et al. 2010). We discuss these frameworks in more detail in Sect. 16.8.

Well-planned bioenergy systems have the potential to provide continuous, large-scale supplies of energy that displace fossil fuels at economically competitive rates without negatively impacting on the environment. Developing such robust systems presents many challenges; in this chapter, we focus on soil health impacts and do not address other environmental impacts or issues such as social impacts, technology development, commercialisation and trade facilitation.

16.3.1 The Role of Bioenergy in Providing Renewable Energy

Bioenergy contributes approximately 10% of the primary energy mix globally, with nearly 9% from woody biomass and the remainder (≈1%) from agricultural crops and by-products plus municipal and industrial wastes (Bauen et al. 2010). Biomass uses range from low technology traditional use of firewood for heating and cooking, to co-firing in large-scale power plants (Berndes et al. 2010) and dedicated heat and power generation (Bernotat and Sandberg 2004). In northern European countries and North America, policy development has led to significant biomass use. For example, in Sweden bioenergy contributes >20% of the national primary energy supply (Alm et al. 2009). By comparison, in Australia, which has abundant supplies of coal and thus low-cost electricity, the contribution of bioenergy is much smaller, comprising 3.6% of the total primary energy in 2007–2008 (ABARE 2010).

The potential for bioenergy to contribute to future national and global energy needs is significant but estimates vary widely. Some consider that bioenergy could sustainably contribute 25–33% of primary energy needs by 2050 replacing fossil fuels in all energy markets (viz. electricity, heat and transport) (Bauen et al. 2010). Much of the uncertainty regarding the capacity to "sustainably" produce feedstock for bioenergy relates to the potential impact of competition for water and land resources, especially as the world population increases. The literature is split between very positive assessments of bioenergy capacity (e.g. Bartle and Abadi 2010; Fischer et al. 2010; Hoogwijk et al. 2009; Wit and Faaij 2010) and those that are negative, predominantly due to concerns over energy or GHG balance (e.g. (Fargione et al. 2008; Pimentel 2009) or conflict with food production (e.g. Pimentel et al. 2009; Schade and Pimentel 2010; Solomon 2010). Studies of likely capacity to produce biomass must, by definition, be hypothetical, based, for example, on assumptions about population growth, technological development and climate change impacts. Importantly these studies determine boundaries of theoretical potential and highlight considerations in developing large-scale production systems. We discuss many of the specific issues in detail below.

The extent to which bioenergy contributes to future global energy requirements will depend on:

- The biophysical capacity to produce biomass feedstock, process and transport energy carriers (Bauen et al. 2010; Berndes et al. 2010);
- Market development (Hansson and Berndes 2009; Hoogwijk et al. 2009);
- Technological development of feedstock conversion including pre-treatment (Uslu et al. 2008) and biorefinery development (Cherubini 2010);
- Social and environmental objectives and limitations applied through policy (Delucchi 2010; O'Connell et al. 2009).

16.3.2 Bioenergy and Climate Change

The major policy driver for the current promotion of bioenergy is its potential to contribute to climate change mitigation. Bioenergy systems can displace the use of fossil fuel, leaving the fossil fuel "in the ground", and may sequester carbon (through storage of carbon in the soil or vegetation). The potential benefit from increasing biomass energy (stationary and transport) and avoiding fossil fuel use is significant. The IPCC estimates that by 2030, increased energy generation efficiency and switching to alternative low-carbon sources including bioenergy could reduce emissions by approximately 1.6–2.5 Gt CO_2-e per annum for light transport (Ribeiro et al. 2007), and between 2.0 and 4.2 Gt CO_2-e per annum for stationary energy production (Sims et al. 2007).

To determine the mitigation value of bioenergy, we can use the *displacement factor*:

$$D_f = \frac{\text{Efficiency of bioenergy system}}{\text{Efficiency of displaced fossil fuel system}} \times \frac{CO_2 \text{ emissions per J fossil fuel}}{CO_2 \text{ emissions per J biofuel}}, \quad (16.1)$$

where D_f is the "amount of carbon emission from fossil fuels that is avoided when biofuels are used instead" (Schlamadinger and Marland 1996) and is a unitless measure. The efficiency of energy production is often summarised by the concept of the *energy return on investment* (EROI). The EROI is determined by the amount of energy gained from a process compared to how much energy was required to grow (or extract) and transform a new unit of energy carrier (e.g. biomass material) into the useable energy in question. Larger EROI indicates a more efficient production of useable energy from primary sources.

Also important are the production-chain emissions, due to: fossil fuel use, for example, in cultivation, harvest, transport, fertiliser manufacture; non-CO_2 emissions, such as nitrous oxide (N_2O) from applied fertiliser; and change in biomass and soil carbon. Co-products can also have a significant impact on the GHG balance of some bioenergy systems: biochar, a co-product of pyrolysis, can be used as a soil amendment improving soil health and yielding mitigation benefits [(Cayuela et al. 2010); also see Chap. 15 and Sect. 16.6.4]; residues from ethanol and biodiesel production can be used as animal feeds, saving emissions associated with feed production (Dale et al. 2010). Land use change (LUC), both direct and indirect, can significantly add to the production-chain emissions of bioenergy systems (see Sect. 16.7).

While climate change is a major driver for the industry, mitigation value must be balanced with other objectives. This becomes a challenge for policy development and implementation, as there are often trade-offs: maximising carbon sequestration and biomass production may reduce potential biodiversity benefits, or adversely impact hydrology. However, there are potential "win–win" scenarios where mitigation, conservation and production objectives can be addressed simultaneously, for example, the development of a mallee industry for bioenergy production in Western Australia (Bartle et al. 2007).

To make a significant contribution to climate change mitigation, a huge quantity of biomass will be required (Bauen et al. 2010; Raunikar et al. 2010). For example, it is estimated that for the USA to produce 30% of their domestic transport fuel demand would require the growing and processing of approximately one billion tonnes of biomass per annum (Perlack et al. 2005). The potential impact on GHG emissions of large-scale bioenergy systems will require carefully developed policy to meet multiple objectives and optimise production.

16.4 Land-Based Bioenergy Systems

There is great variation among bioenergy systems with respect to biomass feedstocks, transformation processes and energy services delivered (Fig. 16.1). In future, algae may become a significant source of biomass for biofuel (Florentinus et al. 2008; Wijffels and Barbosa 2010), but as algae-based systems will have limited impacts on soil health, they are not considered further in this chapter. Rather, we focus on the biomass feedstock production, as this is the stage that predominantly influences soil health.

Biomass for energy systems can be sourced from:

- Crops specifically produced for energy purposes [e.g. sugar cane (*Saccharum* spp.), wheat (*Triticum* spp.), sorghum (*Sorghum bicolor*), rapeseed (*Brassica napus*), short rotation woody crops (SRWC)]. Many of these crops already have competing uses.
- Residues that can be classed as:

  (a) Primary residues, produced during the production of forest products and food crops (e.g. forest thinnings, crop stubble). These resources are generally available in situ and must be collected (i.e., they are a dispersed source of material).

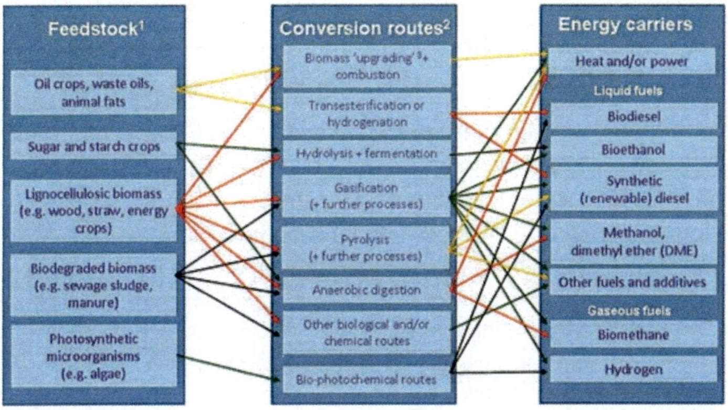

Fig. 16.1 Bioenergy feedstocks can be produced through many different pathways and then converted to meet various energy needs through a multitude of conversion routes (sometimes with multiple processes). The feedstocks will be produced from various primary, secondary and tertiary sources or dedicated energy crops ([1]Different parts of each feedstock can be used in other and/or multiple conversion routes; [2]Each route may also yield co-products (e.g. bagasse for cogeneration as a co-product of sugar production from sugarcane); [3]"Upgrading" includes a densification process such as pelletisation or torrefaction; [adapted from Bauen et al. (2010) with permission]). Only energy related products are considered in this diagram. Other bioproducts (e.g. bioplastics, biochar) are not explicitly incorporated

(b) Secondary residues, generated during the processing of biomass for other products (e.g. bagasse from sugar production from sugarcane, sawdust from the milling of forest products). These resources are generally concentrated at a processing operation.
(c) Tertiary residues, produced from "wastes" (e.g. used cooking oil, organic fraction of municipal solid waste, reclaimed timber products).

Production systems for most of the annual crops used for bioenergy are well developed, as these crops have established uses for food and feed. Some biomass sources (e.g. residues from forest harvest operations) are obtained through modification of existing systems. Additionally, bioenergy demand is encouraging the development of new crops (such as jatropha (*Jatropha curcas*) and miscanthus (*Miscanthus* spp.)), and new production systems such as short rotation tree crops (including willow (*Salix* spp.), polar (*Populus* spp.) and eucalypts), which are sometimes coppiced.

16.5 Biomass Production and Implications for Soil Health

Soil health is a function of the complex interaction of chemical, physical and biological properties. Achieving significant bioenergy targets is likely to require large areas of land (Berndes et al. 2003) and increased production intensity (Pimentel et al. 2009), to increase biomass yields (Fischer et al. 2009). The impacts of bioenergy systems on soil health will depend on:

- LUC – that is, replacement of the previous land use (Luyssaert et al. 2008);
- The type of biomass production system introduced (e.g. annual or perennial);
- Management practices applied to the biomass system.

Impacts of bioenergy systems based on conventional annual crops and forestry are similar to other annual cropping and forestry systems, reviewed in Chaps. 9 and 12, respectively, and we do not address these topics further in this chapter. Generally annual plants require greater inputs and more frequent soil disturbance than perennial systems, which may be replanted after harvest at 5–10 years, or even less frequently if coppiced. Thus, annual systems are likely to pose a greater risk than perennials to soil health.

The anticipated increase in scale and intensity of biomass productions systems will increase pressure on natural resources and test the resilience of the biophysical systems. Options for increased biomass production, and management of their impacts on soil health, are discussed in Sect. 16.6.

In this section, we outline the anticipated impacts of biomass production systems on soil properties and processes that can influence soil health. While there has been little research reported on this topic, we can predict the probable impacts of these bioenergy systems from knowledge of the factors influencing soil health.

16.5.1 Soil Organic Matter

Soil organic matter (SOM) strongly influences many soil properties and as such is a primary indicator of soil health (see Chap. 5). The amount of SOM in soil is a function of climate, topography, parent material, biology and time (Rice 2005). SOM is generally higher in cool climate forests and wetlands where growth is considerable but soil biota activity limited. Conversely, SOM is lower in the wet tropics where it is rapidly mineralised, and lowest in dry environments where plant growth (and therefore input of organic matter) is limited by water availability (Bolin et al. 2000).

Maintaining SOM levels is a significant challenge to the sustainability of bioenergy systems where they increase biomass removal and/or have greater soil disturbance compared to conventional cropping systems. However, where bioenergy systems use degraded or marginal land, or involve establishment of perennial tree crops on intensively cultivated land, then SOM may be increased (Lasch et al. 2010).

Carbon is the major constituent of SOM; therefore changes in SOM have significant implications for climate change. Figure 16.2 shows the respective carbon pools and fluxes and potential impact of management in a forest-based bioenergy system.

It has been suggested that loss of soil carbon in the establishment of energy crops, or as a result of residue removal, could negate the climate change benefits of using bioenergy to displace fossil energy sources (Fargione et al. 2008). Cherubini

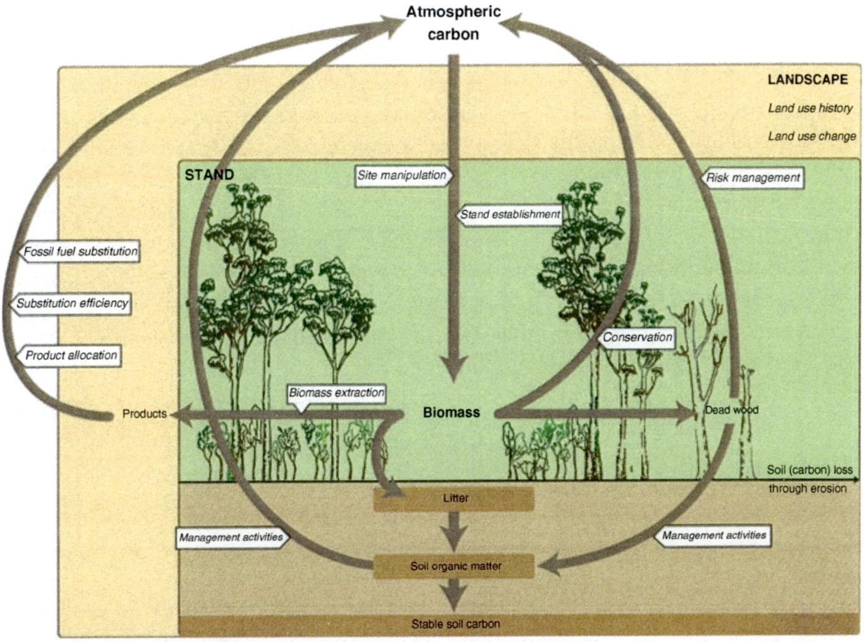

Fig. 16.2 The forms of atmospheric and terrestrial carbon as stocks (*boxes*) and flows (*arrows*) in a forest-based bioenergy system. Management options will impact on the respective flows and stocks of carbon and significantly influence soil health depending on disturbance, additions and subtractions to the soil system. Adapted from Bottcher and Lindner (2010)

and Ulgiati (2010) calculated that the loss of soil carbon contributed approximately 50% of total GHG emissions in the growing of biomass for the production and use of pellets for heat and electricity from corn stover and wheat straw, while Brandão et al. (2011) calculated that loss of soil carbon contributed ≈30% of the emissions due to the production of rapeseed. However, Cowie et al. (2006) determined that for three forest systems in Australia [viz. short rotation eucalypts (*E. globulus*), pine (*P. radiata*) and sawlog eucalypt (*E. grandis*)], the losses of soil carbon were minor compared with the avoided GHG emissions from equivalent fossil fuel use. Similarly, small losses of soil carbon, or even gains, have been reported where perennial low-intensity bioenergy crops including switchgrass (*Panicum virgatum* L.) (Liebig et al. 2005), poplar and miscanthus (Borzecka-Walker et al. 2008) have replaced existing cultivated cropping systems. Impacts of sugarcane on soil carbon vary from losses to gains and are influenced by management; soil carbon is higher where cane is not burned before harvest, and where trash is retained (Cerri et al. 2011; Galdos et al. 2009; Pinheiro et al. 2010).

The loss or gain of soil carbon in a bioenergy system is difficult to assess due to the spatial and temporal variability of soil carbon (Coleman et al. 2004). It will be determined by the LUC, influenced by the species involved (which may differ in their proportional allocation of carbon belowground, and quality of litter

(Pritchard 2011), and the management applied (tillage frequency, proportion of residues removed). Where residues from annual crops are used, soil carbon loss can be significant. For example, Anderson-Teixeira et al. (2009) found that where 25–100% of the corn residue was harvested, soil organic carbon losses of 3–8 Mg ha^{-1} occurred during the first 5 years. Conversely, Anderson-Teixeira et al (2009) reported a small increase in soil organic carbon (1 Mg ha^{-1} year^{-1} in the top 30 cm) in soil sustaining perennial grasses (including miscanthus and switchgrass). These results are consistent with data presented by Dawson and Smith (2007) where the ability of soil to sequester carbon was greater in perennial crops with reduced soil disturbance.

Perennial bioenergy systems that: (1) are established on already cleared agricultural land with minimal soil disturbance and (2) do not remove large amounts of residue will be more likely to retain soil carbon and improve soil health compared with annual cropping systems.

16.5.2 Soil Temperature

Soil temperature is an important factor influencing soil microbial activity (Pingintha et al. 2010). The selection of plant species will influence the amount of radiation reaching the soil (through shading) and the speed of air movement, which will not only affect the soil temperature directly, but also indirectly through effect on the water content. Management of the bioenergy feedstock system such as the removal of straw or logging residues will increase the diurnal temperature fluctuation and impact microbial activity (Blanco-Canqui 2010; Blanco-Canqui and Lal 2009).

16.5.3 Soil Erosion

Soil erosion, by wind or water, has significant impact on soil health through loss of nutrients and soil carbon (Gregg and Izaurralde 2010; Pimentel 2006). This is of particular concern for energy crop production systems where frequent disturbance and removal of residues are likely to increase the risk of erosion (Blanco-Canqui 2010; Delucchi 2010). Furthermore, eroded soil can increase turbidity of streams and rivers (Tilman et al. 2002), and thus combined with eutrophication from nutrient run-off and leaching, bioenergy systems can reduce water quality (Brandão et al. 2011).

16.5.4 Soil Compaction

Operations involving heavy machinery, especially when the soil is wet, can lead to soil compaction (Blanco-Canqui and Lal 2009; Walmsley and Godbold 2010).

Compaction reduces soil porosity and infiltration, thus reducing available water for plant use, and increase waterlogging risk, leading to reduced mineralisation of soil carbon (Cowie et al. 2006). Removal of logging residues in forest-based bioenergy systems may increase soil compaction (Merino et al. 2004), causing reduction in soil health and inhibiting plant growth.

16.5.5 Soil Salinity

An excess of soluble salts, especially sodium, has negative effects on chemical and physical soil properties, and inhibits microbial activity and plant growth (Rengasamy 2006). One of the main drivers for the nascent bioenergy industry in Australia is the potential for an economically viable production system incorporating deep-rooted woody crops to capture and use available soil water to minimise waterlogging and control dryland salinity (Bartle et al. 2007; Crosbie et al. 2007). However, such crops must be carefully sited to avoid negative impacts through reduced streamflow and recharge of aquifers (Benyon et al. 2007; Jackson et al. 2005; Vertessy et al. 2003).

16.5.6 Nutrient Balance and Availability

The production and removal of biomass for bioenergy can significantly impact on chemical fertility through changing the nutrient balance in the soil. The balance of macronutrients (viz. N, P, K, Mg and Ca) and micronutrients (including S, Zn, Fe, Mn and B) is important in all production systems; nutrients removed in harvested biomass or through erosion, run-off and leaching must be replaced to avoid long-term decline in productivity (Blanco-Canqui and Lal 2009; Cowie et al. 2006; Gregg and Izaurralde 2010; Lal 2004). Loss of chemical fertility will limit plant growth, leading to reduced organic matter input and consequently lower microbial activity, nutrient and moisture retention.

Reduced nutrient availability will lead to declining productivity where biomass is continually removed, such as in proposed bioenergy cropping systems including SRWC (Blanco-Canqui 2010; Dickmann 2006; Merino et al. 2004; Sartori et al. 2007) and traditional agricultural crops (Blanco-Canqui and Lal 2009; Muller 2009). The effect of nutrient removal is complicated by the selective nutrient extraction from soil of the respective crops as well as what component of the biomass is removed. Bioenergy systems that remove green foliage (which is particularly high in N, P and K), or bark (which is high in Ca), have a higher risk of nutrient depletion (Cowie et al. 2006). Impacts will differ between species and nutrient removal will be lower in systems using deciduous species or straw harvested after plant maturity.

In their review of the impact of the removal of stover in corn production systems in the USA, Blanco-Canqui and Lal (2009) found that at some sites nitrogen declined and other nutrients remained steady, while at other sites all measured nutrients declined. It is clear that biomass removal will have a significant effect on soil nutrients, particularly nitrogen, and therefore soil health.

LUC can significantly affect the amount and form of nitrogen in the soil. For example, where pasture is converted to a plantation, site preparation will encourage mineralisation of organic matter, releasing nitrogen, which may be lost through leaching in the early stages of plantation growth before the tree roots have fully occupied the site (Kirschbaum et al. 2008; Turner and Lambert 2000).

It is important to consider the fate of the biomass if not removed for bioenergy. In some forest systems residues are windrowed and burned. This practice removes the majority of the carbon, and a large proportion of nitrogen, though the non-volatile nutrients will be retained. In this scenario, removal of residues for bioenergy will have less impact on the soil nutrient (and carbon) status. The retention of sugarcane green trash at harvest can significantly increase soil carbon and nitrogen (Pinheiro et al. 2010) though much of the retained trash decomposes within a year (Robertson and Thorburn 2007).

The ability to definitively determine how much of the residue can be removed without jeopardising soil health in different soil types, crops and management systems remains weak. However, Blanco-Canqui and Lal (2009) suggest that fertility decline is likely if >50% residue is removed from annual crop systems. Ensuring that nutrient-rich leaf material is retained on site (such as by harvesting after leaf fall, or leaving cut branches to dry in the field) is important to maintaining soil fertility (Fung et al. 2002; Walmsley and Godbold 2010).

16.5.7 Biological Activity and Impacts

The presence and activity of soil biota are fundamental to the development and maintenance of soil health (see Chap. 8). In turn, biological activity is influenced by the soil chemical and physical properties, especially porosity, range in moisture content, presence of toxic substances, temperature variation and the availability of nutrients. In addition, SOM provides significant energy for microbial activity. Bioenergy systems that reduce organic matter inputs, for example through increased intensity of biomass harvest, may therefore inhibit microbial activity.

Organic matter is decomposed through comminution and mineralisation, which are dependent on the activity of soil fauna and micro-organisms (Cogle et al. 1995). When land is converted from crop or pasture to forest, the increased shading of the soil surface can reduce surface temperatures, lowering microbial activity and retarding the litter decomposition rate (Grove et al. 2001). This increases the carbon stock in litter while decreasing input to the soil carbon pool. Conversely, reducing tillage disturbance and increasing organic matter and nitrogen (via fertilisation) in soil leads to increased microbial populations (Vargas Gil et al. 2009). There is some "competition" between the respective processes; the balance determines whether microbial activity will be increased or decreased.

Perennial bioenergy systems, with reduced disturbance and increased input of root organic matter, are likely to encourage microbial activity compared with conventionally tilled annual crops. However, even where organic matter is returned to the soil, the impact on biological activity can vary. For example, Smolander et al. (2010) found that the impact of removal of logging residues on microbial processes was greater on low fertility sites.

Understanding the impacts on soil microbes of management options such as plant selection and cultivation, through altered temperature fluxes, retention and incorporation of organic matter, and soil disturbance regimes, is important to maintaining soil health under bioenergy systems.

16.6 Management Options

Realising the anticipated potential for bioenergy will require intensification of biomass production from existing agricultural and forest systems and expansion of the area used for biomass production. Managing environmental impacts is considered one of the most challenging aspects of biomass production on a large scale (Delucchi 2010; Pimentel et al. 2009).

In this section, we consider the impacts on soil health of intensification, including through use of fertiliser, and discuss management options to minimise these impacts. We discuss the implications of expansion of area used for bioenergy in Sect. 16.7.

16.6.1 Use of Marginal and Degraded Land

Currently much of the biomass feedstock for transport fuels is produced from conventional agricultural crops grown on prime agricultural lands. For example, the US ethanol industry is largely based on corn production in the mid-west (Solomon 2010). There is direct competition between the biomass feedstock for energy with food or feed. This leads to concern and debate regarding food and energy security and respective national interests (Pimentel et al. 2009). In addressing these concerns, there is significant opportunity for bioenergy feedstocks to be produced on degraded and marginal lands (Bartle et al. 2007; Delucchi 2010; Tilman et al. 2006; Tilman et al. 2009). While such land is not well suited to conventional crops, it may be satisfactory for lignocellulosic crops such as giant reed (*Arundo donax*), or tree crops such as jatropha, that are reputed to tolerate dry, infertile environments (Jongschaap et al. 2007). Low-input perennial grass systems that increase the opportunity for plant and animal diversity have also been suggested for North America (Tilman et al. 2006). While these species may survive in harsh environments, yields will inevitably be low. Nevertheless, the potential for production of lignocellulosic feedstocks on marginal lands is significant, with estimates of biomass yields capable of meeting up to 10 EJ year^{-1} of global energy requirements (Bauen et al. 2010). It is critical that sustainable land management practices are used to minimise risk to soil health on marginal lands, which can be particularly susceptible to degradation.

There is also potential that use of degraded lands for bioenergy production could deliver substantial net mitigation compared with fossil fuel production and consumption. For example, Wicke et al (2011) demonstrated that degraded lands in Indonesia dominated by the grass weed *Imperata cylindrica* can be used effectively for oil palm production. Use of degraded land often depleted in SOM and nutrients could enhance soil health, as growing plants contribute organic matter and stimulate microbial activity and nutrient cycling. However, establishment of bioenergy crops may be challenging as soil constraints must be overcome.

16.6.2 Intensification

Intensifying production from the current land base, through use of increased inputs, such as fertiliser (discussed below) or irrigation could significantly increase biomass production, particularly in regions such as Africa where yields are well below global averages. However, if larger production per unit time and/or per unit area is to be sustainably achieved, then careful selection of species, inputs and management will be necessary.

Impacts on soil health from intensified production are likely to occur on-site (at a local scale) through direct effects on soil properties, and could include: increase in soil bulk density due to compaction from harvesting operations; nutrient imbalances due to biomass removal; increased mineralisation of soil carbon from site disturbance; and changes to evapotranspiration and soil temperature (e.g. where plantations replace pastures). The magnitude of these direct effects depends on the pre-existing soil conditions, and the previous and new species and production systems. Irrigation offers the capacity to reduce moisture limitations, but can increase the risk to soil health through potential for salinisation if not managed to minimise deep drainage (Rengasamy 2006). Bioenergy feedstocks may be derived from crop residues in irrigated areas (e.g. rice husk, sugarcane trash), but it is unlikely that dedicated energy crops will be grown under irrigation due to high cost of water and higher value of alternative crops.

We know that intensive cropping leads to reduced soil carbon (Dalal and Chan 2001; Tilman et al. 2002). And though this is often attributed simply to soil disturbance (leading to disruption of soil aggregates and increased mineralisation of protected organic matter), there is evidence that soil erosion and decreased input of organic matter are also significant (Cowie et al. 2006; Paul et al. 2002). Thus, cropping generally decreases soil carbon through a combination of reduced input and enhanced loss. Therefore, intensification of biomass production is likely to exacerbate soil carbon decline, thus reducing soil health.

However, there are opportunities, through intensification, for integrated production of biomass for energy and food or feed, while preserving or enhancing soil health (Dale et al. 2010). One example is the use of SRWC such as oil mallees (e.g. *Eucalyptus polybractea*). Oil mallees have been planted in belts across wheat fields and require lower inputs and less frequent soil disturbance than conventional annual crops (Bartle et al. 2007; Hobbs et al. 2009).

16.6.3 Fertiliser

Increased application of fertiliser is one significant management option used in intensive production systems to increase biomass production. Where soils are deficient in specific plant nutrients, for example phosphorus in many Australian soils (Holford 1997), or where harvest or residue removal leads to nutrient imbalances and limitations (Blanco-Canqui and Lal 2009; Cowie et al. 2006; Grigal 2000; Sartori et al. 2007), fertiliser application becomes critical to maintaining plant production. Fertiliser application stimulates plant growth through increased nutrient supply, and can lead to additional biomass input to soil. But this does not necessarily result in higher soil carbon stocks, as SOM mineralisation may also be enhanced (Khan et al. 2007).

Many agricultural systems rely on fertiliser to support increased plant growth; however, globally the efficiency of fertiliser use has declined over the last 50 years (Tilman et al. 2002). For optimum efficiency site-specific management of fertilisers is essential, matching rate and timing of application to plant uptake (Jalali 2007). There is a GHG emissions cost to fertiliser production (Wood and Cowie 2004), and application of nitrogen fertilisers leads to increased flux of N_2O from soils (Stehfest and Bouwman 2006). Thus, increased use of fertilisers in biomass production reduces the climate change mitigation benefits of bioenergy.

Currently most transport biofuels are derived from corn (in the USA), sugarcane (in Brazil) and rapeseed oil (Europe) (Solomon 2010). This biomass is grown in intensively managed production systems with significant inputs, including high fertiliser rates, leading to high yields. The expectation of increased use of lignocellulosic components of biomass for energy (Richard 2010) means that alternative crops and harvest systems will be used in the future to provide feedstocks (e.g. Bartle and Abadi 2010). These crops and their management may differ significantly from current large-scale bioenergy production; the perennial grass and tree crop systems being developed have reduced disturbance (i.e. less frequent tillage) and potentially lower fertiliser requirements. However, fast growing, short rotation biomass plantations will require higher fertiliser inputs than conventional forestry to maximise biomass production and minimise decline in nutrient status. This increased fertiliser use could jeopardise future productivity and soil carbon stocks (Turner and Lambert 2000).

In the search for sustainable biomass production systems, mixed species, in contrast to monocultures, may be beneficial. For example, mixing species and including nitrogen fixers may have greater productivity and reduced need for fertiliser application (Richards et al. 2010). Finding the optimal species mix, while conceptually straightforward (Forrester et al. 2005), will require field trials of potentially promising species and management systems. Another option is to use organic amendments replacing synthetic fertiliser, ensuring the recycling of nutrients from processing and post-consumer residues, including biosolids (Wang et al. 2008).

16.6.4 Biochar

Biochar is a co-product of the bioenergy technology pyrolysis in which biomass is heated in the absence of oxygen. Besides bioenergy, this process can simultaneously produce biochar that can be used beneficially as a soil amendment. Biochar is highly recalcitrant, lasting potentially hundreds to thousands of years in soil (Singh and Cowie 2010). The potential for biochar as a soil amendment is discussed in detail by Waters et al. (See Chap. 15).

The biomass feedstock and pyrolysis conditions determine the physicochemical characteristics of the biochar, and therefore its value as a soil amendment and fertiliser (see Chap. 15). The application of biochar, depending on its properties, can not only increase soil carbon and cation exchange capacity, but also enhance physical properties such as aggregate stability and water-holding capacity. In turn, these changes to physicochemical properties can enhance biological activity, contributing to nutrient cycling and disease suppression (Warnock et al. 2007).

In addition to the benefits of delayed decomposition of organic matter and the supply of renewable energy, biochar can contribute to climate change mitigation through reduction in fertiliser requirements, decreased N_2O emissions from soil and enhanced plant growth (see Chap. 15). Woolf et al. (2010) estimate that the production and use of biochar could reduce global GHG emissions by 12% of current anthropogenic CO_2-e (some 1.8 Gt CO_2-e per annum), with 50% of the reduction from carbon sequestration, 30% from replacement of fossil fuel energy and 20% from avoided emissions of methane and N_2O.

In previous sections, we have discussed the risk that production of biomass crops, and collection of in-field residues for bioenergy, could reduce soil health. That is, we have highlighted the trade-off between maximising organic inputs to soil, to support maintenance of soil health, and removal of biomass for use as a renewable energy source to mitigate climate change. However, through pyrolysis there is potential to simultaneously produce renewable fuel, and biochar that can be used to maintain and/or enhance soil health. The net mitigation of a biochar system could be greater than if the same biomass had been used solely for bioenergy (Woolf et al. 2010). However, this assertion requires confirmation through further research to quantify, for example, the impact of different biochars on N_2O emissions. Furthermore, the conclusions of Woolf et al. (2010) may be true only for specific feedstocks, and where less GHG-intensive fossil energy sources are used. For example, if electricity is supplied from GHG-intensive brown coal at 1.2 kg CO_2-e per kWh (Commonwealth of Australia 2010), then greater mitigation may be achieved using all available biomass for energy. However, where natural gas is the fossil energy source, releasing 0.7 kg CO_2-e per kWh (Integrated Sustainability Analysis 2006), use for biochar production may achieve greater reduction in GHG emissions.

16.7 Indirect Impacts of Bioenergy Systems on Soil Health

Policies that encourage bioenergy are likely to lead to the expansion of land devoted to biomass production. This will entail LUC, that is, conversion from existing uses to biomass production. In already cultivated land, there may be minimal impact where one annual crop is replaced by another. However, LUC could result in conversion of pasture or forest to annual crops, which will directly impact on soil health and especially soil carbon as discussed above (see Chap. 9). Furthermore, the replacement of conventional land uses by biomass production may lead to iLUC: relocation of the displaced industry may result *indirectly* in conversion from forest to agricultural uses. The clearing of forested areas can lead to significant release of carbon into the atmosphere (Bolin et al. 2000; Delucchi 2010; Schlamadinger et al. 2007) and can significantly impact on soil health.

Indirect LUC is an issue associated with the development of various large-scale bioenergy systems including the production of palm oil in tropical countries (Laurance et al. 2010). Another controversial example of iLUC is deforestation in Brazil that is attributed to increased crop production for bioenergy (Fargione et al. 2008; Morton et al. 2006; Pimentel and Patzek 2007). In this situation, a global driver, which is strong markets for biofuels feedstocks, such as soybeans, and processed biofuels (ethanol), leads to local-scale changes with implications for soil health. For example, in Brazil sugarcane and soybeans are replacing pasture, and consequently cattle ranching displaced from pastureland could encroach on forest areas. The resulting deforestation can lead to large SOM loss through oxidation and erosion. Release of soil carbon to the atmosphere can negate the benefits from displacing mineral petroleum use (Fargione et al. 2008; Searchinger et al. 2008). In Brazil, better integration of existing land uses and co-utilisation of feedstocks (e.g. steam cooked bagasse pulp) for energy and feed production are proposed ways of integrating production systems and reducing iLUC (Sparovek et al. 2007, 2009).

The GHG emissions implications of iLUC and attribution to bioenergy are difficult to quantify (Plevin et al. 2010). Nevertheless, policymakers are developing mechanisms that will include estimates of iLUC when calculating the mitigation benefits of bioenergy (European Parliament 2009). Policies have also been introduced to minimise the risk of LUC. One such mechanism is the use of legislation to minimise clearing of native vegetation, which may be implemented for multiple reasons including the protection of biodiversity rather than specifically to address climate change (e.g. the Native Vegetation Act (2003) in New South Wales, Australia). A market-based incentive remains critical to promote development of a large-scale bioenergy industry with minimum GHG footprint (Stern 2007).

16.8 Managing and Regulating the Impacts of Bioenergy Production

Bioenergy systems will need to deliver on current policy-defined expectations such as reducing the emissions of GHGs or minimising environmental impacts and enhancing social outcomes (e.g. enhancing regional development and increasing trade opportunities). Meeting GHG targets and obligations is particularly important if a price (that is effectively "artificial") for carbon is established to address the externalities outlined by Stern (2007).

There are many mechanisms being established around the world that aim to improve sustainability of bioenergy systems. Most sustainability schemes are underpinned by agreed *principles* that are expounded as *criteria* and measured with stated *indicators*. Many of these developing systems are based on the earlier work of Cramer et al. (2006), who followed the "triple P approach" focussing on people, planet and profit to develop six themes for consideration in developing sustainability criteria (Cramer et al. 2006):

- Social well-being;
- Competition with food, local energy supply, medicines and building materials;
- Environment;
- Biodiversity;
- GHG balance;
- Economic prosperity.

To gain market access, bioenergy projects will need to demonstrate compliance with the criteria developed in applicable sustainability schemes. The formalisation of the *criteria* and *indicator* approach is achieved via the implementation of certification schemes that involve third-party verification of compliance. Although the approach of developing sustainability systems based on clear *principles* and *criteria* adds costs to biomass production, these schemes should increase consumer confidence (O'Connell et al. 2009).

We expect that over the next 10 years, there will be ongoing discussion and refinement of existing criteria-based systems for the production of sustainable biomass for energy production. Currently, initiatives include some specific to certain industries (e.g. the Roundtable on Sustainable Palm Oil (RSPO) production; the Roundtable on Responsible Soy; and the Better Sugarcane Initiative). These initiatives, detailed by O'Connell et al. (2009) and Vis et al. (2008), are driven by proponents wanting to support sustainable development of their industries. Larger independent systems are also being developed. These initiatives, including the United Nations Commission on Sustainable Development-sponsored Global

Bioenergy Partnership (GBEP) and the Swiss-led Roundtable on Sustainable Biofuels (RSB), are not feedstock specific. The GBEP is particularly targeted at developing nations (United Nations 2008) and like the RSB aims to identify key *principles* and *criteria* for consideration via a systematic assessment as outlined by van Dam et al. (2008). Using the RSB as an example, Table 16.1 shows the link between the respective *principles* and the three tenets of sustainability espoused

Table 16.1 The Roundtable for Sustainable Biofuels (RSB) principles as an example of the key issues in development of sustainability certification systems for bioenergy production

Roundtable for Sustainable Biofuel Principle	Sustainability tenet
Legality	Social
Biofuel operations shall follow all applicable laws and regulations	
Planning, monitoring and continuous improvement	Economic
Sustainable biofuel operations shall be planned, implemented and continuously improved through an open, transparent and consultative impact assessment and management process and an economic viability analysis	Environmental
Greenhouse gas emissions	Environmental
Biofuels shall contribute to climate change mitigation by significantly reducing life cycle GHG emissions as compared to fossil fuels	
Human and labour rights	Social
Biofuel operations shall not violate human rights or labour rights, and shall promote decent work and the well-being of workers	
Rural and social	Social
In regions of poverty, biofuel operations shall contribute to the social and economic development of local, rural and indigenous people and communities	Economic
Local food security	Social
Biofuel operations shall ensure the human right to adequate food and improve food security in food insecure regions	
Conservation	Environmental
Biofuel operations shall avoid negative impacts on biodiversity, ecosystems and other conservation values	
Soil	Environmental
Biofuel operations shall implement practices that seek to reverse soil degradation and/or maintain soil health	
Water	Environmental
Biofuel operations shall maintain or enhance the quality and quantity of surface and ground water resources, and respect prior formal or customary water rights	
Air	Environmental
Air pollution from biofuel operations shall be minimised along the supply chain	
Use of technology, inputs and management of waste	Social
The use of technologies in biofuel operations shall seek to maximise production efficiency and social and environmental performance, and minimise the risk of damages to the environment and people	Environmental
Land rights	Social
Biofuel operations shall respect land rights and land use rights	

by Cramer et al. (2006). Soil health is nominated specifically as an issue for sustainability of bioenergy systems, in Principle 8 Soil in Table 16.1, and criteria for assessment of compliance have been agreed (Roundtable on Sustainable Biofuels 2009).

Currently many sustainability schemes are voluntary and are used in product differentiation. However, in line with increased societal expectations, some jurisdictions are legislating the production and consumption of bioenergy, especially biofuels that meet nominated sustainability criteria. For example, in the European Union recent legislation requires biofuels to meet thresholds for minimum GHG mitigation and precludes use of biomass from biodiverse lands (European Parliament 2009); in Australia, the State of New South Wales legislated the mandatory use of ethanol and biodiesel, and the supply of these biofuels must comply with sustainability criteria (New South Wales 2007). Maintaining free trade between countries may become an issue in the application of criteria-based systems that could be construed as blocking the exchange of goods (Buchholz et al. 2009; Delzeit and Holm-Muller 2009; Lewandowski and Faaij 2006). van Dam et al. (2008) discuss, in detail, the interaction between the establishment of sustainability criteria and trade especially with respect to the General Agreement on Tariffs and Trade (GATT) and the World Trade Organization (WTO).

While current policy direction aims to improve sustainability of bioenergy systems, many challenges remain in meeting the demand for renewable energy while addressing environmental and social objectives at local, regional, national and international scales (Florin and Bunting 2009; Reinhard and Zah 2009). The application of the respective systems in different countries (with varying levels of governance), large variation in climate and edaphic conditions, multiple land use, crop and product outcomes will challenge the application and compliance of developed sustainability systems. Reports of the successful application of systems are scarce (e.g. Froger et al. 2010), but with the need to develop confidence and allow trade of biomass for energy and biofuels, significant effort is underway.

It is interesting to note that the sustainability requirements being demanded of bioenergy exceed the general expectations of other land uses in many cases. Maintenance of soil health is generally not legislated for food and fibre production systems, nor is avoidance of iLUC. O'Connell et al. (2009) indicate *there are outstanding issues to be resolved around the burden of proof of sustainability being placed on bioenergy especially in situations where land and water resources are shared between many production systems; the drivers for LUC are many-fold and the causal linkages to indirect effects are not clear*. The introduction of sustainability certification is critical but challenging, and significant development, testing and discussion are required to obtain solutions that allow for local interpretation of globally agreed principles and criteria (Florin and Bunting 2009; Hennenberg et al. 2010; O'Connell et al. 2009; Sparovek et al. 2007). It is likely that the debate over sustainability of biomass production for bioenergy will lead to calls for certification of food and fibre production systems that should include expectations of maintenance of soil health.

16.9 Summary and Conclusions

As the world searches for low-carbon renewable energy, there will be significant opportunities for biomass to provide heat, electricity and transport fuels. Bioenergy can contribute to climate change mitigation through increasing sequestration of carbon (e.g. in forests) and through reducing use of fossil fuels. With various feedstocks and process routes and multiple products, there are many bioenergy options that can produce not only energy but also other products, such as biochar, that may enhance the economic, social and environmental role of bioenergy. Bioenergy could supply greater than 30% of world primary energy demands by 2050. However, questions remain concerning the capacity and sustainability of large-scale intensive biomass energy systems (Muller 2009), particularly over implications for soil health.

Soil health is a function of the interaction between the inherent soil type, climate and the management of the bioenergy system. The impacts on soil health will be determined by the current land use, the bioenergy system that replaces it and the management practices applied. Biomass for bioenergy can be obtained from residues from conventional crop and forestry systems which, depending on the pre-existing land use and subsequent management of the bioenergy crop, may have negative impacts on soil health. Alternatively biomass can be produced from purpose-grown energy crops. Where these can be grown on marginal and degraded land, they may improve soil health. However, if they are grown on land that was previously used for cropping or grazing, their conversion to bioenergy crops is likely to lead to iLUC. This is likely to have negative impacts on soil health, with further negative consequences for managing climate change. To meet demand, it is likely that the area devoted to biomass production will expand, using productive land as well as marginal and/or degraded land, which will impact directly on physical, chemical and biological aspects of soil health. Where degraded lands are revegetated, that may have a positive impact on soil health.

If bioenergy is to effectively contribute to climate change mitigation, then it needs to have a positive energy balance, displace the use of fossil fuels for energy and potentially sequester carbon in biomass and soil. Understanding the role of soil carbon and the impact of land use, crop selection and management on soil health is critical to the long-term sustainability of existing and proposed bioenergy feedstock systems. As an important determinant of soil health, the risk of depletion of soil carbon stocks in biomass systems, due to the larger proportion of organic matter and nutrients that are removed from the site, compared with conventional grain, forage and timber production systems, needs to be carefully managed. By avoiding significant disturbance, balancing nutrient inputs and removals, and protecting soil from erosion, soil carbon can be maintained or enhanced. It is likely that in many systems (e.g. where woody plantations replace pastures), there may be an initial decline in soil carbon. But in well-managed systems, this decline will be significantly smaller than the GHG mitigation benefit of avoided fossil fuel emissions.

Other physicochemical and biological aspects of soil health need to be considered as bioenergy feedstock production increases in scale and intensity. The impact on the hydrological cycle can be significant; from affecting the infiltration of water into the soil and the soil water-holding capacity, to regional-scale impacts on streams and rivers. The increased use of fertiliser may lead to pollution problems such as eutrophication, and release of N_2O – a potent GHG. This, coupled with the extraction of biomass, can impact on the nutrient balance in the soil and pH leading to a decline in soil health. Many impacts are interrelated and can lead to significant long-term changes to the soil health from which it may take decades to recover. Maintenance of soil fertility will remain challenging, especially in systems where the majority of the biomass is removed for processing.

The expansion of biomass production for energy needs to be managed, to reduce potential negative impacts through:

- Minimising the extent of direct LUC;
- Reducing the risk of indirect LUC;
- Judicious selection of suitable crops;
- Adaptive management with consideration of economic goals in conjunction with social and environmental values.

Government, industry and land managers have specific roles: in providing guidelines for development; balanced system selection and responsive, site-specific management, respectively. To give confidence to industry and society that bioenergy is not only helping to meet energy needs but also reaching other goals, several sustainability frameworks are being developed. The most robust of these rely on the development of clear principles that are supported by succinct criteria and reported using agreed indicators. We expect that these systems will continue to develop and be tested over the next decade. They provide a clear message to bioenergy feedstock producers and processors that for large-scale production to occur, society needs to be confident that environmental and social goals are considered. This is particularly true in relation to soil health.

Maintaining soil health in biomass production systems is fundamental to sustaining yields and meeting goals such as climate change mitigation and regional development, while minimising on- and off-site impacts. In developing sustainable biomass production systems, maintaining soil health will be a challenge, particularly as the intensity of production is increased to meet the anticipated demand for biomass for energy and other uses.

References

ABARE (2010) Energy in Australia 2010. Australian Bureau of Agricultural and Resource Economics, Canberra

Alm M, Cato J, Lundmark L, Myrman J, Raab U, Karlberg M, Karlsson T, Jöhnemark M, Bohnstedt S, Petersson K, Sahlin K, Berggren T, Bäck M, Sahlin M, Linder E, Bäck M, Wagman N, Sahlin M, Lindberg C, Parikka M, Kärrmarck U, Veibäck E, Östensson E (2009) Energy in Sweden: facts and figures. Swedish Energy Agency, Stockholm

Anderson-Teixeira KJ, Davis SC, Masters MD, Delucia EH (2009) Changes in soil organic carbon under biofuel crops. Glob Change Biol Bioenergy 1(1):75–96

Baffes J, Haniotis T (2010) Placing the 2006/08 commodity price boom into perspective. The World Bank, Washington

Bartle J, Abadi A (2010) Toward sustainable production of second generation bioenergy feedstocks. Energy Fuels 24:2–9

Bartle J, Olsen G, Cooper D, Hobbs T (2007) Scale of biomass production from new woody crops for salinity control in dryland agriculture in Australia. Int J Global Energy Issues 27(2):115–137

Bauen A, Berndes G, Junginger M, Londo M, Vuille F, Ball R, Bole T, Chudziak C, Faaij A, Mozaffarian H (2010) Bioenergy – a sustainable and reliable energy source. A review of status and prospects. IEA Bioenergy, Petten

Benyon R, England J, Eastham J, Polglase P, White D (2007) Tree water use in forestry compared to other dry-land agricultural crops in the Victorian context. CSIRO, Canberra

Berndes G, Hoogwijk M, van den Broek R (2003) The contribution of biomass in the future global energy supply: a review of 17 studies. Biomass Bioenergy 25(1):1–28

Berndes G, Hansson J, Egeskog A, Johnsson F (2010) Strategies for 2nd generation biofuels in EU – co-firing to stimulate feedstock supply development and process integration to improve energy efficiency and economic competitiveness. Biomass Bioenergy 34(2):227–236

Bernotat K, Sandberg T (2004) Biomass fired small-scale CHP in Sweden and the Baltic States: a case study on the potential of clustered dwellings. Biomass Bioenergy 27(6):521–530

Blanco-Canqui H (2010) Energy crops and their implications on soil and environment. Agron J 102(2):403–419

Blanco-Canqui H, Lal R (2009) Crop residue removal impacts on soil productivity and environmental quality. Crit Rev Plant Sci 28(3):139–163

Bolin B, Sukumar R, Ciais P, Cramer W, Jarvis P, Kheshgi H, Nobre C, Semenov S, Steffen W (2000) Global perspective. In: Watson R, Noble IR, Bolin B, Ravindranath NH, Verado DJ, Dokken DJ (eds) Land use, land-use change, and forestry: a special report of the Intergovernmental Panel on Climate Change. Cambridge University Press, Cambridge, pp 23–51

Borzecka-Walker M, Faber A, Borek R (2008) Evaluation of carbon sequestration in energetic crops (Miscanthus and coppice willow). Int Agrophys 22(3):185–190

Bottcher H, Lindner M (2010) Managing forest plantations for carbon sequestration today and in the future. In: Bauhus J, Van der Meer P, Kanninen M (eds) Ecosystem goods and services from plantation forests. Earthscan, London, pp 43–76

Brandão M, Milài Canals L, Clift R (2011) Soil organic carbon changes in the cultivation of energy crops: implications for GHG balances and soil quality for use in LCA. Biomass Bioenergy. doi:10.1016/j.biombioe.2009.10.019:14

Buchholz T, Luzadis VA, Volk TA (2009) Sustainability criteria for bioenergy systems: results from an expert survey. J Cleaner Prod 17(suppl 1):s86–s98

Capon T, Harris M, Reeson A (2010) Soil carbon sequestration market based instruments (MBIs): a literature review. University of Sydney, Sydney

Cayuela ML, Oenema O, Kuikman PJ, Bakker RR, van Groenigen JW (2010) Bioenergy by-products as soil amendments? Implications for carbon sequestration and greenhouse gas emissions. Glob Change Biol Bioenergy 2(4):201–213

Cerri CC, Galdos MV, Maia SMF, Bernoux M, Feigl BJ, Powlson D, Cerri CEP (2011) Effect of sugarcane harvesting systems on soil carbon stocks in Brazil: an examination of existing data. Eur J Soil Sci 62(1):23–28

Cherubini F (2010) The biorefinery concept: using biomass instead of oil for producing energy and chemicals. Energy Convers Manage 51(7):1412–1421

Cherubini F, Ulgiati S (2010) Crop residues as raw materials for biorefinery systems – a LCA case study. Appl Energy 87(1):47–57

Cogle AL, Reddy MVR, Rao KPC, Smith GD, McGarry D, Yule DF (1995) The role of biological practices and the soil biota in management of sealing, crusting and hardsetting soils. In: So HB,

Smith GD, Raine SR, Schafer BM, Loch RJ (eds) Sealing, crusting and hardsetting soils: productivity and conservation. Australian Society of Soil Science (Queensland Branch), Brisbane, pp 305–324

Coleman MD, Isebrands JG, Tolsted DN, Tolbert VR (2004) Comparing soil carbon of short rotation poplar plantations with agricultural crops and woodlots in North Central United States. Environ Manage 33(Suppl 1):S299–S308

Commonwealth of Australia (2010) National greenhouse accounts (NGA) factors. Department of Climate Change and Energy Efficiency, Canberra

Cowie AL, Smith P, Johnson D (2006) Does soil carbon loss in biomass production systems negate the greenhouse benefits of bioenergy? Mitig Adapt Strateg Glob Change 11(5/6):979–1002

Cramer J, Wissema E, Lammers E, Dijk D, Jager H, van Bennekom S, Breunesse E, Horster R, van Leenders C, Wolters W, Kip H, Stam H, Faaij A, Kwant K (2006) Criteria for sustainable biomass production. Project group 'Sustainable production of biomass' in the Energy Transition Task Force, Amsterdam

Crosbie RS, Hughes JD, Friend J, Baldwin BJ (2007) Monitoring the hydrological impact of land use change in a small agricultural catchment affected by dryland salinity in central NSW, Australia. Agric Water Manage 88(1/3):43–53

Dalal RC, Chan KY (2001) Soil organic matter in rainfed cropping systems of the Australian cereal belt. Aust J Soil Res 39:435–464

Dale BE, Bals BD, Kim S, Eranki P (2010) Biofuels done right: land efficient animal feeds enable large environmental and energy benefits. Environ Sci Technol 44(22):8385–8389

Dawson JJC, Smith P (2007) Carbon losses from soil and its consequences for land-use management. Sci Total Environ 382(2–3):165–190

Delucchi MA (2010) Impacts of biofuels on climate change, water use, and land use. Ann NY Acad Sci 1195:28–45

Delzeit R, Holm-Muller K (2009) Steps to discern sustainability criteria for a certification scheme of bioethanol in Brazil: approach and difficulties. Energy 34(5):662–668

Dickmann DI (2006) Silviculture and biology of short-rotation woody crops in temperate regions: then and now. Biomass Bioenergy 30(8/9):696–705

Doran JW, Zeiss MR (2000) Soil health and sustainability: managing the biotic component of soil quality. Appl Soil Ecol 15(1):3–11

Duer H, Christensen PO (2010) Socio-economic aspects of different biofuel development pathways. Biomass Bioenergy 34(2):237–243

European Parliament (2009) Directive 2009/28/EC of the European Parliament and of the Council of 23 April 2009 on the promotion of the use of energy from renewable sources and amending and subsequently repealing Directives 2001/77/EC and 2003/30/E. Official J Eur Union L 140:47

Fargione J, Hill J, Tilman D, Polasky S, Hawthorne P (2008) Land clearing and the biofuel carbon debt. Science 319(5867):1235–1238

Fischer G, Hizsnyik E, Prieler S, Shah M, van Velthuizen H (2009) Biofuels and food security. International Institute for Applied Systems Analysis, Vienna

Fischer G, Prieler S, Velthuizen Hv, Berndes G, Faaij A, Londo M, Wit Md (2010) Biofuel production potentials in Europe: sustainable use of cultivated land and pastures, Part II: Land use scenarios. (Special Issue: a roadmap for biofuels in Europe.). Biomass Bioenergy 34(2):173–187

Florentinus A, Hamelinck C, de Lint S, van Iersel S (2008) Worldwide potential of aquatic biomass. Ecofys, Utrecht

Florin MV, Bunting C (2009) Risk governance guidelines for bioenergy policies. J Cleaner Prod 17(suppl 1):S106–S108

Forrester DI, Bauhus J, Cowie AL (2005) On the success and failure of mixed-species tree plantations: lessons learned from a model system of *Eucalyptus globulus* and *Acacia mearnsii*. For Ecol Manage 209(1–2):147–155

Friedlingstein P (2008) A steep road to climate stabilization. Nature 451(7176):297–298

Fritsche UR, Hennenberg KJ, Hermann A, Hünecke K, Herrera R, Fehrenbach H, Roth E, Hennecke A, Giegrich J (2010) Development of strategies and sustainability standards for the certification of biomass for international trade. Federal Environment Agency, Dessau

Froger E, Paz A, Vissers P (2010) Selection of a sustainability standard for pilot assessments of Jatropha producers in Mozambique. Partners for Innovation, Amsterdam

Fung PYH, Kirschbaum MUF, Raison RJ, Stucley C (2002) The potential for bioenergy production from Australian forests, its contribution to national greenhouse targets and recent developments in conversion processes. Biomass Bioenergy 22(4):223–236

Galdos MV, Cerri CC, Cerri CEP (2009) Soil carbon stocks under burned and unburned sugarcane in Brazil. Geoderma 153(3/4):347–352

Gregg JS, Izaurralde RC (2010) Effect of crop residue harvest on long-term crop yield, soil erosion and nutrient balance: trade-offs for a sustainable bioenergy feedstock. Biofuels 1(1):69–83

Gregory JM, Jones CD, Cadule P, Friedlingstein P (2009) Quantifying carbon cycle feedbacks. J Climate 22(19):5232–5250

Grigal DF (2000) Effects of extensive forest management on soil productivity. For Ecol Manage 138(1–3):167–185

Grove TS, O'Connell AM, Mendham D, Barrow NJ, Rance SJ (2001) Sustaining the productivity of tree crops on agricultural land in South-Western Australia. Rural Industries Research & Development Corporation, Canberra

Halleux H, Lassaux S, Renzoni R, Germain A (2008) Comparative life cycle assessment of two biofuels ethanol from sugar beet and rapeseed methyl ester. Int J Life Cycle Assess 13(3):184–190

Hansson J, Berndes G (2009) Future bioenergy trade in the EU: modelling trading options from a cost-effectiveness perspective. J Cleaner Prod 17(suppl 1):s27–s36

Hennenberg KJ, Dragisic C, Haye S, Hewson J, Semroc B, Savy C, Wiegmann K, Fehrenbach H, Fritsche UR (2010) The power of bioenergy-related standards to protect biodiversity. Conserv Biol 24(2):412–423

Hobbs TJ, Bartle J, Bennell M, George BH (2009) Prioritisation of regional woody crop industries. In: Hobbs T (ed) Regional industry potential for woody biomass crops in lower rainfall southern Australia. Rural Industries Research & Development Corporation, Canberra, pp 5–38

Holford ICR (1997) Soil phosphorus: its measurement, and its uptake by plants. Aust J Soil Res 35:227–240

Hoogwijk M, Faaij A, Bd V, Turkenburg W (2009) Exploration of regional and global cost-supply curves of biomass energy from short-rotation crops at abandoned cropland and rest land under four IPCC SRES land-use scenarios. Biomass Bioenergy 33(1):26–43

Integrated Sustainability Analysis (2006) Life-cycle energy balance and greenhouse gas emissions of nuclear energy in Australia. The University of Sydney, Sydney

International Energy Agency (2009) World energy outlook 2009. OECD/IEA, Paris

IPCC (2001) Working Group 1: the scientific basis. Intergovernmental Panel on Climate Change, Cambridge

Jackson RB, Jobbagy EG, Avissar R, Roy SB, Barrett DJ, Cook CW, Farley KA, le Maitre DC, McCarl BA, Murray BC (2005) Trading water for carbon with biological carbon sequestration. Science 310(5756):1944–1947

Jalali M (2007) Site-specific potassium application based on the fertilizer potassium availability index of soil. Precision Agric 8(4–5):199–211

Jongschaap REE, Corre WJ, Bindrabin PS, Brandanberg WA (2007) Claims and facts on *Jatropha curcas* L. Plant Research International B.V, Wageningen

Khan SA, Mulvaney RL, Ellsworth TR, Boast CW (2007) The myth of nitrogen fertilization for soil carbon sequestration. J Environ Qual 36(6):1821–1832

Kirschbaum MUF, Guo LB, Gifford RM (2008) Why does rainfall affect the trend in soil carbon after converting pastures to forests? A possible explanation based on nitrogen dynamics. For Ecol Manage 255(7):2990–3000

Lal R (2004) Soil carbon sequestration to mitigate climate change. Geoderma 123(1–2):1–22

Lal R, Follett RF, Kimble JM (2003) Achieving soil carbon sequestration in the United States: a challenge to the policy makers. Soil Sci 168(12):827–845

Lasch P, Kollas C, Rock J, Suckow F (2010) Potentials and impacts of short-rotation coppice plantation with aspen in Eastern Germany under conditions of climate change. Reg Environ Change 10(2):83–94

Lattimore B, Smith CT, Titus BD, Stupak I, Egnell G (2009) Environmental factors in woodfuel production: opportunities, risks, and criteria and indicators for sustainable practices. Biomass Bioenergy 33(10):1321–1342

Laurance WF, Koh LP, Butler R, Sodhi NS, Bradshaw CJA, Neidel JD, Consunji H, Mateo Vega J (2010) Improving the performance of the roundtable on sustainable palm oil for nature conservation. Conserv Biol 24(2):377–381

Lewandowski I, Faaij APC (2006) Steps towards the development of a certification system for sustainable bio-energy trade. Biomass Bioenergy 30(2):83–104

Liebig MA, Johnson HA, Hanson JD, Frank AB (2005) Soil carbon under switchgrass stands and cultivated cropland. Biomass Bioenergy 28(4):347–354

Luyssaert S, Schulze ED, Borner A, Knohl A, Hessenmoller D, Law BE, Ciais P, Grace J (2008) Old-growth forests as global carbon sinks. Nature 455(7210):213–215

Merino A, Fernandez-Lopez A, Solla-Gullon F, Miguel Edeso J (2004) Soil changes and tree growth in intensively managed *Pinus radiata* in northern Spain. For Ecol Manage 196(2/3):393–404

Morton DC, DeFries RS, Shimabukuro YE, Anderson LO, Arai E, del Bon Espirito-Santo F, Freitas R, Morisette J (2006) Cropland expansion changes deforestation dynamics in the southern Brazilian Amazon. Proc Natl Acad Sci USA 103(39):14637–14641

Muller A (2009) Sustainable agriculture and the production of biomass for energy use. Clim Change 94(3):319–331

New South Wales (2003) Native Vegetation Act 2003. Parliament of New South Wales, Sydney

New South Wales (2007) Biofuels Act 2007. Parliament of New South Wales, Sydney

O'Connell D, Braid A, Raison J, Handberg K, Cowie A, Rodriguez L, George BH (2009) Sustainable production of bioenergy: a review of global bioenergy sustainability frameworks and assessment system. Rural Industries Research and Development Corporation, Canberra

Parliament E (2009) Directive 2009/28/EC of the European Parliament and of the Council of 23 April 2009 on the promotion of the use of energy from renewable sources and amending and subsequently repealing Directives 2001/77/EC and 2003/30/E. Off J Eur Union L 140:47

Paul KI, Polglase PJ, Nyakuengama JG, Khanna PK (2002) Change in soil carbon following afforestation. For Ecol Manage 168(1–3):241–257

Perlack RD, Wright LL, Turhollow AF, Graham RL, Stokes BJ, Erbach DC (2005) Biomass as feedstock for a bioenergy and bioproducts industry: the technical feasibility of a billion-ton annual supply. U.S. Department of Energy and U.S Department of Agriculture, Oak Ridge

Pimentel D (2006) Soil erosion: a food and environmental threat. Environ Dev Sustain 8(1):119–137

Pimentel D (2009) Biofuel food disasters and cellulosic ethanol problems. Bull Sci Technol Soc 29(3):205–212

Pimentel D, Patzek T (2007) Ethanol production: energy and economic issues related to U.S. and Brazilian sugarcane. Nat Resour Res 16(3):235–242

Pimentel D, Marklein A, Toth M, Karpoff M, Paul G, McCormack R, Kyriazis J, Krueger T (2008) Biofuel impacts on world food supply: use of fossil fuel, land and water resources. Energies 1(2):41–78

Pimentel D, Marklein A, Toth MA, Karpoff MN, Paul GS, McCormack R, Kyriazis J, Krueger T (2009) Food versus biofuels: environmental and economic costs. Hum Ecol 37(1):1–12

Pingintha N, Leclerc MY, Beasley JP Jr, Zhang G, Senthong C (2010) Assessment of the soil CO_2 gradient method for soil CO_2 efflux measurements: comparison of six models in the calculation of the relative gas diffusion coefficient. Tellus Ser B 62(1):47–58

Pinheiro EFM, Lima E, Ceddia MB, Urquiaga S, Alves BJR, Boddey RM (2010) Impact of pre-harvest burning versus trash conservation on soil carbon and nitrogen stocks on a sugarcane plantation in the Brazilian Atlantic forest region. Plant Soil 333(1/2):71–80

Plevin RJ, O'Hare M, Jones AD, Torn MS, Gibbs HK (2010) Greenhouse gas emissions from biofuels' indirect land use change are uncertain but may be much greater than previously estimated. Environ Sci Technol 44(21):8015–8021

Pritchard SG (2011) Soil organisms and global climate change. Plant Pathol 60(1):82–99

Raison RJ, Khanna PK (2011) Possible impacts of climate change on forest soil health. In: Singh B, Cowie AL, Chan KY (eds) Soil health and climate change. Springer, Heidelberg

Raunikar R, Buongiorno J, Turner JA, Zhu SS (2010) Global outlook for wood and forests with the bioenergy demand implied by scenarios of the Intergovernmental Panel on Climate Change. For Policy Econ 12(1):48–56

Reinhard J, Zah R (2009) Global environmental consequences of increased biodiesel consumption in Switzerland: consequential life cycle assessment. J Cleaner Prod 17(suppl 1):S46–S56

Rengasamy P (2006) World salinization with emphasis on Australia. J Exp Bot 57(5):1017–1023

Ribeiro SK, Kobayashi S, Beuthe M, Gasca J, Greene D, Lee DS, Muromachi Y, Newton PJ, Plotkin S, Sperling D, Wit R, Zhou PJ, Hata H, Sims R, Skjolsvik KO (2007) Transport and its infrastructure. In: Metz B, Davidson OR, Bosch PR, Dave R, Meyer LA (eds) Climate change 2007: mitigation. Contribution of Working Group III to the Fourth Assessment Report of the Intergovernmental Panel on Climate Change. Cambridge University Press, Cambridge, p 64

Rice CW (2005) Carbon cycle in soils: dynamics and management. In: Daniel H (ed) Encyclopedia of soils in the environment. Elsevier, Oxford, pp 164–170

Richard TL (2010) Challenges in scaling up biofuels infrastructure. Science 329(5993):793–796

Richards AE, Forrester DI, Bauhus J, Scherer-Lorenzen M (2010) The influence of mixed tree plantations on the nutrition of individual species: a review. Tree Physiol 30(9):1192–1208

Robertson FA, Thorburn PJ (2007) Management of sugarcane harvest residues: consequences for soil carbon and nitrogen. Aust J Soil Res 45(1):13–23

Roundtable on Sustainable Biofuels (2009) RSB principles and criteria for sustainable biofuel production. Ecole Polytechnique Federale De Lausanne, Lausanne

Sartori F, Lal R, Ebinger MH, Eaton JA (2007) Changes in soil carbon and nutrient pools along a chronosequence of poplar plantations in the Columbia Plateau, Oregon, USA. Agric Ecosyst Environ 122(3):325–339

Schade C, Pimentel D (2010) Population crash: prospects for famine in the twenty-first century. Environ Dev Sustain 12(2):245–262

Schlamadinger B, Marland G (1996) The role of forest and bioenergy strategies in the global carbon cycle. Biomass Bioenergy 10(5/6):275–300

Schlamadinger B, Bird N, Johns T, Brown S, Canadell J, Ciccarese L, Dutschke M, Fiedler J, Fischlin A, Fearnside P, Forner C, Freibauer A, Frumhoff P, Hoehne N, Kirschbaum MUF, Labat A, Marland G, Michaelowa A, Montanarella L, Moutinho P, Murdiyarso D, Pena N, Pingoud K, Rakonczay Z, Rametsteiner E, Rock J (2007) A synopsis of land use, land-use change and forestry (LULUCF) under the Kyoto Protocol and Marrakech Accords. (Options for including agriculture and forestry activities in a post-2012 international climate agreement). Environ Sci Policy 10(4):271–282

Searchinger T, Heimlich R, Houghton RA, Dong FX, Elobeid A, Fabiosa J, Tokgoz S, Hayes D, Yu TH (2008) Use of U.S. croplands for biofuels increases greenhouse gases through emissions from land-use change. Science 319(5867):1238–1240

Sims REH, Schock RN, Adegbululgbe A, Fenhann J, Konstantinaviciute I, Moomaw W, Nimir HB, Schlamadinger B, Torres-Martínez J, Turner C, Uchiyama Y, Vuori SJV, Wamukonya N, Zhang X (2007) Energy supply. Cambridge University Press, Cambridge

Singh BP, Cowie A (2010) The mean turnover time of biochar in soil varies depending on biomass source and pyrolysis temperature. In: 19th world congress of soil science, soil solutions for a changing world, Brisbane, 1–6 Aug 2010. http://www.iuss.org/19th%20WCSS/symposium/pdf/0326.pdf. Accessed 1 December 2010

Smolander A, Kitunen V, Tamminen P, Kukkola M (2010) Removal of logging residue in Norway spruce thinning stands: long-term changes in organic layer properties. Soil Biol Biochem 42(8):1222–1228

Solomon BD (2010) Biofuels and sustainability. Ann NY Acad Sci 1185:119–134

Solomon S, Qin D, Manning M, Alley RB, Berntsen T, Bindoff NL, Chen Z, Chidthaisong A, Gregory JM, Hegerl GC, Heimann M, Hewitson B, Hoskins BJ, Joos F, Jouzel J, Kattsov V, Lohmann U, Matsuno T, Molina M, Nicholls N, Overpeck J, Raga G, Ramaswamy V, Ren J, Rusticucci M, Somerville R, Stocker TF, Whetton P, Wood RA, Wratt D (2007) Technical summary. Cambridge

Sparovek G, Berndes G, Egeskog A, de Freitas FLM, Gustafsson S, Hansson J (2007) Sugarcane ethanol production in Brazil: an expansion model sensitive to socioeconomic and environmental concerns. Biofuels, Bioprod Biorefin 1(4):270–282

Sparovek G, Barretto A, Berndes G, Martins S, Maule R (2009) Environmental, land-use and economic implications of Brazilian sugarcane expansion 1996–2006. Mitig Adapt Strateg Glob Change 14(3):285–298

Stehfest E, Bouwman L (2006) N_2O and NO emission from agricultural fields and soils under natural vegetation: summarizing available measurement data and modeling of global annual emissions. Nutr Cycl Agroecosyst 74(3):207–228

Stern NH (2007) The economics of climate change: the Stern review. Cambridge University Press, Cambridge

Tilman D, Cassman KG, Matson PA, Naylor R, Polasky S (2002) Agricultural sustainability and intensive production practices. Nature 418(6898):671–677

Tilman D, Hill J, Lehman C (2006) Carbon-negative biofuels from low-input high-diversity grassland biomass. Science 314(5805):1598–1600

Tilman D, Socolow R, Foley JA, Hill J, Larson E, Lynd L, Pacala S, Reilly J, Searchinger T, Somerville C, Williams R (2009) Beneficial biofuels – the food, energy, and environment trilemma. Science 325(5938):270–271

Turner J, Lambert M (2000) Change in organic carbon in forest plantation soils in eastern Australia. For Ecol Manage 133(3):231–247

Tyner WE, Taheripour F, Zhuang Q, Birur D, Baldos U (2010) Land use changes and consequent CO_2 emissions due to US corn ethanol production: a comprehensive analysis. Purdue University, West Lafayette

United Nations (2008) Partnerships for sustainable development. United Nations, Rome

Uslu A, Faaij APC, Bergman PCA (2008) Pre-treatment technologies, and their effect on international bioenergy supply chain logistics. Techno-economic evaluation of torrefaction, fast pyrolysis and pelletisation. Energy 33(8):1206–1223

van Dam J, Junginger M, Faaij A, Jürgens I, Best G, Fritsche U (2008) Overview of recent developments in sustainable biomass certification. Biomass Bioenergy 32(8):749–780

van Dam J, Junginger M, Faaij APC (2010) From the global efforts on certification of bioenergy towards an integrated approach based on sustainable land use planning. Renewable Sustainable Energy Rev 14(9):2445–2472

Vargas Gil S, Becker A, Oddino C, Zuza M, Marinelli A, March G (2009) Field trial assessment of biological, chemical, and physical responses of soil to tillage intensity, fertilization, and grazing. Environ Manage 44(2):378–386

Vertessy RA, Zhang L, Dawes W (2003) Plantations, river flows and river salinity. Aust Forest 66:55–61

Vis MW, Vos J, van den Berg D (2008) Sustainability criteria and certification systems for biomass production. Biomass Technology Group, Enschede

Vivid Economics (2010) The implicit price of carbon in the electricity sector of six major economies. The Climate Institute, Sydney

Walmsley JD, Godbold DL (2010) Stump harvesting for bioenergy – a review of the environmental Impacts. Forestry 83(1):17–38

Wang H, Brown SL, Magesan GN, Slade AH, Quintern M, Clinton PW, Payn TW (2008) Technological options for the management of biosolids. Environ Sci Pollut Res 15(4):308–317

Warnock D, Lehmann J, Kuyper T, Rillig M (2007) Mycorrhizal responses to biochar in soil – concepts and mechanisms. Plant Soil 300(1):9–20

Wicke B, Sikkema R, Dornburg V, Faaij A (2011) Exploring land use changes and the role of palm oil production in Indonesia and Malaysia. Land Use Policy 28(1):193–206

Wijffels RH, Barbosa MJ (2010) An outlook on microalgal biofuels. Science 329(5993):796–799

Wit MD, Faaij A (2010) European biomass resource potential and costs. Biomass Bioenergy 34(2):188–202

Wood S, Cowie AL (2004) A review of greenhouse gas emission factors for fertiliser production. Task 38. IEA Bioenergy, Sydney

Woolf D, Amonette JE, Street-Perrott FA, Lehmann J, Joseph S (2010) Sustainable biochar to mitigate global climate change. Nat Commun 1(5):56

Index

B.P. Singh et al. (eds.), *Soil Health and Climate Change*, Soil Biology 29,
DOI 10.1007/978-3-642-20256-8, © Springer-Verlag Berlin Heidelberg 2011